NanoScience and Technology

NanoScience and Technology

Series Editors:
P. Avouris   B. Bhushan   D. Bimberg   K. von Klitzing   H. Sakaki   R. Wiesendanger

The series NanoScience and Technology is focused on the fascinating nano-world, mesoscopic physics, analysis with atomic resolution, nano and quantum-effect devices, nanomechanics and atomic-scale processes. All the basic aspects and technology-oriented developments in this emerging discipline are covered by comprehensive and timely books. The series constitutes a survey of the relevant special topics, which are presented by leading experts in the field. These books will appeal to researchers, engineers, and advanced students.

**Magnetic Microscopy
of Nanostructures**
Editors: H. Hopster and H.P. Oepen

Applied Scanning Probe Methods I
Editors: B. Bhushan, H. Fuchs,
S. Hosaka

The Physics of Nanotubes
Fundamentals of Theory, Optics
and Transport Devices
Editors: S.V. Rotkin and S. Subramoney

**Single Molecule Chemistry
and Physics**
An Introduction
By C. Wang, C. Bai

**Atomic Force Microscopy, Scanning
Nearfield Optical Microscopy
and Nanoscratching**
Application to Rough
and Natural Surfaces
By G. Kaupp

Applied Scanning Probe Methods II
Scanning Probe Microscopy
Techniques
Editors: B. Bhushan, H. Fuchs

Applied Scanning Probe Methods III
Characterization
Editors: B. Bhushan, H. Fuchs

Applied Scanning Probe Methods IV
Industrial Application
Editors: B. Bhushan, H. Fuchs

Nanocatalysis
Editors: U. Heiz, U. Landman

**Roadmap
of Scanning Probe Microscopy**
Editors: S. Morita

**Nanostructures –
Fabrication and Analysis**
Editor: H. Nejo

Applied Scanning Probe Methods V
Scanning Probe Microscopy Techniques
Editors: B. Bhushan, H. Fuchs,
S. Kawata

Applied Scanning Probe Methods VI
Characterization
Editors: B. Bhushan, S. Kawata

Applied Scanning Probe Methods VII
Biomimetics and Industrial Applications
Editors: B. Bhushan, H. Fuchs

Applied Scanning Probe Methods VIII
Scanning Probe Microscopy Techniques
Editors: B. Bhushan, H. Fuchs,
M. Tomitori

Applied Scanning Probe Methods IX
Characterization
Editors: B. Bhushan, H. Fuchs,
M. Tomitori

Applied Scanning Probe Methods X
Biomimetics and Industrial Applications
Editors: B. Bhushan, H. Fuchs,
M. Tomitori

Bharat Bhushan
Harald Fuchs
Masahiko Tomitori

Applied Scanning Probe Methods X

Biomimetics and Industrial Applications

With 306 Figures and 9 Tables
Including 81 Color Figures

Springer

Editors:

Professor Bharat Bhushan
Nanotribology Laboratory for Information
Storage and MEMS/NEMS (NLIM)
W 390 Scott Laboratory, 201 W. 19th Avenue
The Ohio State University, Columbus
Ohio 43210-1142, USA
e-mail: Bhushan.2@osu.edu

Professor Dr. Harald Fuchs
Institute of Physics, FB 16
University of Münster
Wilhelm-Klemm-Str. 10
48149 Münster, Germany
e-mail: fuchsh@uni-muenster.de

Professor Dr. Masahiko Tomitori
Advanced Institute of Science & Technology
School of Materials Science
Asahidai, Nomi 1-1
923-1292 Ishikawa, Japan
e-mail: tomitori@jaist.ac.jp

Series Editors:

Professor Dr. Phaedon Avouris
IBM Research Division
Nanometer Scale Science & Technology
Thomas J. Watson Research Center, P.O. Box 218
Yorktown Heights, NY 10598, USA

Professor Bharat Bhushan
Nanotribology Laboratory for Information
Storage and MEMS/NEMS (NLIM)
W 390 Scott Laboratory, 201 W. 19th Avenue
The Ohio State University, Columbus
Ohio 43210-1142, USA

Professor Dr. Dieter Bimberg
TU Berlin, Fakutät Mathematik,
Naturwissenschaften,
Institut für Festkörperphysik
Hardenbergstr. 36, 10623 Berlin, Germany

Professor Dr., Dres. h. c. Klaus von Klitzing
Max-Planck-Institut für Festkörperforschung
Heisenbergstrasse 1, 70569 Stuttgart, Germany

Professor Hiroyuki Sakaki
University of Tokyo
Institute of Industrial Science,
4-6-1 Komaba, Meguro-ku, Tokyo 153-8505, Japan

Professor Dr. Roland Wiesendanger
Institut für Angewandte Physik
Universität Hamburg
Jungiusstrasse 11, 20355 Hamburg, Germany

ISBN 978-3-540-74084-1
DOI 10.1007/978-3-540-74085-8

e-ISBN 978-3-540-74085-8

NanoScience and Technology ISSN 1434-4904

Library of Congress Control Number: 2007937294

© 2008 Springer-Verlag Berlin Heidelberg

This work is subject to copyright. All rights are reserved, whether the whole or part of the material is concerned, specifically the rights of translation, reprinting, reuse of illustrations, recitation, broadcasting, reproduction on microfilm or in any other way, and storage in data banks. Duplication of this publication or parts thereof is permitted only under the provisions of the German Copyright Law of September 9, 1965, in its current version, and permission for use must always be obtained from Springer. Violations are liable to prosecution under the German Copyright Law.

The use of general descriptive names, registered names, trademarks, etc. in this publication does not imply, even in the absence of a specific statement, that such names are exempt from the relevant protective laws and regulations and therefore free for general use.

Typesetting: LE-TEX Jelonek, Schmidt & Vöckler GbR, Leipzig
Production: LE-TEX Jelonek, Schmidt & Vöckler GbR, Leipzig
Cover design: WMXDesign GmbH, Heidelberg

Printed on acid-free paper

9 8 7 6 5 4 3 2 1

springer.com

Preface

The success of the Springer Series Applied Scanning Probe Methods I–VII and the rapidly expanding activities in scanning probe development and applications worldwide made it a natural step to collect further specific results in the fields of development of scanning probe microscopy techniques (Vol. VIII), characterization (Vol. IX), and biomimetics and industrial applications (Vol. X). These three volumes complement the previous set of volumes under the subject topics and give insight into the recent work of leading specialists in their respective fields. Following the tradition of the series, the chapters are arranged around techniques, characterization and biomimetics and industrial applications.

Volume VIII focuses on novel scanning probe techniques and the understanding of tip/sample interactions. Topics include near field imaging, advanced AFM, specialized scanning probe methods in life sciences including new self sensing cantilever systems, combinations of AFM sensors and scanning electron and ion microscopes, calibration methods, frequency modulation AFM for application in liquids, Kelvin probe force microscopy, scanning capacitance microscopy, and the measurement of electrical transport properties at the nanometer scale.

Vol. IX focuses on characterization of material surfaces including structural as well as local mechanical characterization, and molecular systems. The volume covers a broad spectrum of STM/AFM investigations including fullerene layers, force spectroscopy for probing material properties in general, biological films .and cells, epithelial and endothelial layers, medical related systems such as amyloidal aggregates, phospholipid monolayers, inorganic films on aluminium and copper oxides, tribological characterization, mechanical properties of polymer nanostructures, technical polymers, and nearfield optics.

Volume X focuses on biomimetics and industrial applications such as investigation of structure of gecko feet, semiconductors and their transport phenomena, charge distribution in memory technology, the investigation of surfaces treated by chemical-mechanical planarization, polymeric solar cells, nanoscale contacts, cell adhesion to substrates, nanopatterning, indentation application, new printing techniques, the application of scanning probes in biology, and automatic AFM for manufacturing.

As a result, Volumes VIII to X of Applied Scanning Probes microscopies cover a broad and impressive spectrum of recent SPM development and application in many fields of technology, biology and medicine, and introduce many technical concepts and improvements of existing scanning probe techniques.

We are very grateful to all our colleagues who took the efforts to prepare manuscripts and provided them in timely manner. Their activity will help both students and established scientists in research and development fields to be informed about the latest achievements in scanning probe methods. We would like to cordially thank Dr. Marion Hertel, Senior Editor Chemistry and Mrs. Beate Siek of Springer for their continuous professional support and advice which made it possible to get this volume to the market on time.

Bharat Bhushan
Harald Fuchs
Masahiko Tomitori

Contents – Volume X

27 **Gecko Feet: Natural Attachment Systems for Smart Adhesion—Mechanism, Modeling, and Development of Bio-Inspired Materials**
Bharat Bhushan, Robert A. Sayer 1

27.1	Introduction	1
27.2	Tokay Gecko	2
27.2.1	Construction of Tokay Gecko	2
27.2.2	Other Attachment Systems	5
27.2.3	Adaptation to Surface Roughness	7
27.2.4	Peeling	8
27.2.5	Self-Cleaning	10
27.3	Attachment Mechanisms	12
27.3.1	Van der Waals Forces	12
27.3.2	Capillary Forces	13
27.4	Experimental Adhesion Test Techniques and Data	14
27.4.1	Adhesion Under Ambient Conditions	15
27.4.2	Effects of Temperature	17
27.4.3	Effects of Humidity	18
27.4.4	Effects of Hydrophobicity	18
27.5	Adhesion Modeling	19
27.5.1	Spring Model	21
27.5.2	Single Spring Contact Analysis	21
27.5.3	The Multilevel Hierarchical Spring Analysis	23
27.5.4	Adhesion Results for the Gecko Attachment System Contacting a Rough Surface	26
27.5.5	Capillarity Effects	30
27.5.6	Adhesion Results that Account for Capillarity Effects	31
27.6	Modeling of Biomimetic Fibrillar Structures	34
27.6.1	Fiber Model	34
27.6.2	Single Fiber Contact Analysis	34
27.6.3	Constraints	35
27.6.4	Numerical Simulation	39
27.6.5	Results and Discussion	41

27.7	Fabrication of Biomimetric Gecko Skin	48
27.7.1	Single-Level Hierarchical Structures	49
27.7.2	Multilevel Hierarchical Structures	53
27.8	Closure	55
Appendix		56
References		59

28 Carrier Transport in Advanced Semiconductor Materials
Filippo Giannazzo, Patrick Fiorenza, Vito Raineri 63

28.1	Majority Carrier Distribution in Semiconductors: Imaging and Quantification	64
28.1.1	Basic Principles of SCM	64
28.1.2	Carrier Imaging Capability by SCM	67
28.1.3	Quantification of SCM Raw Data	70
28.1.4	Basic Principles of SSRM	78
28.1.5	Carrier Imaging Capability by SSRM	81
28.1.6	Quantification of SSRM Raw Data	81
28.1.7	Drift Mobility by SCM and SSRM	85
28.2	Carrier Transport Through Metal–Semiconductor Barriers by C-AFM	88
28.3	Charge Transport in Dielectrics by C-AFM	93
28.3.1	Direct Determination of Breakdown	97
28.3.2	Weibull Statistics by C-AFM	99
28.4	Conclusion	101
References		101

29 Visualization of Fixed Charges Stored in Condensed Matter and Its Application to Memory Technology
Yasuo Cho 105

29.1	Introduction	105
29.2	Principle and Theory for SNDM	106
29.3	Microscopic Observation of Area Distribution of the Ferroelectric Domain Using SNDM	107
29.4	Visualization of Stored Charge in Semiconductor Flash Memories Using SNDM	109
29.5	Higher-Order SNDM	110
29.6	Noncontact SNDM	111
29.7	SNDM for 3D Observation of Nanoscale Ferroelectric Domains	112

29.8	Next-Generation Ultra-High-Density Ferroelectric Data Storage Based on SNDM	114
29.8.1	Overview of Ferroelectric Data Storage	114
29.8.2	SNDM Nanodomain Engineering System and Ferroelectric Recording Medium	116
29.8.3	Nanodomain Formation in a $LiTaO_3$ Single Crystal	117
29.8.4	High-Speed Switching of Nanoscale Ferroelectric Domains in Congruent Single-Crystal $LiTaO_3$	120
29.8.5	Prototype of a High-Density Ferroelectric Data Storage System	122
29.8.6	Realization of 10 Tbit/in.^{2} Memory Density	126
29.9	Outlook	128
References		129

30 Applications of Scanning Probe Methods in Chemical Mechanical Planarization
Toshi Kasai, Bharat Bhushan 131

30.1	Overview of CMP Technology and the Need for SPM	131
30.1.1	CMP Technology and Its Key Elements	131
30.1.2	Various CMP Processes and the Need for SPM	134
30.2	AFP for the Evaluation of Dishing and Erosion	137
30.3	Surface Planarization and Roughness Characterization in CMP Using AFM	141
30.4	Use of Modified Atomic Force Microscope Tips for Fundamental Studies of CMP Mechanisms	144
30.5	Conclusions	149
References		149

31 Scanning Probe Microscope Application for Single Molecules in a π-Conjugated Polymer Toward Molecular Devices Based on Polymer Chemistry
Ken-ichi Shinohara 153

31.1	Introduction	153
31.2	Chiral Helical π-Conjugated Polymer	154
31.2.1	Helical Chirality of a π-Conjugated Main Chain Induced by Polymerization of Phenylacetylene with Chiral Bulky Groups	156
31.2.2	Direct Measurement of the Chiral Quaternary Structure in a π-Conjugated Polymer	158
31.2.3	Direct Measurement of Structural Diversity in Single Molecules of a Chiral Helical π-Conjugated Polymer	163

31.2.4	Dynamic Structure of Single Molecules in a Chiral Helical π-Conjugated Polymer by a High-Speed AFM	166
31.3	Supramolecular Chiral π-Conjugated Polymer	169
31.3.1	Simultaneous Imaging of Structure and Fluorescence of a Supramolecular Chiral π-Conjugated Polymer	169
31.3.2	Dynamic Structure of a Supramolecular Chiral π-Conjugated Polymer by a High-Speed AFM	177
References		181

32 Scanning Probe Microscopy on Polymer Solar Cells
Joachim Loos, Alexander Alexeev 183

32.1	Brief Introduction to Polymer Solar Cells	184
32.2	Sample Preparation and Characterization Techniques	188
32.3	Morphology Features of the Photoactive Layer	190
32.3.1	Influence of Composition and Solvents on the Morphology of the Active Layer	190
32.3.2	Influence of Annealing	193
32.3.3	All-Polymer Solar Cells	199
32.4	Nanoscale Characterization of Properties of the Active Layer	201
32.4.1	Local Optical Properties As Measured by Scanning Near-Field Optical Microscopy	201
32.4.2	Characterization of Nanoscale Electrical Properties	203
32.5	Summary and Outlook	212
References		213

33 Scanning Probe Anodization for Nanopatterning
Hiroyuki Sugimura 217

33.1	Introduction	217
33.2	Electrochemical Origin of SPM-Based Local Oxidation	218
33.3	Variation in Scanning Probe Anodization	223
33.3.1	Patternable Materials in Scanning Probe Anodization	223
33.3.2	Environment Control in Scanning Probe Anodization	226
33.3.3	Electrochemical Scanning Surface Modification Using Cathodic Reactions	229
33.4	Progress in Scanning Probe Anodization	232
33.4.1	From STM-Based Anodization to AFM-Based Anodization	232
33.4.2	Versatility of AFM-Based Scanning Probe Anodization	233
33.4.3	In Situ Characterization of Anodized Structures by AFM-Based Methods	233

33.4.4	Technical Development of Scanning Probe Anodization	237
33.5	Lithographic Applications of Scanning Probe Anodization	239
33.5.1	Device Prototyping	239
33.5.2	Pattern Transfer from Anodic Oxide to Other Materials	240
33.5.3	Integration of Scanning Probe Lithography with Other High-Throughput Lithographies	247
33.5.4	Chemical Manipulation of Nano-objects by the Use of a Nanotemplate Prepared by Scanning Probe Anodization	248
33.6	Conclusion	251
References		251

34 Tissue Engineering: Nanoscale Contacts in Cell Adhesion to Substrates
Mario D'Acunto, Paolo Giusti, Franco Maria Montevecchi, Gianluca Ciardelli 257

34.1	Tissue Engineering: A Brief Introduction	257
34.2	Fundamental Features of Cell Motility and Cell–Substrates Adhesion	261
34.2.1	Biomimetic Scaffolds, Roughness, and Contact Guidance for Cell Adhesion and Motility	268
34.3	Experimental Strategies for Cell–ECM Adhesion Force Measurements	271
34.4	Conclusions	279
34.5	Glossary	279
References		280

35 Scanning Probe Microscopy in Biological Research
Tatsuo Ushiki, Kazushige Kawabata 285

35.1	Introduction	285
35.2	SPM for Visualization of the Surface of Biomaterials	286
35.2.1	Advantages of AFM in Biological Studies	286
35.2.2	AFM of Biomolecules	287
35.2.3	AFM of Isolated Intracellular and Extracellular Structures	289
35.2.4	AFM of Tissue Sections	292
35.2.5	AFM of Living Cells and Their Movement	292
35.2.6	Combination of AFM with Scanning Near-Field Optical Microscopy for Imaging Biomaterials	294
35.3	SPM for Measuring Physical Properties of Biomaterials	296
35.3.1	Evaluation Methods of Viscoelasticity	296
35.3.2	Examples for Viscoelasticity Mapping Measurements	299

35.3.3	Combination of Viscoelasticity Measurement with Other Techniques	302
35.4	SPM as a Manipulation Tool in Biology	304
35.5	Conclusion	306
References		306

36	**Novel Nanoindentation Techniques and Their Applications** *Jiping Ye*	309
36.1	Introduction	309
36.2	Basic Principles of Contact	311
36.2.1	Meyer's Law	311
36.2.2	Elastic Contact Solution	312
36.3	Tip Rigidity and Geometry	313
36.4	Hardness and Modulus Measurements	314
36.4.1	Analysis Method	314
36.4.2	Practical Application Aspects	316
36.4.3	Recent Applications	320
36.5	Yield Stress and Modulus Measurements	324
36.5.1	Analysis Method	324
36.5.2	Recent Applications	326
36.6	Work-Hardening Rate and Exponent Measurements	329
36.6.1	Analysis Method	329
36.6.2	Practical Application Aspects	333
36.6.3	Recent Applications	335
36.7	Viscoelastic Compliance and Modulus	336
36.7.1	Analysis Method	336
36.7.2	Practical Application Aspects	339
36.8	Other Mechanical Characteristics	342
36.9	Outlook	343
References		343

37	**Applications to Nano-Dispersion Macromolecule Material Evaluation in an Electrophotographic Printer** *Yasushi Kadota*	347
37.1	Introduction	347
37.2	Electrophotographic Processes	348
37.2.1	Principle and Characteristics of an Electrophotographic System	348

37.2.2	Microcharacteristic and Analysis Technology for Functional Components	349
37.3	SPM Applications to Electrophotographic Systems	352
37.3.1	Measurement of Electrostatic Charge of Toner	352
37.3.2	Measurement of the Adhesive Force Between a Particle and a Substrate	353
37.3.3	Observation of a Nanodispersion Macromolecule Interface —Toner Adhesion to a Fusing Roller	355
37.4	Current Technology Subjects	357
References		357

38   Automated AFM as an Industrial Process Metrology Tool for Nanoelectronic Manufacturing
Tianming Bao, David Fong, Sean Hand 359

38.1	Introduction	359
38.2	Dimensional Metrology with AFM	361
38.2.1	Dimensional Metrology	361
38.2.2	AFM Scanning Technology	362
38.2.3	AFM Probe Technology	367
38.2.4	AFM Metrology Capability	367
38.3	Applications in Semiconductors— Logic and Memory Integrated Circuits	370
38.3.1	Shallow Trench Isolation Resist Pattern	370
38.3.2	STI Etch	372
38.3.3	STI CMP	375
38.3.4	Gate Resist Pattern	378
38.3.5	Gate Etch	379
38.3.6	FinFET Gate Formation	383
38.3.7	Gate Sidewall Spacer	385
38.3.8	Strained SiGe Source/Drain Recess	385
38.3.9	Pre-metal Dielectric CMP	386
38.3.10	Contact and Via Photo Pattern	387
38.3.11	Contact Etch	387
38.3.12	Contact CMP	389
38.3.13	Metal Trench Photo Pattern	390
38.3.14	Metal Trench Etch	390
38.3.15	Via Etch	392
38.3.16	Via Etch	394
38.3.17	Roughness	396
38.3.18	LWR, LER, and SWR	397
38.3.19	DRAM DT Capacitor	397
38.3.20	Ferroelectric RAM Capacitor	398
38.3.21	Optical Proximity Correction	398

38.4	Applications in Photomask	399
38.4.1	Photomask Pattern and Etch	399
38.4.2	Photomask Defect Review and Repair	400
38.5	Applications in Hard Disk Manufacturing	401
38.5.1	Magnetic Thin-Film Recording Head	401
38.5.2	Slider for Hard Drive	405
38.6	Applications in Microelectromechanical System Devices	406
38.6.1	Contact Image Sensor	406
38.6.2	Digital Light Processor Mirror Device	408
38.7	Challenge and Potential Improvement	408
38.8	Conclusion	409
References		411
Subject Index		413

Contents – Volume VIII

1	**Background-Free Apertureless Near-Field Optical Imaging** *Pietro Giuseppe Gucciardi, Guillaume Bachelier,* *Stephan J. Stranick, Maria Allegrini*	1
1.1	Introduction	1
1.2	Principles of Apertureless SNOM	3
1.2.1	The Homodyne Apertureless SNOM Concept	5
1.2.2	The Heterodyne and Pseudo-Heterodyne Apertureless SNOM Concepts	8
1.3	Interpretation of the Measured Near-Field Signal in the Presence of a Background	9
1.3.1	Noninterferometric Detection	9
1.3.2	Interferometric Detection	12
1.3.3	Artifacts in Apertureless SNOM and Identification Criteria	14
1.3.4	New Techniques for Background Removal	17
1.4	Applications of Elastic-Scattering Apertureless SNOM	17
1.4.1	Material-Specific Imaging	18
1.4.2	Phase Mapping in Metallic Nanostructures and Optical Waveguides	19
1.4.3	Tip-Induced Resonances in Polaritonic Samples	22
1.4.4	Applications to Identification of Biosamples	24
1.4.5	Subsurface Imaging and Superlensing	25
1.5	Conclusions	27
References		27
2	**Critical Dimension Atomic Force Microscopy** **for Sub-50-nm Microelectronics Technology Nodes** *Hao-Chih Liu, Gregory A. Dahlen, Jason R. Osborne*	31
2.1	Introduction	32
2.1.1	AFM for Semiconductor and Data Storage Industries	32
2.1.2	Scanning Modes: Tapping Versus Deep Trench and CD Mode . . .	32
2.1.3	Specialty Probes	34
2.2	Reference Metrology System and Semiconductor Production . . .	34

2.2.1	Requirements for Metrology Tools	37
2.2.2	AFM as a Reference Metrology System	37
2.2.3	AFM as an In-Line Metrology System	39
2.3	Image Analysis for Accurate Metrology	40
2.3.1	Background	40
2.3.2	Conventional Tip Characterization and Image Reconstruction	41
2.3.3	CD Tip Shape Parameters	44
2.3.4	CD Tip Shape Characterization Techniques	44
2.3.5	CD Image (Reentrant) Reconstruction Algorithms	47
2.4	Metrology Applications	52
2.4.1	Examples within Process Control	52
2.4.2	"Fingerprinting" of Sample Features	54
2.5	Developments in Probe Fabrication	60
2.5.1	Tip–Sample Interactions: Tip Shape, Stiffness, and Tip Wear	61
2.5.2	Tip Wear and Surface Modification	64
2.5.3	Application-Oriented Probe Designs	67
2.6	Outlook: CD AFM Technologies for 45-/32-/22-nm Nodes	70
2.6.1	Measuring Sub-50-nm Devices: System Requirements	70
2.6.2	Probe Technology for 45-/32-nm Structures	71
References		73

3	**Near Field Probes: From Optical Fibers to Optical Nanoantennas** *Eugenio Cefalì, Salvatore Patanè, Salvatore Spadaro, Renato Gardelli, Matteo Albani, Maria Allegrini*	77
3.1	Introduction	77
3.2	Conventional Microscopy and Near-Field Optical Techniques	78
3.3	The Probe	84
3.3.1	Aperture SNOM Probes	85
3.3.2	The Apertureless Probe: Optical Nanoantennas	118
3.4	Applications and Perspectives	127
References		129

4	**Carbon Nanotubes as SPM Tips: Mechanical Properties of Nanotube Tips and Imaging** *Sophie Marsaudon, Charlotte Bernard, Dirk Dietzel, Cattien V. Nguyen, Anne-Marie Bonnot, Jean-Pierre Aimé, Rodolphe Boisgard*	137
4.1	Introduction	138

4.2	CNT Tip Fabrication	140
4.2.1	MWCNTs and Fusing	141
4.2.2	SWCNTs and Direct Growth	144
4.2.3	Controlling and Tailoring the Properties of CNT Tips	148
4.3	Understanding the Mechanical Properties of CNT Tips: A Dynamical SPM Frequency Modulation Study	149
4.3.1	Mechanical Properties of CNTs	149
4.3.2	Mechanical Properties of CNT Tips	150
4.3.3	Mechanical Properties of CNT Tips in Dynamical Experiments: Competition Between Elasticity and Adhesion	151
4.3.4	Experimental Signals	155
4.3.5	Mechanical Properties of MWCNTs	158
4.3.6	Mechanical Properties of SWCNTs: Main Adhesive Contribution	163
4.3.7	Comparison of Mechanical Properties of CNTs	166
4.3.8	Special cases	170
4.4	Imaging	174
4.4.1	Literature Tour	174
4.4.2	Using the Mechanical Properties of CNTs for Imaging	174
4.5	Conclusion	176
References		178

5 Scanning Probes for the Life Sciences
Andrea M. Ho, Horacio D. Espinosa 183

5.1	Introduction	183
5.2	Microarray Technology	184
5.2.1	Microcontact Printing	185
5.2.2	Optical Lithography	186
5.2.3	Protein Arrays	188
5.3	Nanoarray Technology	189
5.3.1	The Push for Nanoscale Detection	189
5.3.2	Probe-Based Patterning	191
5.3.3	Alternative Patterning Methods	202
5.4	Nanoscale Deposition Mechanisms	204
5.5	AFM Parallelization	207
5.5.1	One-Dimensional Arrays	208
5.5.2	Two-Dimensional Arrays	209
5.6	Future Prospects for Nanoprobes	212
References		214

6	**Self-Sensing Cantilever Sensor for Bioscience**	
	Hayato Sone, Sumio Hosaka	219
6.1	Introduction	219
6.2	Basics of the Cantilever Mass Sensor	220
6.3	Finite Element Method Simulation of the Cantilever Vibration	223
6.4	Detection of Cantilever Deflection	226
6.4.1	Using a Position Sensor	226
6.4.2	Using a Piezoresistive Sensor	227
6.5	Self-Sensing Systems	232
6.5.1	Vibration Systems	232
6.5.2	Vibration-Frequency Detection Systems	232
6.6	Applications	233
6.6.1	Water Molecule Detection in Air	233
6.6.2	Antigen and Antibody Detection in Water	238
6.7	Prospective Applications	244
	References	244

7	**AFM Sensors in Scanning Electron and Ion Microscopes: Tools for Nanomechanics, Nanoanalytics, and Nanofabrication**	
	Vinzenz Friedli, Samuel Hoffmann, Johann Michler, Ivo Utke	247
7.1	Introduction	248
7.2	Description of Standalone Techniques	250
7.2.1	FEB/FIB Nanofabrication	250
7.2.2	Cantilever as a Static Force Sensor	255
7.2.3	Cantilever as a Resonating Mass Sensor	255
7.2.4	Nanomanipulation	256
7.3	Fundamentals of Cantilever-Based Sensors	257
7.3.1	Static Operation—Force Sensors	257
7.3.2	Dynamic Operation—Mass Sensors	258
7.3.3	Sensor Scaling	263
7.3.4	Cantilever Calibration	264
7.3.5	Temperature Stability	265
7.3.6	Piezoresistive Detection	267
7.4	Analytics at the Nanoscale	268
7.4.1	Nanomechanics	268
7.4.2	Cantilever-Based Gravimetry	276
7.4.3	Atomic Force Microscopy in a SEM	282
7.5	Perspectives and Outlook	283
	References	284

8	**Cantilever Spring-Constant Calibration in Atomic Force Microscopy**	
	Peter J. Cumpson, Charles A. Clifford, Jose F. Portoles,	
	James E. Johnstone, Martin Munz	289
8.1	Introduction	289
8.2	Applications of AFM	291
8.3	Force Measurements and Spring-Constant Calibration	292
8.3.1	Theoretical Methods	292
8.3.2	V-Shaped Cantilevers and the Parallel-Beam Approximation	293
8.3.3	Dynamic Experimental Methods	295
8.3.4	Thermal Methods	297
8.4	Repeatability in AFM Force Measurements	299
8.4.1	z-Axis Displacement Repeatability	300
8.4.2	Cantilever Deflection Repeatability	300
8.5	Microfabricated Devices for AFM Force Calibration	302
8.6	Lateral Force Calibration	308
References		312

9	**Frequency Modulation Atomic Force Microscopy in Liquids**	
	Suzanne P. Jarvis, John E. Sader, Takeshi Fukuma	315
9.1	Introduction	315
9.2	Instrumentation	318
9.2.1	Basic Setup for FM-AFM	318
9.2.2	Cantilever Excitation in a Liquid	319
9.2.3	Cantilever-Deflection Detection	320
9.3	Applications	326
9.3.1	Nonbiological Systems	326
9.3.2	Biological Systems	327
9.4	Theoretical Framework for Quantitative FM-AFM Force	
	Measurements	332
9.4.1	Decomposition of Interaction Force	332
9.4.2	Governing Equations	335
9.4.3	Fundamental Conditions on Interaction Force	336
9.4.4	Determination of Forces	337
9.4.5	Validation of Formulas	339
9.5	Operation in a Fluid	340
9.5.1	Governing Equations and Resonance Frequency in a Fluid	340
9.5.2	Validation of FM-AFM Force Measurements in a Liquid	342

9.6	Phase Detuning in FM-AFM	343
9.6.1	Governing Equations for Arbitrary Phase Shift	344
9.6.2	Coupling of Conservative and Dissipative Forces	345
9.6.3	Operation of FM-AFM Away From the Resonance Frequency	346
9.6.4	Calibration of 90° Phase Shift	346
9.7	Future Prospects	348
References		349

10	**Kelvin Probe Force Microscopy: Recent Advances and Applications** *Yossi Rosenwaks, Oren Tal, Shimon Saraf, Alex Schwarzman, Eli Lepkifker, Amir Boag*	351
10.1	Kelvin Probe Force Microscopy	351
10.2	Sensitivity and Spatial Resolution in KPFM	354
10.2.1	Tip–Sample Electrostatic Interaction	354
10.2.2	A Fast Algorithm for Calculating the Electrostatic Force	356
10.2.3	Noise in KPFM Images	359
10.2.4	Deconvolution of KPFM Images	360
10.3	Measurement of Semiconductor Surface States	362
10.3.1	Surface Charge and Band Bending Measurements	362
10.3.2	Measuring the Energy Distribution of the Surface States	365
10.3.3	Organic Semiconductors: Bulk Density of States	368
References		374

11	**Application of Scanning Capacitance Microscopy to Analysis at the Nanoscale** *Štefan Lányi*	377
11.1	Introduction	377
11.2	Capacitance Microscopes	379
11.2.1	Resolution of Capacitance Transducers	382
11.2.2	Stray Capacitance of the Probe	387
11.2.3	Sensitivity of the Probe	392
11.2.4	SCM Operation Modes	399
11.3	Looking at the Invisible—Capacitance Contrast	400
11.4	Semiconductor Analysis	401
11.5	Other Semiconductor Structures	405
11.6	Looking Deeper	406
11.6.1	Impedance Spectroscopy	407
11.6.2	Deep Level Transient Spectroscopy	408

11.7	Optimising the Experimental Conditions	414
11.8	Conclusions	416
References		417

12 Probing Electrical Transport Properties at the Nanoscale by Current-Sensing Atomic Force Microscopy
Laura Fumagalli, Ignacio Casuso, Giorgio Ferrari, Gabriel Gomila 421

12.1	Introduction	421
12.2	Fundamentals of Electrical Transport Properties: Resistance, Impedance and Noise	424
12.3	Experimental Setups for CS-AFM	429
12.3.1	Conductive Probes	429
12.3.2	Operating Modes of AFM	431
12.3.3	Current Detection Instrumentation	432
12.4	Conductive Atomic Force Microscopy	434
12.5	Nanoscale Impedance Microscopy	437
12.6	Electrical Noise Microscopy	443
12.7	Conclusions	445
References		446

Subject Index 451

Contents – Volume IX

13 Ultrathin Fullerene-Based Films via STM and STS
Luca Gavioli, Cinzia Cepek 1

13.1 Introduction 1
13.2 Basic Principles of STM and STS 2
13.3 Survey of Fullerene-Based Systems 4
13.3.1 Bulk Properties 4
13.3.2 Electronic Structure 6
13.3.3 Alkali-Metal-Doped C_{60} 8
13.3.4 Interaction of C_{60} with Surfaces 10
13.4 Summary 17
References 18

14 Quantitative Measurement of Materials Properties with the (Digital) Pulsed Force Mode
Alexander M. Gigler, Othmar Marti 23

14.1 Introduction 23
14.2 Modes of Intermittent Operation 24
14.2.1 Destructive Versus Nondestructive Measurements 25
14.2.2 The Pulsed Force Mode 26
14.2.3 Operating Principle of the Pulsed Force Mode 26
14.2.4 Analog Pulsed Force Mode 32
14.2.5 Digital Pulsed Force Mode 32
14.3 Contact Mechanics Relevant for Pulsed Force Mode Investigations 33
14.3.1 Hertz Model 33
14.3.2 Sneddon's Extensions to the Hertz Model 36
14.4 Models Incorporating Adhesion 37
14.4.1 Data Processing 39
14.4.2 Polymers 42
14.4.3 Other Applications 47

14.4.4	Pulsed Force Mode and Friction Measurements	47
14.4.5	Cell Mechanics	50
14.5	Summary	51
References		52

15 Advances in SPMs for Investigation and Modification of Solid-Supported Monolayers
Bruno Pignataro 55

15.1	Introduction	55
15.2	SSMs and Their Preparation	57
15.2.1	Self-Assembled Monolayers	57
15.2.2	LB Monolayers	60
15.3	Fundamental and Technological Applications of SSMs	61
15.4	Characterization and Modification of SSMs	64
15.4.1	Characterization of SSMs	64
15.4.2	Modification of SSMs	67
15.5	Latest Advances in SPMs and Applications for Imaging of SSMs	67
15.5.1	Dynamic SFM in the Attractive Regime	69
15.5.2	Dynamic SFM at Different Level of Interaction Forces	71
15.6	Nanopatterning by SPMs and SSMs	76
15.6.1	Addition Nanolithography	76
15.6.2	Elimination and Substitution Nanolithography	78
15.6.3	Nanoelectrochemical Lithography	79
15.6.4	3D Nanolithography	82
15.7	Conclusions and Perspectives	84
References		85

16 Atomic Force Microscopy Studies of the Mechanical Properties of Living Cells
Félix Rico, Ewa P. Wojcikiewicz, Vincent T. Moy 89

16.1	Introduction	89
16.2	Principle of Operation	90
16.2.1	AFM Imaging	92
16.2.2	Force Measurements	92
16.3	Cell Viscoelasticity	93
16.3.1	AFM Tip Geometries	94
16.3.2	Elasticity: Young's Modulus	94
16.3.3	Viscoelasticity: Complex Shear Modulus	96

16.3.4	Cell Adhesion	98
16.4	Concluding Remarks and Future Directions	104
References		105

17	**Towards a Nanoscale View of Microbial Surfaces Using the Atomic Force Microscope** *Claire Verbelen, Guillaume Andre, Xavier Haulot, Yann Gilbert, David Alsteens, Etienne Dague and Yves F. Dufrêne*	111
17.1	Introduction	111
17.2	Imaging	112
17.2.1	Sample Preparation	112
17.2.2	Visualizing Membrane Proteins at Subnanometer Resolution	112
17.2.3	Live-Cell Imaging	113
17.3	Force Spectroscopy	116
17.3.1	Customized Tips	116
17.3.2	Probing Nanoscale Elasticity and Surface Properties	117
17.3.3	Stretching Cell Surface Polysaccharides and Proteins	119
17.3.4	Nanoscale Mapping and Functional Analysis of Molecular Recognition Sites	120
17.4	Conclusions	123
References		124

18	**Cellular Physiology of Epithelium and Endothelium** *Christoph Riethmüller, Hans Oberleithner*	127
18.1	Introduction	127
18.2	Epithelium	128
18.2.1	Transport Through a Septum	128
18.2.2	In the Kidney	130
18.3	Endothelium	136
18.3.1	Paracellular Gaps	137
18.3.2	Cellular Drinking	139
18.3.3	Wound Healing	142
18.3.4	Transmigration of Leukocytes	143
18.4	Technical Remarks	144
18.5	Summary	145
References		145

19	**Application of Atomic Force Microscopy to the Study of Expressed Molecules in or on a Single Living Cell** *Hyonchol Kim, Hironori Uehara, Rehana Afrin, Hiroshi Sekiguchi, Hideo Arakawa, Toshiya Osada, Atsushi Ikai*	149
19.1	Introduction	150
19.2	Methods of Manipulation To Study Molecules in or on a Living Cell Using an AFM	151
19.2.1	AFM Tip Preparation To Manipulate Receptors on a Cell Surface .	151
19.2.2	Analysis of Molecular Interactions Where Multiple Bonds Formed	153
19.2.3	Measurement of Single-Molecule Interaction Strength on Soft Materials	155
19.3	Observation of the Distribution of Specific Receptors on a Living Cell Surface	156
19.3.1	Distribution of Fibronectin Receptors on a Living Fibroblast Cell .	156
19.3.2	Distribution of Vitronectin Receptors on a Living Osteoblast Cell .	159
19.3.3	Quantification of the Number of Prostaglandin Receptors on a Chinese Hamster Ovary Cell Surface	161
19.4	Further Application of the AFM to the Study of Single-Cell Biology	164
19.4.1	Manipulation of Expressed mRNAs in a Living Cell Using an AFM	164
19.4.2	Manipulation of Membrane Receptors on a Living Cell Surface Using an AFM	170
References	. . . .	173
20	**What Can Atomic Force Microscopy Say About Amyloid Aggregates?** *Amalisa Relini, Ornella Cavalleri, Claudio Canale, Tiziana Svaldo-Lanero, Ranieri Rolandi, Alessandra Gliozzi*	177
20.1	Introduction	178
20.2	Techniques and Methods Used To Study Amyloid Aggregates . . .	181
20.2.1	Optical Methods	182
20.2.2	Electron Microscopy	183
20.2.3	X-ray Diffraction	185
20.2.4	Nuclear Magnetic Resonance	185
20.2.5	Atomic Force Microscopy	186
20.3	Monitoring the Aggregation Process by AFM	188
20.4	Effect of Surfaces on the Aggregation Process	190
20.5	Interaction with Model Membranes	193
20.6	Physical Properties of Fibrils Obtained by AFM	197
References	. . . .	201

21	**Atomic Force Microscopy: Interaction Forces Measured in Phospholipid Monolayers, Bilayers and Cell Membranes**	
	Zoya Leonenko, David Cramb, Matthias Amrein, Eric Finot	207
21.1	Introduction	207
21.2	Phase Transitions of Lipid Bilayers in Water	209
21.2.1	Morphology Change During Lamellar Phase Transition	210
21.2.2	Change in Forces During Phase Transition	212
21.3	Force Measurements on Pulmonary Surfactant Monolayers in Air	219
21.3.1	Adhesion Measurements: Monolayer Stiffness and Function	221
21.3.2	Repulsive Forces: The Interaction of Charged Airborne Particles with Surfactant	222
21.4	Interaction Forces Measured on Lung Epithelial Cells in Buffer	224
21.4.1	Cell Culture/Force Measurement Setup	225
21.4.2	Mechanical Properties	227
21.5	Conclusions	230
References		231

22	**Self-Assembled Monolayers on Aluminum and Copper Oxide Surfaces: Surface and Interface Characteristics, Nanotribological Properties, and Chemical Stability**	
	E. Hoque, J. A. DeRose, B. Bhushan, H. J. Mathieu	235
22.1	Introduction	236
22.2	Substrate Preparation	238
22.2.1	Aluminum	238
22.2.2	Copper	239
22.3	Phosphonic Acid and Silane Based SAMs on Al and Cu	239
22.3.1	SAM Preparation	239
22.3.2	Surface and Interface Characterization	240
22.3.3	Nanotribological Properties	261
22.3.4	Chemical Stability	268
22.4	Summary	277
References		279

23	**High Sliding Velocity Nanotribological Investigations of Materials for Nanotechnology Applications**	
	Nikhil S. Tambe, Bharat Bhushan	283
23.1	Bridging Science and Engineering for Nanotribological Investigations	283

23.1.1	Microtribology/Nanotribology	284
23.1.2	Historical Perspective for Velocity Dependence of Friction	284
23.2	Need for Speed: Extending the AFM Capabilities for High Sliding Velocity Studies	285
23.2.1	Modifications to the Commercial AFM Setup	287
23.2.2	Friction Investigations on the Microscale/Nanoscale at High Sliding Velocities	298
23.3	Microscale/Nanoscale Friction and Wear Studies at High Sliding Velocities	300
23.3.1	Nanoscale Friction Mapping: Understanding Normal Load and Velocity Dependence of Friction Force	301
23.3.2	Nanoscale Wear Mapping: Wear Studies at High Sliding Velocities	303
23.4	Closure	308
References		309

24	**Measurement of the Mechanical Properties of One-Dimensional Polymer Nanostructures by AFM** *Sung-Kyoung Kim, Haiwon Lee*	311
24.1	Introduction	311
24.2	AFM-Based Techniques for Measuring the Mechanical Properties of 1D Polymer Nanostructures	312
24.3	Mechanical Properties of Electrospun Polymer Nanofibers	318
24.4	Test of Reliability of AFM-Based Measurements	323
References		327

25	**Evaluating Tribological Properties of Materials for Total Joint Replacements Using Scanning Probe Microscopy** *Sriram Sundararajan, Kanaga Karuppiah Kanaga Subramanian*	329
25.1	Introduction	329
25.1.1	Total Joint Replacements	329
25.1.2	Social and Economic Significance	330
25.2	Problems Associated with Total Joint Replacements	330
25.2.1	Tribology	332
25.2.2	Materials	332
25.2.3	Lubrication in Joints—the Synovial Fluid	333
25.3	Conventional Tribological Testing of Material Pairs for Total Joint Replacements	334

25.3.1	Wear Tests	334
25.3.2	Friction Tests	334
25.4	Scanning Probe Microscopy as a Tool to Study Tribology of Total Joint Replacements	334
25.4.1	Nanotribology of Ultrahigh Molecular Weight Polyethylene	335
25.4.2	Fretting Wear of Cobalt–Chromium Alloy	343
25.5	Summary and Future Outlook	347
References		348

26	**Near-Field Optical Spectroscopy of Single Quantum Constituents** *Toshiharu Saiki*	351
26.1	Introduction	351
26.2	General Description of NSOM	353
26.3	NSOM Aperture Probe	354
26.3.1	Basic Process of Aperture Probe Fabrication	354
26.3.2	Tapered Structure and Optical Throughput	355
26.3.3	Fabrication of a Double-Tapered Aperture Probe	356
26.3.4	Evaluation of Transmission Efficiency and Collection Efficiency	357
26.3.5	Evaluation of Spatial Resolution with Single QDs	358
26.4	Single-Quantum-Constituent Spectroscopy	360
26.4.1	Light–Matter Interaction at the Nanoscale	361
26.4.2	Real-Space Mapping of an Exciton Wavefunction Confined in a QD	363
26.4.3	Carrier Localization in Cluster States in GaNAs	367
26.5	Perspectives	370
References		371

Subject Index	373

Contents – Volume I

Part I Scanning Probe Microscopy

1 Dynamic Force Microscopy
André Schirmeisen, Boris Anczykowski, Harald Fuchs 3

2 Interfacial Force Microscopy: Selected Applications
Jack E. Houston 41

3 Atomic Force Microscopy with Lateral Modulation
Volker Scherer, Michael Reinstädtler, Walter Arnold 75

4 Sensor Technology for Scanning Probe Microscopy
Egbert Oesterschulze, Rainer Kassing 117

5 Tip Characterization for Dimensional Nanometrology
John S. Villarrubia 147

Part II Characterization

6 Micro/Nanotribology Studies Using Scanning Probe Microscopy
Bharat Bhushan 171

7 Visualization of Polymer Structures with Atomic Force Microscopy
Sergei Magonov 207

8 Displacement and Strain Field Measurements from SPM Images
Jürgen Keller, Dietmar Vogel, Andreas Schubert, Bernd Michel . . 253

9 AFM Characterization of Semiconductor Line Edge Roughness
Ndubuisi G. Orji, Martha I. Sanchez, Jay Raja,
Theodore V. Vorburger 277

10 Mechanical Properties of Self-Assembled Organic Monolayers: Experimental Techniques and Modeling Approaches
Redhouane Henda 303

| 11 | Micro-Nano Scale Thermal Imaging Using Scanning Probe Microscopy
Li Shi, Arun Majumdar . . . . 327 |
|---|---|
| 12 | The Science of Beauty on a Small Scale. Nanotechnologies Applied to Cosmetic Science
Gustavo Luengo, Frédéric Leroy . . . . 363 |

Part III Industrial Applications

| 13 | SPM Manipulation and Modifications and Their Storage Applications
Sumio Hosaka . . . . 389 |
|---|---|
| 14 | Super Density Optical Data Storage by Near-Field Optics
Jun Tominaga . . . . 429 |
| 15 | Capacitance Storage Using a Ferroelectric Medium and a Scanning Capacitance Microscope (SCM)
Ryoichi Yamamoto . . . . 439 |
| 16 | Room-Temperature Single-Electron Devices formed by AFM Nano-Oxidation Process
Kazuhiko Matsumoto . . . . 459 |

Subject Index 469

Contents – Volume II

1. **Higher Harmonics in Dynamic Atomic Force Microscopy**
   Robert W. Stark, Martin Stark 1

2. **Atomic Force Acoustic Microscopy**
   Ute Rabe 37

3. **Scanning Ion Conductance Microscopy**
   Tilman E. Schäffer, Boris Anczykowski, Harald Fuchs 91

4. **Spin-Polarized Scanning Tunneling Microscopy**
   Wulf Wulfhekel, Uta Schlickum, Jürgen Kirschner 121

5. **Dynamic Force Microscopy and Spectroscopy**
   Ferry Kienberger, Hermann Gruber, Peter Hinterdorfer 143

6. **Sensor Technology for Scanning Probe Microscopy and New Applications**
   Egbert Oesterschulze, Leon Abelmann, Arnout van den Bos, Rainer Kassing, Nicole Lawrence, Gunther Wittstock, Christiane Ziegler 165

7. **Quantitative Nanomechanical Measurements in Biology**
   Małgorzata Lekka, Andrzej J. Kulik 205

8. **Scanning Microdeformation Microscopy: Subsurface Imaging and Measurement of Elastic Constants at Mesoscopic Scale**
   Pascal Vairac, Bernard Cretin 241

9. **Electrostatic Force and Force Gradient Microscopy: Principles, Points of Interest and Application to Characterisation of Semiconductor Materials and Devices**
   Paul Girard, Alexander Nikolaevitch Titkov 283

10. **Polarization-Modulation Techniques in Near-Field Optical Microscopy for Imaging of Polarization Anisotropy in Photonic Nanostructures**
    Pietro Giuseppe Gucciardi, Ruggero Micheletto, Yoichi Kawakami, Maria Allegrini 321

11   **Focused Ion Beam as a Scanning Probe: Methods and Applications**
Vittoria Raffa, Piero Castrataro, Arianna Menciassi, Paolo Dario .   361

Subject Index 413

Contents – Volume III

12	**Atomic Force Microscopy in Nanomedicine** *Dessy Nikova, Tobias Lange, Hans Oberleithner,* *Hermann Schillers, Andreas Ebner, Peter Hinterdorfer*	1
13	**Scanning Probe Microscopy:** **From Living Cells to the Subatomic Range** *Ille C. Gebeshuber, Manfred Drack, Friedrich Aumayr,* *Hannspeter Winter, Friedrich Franek*	27
14	**Surface Characterization and Adhesion and Friction Properties** **of Hydrophobic Leaf Surfaces and Nanopatterned Polymers** **for Superhydrophobic Surfaces** *Zachary Burton, Bharat Bhushan*	55
15	**Probing Macromolecular Dynamics and the Influence** **of Finite Size Effects** *Scott Sills, René M. Overney*	83
16	**Investigation of Organic Supramolecules by Scanning Probe Microscopy** **in Ultra-High Vacuum** *Laurent Nony, Enrico Gnecco, Ernst Meyer*	131
17	**One- and Two-Dimensional Systems: Scanning Tunneling Microscopy** **and Spectroscopy of Organic and Inorganic Structures** *Luca Gavioli, Massimo Sancrotti*	183
18	**Scanning Probe Microscopy Applied to Ferroelectric Materials** *Oleg Tikhomirov, Massimiliano Labardi, Maria Allegrini*	217
19	**Morphological and Tribological Characterization of Rough Surfaces** **by Atomic Force Microscopy** *Renato Buzio, Ugo Valbusa*	261
20	**AFM Applications for Contact and Wear Simulation** *Nikolai K. Myshkin, Mark I. Petrokovets, Alexander V. Kovalev* . .	299
21	**AFM Applications for Analysis of Fullerene-Like Nanoparticles** *Lev Rapoport, Armen Verdyan*	327

22	**Scanning Probe Methods in the Magnetic Tape Industry**	
	James K. Knudsen	343

Subject Index 371

Contents – Volume IV

23	Scanning Probe Lithography for Chemical, Biological and Engineering Applications *Joseph M. Kinsella, Albena Ivanisevic*	1
24	Nanotribological Characterization of Human Hair and Skin Using Atomic Force Microscopy (AFM) *Bharat Bhushan, Carmen LaTorre*	35
25	Nanofabrication with Self-Assembled Monolayers by Scanning Probe Lithography *Jayne C. Garno, James D. Batteas*	105
26	Fabrication of Nanometer-Scale Structures by Local Oxidation Nanolithography *Marta Tello, Fernando García, Ricardo García*	137
27	Template Effects of Molecular Assemblies Studied by Scanning Tunneling Microscopy (STM) *Chen Wang, Chunli Bai*	159
28	Microfabricated Cantilever Array Sensors for (Bio-)Chemical Detection *Hans Peter Lang, Martin Hegner, Christoph Gerber*	183
29	Nano-Thermomechanics: Fundamentals and Application in Data Storage Devices *B. Gotsmann, U. Dürig*	215
30	Applications of Heated Atomic Force Microscope Cantilevers *Brent A. Nelson, William P. King*	251

Subject Index 277

Contents – Volume V

1. **Integrated Cantilevers and Atomic Force Microscopes**
   Sadik Hafizovic, Kay-Uwe Kirstein, Andreas Hierlemann 1

2. **Electrostatic Microscanner**
   Yasuhisa Ando 23

3. **Low-Noise Methods for Optical Measurements of Cantilever Deflections**
   Tilman E. Schäffer 51

4. **Q-controlled Dynamic Force Microscopy in Air and Liquids**
   Hendrik Hölscher, Daniel Ebeling, Udo D. Schwarz 75

5. **High-Frequency Dynamic Force Microscopy**
   Hideki Kawakatsu 99

6. **Torsional Resonance Microscopy and Its Applications**
   Chanmin Su, Lin Huang, Craig B. Prater, Bharat Bhushan 113

7. **Modeling of Tip-Cantilever Dynamics in Atomic Force Microscopy**
   Yaxin Song, Bharat Bhushan 149

8. **Combined Scanning Probe Techniques for In-Situ Electrochemical Imaging at a Nanoscale**
   Justyna Wiedemair, Boris Mizaikoff, Christine Kranz 225

9. **New AFM Developments to Study Elasticity and Adhesion at the Nanoscale**
   Robert Szoszkiewicz, Elisa Riedo 269

10. **Near-Field Raman Spectroscopy and Imaging**
    Pietro Giuseppe Gucciardi, Sebastiano Trusso, Cirino Vasi,
    Salvatore Patanè, Maria Allegrini 287

Subject Index 331

Contents – Volume VI

11 **Scanning Tunneling Microscopy of Physisorbed Monolayers: From Self-Assembly to Molecular Devices**
Thomas Müller 1

12 **Tunneling Electron Spectroscopy Towards Chemical Analysis of Single Molecules**
Tadahiro Komeda 31

13 **STM Studies on Molecular Assembly at Solid/Liquid Interfaces**
Ryo Yamada, Kohei Uosaki 65

14 **Single-Molecule Studies on Cells and Membranes Using the Atomic Force Microscope**
Ferry Kienberger, Lilia A. Chtcheglova, Andreas Ebner, Theeraporn Puntheeranurak, Hermann J. Gruber, Peter Hinterdorfer 101

15 **Atomic Force Microscopy of DNA Structure and Interactions**
Neil H. Thomson 127

16 **Direct Detection of Ligand–Protein Interaction Using AFM**
Małgorzata Lekka, Piotr Laidler, Andrzej J. Kulik 165

17 **Dynamic Force Microscopy for Molecular-Scale Investigations of Organic Materials in Various Environments**
Hirofumi Yamada, Kei Kobayashi 205

18 **Noncontact Atomic Force Microscopy**
Yasuhiro Sugawara 247

19 **Tip-Enhanced Spectroscopy for Nano Investigation of Molecular Vibrations**
Norihiko Hayazawa, Yuika Saito 257

20 **Investigating Individual Carbon Nanotube/Polymer Interfaces with Scanning Probe Microscopy**
Asa H. Barber, H. Daniel Wagner, Sidney R. Cohen 287

Subject Index 325

Contents – Volume VII

21	**Lotus Effect: Roughness-Induced Superhydrophobicity** *Michael Nosonovsky, Bharat Bhushan*	1
22	**Gecko Feet: Natural Attachment Systems for Smart Adhesion** *Bharat Bhushan, Robert A. Sayer*	41
23	**Novel AFM Nanoprobes** *Horacio D. Espinosa, Nicolaie Moldovan, K.-H. Kim*.	77
24	**Nanoelectromechanical Systems – Experiments and Modeling** *Horacio D. Espinosa, Changhong Ke*	135
25	**Application of Atom-resolved Scanning Tunneling Microscopy in Catalysis Research** *Jeppe Vang Lauritsen, Ronny T. Vang, Flemming Besenbacher*. . .	197
26	**Nanostructuration and Nanoimaging of Biomolecules for Biosensors** *Claude Martelet, Nicole Jaffrezic-Renault, Yanxia Hou, Abdelhamid Errachid, François Bessueille*	225
27	**Applications of Scanning Electrochemical Microscopy (SECM)** *Gunther Wittstock, Malte Burchardt, Sascha E. Pust*	259
28	**Nanomechanical Characterization of Structural and Pressure-Sensitive Adhesives** *Martin Munz, Heinz Sturm*	301
29	**Development of MOEMS Devices and Their Reliability Issues** *Bharat Bhushan, Huiwen Liu*	349

Subject Index 367

List of Contributors – Volume X

Alexander Alexeev
NT-MDT Co., Post Box 158, Building 317, Zelenograd, Moscow 124482, Russia
e-mail: alexander@ntmdt.ru

Tianming Bao
Veeco Instruments Inc, 14990 Blakehill Dr, Frisco, TX 75035, USA
e-mail: tbao@veeco.com

Bharat Bhushan
Nanotribology Lab for Information Storage and MEMS/NEMS (NLIM)
The Ohio State University
201 W. 19th Avenue, W390 Scott Laboratory, Columbus, Ohio 43210-1142, USA
e-mail: bhushan.2@osu.edu

Yasuo Cho
Research Institute of Electrical Communication, Tohoku University
2-1-1 Katahira Aoba-ku Sendai, 980-8577, Japan
e-mail: cho@riec.tohoku.ac.jp

Gianluca Ciardelli
Department of Mechanics, Politecnico di Torino
Corso Duca degli Abruzzi 24, 10129 Torino, Italy
e-mail: gianluca.ciardelli@polito.it

Mario D'Acunto
Department of Chemical Engineering and Materials Science, University of Pisa
via Diotisalvi 2, I-56126 Pisa, Italy
e-mail: m.dacunto@ing.unipi.it

Patrick Fiorenza
CNR-IMM, Stradale Primosole 50, I-95121 Catania, Italy
e-mail: patrick.fiorenza@imm.cnr.it

David Fong
Veeco Instruments Inc, 112 Robin Hill Rd, Santa Barbara, CA 93117, USA
e-mail: david.fong@veeco.com

Filippo Giannazzo
CNR-IMM, Stradale Primosole 50, I-95121 Catania, Italy
e-mail: filippo.giannazzo@imm.cnr.it

Paolo Giusti
Department of Chemical Engineering and Materials Science, University of Pisa
via Diotisalvi 2, I-56126 Pisa, Italy
e-mail: giusti@ing.unipi.it

Sean Hand
Veeco Instruments Inc, 112 Robin Hill Rd, Santa Barbara, CA 93117, USA

Yasushi Kadota
Ricoh, 1-3-6 Nakamagame, Ohta-ku, Tokyo 143-8555, Japan
e-mail: yasushi.kadota@nts.ricoh.co.jp

Toshi Kasai
Cabot Microelectronics
870 N. Commons Drive, Aurora, IL 60504, USA
e-mail: toshi_kasai@cabotcmp.com

Kazushige Kawabata
Division of Biological Sciences, Graduate School of Science, Hokkaido University
North 10 West 8, Kita-ku, Sapporo 060-0810, Japan
e-mail: kaw@sci.hokudai.ac.jp

Joachim Loos
Department of Chemical Engineering and Chemistry
Eindhoven University of Technology
P.O. Box 513, 5600 MB Eindhoven, The Netherlands
e-mail: j.loos@tue.nl

Franco Montevecchi
Department of Mechanics, Politecnico di Torino
Corso Duca degli Abruzzi 24, I-10129 Torino, Italy
e-mail: franco.montevecchi@polito.it

Vito Raineri
CNR-IMM, Stradale Primosole 50, I-95121 Catania, Italy
e-mail: vito.raineri@imm.cnr.it

Robert A. Sayer
Nanotribology Laboratory for Information Storage and MEMS/NEMS (NLIM)
The Ohio State University
201 West 19th Avenue, Columbus, OH 43210-1142, USA

Ken-ichi Shinohara
School of Materials Science, Japan Advanced Institute of Science and Technology
Asahi-dai, Nomi, Ishikawa 923-1292, Japan
e-mail: shinoken@jaist.ac.jp

Hiroyuki Sugimura
Kyoto University, Sakyo, Kyoto 606-8501, Japan
e-mail: hiroyuki-sugimura@mtl.kyoto-u.ac.jp

Tatsuo Ushiki
Division of Microscopic Anatomy
Graduate School of Medical and Dental Sciences, Niigata University
757 Asahimachi-dori, Chuo-ku, Niigata 951-8510, Japan
e-mail: t-ushiki@med.niigata-u.ac.jp

Jiping Ye
Nano Analysis Section, Research Department, Nissan ARC
1 Natsushima-cho, Yokosuka 237-0061, Japan
e-mail: ye@nissan-arc.co.jp

List of Contributors – Volume VIII

Jean-Pierre Aimé
CPMOH, Université Bordeaux 1
351 Cours de la Libération, 33405 Talence cedex, France
e-mail: jp.aime@cpmoh.u-bordeaux1.fr

Matteo Albani
Dipartimento di Ingegneria dell'Informazione, Università di Siena
Via Roma 56, 53100 Siena, Italy
e-mail: matteo.albani@ing.unisi.it

Maria Allegrini
Dipartimento di Fisica "Enrico Fermi", Università di Pisa
Largo Bruno Pontecorvo 3, 56127 Pisa, Italy
e-mail: maria.allegrini@df.unipi.it

Guillaume Bachelier
Laboratoire de Spectrométrie Ionique et Moléculaire (LASIM)
Université Claude Bernard – Lyon 1, Bât. A. Kastler
43 Boulevard du 11 Novembre 1918, 69622 Villeurbanne Cedex, France
e-mail: guillaume.bachelier@lasim.univ-lyon1.fr

Charlotte Bernard
CPMOH, Université Bordeaux 1
351 Cours de la Libération, 33405 Talence cedex, France
e-mail: c.bernard@cpmoh.u-bordeaux1.fr

Bharat Bhushan
Nanotribology Lab for Information Storage and MEMS/NEMS (NLIM)
The Ohio State University
201 W. 19th Avenue, W390 Scott Laboratory, Columbus, Ohio 43210-1142
USA
e-mail: bhushan.2@osu.edu

Amir Boag
School of Electrical Engineering, Faculty of Engineering, Tel Aviv University
Ramat-Aviv, 69978, Israel
e-mail: boag@eng.tau.ac.il

Rodolphe Boisgard
CPMOH, Université Bordeaux 1
351 Cours de la Libération, 33405 Talence cedex, France
e-mail: r.boisgard@cpmoh.u-bordeaux1.fr

Anne Marie Bonnot
LEPES
25 av. des Martyrs, 38042 Grenoble cedex, France
e-mail: bonnot@grenoble.cnrs.fr

P. Carl
Institute of Physiology II, University of Münster
Robert-Koch-Straße 27b, 48149 Münster, Germany
e-mail: pcarl@uni-muenster.de

Ignacio Casuso
Laboratory of Nanobioengineering, Barcelona Science Park
and Department of Electronics, University of Barcelona
C/ Marti i Franque 1, 08028-Barcelona, Spain
e-mail: icasuso@pcb.ub.es

Eugenio Cefalì
Dipartimento di Fisica della Materia e Tecnologie Fisiche Avanzate
Università di Messina
Salita Sperone n. 31, 98166 Messina, Italy
e-mail: eugcef@libero.it

Charles Clifford
Analytical Science Team, Quality of Life Division, National Physical Laboratory
Teddington, Middlesex TW11 0LW, UK
e-mail: charles.clifford@npl.co.uk

Peter J. Cumpson
Analytical Science Team, Quality of Life Division, National Physical Laboratory
Teddington, Middlesex TW11 0LW, UK
e-mail: peter.cumpson@npl.co.uk

Gregory A. Dahlen
Research Scientist
112 Robin Hill Road, Santa Barbara, CA 93117, USA
e-mail: gdahlen@veeco.com

Dirk Dietzel
Physikalisches Institut, Universität Münster
Wilhelm-Klemm-Straße 10, 48149 Münster, Germany
e-mail: dirk.dietzel@uni-muenster.de

Robert Eibl
Plainburgstraße 8, 83457 Bayerisch Gmain, Germany
e-mail: robert_eibl@yahoo.com

Horacio D. Espinosa
Department of Mechanical Engineering, Northwestern University
2145 Sheridan Rd., Evanston, IL 60208-3111, USA
e-mail: espinosa@northwestern.edu

Giorgio Ferrari
Dipartimento di Elettronica e Informazione, Politecnico di Milano
P.za L. da Vinci 32, 20133 Milano, Italy
e-mail: ferrari@elet.polimi.it

Vinzenz Friedli
EMPA – Materials Science and Technology
Feuerwerker Strasse 39, CH-3602 Thun, Switzerland
e-mail: vinzenz.friedli@empa.ch

Takeshi Fukuma
Department of Physics, Kanazawa University
Kakuma-machi, Kanazawa, 920-1192, Japan
email: takeshi.fukuma@tcd.ie

Laura Fumagalli
Laboratory of Nanobioengineering Barcelona Science Park
and Department of Electronics, University of Barcelona
C/ Marti i Franque 1, 08028-Barcelona, Spain
e-mail: Laura.fumagalli@polimi.it

Renato Gardelli
Dipartimento di Fisica della Materia e Tecnologie Fisiche Avanzate
Università di Messina
Salita Sperone n. 31, 98166 Messina, Italy
e-mail: rgardelli@unime.it

G. Gomila
Laboratory of Nanobioengineering, Barcelona Science Park
and Department of Electronics, University of Barcelona
C/ Marti i Franque 1, 08028-Barcelona, Spain
e-mail: ggomila@el.ub.es

Pietro Giuseppe Gucciardi
CNR – Istituto per i Processi Chimico-Fisici, Sezione Messina
Via La Farina 237, 98123 Messina, Italy
e-mail: gucciardi@me.cnr.it

Andrea Ho
Department of Mechanical Engineering, Northwestern University
2145 Sheridan Road, Evanston, IL 60208-3111, USA

Samuel Hoffmann
EMPA – Materials Science and Technology
Feuerwerker Strasse 39, CH-3602 Thun, Switzerland
e-mail: samuel.hoffmann@empa.ch

Sumio Hosaka
Department of Production Science and Technology, Gunma University
1-5-1 Tenjin-cho, Kiryu, Gunma, 376-8515, Japan

Suzanne P. Jarvis
Conway Institute of Biomolecular and Biomedical Research
University College Dublin, Belfield, Dublin 4, Ireland
e-mail: suzi.jarvis@ucd.ie

James Johnstone
DEPC, National Physical Laboratory
Hampton Road, Teddington TW11 0LW, UK
e-mail: james.johnstone@npl.co.uk

P.M. Koenraad
Semiconductor Physics Group
Department of Applied Physics, Eindhoven University of Technology
P.O. Box 513, 5600 MB Eindhoven, The Netherlands
e-mail: P.M.Koenraad@tue.nl

Stefan Lanyi
Institute of Physics, Slovakian Academy o Sciences
Dubravska cesta 9, SK-845 11 Bratislava, Slovakia
e-mail: lanyi@savba.sk

Eli Lepkifker
School of Electrical Engineering, Faculty of Engineering, Tel Aviv University
Ramat-Aviv, 69978, Israel
e-mail: lepkifke@post.tau.ac.il

Hao-Chih Liu
SPM Probe Scientist
112 Robin Hill Road, Santa Barbara, CA 93117, USA
e-mail: hcliu@veeco.com

Sophie Marsaudon
CPMOH, Université Bordeaux 1
351 Cours de la Libération, 33405 Talence cedex, France
e-mail: s.marsaudon@cpmoh.u-bordeaux1.fr

Johann Michler
EMPA – Materials Science and Technology
Feuerwerker Strasse 39, CH-3602 Thun, Switzerland
e-mail: johann.michler@empa.ch

Martin Munz
Analytical Science Team, Quality of Life Division, National Physical Laboratory
Teddington, Middlesex TW11 0LW, UK
e-mail: martin.munz@npl.co.uk

Cattien V. Nguyen
Research Scientist, Eloret Corp./NASA Ames Research Center
MS 229-1, Moffett Field, CA 94035-1000, USA
e-mail: cvnguyen@mail.arc.nasa.gov

Jason R. Osborne
Eng. Manager, Advanced Development Group
112 Robin Hill Road, Santa Barbara, CA 93117, USA
e-mail: josborne@veeco.com

Salvatore Patanè
Dipartimento di Fisica della Materia e Tecnologie Fisiche Avanzate
Università di Messina
Salita Sperone n. 31, 98166 Messina, Italy
e-mail: salvatore.patane@unime.it

Jose Portoles
Laboratory of Biophysics and Surface Analysis, University of Nottingham
Nottingham, NG72RD, UK
e-mail: paxjp1@nottingham.ac.uk

Yossi Rosenwaks
School of Electrical Engineering, Faculty of Engineering, Tel Aviv University
Ramat-Aviv, 69978, Israel
e-mail: yossir@eng.tau.ac.il

John E. Sader
Department of Mathematics and Statistics, The University of Melbourne
Victoria 3010, Australia
e-mail: jsader@unimelb.edu.au

Shimon Saraf
School of Electrical Engineering, Faculty of Engineering, Tel Aviv University
Ramat-Aviv, 69978, Israel
e-mail: saraf@eng.tau.ac.il

Alex Schwarzman
School of Electrical Engineering, Faculty of Engineering, Tel Aviv University
Ramat-Aviv, 69978, Israel

Hayato Sone
Department of Production Science and Technology, Gunma University
1-5-1 Tenjin-cho, Kiryu, Gunma, 376-8515, Japan
e-mail: sone@el.gunma-u.ac.jp

Salvatore Spadaro
Dipartimento di Fisica della Materia e Tecnologie Fisiche Avanzate
Università di Messina
Salita Sperone n. 31, 98166 Messina, Italy
e-mail: salvatore.spadaro@ortica.unime.it

Stephan J. Stranick
National Institute of Standards and Technology Surface
and Microanalysis Science Division (837) 100 Bureau Drive
Gaithersburg, MD 20899-8372, USA
e-mail: stephan.stranick@nist.gov

Oren Tal
School of Electrical Engineering, Faculty of Engineering, Tel Aviv University
Ramat-Aviv, 69978, Israel
e-mail: tal@Physics.LeidenUniv.nl

Ivo Utke
EMPA – Materials Science and Technology
Feuerwerker Strasse 39, CH-3602 Thun, Switzerland
e-mail: ivo.utke@empa.ch

List of Contributors – Volume IX

Rehana Afrin
Department of Life Science, Tokyo Institute of Technology
Nagatsuta, Midori-ku, Yokohama 226-8501, Japan
e-mail: rafrin@bio.titech.ac.jp

David Alsteens
Unité de Chimie des Interfaces, Université catholique de Louvain
Croix du Sud 2/18, 1348 Louvain-la-Neuve, Belgium
e-mail: alsteens@cifa.ucl.ac.be

Matthias Amrein
Department of Cell Biology and Anatomy, Faculty of Medicine
University of Calgary, Canada
e-mail: mamrein@ucalgary.ca

Guillaume Andre
Unité de Chimie des Interfaces, Université catholique de Louvain
Croix du Sud 2/18, 1348 Louvain-la-Neuve, Belgium
e-mail: andre@cifa.ucl.ac.be

Hideo Arakawa
Bio Nanomaterials Group, National Institute for Materials Science
1-1, Namiki, Tsukuba 305-0044, Japan
e-mail : ARAKAWA.Hideo@nims.go.jp

Bharat Bhushan
Nanotribology Lab for Information Storage and MEMS/NEMS (NLIM)
The Ohio State University
201 W. 19th Avenue, W390 Scott Laboratory, Columbus, Ohio 43210-1142, USA
e-mail: bhushan.2@osu.edu

Claudio Canale
Physics Department, University of Genoa
Via Dodecaneso 33, I-16146 Genoa, Italy
e-mail: canale@fisica.unige.it

Ornella Cavalleri
Physics Department, University of Genoa
Via Dodecaneso 33, I-16146 Genoa, Italy
e-mail: cavalleri@fisica.unige.it

Cinzia Cepek
Laboratorio Nazionale TASC-INFM
Strada Statale 14, km 163.5 Basovizza, I-34012 Trieste, Italy
e-mail: cepek@tasc.infm.it

David T. Cramb
Department of Chemistry, University of Calgary, Canada
e-mail: dcramb@ucalgary.ca

Etienne Dague
Unité de Chimie des Interfaces, Université catholique de Louvain
Croix du Sud 2/18, 1348 Louvain-la-Neuve, Belgium
e-mail: dague@cifa.ucl.ac.be

James A. DeRose
EMPA
Ueberlandstrasse 129, CH-8600 Duebendorf, Switzerland
e-mail: james.derose@empa.ch

Y. F. Dufrene
Unité de Chimie des Interfaces, Université catholique de Louvain
Croix du Sud 2/18, 1348 Louvain-la-Neuve, Belgium
e-mail: dufrene@cifa.ucl.ac.be

Eric Finot
Institut Carnot de Bourgogne, UMR 5209 CNRS
Université de Bourgogne, France
e-mail: efinot@u-bourgogne.fr

Luca Gavioli
Dipartimento di Matematica e Fisica, Università Cattolica del Sacro Cuore
Via dei Musei 41, I-25121 Brescia, Italy
e-mail: l.gavioli@dmf.bs.unicatt.it

Alexander Gigler
Sektion Kristallographie, Ludwig-Maximilians-Universität München
Theresienstraße 41/II, 80333 München, Germany
e-mail: gigler@lrz.uni-muenchen.de

Yann Gilbert
Unité de Chimie des Interfaces, Université catholique de Louvain
Croix du Sud 2/18, 1348 Louvain-la-Neuve, Belgium
e-mail: gilbert@cifa.ucl.ac.be

Alessandra Gliozzi
Physics Department, University of Genoa
Via Dodecaneso 33, I-16146 Genoa, Italy
e-mail: gliozzi@fisica.unige.it

Xavier Haulot
Unité de Chimie des Interfaces, Université catholique de Louvain
Croix du Sud 2/18, 1348 Louvain-la-Neuve, Belgium
e-mail: haulot@cifa.ucl.ac.be

Enamul Hoque
Department of Chemistry and Materials Science, Washington State University
Pullman, WA 99164-4630, USA
e-mail: ehoque@ad.wsu.edu

Atsushi Ikai
Department of Life Science, Tokyo Institute of Technology
Nagatsuta, Midori-ku, Yokohama 226-8501, Japan
e-mail: aikai@bio.titech.ac.jp

K. S. Kanaga Karuppiah
Department of Mechanical Engineering, 2025 H. M. Black Engineering Building
Iowa State University, Ames, IA 50011-2161, USA
e-mail: kskarup@iastate.edu

Hyonchol Kim
Division of Biosystems, Institute of Biomaterials and Bioengineering
Tokyo Medical and Dental University
2-3-10, Kanda-surugadai, Chiyoda-ku, Tokyo, 101-0062, Japan
e-mail: kim.bmi@tmd.ac.jp

Sung-Kyoung Kim
Department of Chemistry, Hanyang University
17 Haengdangdong, Seoul 133-791, Korea
e-mail: s_ksk92@hanmail.net

Tiziana Svaldo-Lanero
Physics Department, University of Genoa
Via Dodecaneso 33, I-16146 Genoa, Italy
e-mail: svaldo@fisica.unige.it

Haiwon Lee
Department of Chemistry, Hanyang University
17 Haengdangdong, Seoul 133-791, Korea
e-mail: haiwon@hanyang.ac.kr

Zoya Leonenko
Department of Cell Biology and Anatomy, Faculty of Medicine
University of Calgary, Canada
e-mail: zleonenk@ucalgary.ca

Othmar Marti
Experimental Physics, Ulm University
89069 Ulm, Germany
e-mail: othmar.marti@uni-ulm.de

Hans Jörg Mathieu
EPFL – IMX, Station 12, 1015 Lausanne, Switzerland
e-mail: HansJoerg.Mathieu@EPFL.ch

Vincent T. Moy
Department of Physiology & Biophysics, University of Miami
1600 NW 12th Ave., RMSB 5077, Miami, FL 33136, USA
e-mail: vmoy@miami.edu

H. Oberleithner
University of Münster, Institute of Physiology II
Robert-Koch-Straße 27b, 48149 Münster, Germany
e-mail: oberlei@uni-muenster.de

Toshiya Osada
Department of Life Science, Tokyo Institute of Technology
Nagatsuta, Midori-ku, Yokohama 226-8501, Japan
e-mail: tosada@bio.titech.ac.jp

Bruno Pignataro
Dipartimento di Chimica Fisica "F. Accascina", Università di Palermo
V.le delle Scienze – Parco d'Orleans II, 90128 Palermo, Italy
e-mail: bruno.pignataro@unipa.it

Annalisa Relini
Physics Department, University of Genoa
Via Dodecaneso 33, I-16146 Genoa, Italy
e-mail: relini@fisica.unige.it

Félix Rico
Department of Physiology & Biophysics, University of Miami
1600 NW 12th Ave, RMSB 5077, Miami, FL 33136, USA
e-mail: frico@med.miami.edu

C. Riethmüller
Institute of Physiology II, University of Münster
Robert-Koch-Straße 27b, 48149 Münster, Germany
e-mail: chrth@uni-muenster.de

Ranieri Rolandi
Physics Department, University of Genoa
Via Dodecaneso 33, I-16146 Genoa, Italy
e-mail: rolandi@fisica.unige.it

Toshiharu Saiki
Department Electronics and Electrical Engineering, Keio University
3-14-1 Hiyoshi, Kohoku-ku, Yokohama 223-8522, Japan
e-mail: saiki@elec.keio.ac.jp

Hiroshi Sekiguchi
Department of Life Science, Tokyo Institute of Technology
Nagatsuta, Midori-ku, Yokohama 226-8501, Japan
e-mail: hsekiguc@bio.titech.ac.jp

Sriram Sundararajan
Department of Mechanical Engineering, 2025 H. M. Black Engineering Building
Iowa State University, Ames IA 50011-2161, USA
e-mail: srirams@iastate.edu

Nikhil S. Tambe
Material Systems Technologies, GE Global Research, #122, EPIP, Phase 2
Hoodi Village, Whitefield Road, Bangalore 560066, India
e-mail: nikhil.tambe@ge.com

Hironori Uehara
Department of Life Science, Tokyo Institute of Technology
Nagatsuta, Midori-ku, Yokohama 226-8501, Japan
e-mail: huehara@bio.titech.ac.jp

Claire Verbelen
Unité de Chimie des Interfaces, Université catholique de Louvain
Croix du Sud 2/18, 1348 Louvain-la-Neuve, Belgium
e-mail: verbelen@cifa.ucl.ac.be

Ewa P. Wojocikiewicz
Department of Physiology & Biophysics, University of Miami
1600 NW 12th Ave, RMSB 5077, Miami, FL 33136, USA
e-mail: e.wojcikiewicz@miami.edu

27 Gecko Feet: Natural Attachment Systems for Smart Adhesion—Mechanism, Modeling, and Development of Bio-Inspired Materials

Bharat Bhushan · Robert A. Sayer

Abstract. Several creatures, including insects, spiders, and lizards, have developed a unique clinging ability that utilizes dry adhesion. Geckos, in particular, have developed the most complex adhesive structures capable of smart adhesion—the ability to cling to different smooth and rough surfaces and detach at will. These animals make use of about three million microscale hairs (setae) (about 14000/mm^2) that branch off into hundreds of nanoscale spatulae (about a billion spatulae). This hierarchical surface construction gives the gecko the adaptability to create a large real area of contact with surfaces. Modeling of the gecko attachment system as a hierarchical spring model has provided insight into adhesion enhancement generated by this system. van der Waals forces are the primary mechanism utilized to adhere to surfaces, and capillary forces are a secondary effect that can further increase adhesion force. Preload applied to the setae increases adhesion force. Although a gecko is capable of producing on the order of 20 N of adhesion force, it retains the ability to remove its feet from an attachment surface at will. The adhesion strength of gecko setae is dependent on the orientation; maximum adhesion occurs at 30°. During walking a gecko is able to peel its foot from surfaces by changing the angle at which its setae contact a surface. A man-made fibrillar structure capable of replicating gecko adhesion has the potential for use in dry, superadhesive tapes that would be of use in a wide range of applications. These adhesives could be created using microfabrication/nanofabrication techniques or self-assembly.

Key words: Gecko feet, Adhesion, Surface energy, Nanostructures, Robots

27.1
Introduction

Almost 2500 years ago, the ability of the gecko to "run up and down a tree in any way, even with the head downwards" was observed by Aristotle (1918). This phenomenon is not limited to geckos, but occurs in several animals and insects as well. This dynamic attachment ability will be referred to as reversible adhesion or smart adhesion (Bhushan et al. 2006). Many insects (e.g., flies and beetles) and spiders have been the subject of investigation. However, the attachment pads of geckos have been the most widely studied owing to the fact that they exhibit the most versatile and effective adhesive known in nature. As a result, the vast majority of this chapter will be concerned with gecko feet.

Although there are over 1000 species of geckos (Kluge 2001; Han et al. 2004) that have attachment pads of varying morphology (Ruibal and Ernst 1965), the Tokay gecko (*Gekko gecko*) has been the main focus of scientific research (Hiller 1968; Irschick et al. 1996; Autumn 2006). The Tokay gecko is the second-largest gecko species, attaining respective lengths of approximately 0.3–0.4 and 0.2–0.3 m for males and females. They have a distinctive blue or gray body with orange or red spots and can weigh up to 300 g (Tinkle 1992). These geckos have been the most widely investigated species of gecko owing to the availability and size of these creatures.

Even though the adhesive ability of geckos has been known since the time of Aristotle, little was understood about this phenomenon until the late nineteenth century when microscopic hairs covering the toes of the gecko were first noted. The development of electron microscopy in the 1950s enabled scientists to view a complex hierarchical morphology that covers the skin on the gecko's toes. Over the past century and a half, scientific studies have been conducted to determine the factors that allow the gecko to adhere to and detach from surfaces at will, including surface structure (Ruibal and Ernst 1965; Russell 1975, 1986; Williams and Peterson 1982; Schleich and Kästle 1986; Irschick et al. 1996; Autumn and Peattie 2002; Arzt et al. 2003), the mechanisms of adhesion (Wagler 1830; Simmermacher 1884; Schmidt 1904; Hora 1923; Dellit 1934; Ruibal and Ernst 1965; Hiller 1968; Gennaro 1969; Stork 1980; Autumn et al. 2000, 2002; Bergmann and Irschick 2005; Huber et al. 2005b), and adhesion strength (Hiller 1968; Irschick et al. 1996; Autumn et al. 2000; Arzt et al. 2003; Huber et al. 2005a,b). Recent work in modeling the gecko attachment system as a system of springs (Bhushan et al. 2006; Kim and Bhushan 2007a–d) has provided valuable insight into adhesion enhancement. van der Waals forces are widely accepted in the literature as the dominant adhesive mechanism utilized by hierarchical attachment systems. Capillary forces created by humidity naturally present in the air can further increase the adhesive force generated by the spatulae. Both experimental and theoretical work support these adhesive mechanisms.

There is great interest among the scientific community to further study the characteristics of gecko feet in the hope that this information could be applied to the production of microsurfaces/nanosurfaces capable of recreating the adhesion forces generated by these lizards (Bhushan, 2007b). Common man-made adhesives such as tape or glue involve the use of wet adhesives that permanently attach two surfaces. However, replication of the characteristics of gecko feet would enable the development of a superadhesive polymer tape capable of clean, dry adhesion (Geim et al. 2003; Sitti 2003; Sitti and Fearing 2003a; Northen and Turner 2005, 2006; Yurdumakan et al. 2005; Zhao et al. 2006; Bhushan 2007a, Bhushan and Sayer 2007; Gorb et al. 2007). These reusable adhesives have potential for use in everyday objects such as tapes, fasteners, and toys and in high technology such as microelectric and space applications. Replication of the dynamic climbing and peeling ability of geckos could find use in the treads of wall-climbing robots (Sitti and Fearing 2003b; Menon et al. 2004; Autumn et al. 2005).

27.2
Tokay Gecko

27.2.1
Construction of Tokay Gecko

The explanation for the adhesive properties of gecko feet can be found in the surface morphology of the skin on the toes of the gecko. The skin is composed of a complex hierarchical structure of lamellae, setae, branches, and spatulae (Ruibal and Ernst

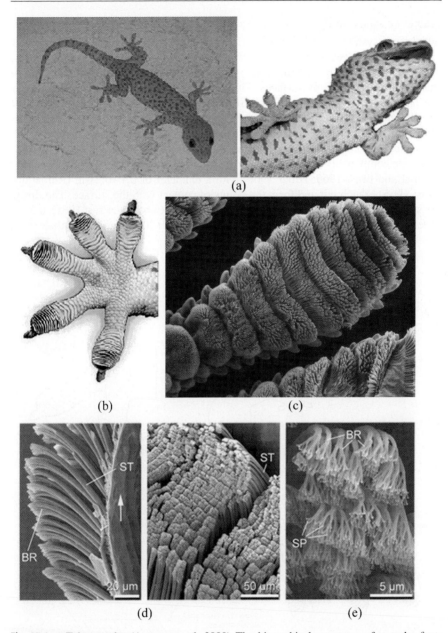

Fig. 27.1. a Tokay gecko (Autumn et al. 2000) The hierarchical structures of a gecko foot; **b** a gecko foot (Autumn et al. 2000) and **c** a gecko toe (Autumn 2006). Each toe contains hundreds of thousands of setae and each seta contains hundreds of spatulae. Scanning electron microscope (SEM) micrographs of **d** the setae (Gao et al. 2005) and **e** the spatulae (Gao et al. 2005). *ST* seta, *SP* spatula, *BR* branch

1965). As shown in Figs. 27.1 and 27.2 and summarized in Table 27.1, the gecko attachment system consists of an intricate hierarchy of structures beginning with lamellae, soft ridges that are 1–2 mm in length (Ruibal and Ernst 1965) that are located on the attachment pads (toes) that compress easily so that contact can be made with rough, bumpy surfaces. Tiny curved hairs known as setae extend from the lamellae with a density of approximately 14,000/mm^2 (Schleich and Kästle 1986). These setae are typically 30–130 μm in length and 5–10 μm in diameter (Ruibal and Ernst 1965; Hiller 1968; Russell 1975; Williams and Peterson 1982) and are composed primarily of β-keratin (Maderson 1964; Russell 1986) with some α-keratin components (Rizzo et al. 2006). At the end of each seta, 100–1000 spatulae (Ruibal and Ernst 1965; Hiller 1968) with a diameter of 0.1–0.2 μm (Ruibal and Ernst 1965) branch out and form the points of contact with the surface. The tips of the spatulae are approximately 0.2–0.3 μm in width (Ruibal and Ernst 1965), 0.5 μm in length, and 0.01 μm in thickness (Persson and Gorb 2003) and garner their name from their resemblance to a spatula.

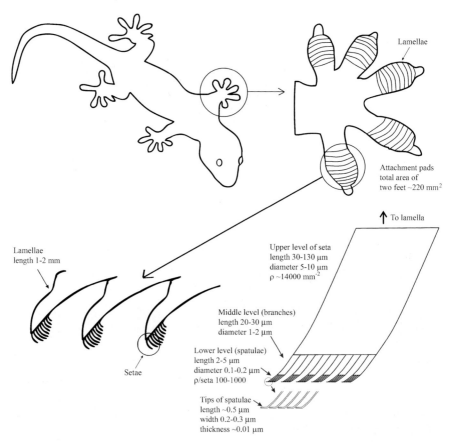

Fig. 27.2. A Tokay gecko including the overall body, one foot, a cross-sectional view of the lamellae, and an individual seta. ρ represents number of spatulae

Table 27.1. Surface characteristics of Tokay gecko feet

Component	Size	Density	Adhesive force
Seta	30–130[a–d]/5–10[a–d] length/diameter (μm)	~ 14000[f,g] setae/mm^2	194 μN[h] (in shear) ~ 20 μN[h] (normal)
Branch	20–30[a]/1–2[a] length/diameter (μm)	–	–
Spatula	2–5[a]/0.1–0.2[a,e] length/diameter (μm)	100–1000[c,d] spatulae per seta	–
Tip of spatula	~ 0.5[a,e]/0.2–0.3[a,d]/ ~ 0.01[e] length/width/thickness (μm)	– –	11 nN[i] (normal)

Young's modulus of surface material, keratin 1–20 GPa (Russell 1986; Bertram and Gosline 1987)
[a]Ruibal and Ernst (1965)
[b]Hiller (1968)
[c]Russell (1975)
[d]Williams and Peterson (1982)
[e]Persson and Gorb (2003)
[f]Schleich and Kästle (1986)
[g]Autumn and Peattie (2002)
[h]Autumn et al. (2000)
[i]Huber et al. (2005a)

The attachment pads on two feet of the Tokay gecko have an area of about 220 mm^2. About three million setae on their toes can produce a clinging ability of about 20 N [vertical force required to pull a lizard down a nearly vertical (85°) surface] (Irschick et al. 1996) and allow them to climb vertical surfaces at speeds over 1 m/s with capability to attach and detach their toes in milliseconds. In isolated setae, a 2.5-μN preload yielded adhesion of 20–40 μN and thus the adhesion coefficient, which represents the strength of adhesion as a function of preload, ranges from 8 to 16 (Autumn et al. 2002).

27.2.2
Other Attachment Systems

Attachment systems in other creatures such as insects and spiders have similar structures to that of gecko skin. The microstructures utilized by beetles, flies, spiders, and geckos can be seen in Fig. 27.3a. As the size (mass) of the creature increases, the radius of the terminal attachment elements decreases. This allows a greater number of setae to be packed into an area, hence increasing the linear dimension of contact and the adhesion strength. Arzt et al. (2003) determined that the density of the terminal attachment elements ρ_A per square meter strongly increases with increasing body mass m in kilograms. In fact, a master curve can be fit for all the different species (Fig. 27.3b):

$$\log \rho_A = 13.8 + 0.669 \log m \,. \tag{27.1}$$

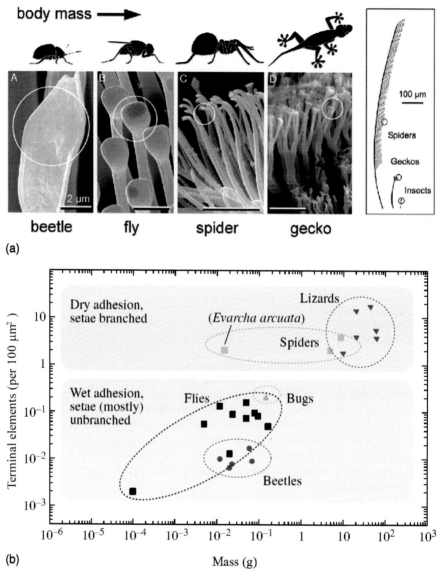

Fig. 27.3. a Terminal elements of the hairy attachment pads of a beetle, fly, spider, and gecko shown at two different scales (Arzt et al. 2003) and **b** the dependence of terminal element density on body mass (Federle 2006). (The data are from Artz et al. 2003 and Kesel et al. 2003)

The correlation coefficient of the master curve is equal to 0.919. Flies and beetles have the largest attachment pads and the lowest density of terminal attachment elements. Spiders have highly refined attachment elements that cover the leg. Geckos have both the highest body mass and the greatest density of terminal elements (spatulae).

Spiders and geckos can generate high dry adhesion, whereas beetles and flies increase adhesion by secreting liquid at the contacting surface.

27.2.3
Adaptation to Surface Roughness

Typical rough, rigid surfaces are only able to make intimate contact with a mating surface over a very small portion of the perceived apparent area of contact. In fact, the real area of contact is typically 2–6 orders of magnitude less than the apparent area of contact (Bhushan 2002, 2005). Autumn et al. (2002) proposed that divided contacts serve as a means for increasing adhesion. A surface energy approach can be used to calculate adhesion force in dry environments in order to calculate the effect of division of contacts. If the tip of a spatula is considered to be a hemisphere with radius R, the adhesion force of a single contact F_{ad} based on the Johnson–Kendall–Roberts (JKR) theory is given as (Johnson et al. 1971):

$$F_{ad} = \frac{3}{2}\pi W_{ad} R, \qquad (27.2)$$

where W_{ad} is the work of adhesion (units of energy per unit area). Equation (27.2) shows that adhesion force of a single contact is proportional to a linear dimension of the contact. For a constant area divided into a large number of contacts or setae, n, the radius of a divided contact, R_1, is given by $R_1 = R/\sqrt{n}$ (self-similar scaling) (Arzt et al. 2003). Therefore, the adhesion force of (27.2) can be modified for multiple contacts such that

$$F'_{ad} = \frac{3}{2}\pi W_{ad} \left(\frac{R}{\sqrt{n}}\right) n = \sqrt{n} F_{ad}, \qquad (27.3)$$

where F'_{ad} is the total adhesion force from the divided contacts. Thus, the total adhesion force is simply the adhesion force of a single contact multiplied by the square root of the number of contacts.

For a contact in the humid environment, the meniscus (or capillary) forces further increase the adhesion force (Bhushan 1999, 2002, 2005). The attractive meniscus force (F_m) consists of a contribution from both Laplace pressure and surface tension (Orr et al. 1975; Bhushan 2002). The contribution from Laplace pressure is directly proportional to the meniscus area. The other contribution is from the vertical component of surface tension around the circumference. This force is proportional to the circumference as is the case for the work of adhesion (Bhushan, 2002). Going through the analysis presented earlier, one can show that the contribution from the vertical component of surface tension increases as a surface is divided into a larger number of contacts. It increases linearly with the square root of the number of contacts n (self-similar scaling) (Bhushan 2007b; Kim and Bhushan 2007d):

$$\left(F'_m\right)_{\text{surface tension}} = \sqrt{n}\, \left(F_m\right)_{\text{surface tension}}, \qquad (27.4)$$

where F'_m is the force from the divided contacts and F_m is the force of an undivided contact.

The models just presented only consider contact with a flat surface. On natural rough surfaces the compliance and adaptability of setae are the primary sources of high adhesion. Intuitively, the hierarchical structure of gecko setae allows for greater contact with a natural rough surface than a nonbranched attachment system. Modeling of the contact between gecko setae and rough surfaces is discussed in detail in Sect. 27.5.

Material properties also play an important role in adhesion. A soft material is able to achieve greater contact with a mating surface than a rigid material. Although, gecko skin is primarily comprised of β-keratin, a stiff material with a Young's modulus in the range of 1–20 GPa (Russell 1986; Bertman and Gosline 1987), the effective modulus of the setal arrays on gecko feet is about 100 kPa (Autumn et al. 2006), which is approximately 4 orders of magnitude lower than for the bulk material. Nature has selected a relatively stiff material to avoid clinging to adjacent setae. Division of contacts, as discussed earlier, provides high adhesion. By combining optimal surface structure and material properties, mother nature has created an evolutionary superadhesive.

27.2.4
Peeling

Although geckos are capable of producing large adhesion forces, they retain the ability to remove their feet from an attachment surface at will by peeling action. The orientation of the spatulae facilitates peeling. Autumn et al. (2000) were the first to experimentally show that the adhesion force of gecko setae is dependent on the three-dimensional orientation as well as the preload applied during attachment (see Sect. 27.4.1.1). Owing to this fact, geckos have developed a complex foot motion during walking. First the toes are carefully uncurled during attachment. The maximum adhesion occurs at an attachment angle of 30°—the angle between a seta and the mating surface. The gecko is then able to peel its foot from surfaces one row of setae at a time by changing the angle at which its setae contact a surface. At an attachment angle greater than 30° the gecko will detach from the surface.

Shah and Sitti (2004) determined the theoretical preload required for adhesion as well as the adhesion force generated for setal orientations of 30°, 40°, 50°, and 60°. We consider a solid material (elastic modulus E, Poisson's ratio ν) in contact with the rough surface whose profile is given by

$$f(x) = H \sin^2\left(\frac{\pi x}{\chi}\right), \qquad (27.5)$$

where H is the amplitude and χ is the wavelength of the roughness profile. For a solid adhesive block to achieve intimate contact with the rough surface neglecting surface forces, it is necessary to apply a compressive stress, σ_c (Jagota and Bennison 2002):

$$\sigma_c = \frac{\pi E H}{2\chi (1 - \nu^2)}. \qquad (27.6)$$

Equation (27.6) can be modified to account for fibers oriented at an angle θ. The preload required for contact is summarized in Fig. 27.4a. As the orientation angle

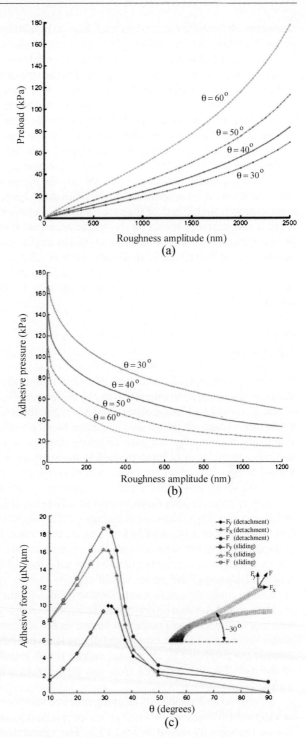

Fig. 27.4. Contact mechanics results for the affect of fiber orientation on **a** preload and **b** adhesive force for roughness amplitudes ranging from 0 to 2500 nm (Shah and Sitti 2004). **c** Finite-element analysis of the adhesive force of a single seta as a function of pull direction (Gao et al. 2005)

decreases, so does the required preload. Similarly, adhesion strength is influenced by fiber orientation. As seen in Fig. 27.4b, the greatest adhesion force occurs at $\theta = 30°$.

Gao et al. (2005) created a finite-element model of a single gecko seta in contact with a surface. A tensile force was applied to the seta at various angles, θ, as shown in Fig. 27.4c. For forces applied at an angle less than 30°, the dominant failure mode was sliding. In contrast, the dominant failure mode for forces applied at angles greater than 30° was detachment. This verifies the results of Autumn et al. (2000) that detachment occurs at attachment angles greater than 30°.

Tian et al. (2006) have suggested that during detachment, the angular dependence of adhesion and that of friction play a role. The pulling force of a spatula along its shaft with an angle between 0 and 90° to the substrate has a normal adhesive force produced at the spatula–substrate bifurcation zone, and a lateral friction force contribution from the part of the spatula still in contact with the substrate. High net friction and adhesion forces on the whole gecko are obtained by rolling down and gripping the toes inward to realize small pulling angles of the large number of spatulae in contact with the substrate. To detach, the high adhesion/friction is rapidly reduced to a very low value by rolling the toes upward and downward, which, mediated by the lever function of the setal shaft, peels the spatula off perpendicularly from the substrate.

27.2.5
Self-Cleaning

Natural contaminants (dirt and dust) as well as man-made pollutants are unavoidable and have the potential to interfere with the clinging ability of geckos. Particles found in the air consist of particulates that are typically less than 10 μm in diameter, while those found on the ground can often be larger (Hinds 1982; Jaenicke 1998). Intuitively, it seems that the great adhesion strength of gecko feet would cause dust and other particles to become trapped in the spatulae and that they would have no way of being removed without some sort of manual cleaning action on behalf of the gecko. However, geckos are not known to groom their feet like beetles (Stork 1983) nor do they secrete sticky fluids to remove adhering particles like ants (Federle et al. 2002) and tree frogs (Hanna and Barnes 1991), yet they retain adhesive properties. One potential source of cleaning is during the time when the lizards undergo molting, or the shedding of the superficial layer of epidermal cells. However, this process only occurs approximately once per month (Van der Kloot 1992). If molting were the sole source of cleaning, the gecko would rapidly lose its adhesive properties as it is exposed to contaminants in nature (Hansen and Autumn 2005).

Hansen and Autumn (2005) tested the hypothesis that gecko setae become cleaner with repeated use—a phenomenon known as self-cleaning. The cleaning ability of gecko feet was first tested experimentally by applying 2.5-μm-radius silica–alumina ceramic microspheres to clean setal arrays. It was found that a significant fraction of the particles was removed from the setal arrays with each step taken by the gecko.

In order to understand this cleaning process, substrate–particle interactions must be examined. The interaction energy between a dust particle and a wall and spatulae can be modeled as shown in Fig. 27.5. The interaction between a spherical dust

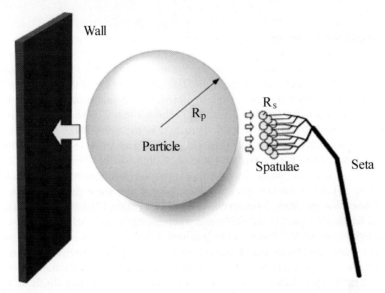

Fig. 27.5. Model of interactions between gecko spatulae of radius R_s, a spherical dirt particle of radius R_p, and a planar wall that enable self-cleaning (Hansen and Autumn 2005)

particle and the wall, W_{pw}, can be expressed as (Israelachvili 1992)

$$W_{pw} = \frac{-H_{pw} R_p}{6 D_{pw}}, \qquad (27.7)$$

where p and w refer to the particle and wall, respectively. H is the Hamaker constant, R_p is the radius of the particle, and D_{pw} is the separation distance between the particle and the wall. Similarly, the interaction energy between a spherical dust particle and a spatula, s, assuming that the spatula tip is spherical is (Israelachvili 1992)

$$W_{ps} = \frac{-H_{ps} R_p R_s}{6 D_{ps}(R_p + R_s)}. \qquad (27.8)$$

The ratio of the two interaction energies, N, can be expressed as

$$N = \frac{W_{pw}}{W_{ps}} = \left(1 + \frac{R_p}{R_s}\right) \frac{H_{pw} D_{ps}}{H_{ps} D_{pw}}. \qquad (27.9)$$

When the energy required to separate a particle from the wall is greater than that required to separate it from a spatula, self-cleaning will occur. For small contaminants ($R_p < 0.5$ μm), there are not enough spatulae available to adhere to the particle. For larger contaminants, the curvature of the particle makes it impossible for enough spatulae to adhere to it. As a result, Hansen and Autumn (2005) concluded that self-cleaning should occur for all spherical spatulae interacting with all spherical particles.

27.3
Attachment Mechanisms

When asperities of two solid surfaces are brought into contact with each other, chemical and/or physical attractions occur. The force developed that holds the two surfaces together is known as adhesion. In a broad sense, adhesion is considered to be either physical or chemical in nature (Bikerman 1961; Zisman 1963; Houwink and Salomon 1967; Israelachvili 1992; Bhushan 1996, 1999, 2002, 2005, 2007a). Chemical interactions such as electrostatic attraction charges (Schmidt 1904) as well as intermolecular forces (Hiller 1968) including van der Waals and capillary forces have all been proposed as potential adhesion mechanisms in gecko feet. Others have hypothesized that geckos adhere to surfaces through the secretion of sticky fluids (Wagler 1830; Simmermacher 1884), suction (Simmermacher 1884), increased frictional force (Hora 1923), and microinterlocking (Dellit 1934).

Through experimental testing and observations conducted over the last century and a half many potential adhesive mechanisms have been eliminated. Observation has shown that geckos lack glands capable of producing sticky fluids (Wagler 1830; Simmermacher 1884), thus ruling out the secretion of sticky fluids as a potential adhesive mechanism. Furthermore, geckos are able to create large adhesive forces normal to a surface. Since friction only acts parallel to a surface, the attachment mechanism of increased frictional force has been ruled out. Dellit (1934) experimentally ruled out suction and electrostatic attraction as potential adhesive mechanisms. Experiments carried out in a vacuum did not show a difference between the adhesion force at low pressures and that in ambient conditions. Since adhesive forces generated during suction are based on pressure differentials, which are insignificant under vacuum, suction was rejected as an adhesive mechanism (Dellit 1934). Additional testing utilized X-ray bombardment to create ionized air in which electrostatic attraction charges would be eliminated. It was determined that geckos were still able to adhere to surfaces in these conditions and, therefore, electrostatic charges could not be the sole cause of attraction (Dellit 1934). Autumn et al. (2000) demonstrated the ability of a gecko to generate large adhesive forces when in contact with a molecularly smooth SiO_2 microelectromechanical semiconductor. Since surface roughness is necessary for microinterlocking to occur, it has been ruled out as a mechanism of adhesion. Two mechanisms, van der Waals forces and capillary forces, remain as the potential sources of gecko adhesion. These attachment mechanisms are described in detail in the following sections.

27.3.1
Van der Waals Forces

Van der Waals bonds are secondary bonds that are weak in comparison with other physical bonds such as covalent, hydrogen, ionic, and metallic bonds. Unlike other physical bonds, van der Waals forces are always present regardless of separation and are effective from very large separations (approximately 50 nm) down to atomic separation (approximately 0.3 nm). The van der Waals force per unit area between two parallel surfaces, f_{vdW}, is given by (Hamaker 1937; Israelachvili and Tabor 1972; Israelachvili 1992)

$$f_{vdW} = \frac{H}{6\pi D^3} \quad \text{for } D < 30 \,\text{nm}, \tag{27.10}$$

where H is the Hamaker constant and D is the separation between surfaces.

Hiller (1968) showed experimentally that the surface energy of a substrate is responsible for gecko adhesion. One potential adhesive mechanism would then be van der Waals forces (Stork 1980; Autumn et al. 2000). Assuming van der Waals forces to be the dominant adhesive mechanism utilized by geckos, one can claculate the adhesion force of a gecko. Typical values of the Hamaker constant range from 4×10^{-20} to 4×10^{-19} J (Israelachvili 1992). In calculation, the Hamaker constant is assumed to be 10^{-19} J, the surface area of a spatula is taken to be 2×10^{-14} m^2 (Ruibal and Ernst 1965; Williams and Peterson 1982; Autumn and Peattie 2002), and the separation between the spatula and the contact surface is estimated to be 0.6 nm. This equation yields the force of a single spatula to be about 0.5 μN. By applying the surface characteristics of Table 27.1, the maximum adhesive force of a gecko is 150–1500 N for varying spatula density of 100–1000 spatulae per seta. If an average value of 550 spatulae per seta is used, the adhesion force of a single seta is approximately 270 μN, which is in agreement with the experimental value obtained by Autumn et al. (2000), which will be discussed in Sect. 27.4.1.1.

Another approach to calculate adhesive force is to assume that spatulae are cylinders that terminate in hemispherical tips. By using (27.2) and assuming that the radius of each spatula is about 100 nm and that the surface energy is expected to be 50 mJ/m^2 (Arzt et al. 2003), one predicts the the adhesion force of a single spatula to be 0.02 μN. This result is an order of magnitude lower than the first approach calculated for the higher value of A. For a lower value of 10^{-20} J for the Hamaker constant, the adhesion force of a single spatula is comparable to that obtained using the surface-energy approach.

Several experimental results favor van der Waals forces as the dominant adhesive mechanism, including temperature testing (Bergman and Irschick 2005) and adhesion force measurements of a gecko seta with both hydrophilic and hydrophobic surfaces (Autumn et al. 2000). These data will be presented in the Sects. 27.4.2–27.4.4.

27.3.2
Capillary Forces

It has been hypothesized that capillary forces that arise from liquid-mediated contact could be a contributing or even the dominant adhesive mechanism utilized by gecko spatulae (Hiller 1968; Stork 1980). Experimental adhesion measurements (presented in Sects. 27.4.3, 27.4.4) conducted on surfaces with different hydrophobicities and at various humidities (Huber et al. 2005b) as well as numerical simulations (Kim and Bhushan 2007d) support this hypothesis as a contributing mechanism. During contact, any liquid that wets or has a small contact angle on surfaces will condense from vapor in the form of an annular-shaped capillary condensate. Owing to the natural humidity present in the air, water vapor will condense to liquid on the surface of bulk materials. During contact this will cause the formation of adhesive bridges (menisci) owing to the proximity of the two surfaces and the affinity of the surfaces for condensing liquid. (Zimon 1969; Fan and O'Brien 1975; Phipps and Rice 1979).

Capillary force can be divided into two components: the Laplace force F_L and the surface tension force F_s such that the total capillary force F_c is given by the sum of the components:

$$F_c = F_L + F_s . \tag{27.11}$$

The Laplace force is caused by the pressure difference across the interface of a curved liquid surface (Fig. 27.6) and depends on pressure difference multiplied by the meniscus area, which can be expressed as (Orr et al. 1975)

$$F_L = -\pi \kappa \gamma R^2 \sin^2 \phi , \tag{27.12}$$

where γ is the surface tension of the liquid, R is the tip radius, ϕ is the filling angle and κ is the mean curvature of the meniscus. From the Kelvin equation (Israelachvili 1992), which is the thermal equilibrium relation, the mean curvature of the meniscus can be determined as

$$\kappa = \frac{\mathcal{R}T}{V\gamma} \ln\left(\frac{p}{p_o}\right) , \tag{27.13}$$

where $\mathcal{R}$ is the universal gas constant, T is the absolute temperature, V is the molecular volume, p_o is the saturated vapor pressure of the liquid at T, and p is the ambient pressure acting outside the curved surface (p/p_o is the relative humidity). Orr et al. (1975) formulated the mean curvature of the meniscus between a sphere and a plane in terms of elliptic integrals. The filling angle ϕ can be calculated from the expression just mentioned and (27.13) using an iteration method. Then the Laplace force is calculated at a given environment using (27.12).

The surface tension of the liquid results in the formation of a curved liquid–air interface. The surface tension force acting on the sphere is (Orr et al. 1975)

$$F_s = 2\pi R \gamma \sin \phi \sin(\theta_1 + \phi) . \tag{27.14}$$

Owing to the fact that surface tension force depends on the radius, division would result in an increase of the surface tension force by the square root of N_c upon division (Bhushan et al. 2007b; Kim and Bhushan 2007d).

Hence, the total capillary force on the sphere is

$$F_c = \pi R \gamma \left[2 \sin \phi \sin(\theta_1 + \phi) - \kappa R \sin^2 \phi\right] . \tag{27.15}$$

27.4
Experimental Adhesion Test Techniques and Data

Experimental measurements of the adhesion force of a single gecko seta (Autumn et al. 2000) and single gecko spatula (Huber et al. 2005a, 2005b) have been made.

The effect of the environment, including temperature (Losos 1990; Bergmann and Irschick 2005) and humidity (Huber et al. 2005b), has been studied. Some of the data have been used to understand the adhesion mechanism utilized by the gecko attachment system—van der Waals or capillary forces. The majority of experimental results point towards van der Waals forces as the dominant mechanism of adhesion (Autumn et al. 2000; Bergmann and Irschick 2005). Recent research suggests that capillary forces can be a contributing adhesive factor (Huber et al. 2005b).

27.4.1
Adhesion Under Ambient Conditions

Two feet of a Tokay gecko are capable of producing about 20 N of adhesion force with a pad area of about 220 mm^2 (Irschick et al. 1996). Assuming that there are about 14,000 setae per square millimeter, the adhesion force from a single hair should be approximately 7 µN. It is likely that the magnitude is actually greater than this value because it is unlikely that all setae are in contact with the mating surface (Autumn et al. 2000). Setal orientation greatly influences adhesive strength. This dependency was first noted by Autumn et al. (2000). It was determined that the greatest adhesion occurs at 30°. In order to determine the adhesion mechanism(s) utilized by gecko feet, it is important to know the adhesion force of a single seta. Hence, the adhesion force of gecko foot-hair has been the focus of several investigations (Autumn et al. 2000; Huber et al. 2005a).

27.4.1.1
Adhesion Force of a Single Seta

Autumn et al. (2000) used both a microelectromechanical force sensor and a wire as a force gauge to determine the adhesion force of a single seta. The microelectromechanical force sensor is a dual-axis atomic force microscope (AFM) cantilever with independent piezoresistive sensors, which allows simultaneous detection of

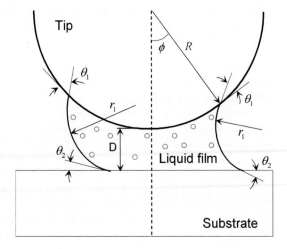

Fig. 27.6. A sphere on a plane at distance D with a liquid film in-between, forming menisci. R is the tip radius, ϕ is the filling angle, θ_1 and θ_2 are contact angles on the sphere and the plane respectively, and r_1 and r_2 are the two principal radii of the curved surface (Kim and Bhushan 2007d)

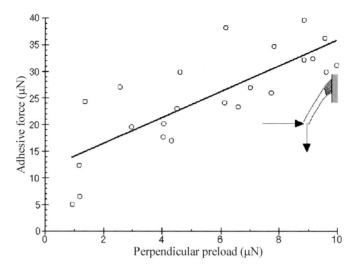

Fig. 27.7. Adhesive force of a single gecko seta as a function of applied preload. The seta was first pushed perpendicularly against the surface and then pulled parallel to the surface (Autumn et al. 2000)

vertical and lateral forces (Chui et al. 1998). The wire force gauge consisted of an aluminum bonding wire that displaced under a perpendicular pull. Autumn et al. (2000) discovered that setal force actually depends on the three-dimensional orientation of the seta as well as the preloading force applied during initial contact. Setae that were preloaded vertically to the surface exhibited only one tenth of the adhesive force ($0.6 \pm 0.7\,\mu N$) compared with setae that were pushed vertically and then pulled horizontally to the surface ($13.6 \pm 2.6\,\mu N$). The dependence of adhesion force of a single gecko spatula on perpendicular preload is illustrated in Fig. 27.7. The adhesion force increases linearly with the preload, as expected (Bhushan 1996, 1999, 2002). The maximum adhesion force of a single gecko foot-hair occurred when the seta was first subjected to a normal preload and then slid $5\,\mu m$ along the contacting surface. Under these conditions, adhesion force measured $194 \pm 25\,\mu N$ (approximately 10 atm adhesive pressure).

27.4.1.2
Adhesion Force of a Single Spatula

Huber et al. (2005a) used atomic force microscopy to determine the adhesion force of individual gecko spatulae. A seta with four spatulae was glued to an AFM tip. The seta was then brought in contact with a surface and a compressive preload of 90 nN was applied. The force required to pull the seta off the surface was then measured. As seen in Fig. 27.8, there are two distinct peaks in the graph—one at 10 nN and the other at 20 nN. The first peak corresponds to one of the four spatulae adhering to the contact surface, while the peak at 20 nN corresponds to two of the four spatulae adhering to the contact surface. The average adhesion force of a single spatula was

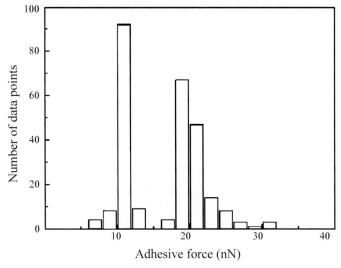

Fig. 27.8. Adhesive force of a single gecko spatula. The peak at 10 nN corresponds to the adhesive force of one spatula and the peak at 20 nN corresponds to the adhesive force of two spatulae (Huber et al. 2005a)

found to be 10.8 ± 1 nN. The measured value is in agreement with the measured adhesive strength of an entire gecko (on the order of 10^9 spatulae on a gecko).

27.4.2
Effects of Temperature

Environmental factors are known to affect several aspects of vertebrate function, including speed of locomotion, digestion rate, and muscle contraction, and as a result several studies have been undertaken to investigate environmental impact on these functions. Relationships between the environment and other properties such as adhesion have been far less studied (Bergmann and Irschick 2005). Only two known studies exist that examine the affect of temperature on the clinging force of the gecko (Losos 1990; Bergmann and Irschick 2005). Losos (1990) examined the adhesion ability of large live geckos at temperatures up to 17 °C. Bergmann and Irschick (2005) expanded upon this research for body temperatures ranging from 15 to 35 °C. The geckos were incubated until their body temperature reached a desired level. The clinging ability of these animals was then determined by measuring the maximum exerted force by the geckos as they were pulled off a custom-built force plate. The clinging force of a gecko for the experimental test range is plotted in Fig. 27.9. It was determined that variation in temperature is not statistically significant for the adhesion force of a gecko. From these results, it was concluded that the temperature independence of adhesion supports the hypothesis of clinging as a passive mechanism (i. e., van der Waals forces). Both studies only measured overall clinging ability on the macroscale. There have not been any investigations into effects of temperature on the clinging ability of a single seta on the microscale and therefore testing in this area would be extremely important.

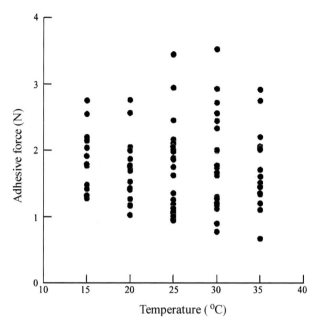

Fig. 27.9. Adhesive force of a gecko as a function of temperature (Bergmann and Irschick 2005)

27.4.3
Effects of Humidity

Huber et al. (2005b) employed similar methods to Huber et al. (2005a) (discussed previously in Sect. 27.4.1.2) in order to determine the adhesion force of a single spatula at varying humidity. Measurements were made using an AFM placed in an airtight chamber. The humidity was adjusted by varying the flow rate of dry nitrogen into the chamber. The air was continuously monitored with a commercially available hygrometer. All tests were conducted at ambient temperature.

As seen in Fig. 27.10, even at low humidity, adhesion force is large. An increase in humidity further increases the overall adhesion force of a gecko spatula. The pull-off force roughly doubled as the humidity was increased from 1.5 to 60%. This humidity effect can be explained in two possible ways: (1) by standard capillarity or (2) by a change of the effective short-range interaction due to absorbed monolayers of water—in other words, the water molecules increase the number of van der Waals bonds that are made. On the basis of these data, van der Waals forces are the primary adhesion mechanism and capillary forces are a secondary adhesion mechanism.

27.4.4
Effects of Hydrophobicity

To further test the hypothesis that capillary forces play a role in gecko adhesion, the spatular pull-off force was determined for contact with both hydrophilic and hydrophobic surfaces. As seen in Fig. 27.11a, the capillary adhesion theory predicts

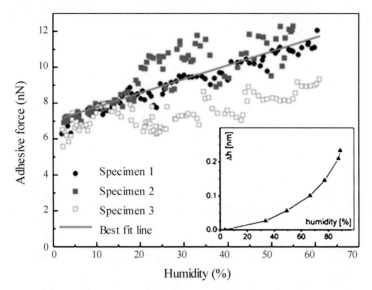

Fig. 27.10. Humidity effects on spatular pull-off force. *Inset*: The increase in water film thickness on a Si wafer with increasing humidity (Huber et al. 2005b)

that a gecko spatula will generate a greater adhesion force when in contact with a hydrophilic surface compared with a hydrophobic surface, while the van der Waals adhesion theory predicts that the adhesion force between a gecko spatula and a surface will be the same regardless of the hydrophobicity of the surface (Autumn et al. 2002). Figure 27.11b shows the adhesion pressure of a whole gecko and the adhesion force of a single seta on hydrophilic and hydrophobic surfaces. The data show that the values of the adhesion force are the same for both surfaces. This supports the van der Waals prediction of Fig. 27.11a. Huber et al. (2005b) found that the hydrophobicity of the attachment surface had an effect on the adhesion force of a single gecko spatula, as shown in Fig. 27.11c. These results show that adhesion force has a finite value for a superhydrophobic surface and increases as the surface becomes hydrophilic. It is concluded that van der Waals forces are the primary mechanism and capillary forces further increase the adhesion force generated.

27.5
Adhesion Modeling

With regard to the natural living conditions of the animals, the mechanics of gecko attachment can be separated into two parts: the mechanics of adhesion of a single contact with a flat surface, and an adaptation of a large number of spatulae to a natural, rough surface. Modeling of the mechanics of adhesion of spatulae to a smooth surface was developed by Autumn et al. (2002), Jagota and Bennison (2002), and Arzt et al. (2003). As discussed previously in Sect. 27.2.3, the adhesion force of multiple contacts F'_{ad} can be increased by dividing the contact into a large

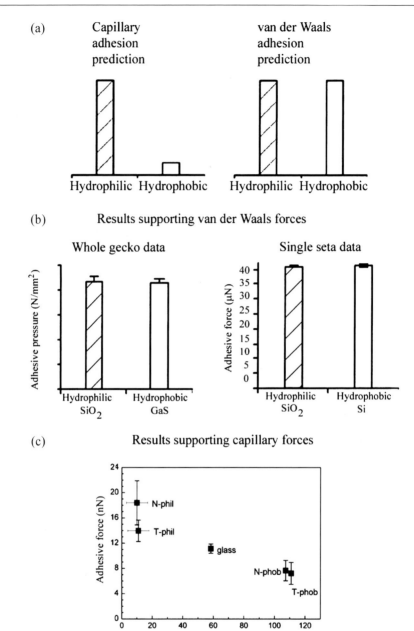

Fig. 27.11. a Capillary and van der Waals adhesion predictions for the relative magnitude of the adhesive force of gecko setae with hydrophilic and hydrophobic surfaces (Autumn et al. 2002). **b** Results of adhesion testing for a whole gecko and single seta with hydrophilic and hydrophobic surfaces (Autumn et al. 2002) and **c** results of adhesive force testing with surfaces with different contact angles (Huber et al. 2005b)

number (n) of small contacts, while the nominal area of the contact remains the same, $F'_{ad} \sim \sqrt{n} F_{ad}$. However, this model only considers contact with a flat surface. On natural, rough surfaces, the compliance and adaptability of setae are the primary sources of high adhesion. Intuitively, the hierarchical structure of gecko setae allows for a greater contact with a natural, rough surface than a nonbranched attachment system (Sitti and Fearing 2003a).

27.5.1
Spring Model

Bhushan et al. (2006) and Kim and Bhushan (2007a–d) have recently approximated a gecko seta in contact with random rough surfaces using a hierarchical spring model. Each level of springs in their model corresponds to a level of seta hierarchy. The upper level of springs corresponds to the thicker part of gecko seta, the middle spring level corresponds to the branches, and the lower level of springs corresponds to the spatulae. The upper level is the thickest branch of the seta. It is 75 μm in length and 5 μm in diameter. The middle level, referred to as a branch, has a length of 25 μm and a diameter of 1 μm. The lower level, called a spatula, is the thinnest branch, with a length of 2.5 μm and a diameter of about 0.1 μm (Table 27.2). Autumn et al. (2000) showed that the optimal attachment angle between the substrate and a gecko seta is 30° in the single seta pull-off experiment. This finding is supported by the adhesion models of setae as cantilever beams (Shah and Sitti 2004; Gao et al. 2005) (see Sect. 27.2.4 for more details). Therefore, θ was fixed at 30° in the studies (Bhushan et al. 2006; Kim and Bhushan 2007a–d).

Table 27.2. Geometrical size, calculated stiffness, and typical densities of branches of seta for Tokay gecko (Kim and Bhushan 2007a)

Level of seta	Length (μm)	Diameter (μm)	Bending stiffness[a] (N/m)	Typical density (no./mm^2)
III upper	75	5	2.908	14×10^3
II middle	25	1	0.126	–
I lower	2.5	0.1	0.0126	$1.4–14 \times 10^6$

[a] For an elastic modulus of 10 GPa with load applied at 60° to the spatula long axis

27.5.2
Single Spring Contact Analysis

In their analysis, Bhushan et al. (2006) and Kim and Bhushan (2007a–d) assumed the tip of the spatula in a single contact to be spherical. The springs on every level of hierarchy have the same stiffness as the bending stiffness of the corresponding branches of the seta. If the beam is oriented at an angle θ to the substrate and the contact load F is aligned normal to the substrate, its components along and tangential to the direction of the beam, $F\cos\theta$ and $F\sin\theta$, give rise to bending and compressive deformations, δ_b and δ_c, respectively, as (Young and Budynas 2001)

$$\delta_b = \frac{F \cos \theta l_m^3}{3EI}, \quad \delta_c = \frac{F \sin \theta l_m}{A_C E}, \qquad (27.16)$$

where $I = \pi R_m^4/4$ and $A_C = \pi R_m^2$ are the moment of inertia and the cross-sectional area of the beam, respectively, l_m and R_m are the length and the radius of seta branches, respectively, and m is the level number. The net displacement, $\delta_\perp$, normal to the substrate is given by

$$\delta_\perp = \delta_c \sin \theta + \delta_b \cos \theta . \qquad (27.17)$$

Using (27.16) and (27.17), one can calculate the stiffness of seta branches k_m as (Glassmaker et al. 2004)

$$k_m = \frac{\pi R_m^2 E}{l_m \sin^2 \theta \left(1 + \frac{4 l_m^2 \cot^2 \theta}{3 R_m^2}\right)} . \qquad (27.18)$$

For an assumed elastic modulus E of seta material of 10 GPa with a load applied at an angle of 60° to spatulae long axis, Kim and Bhushan (2007a) calculated the stiffness of every level of seta as given in Table 27.2.

In the model, both tips of a spatula and asperity summits of the rough surface are assumed to be spherical with a constant radius (Bhushan et al. 2006). As a result, a single spatula adhering to a rough surface was modeled as the interaction between two spherical tips. Because β-keratin has a high elastic modulus (Russell 1986; Bertram and Gosline 1987), the adhesion force between two round tips was calculated according to the Derjaguin–Muller–Toporov (DMT) theory (Derjaguin et al. 1975) as

$$F_{ad} = 2\pi R_c W_{ad} , \qquad (27.19)$$

where R_c is the reduced radius of contact, which is calculated as $R_c = 1/(1/R_1 + 1/R_2)$; R_1 and R_2 are the radii of the contacting surfaces; $R_1 = R_2$, $R_c = R/2$. The work of adhesion W_{ad} is then calculated using the following equation for two flat surfaces separated by a distance D (Israelachvili 1992):

$$W_{ad} = -\frac{H}{12\pi D^2} , \qquad (27.20)$$

where H is the Hamakar constant, which depends on the medium between the two surfaces. Typical values of the Hamakar constant for polymers are 10^{-19} J in air and 3.7×10^{-20} J in water (Israelachvili 1992). For a gecko seta, which is composed of β-keratin, the value of H is assumed to be 10^{-19} J. The work of adhesion of two surfaces in contact separated by an atomic distance $D \approx 0.2$ nm is approximately equal to 66 mJ/m² (Israelachvili 1992). By assuming that the tip radius R is 50 nm, using (27.19), one can calculate the adhesion force of a single contact to be 10 nN (Kim and Bhushan 2007a). This value is identical to the adhesion force of a single spatula measured by Huber et al. (2005a). This adhesion force is used as a critical force in the model for judging whether the contact between the tip and the surface is broken or not during the pull-off cycle (Bhushan et al. 2006). If the elastic force of a single spring is less than the adhesion force, the spring is regarded as having been detached.

27.5.3
The Multilevel Hierarchical Spring Analysis

In order to study the effect of the number of hierarchical levels in the attachment system on attachment ability, models with one level (Bhushan et al. 2006; Kim and Bhushan 2007a, b), two levels (Bhushan et al. 2006; Kim and Bhushan 2007a, b) and three levels (Kim and Bhushan 2007a, b) of hierarchy were simulated (Fig. 27.12). The one-level model has springs with length $l_\mathrm{I} = 2.5\,\mu\mathrm{m}$ and stiffness $k_\mathrm{I} = 0.0126\,\mathrm{N/m}$. The length and stiffness of the springs in the two-level model are $l_\mathrm{I} = 2.5\,\mu\mathrm{m}$, $k_\mathrm{I} = 0.0126\,\mathrm{N/m}$ and $l_\mathrm{II} = 25\,\mu\mathrm{m}$, $k_\mathrm{II} = 0.126\,\mathrm{N/m}$ for levels I and II, respectively. The three-level model has additional upper-level springs with $l_\mathrm{III} = 75\,\mu\mathrm{m}$, $k_\mathrm{III} = 2.908\,\mathrm{N/m}$ on the springs of the two-level model, which is identical to gecko setae. The base of the springs and the connecting plate between the levels are assumed to be rigid. The distance S_I between neighboring structures of level I is $0.35\,\mu\mathrm{m}$, obtained from the average value of the measured spatula density, $8 \times 10^6\,\mathrm{mm}^{-2}$, calculated by multiplying 14,000 setae/mm^2 by an average of 550 spatulae per seta (Schleich and Kästle 1986) (Table 27.2). A 1:10 proportion of the number of springs in the upper level to that in the level below was assumed (Bhushan et al. 2006). This corresponds to one spring at level III being connected to ten springs at level II and each spring at level II also has ten springs at level I. The number of springs at level I considered in the model is calculated by dividing the scan length (2000 μm) with the distance S_I (0.35 μm), which corresponds to 5700.

The spring deflection Δl was calculated as

$$\Delta l = h - l_0 - z, \tag{27.21}$$

where h is the position of the spring base relative to the mean line of the surface; l_0 is the total length of a spring structure, which is $l_0 = l_\mathrm{I}$ for the one-level model, $l_0 = l_\mathrm{I} + l_\mathrm{II}$ for the two-level model, and $l_0 = l_\mathrm{I} + l_\mathrm{II} + l_\mathrm{III}$ for the three-level model; and z is the profile height of the rough surface. The elastic force F_el arising in the springs at a distance h from the surface was calculated for the one-level model as (Bhushan et al. 2006)

$$F_\mathrm{el} = -k_\mathrm{I} \sum_{i=1}^{p} \Delta l_i u_i \quad u_i = \begin{cases} 1 & \text{if contact} \\ 0 & \text{if no contact}, \end{cases} \tag{27.22}$$

where p is the number of springs in level I of the model. For the two-level model, the elastic force was calculated as (Bhushan et al. 2006)

$$F_\mathrm{el} = -\sum_{j=1}^{q}\sum_{i=1}^{p} k_{ji}\left(\Delta l_{ji} - \Delta l_j\right) u_{ji} \quad u_{ji} = \begin{cases} 1 & \text{if contact} \\ 0 & \text{if no contact}, \end{cases} \tag{27.23}$$

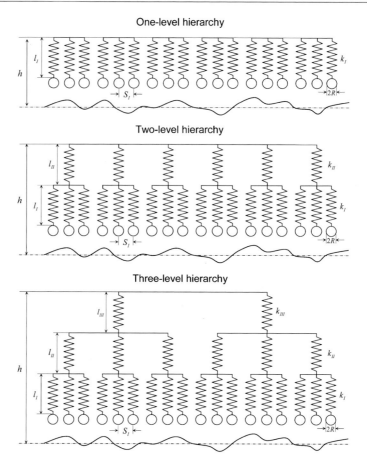

Fig. 27.12. One-, two-, and three-level hierarchical spring models for simulating the effect of hierarchical morphology on the interaction of a seta with a rough surface. l_I, l_{II}, and l_{III} are lengths of the structures, S_I is the space between spatulae, k_I, k_{II}, and k_{III} are stiffnesses of the structures, I, II, and III are level indexes, R is the radius of the tip, and h is the distance between the base of the upper spring of each model and the mean line of the rough profile (Kim and Bhushan 2007a)

where q is the number of springs level II of the model. For the three-level model, the elastic force was calculated as (Kim and Bhushan 2007a)

$$F_{el} = -\sum_{k=1}^{r}\sum_{j=1}^{q}\sum_{i=1}^{p} k_{kji}\left(\Delta l_{kji} - \Delta l_{kj} - \Delta l_j\right) u_{kji} \quad u_{kji} = \begin{cases} 1 & \text{if contact} \\ 0 & \text{if no contact} \end{cases},$$
(27.24)

where r is the number of springs in level III of the model. The spring force when the springs approach the rough surface is calculated using (27.22), (27.23), or (27.24) for the one-, two-, and three-level models, respectively. During pull-off, the same equations are used to calculate the spring force. However, when the applied load is zero, the springs do not detach owing to adhesion attraction given by (27.19). The springs are pulled apart until the net force (pull-off force minus attractive adhesion force) at the interface is zero.

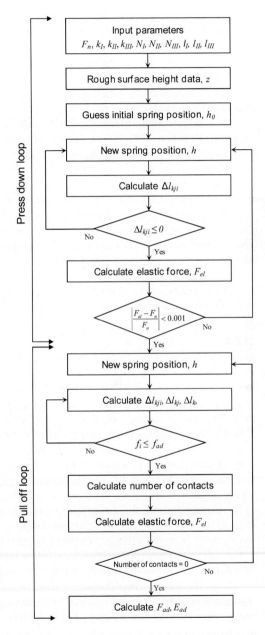

Fig. 27.13. Flow chart for the calculation of the adhesion force (F_{ad}) and the adhesion energy (E_{ad}) for three-level hierarchical spring model. F_n is an applied load, k_I, k_{II}, and k_{III} and l_I, l_{II}, and l_{III} are stiffnesses and lengths of the structures, Δl_{kji}, Δl_{ki}, and Δl_k are the spring deformations on levels I, II, and III, respectively, i, j, and k are spring indexes on each level, f_i is the elastic force of a single spring and f_{ad} is the adhesion force of a single contact (Kim and Bhushan 2007a)

The adhesion force is the lowest value of the elastic force F_{el} when the seta has detached from the contacting surface. The adhesion energy is calculated as

$$W_{ad} = \int_{\overline{D}}^{\infty} F_{el}(D)\, dD\,, \tag{27.25}$$

where D is the distance that the spring base moves away from the contacting surface. The lower limit of the distance $\overline{D}$ is the value of D where F_{el} is first zero when the model is pulled away from the contacting surface. Also although the upper limit of the distance is infinity, in practice, the $F_{el}(D)$ curve is integrated to the upper limit where F_{el} increases from a negative value to zero. Figure 27.13 shows the flow chart for the calculation of the adhesion force and the adhesion energy employed by Kim and Bhushan (2007a).

The random rough surfaces used in the simulations were generated by a computer program (Bhushan 1999, 2002). Two-dimensional profiles of surfaces that a gecko may encounter were obtained using a stylus profiler (Bhushan et al. 2006). These profiles along with the surface selection methods and surface roughness parameters [root mean square (RMS) amplitude σ and correlation length β^*] for scan lengths of 80, 400, and 2000 µm are presented in the Appendix. The roughness parameters are scale-dependent, and, therefore, adhesion values also are expected to be scale-dependent. As the scan length was increased, the measured values of RMS amplitude and correlation length both increased. The range of values of σ from 0.01 to 30 µm and a fixed value of $\beta^* = 200$ µm were used for modeling the contact of a seta with random rough surfaces. The range chosen covers values of roughnesses for relatively smooth, artificial surfaces to natural, rough surfaces. A typical scan length of 2000 µm was also chosen, which is comparable to the length of a gecko lamella.

27.5.4
Adhesion Results for the Gecko Attachment System Contacting a Rough Surface

The multilevel spring model produced many useful results. Those obtained by Kim and Bhushan (2007a) will be discussed in detail in this section. Figure 27.14a shows the calculated spring force–distance curves for the one-, two-, and three-level hierarchical models in contact with rough surfaces of different values of RMS amplitude σ ranging from $\sigma = 0.01$ µm to $\sigma = 10$ µm at an applied load of 1.6 µN, which was derived from the gecko's weight. When the spring model is pressed against the rough surface, contact between the spring and the rough surface occurs at point A; as the spring tip presses into the contacting surface, the force increases up to point B, B′, or B″. During pull-off, the spring relaxes, and the spring force passes an equilibrium state (0 N); tips break free of adhesion forces at point C, C′, or C″ as the spring moves away from the surface. The perpendicular distance from C, C′, or C″ to zero is the adhesion force. Adhesion energy stored during contact can be obtained by calculating the area of the triangle during the unloading part of the curves (27.25).

Using the spring force–distance curves, Kim and Bhushan (2007a) calculated the adhesion coefficient, the number of contacts per unit length, and the adhesion energy per unit length of the one-, two-, and three-level models for an applied load of 1.6 µN and a wide range of RMS roughness values as seen in the left graphs in

Fig. 27.14b. The adhesion coefficient, defined as the ratio of the pull-off force to the applied preload, represents the strength of adhesion with respect to the preload. For the applied load of 1.6 µN, which corresponds to the weight of a gecko, the maximum adhesion coefficient is about 36 when σ is smaller than 0.01 µm. This means that a gecko can generate enough adhesion force to support 36 times its bodyweight. However, if σ is increased to 1 µm, the adhesion coefficient for the three-level model is reduced to 4.7. It is noteworthy that the adhesion coefficient falls below 1 when the contacting surface has a RMS roughness σ greater than 10 µm. This implies that the attachment system is no longer capable of supporting the gecko's weight. Autumn et al. (2000, 2002) showed that in isolated gecko setae contacting with the surface of a single crystalline silicon wafer, a 2.5-µN preload yielded adhesion of 20–40 µN and thus a value of the adhesion coefficient of 8–16, which supports the simulation results of Kim and Bhushan (2007a).

Figure 27.14b (top left) shows that the adhesion coefficient for the two-level model is lower than that for the three-level model, but there is only a small difference

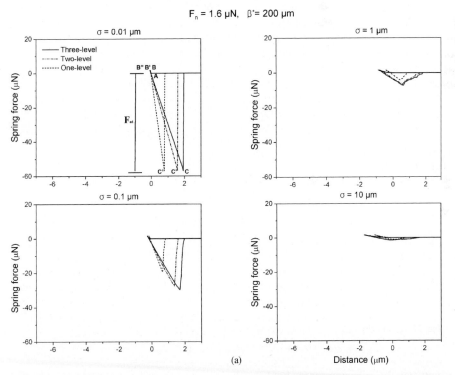

Fig. 27.14. a Force–distance curves of one-, two-, and three-level models in contact with rough surfaces with different σ values for an applied load of 1.6 µN. **b** The adhesion coefficient, the number of contacts, and the adhesion energy per unit length of profile for one-level and multilevel models with an increase of σ value (*left*), and relative increases between multilevel and one-level models (*right*) for an applied load of 1.6 µN. The value of k_{III} in the analysis is 2.908 N/m (Kim and Bhushan 2007a)

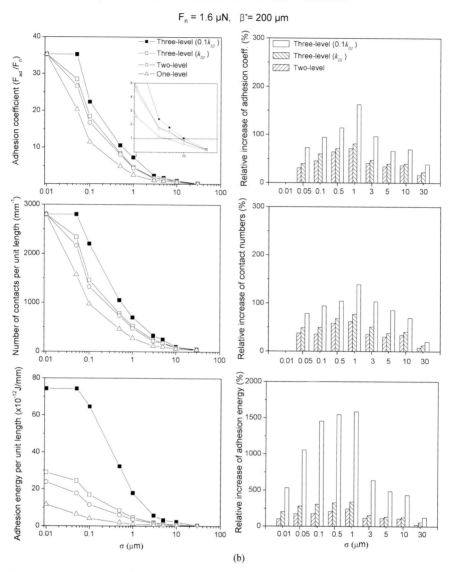

Fig. 27.14. (continued)

between the adhesion forces between the two- and three-level models, because the stiffness of level III for the three-level model is calculated to be higher than those of levels I and II. In order to show the effect of stiffness, the results for the three-level model with springs in level III of which the stiffness is 10 times smaller than that of original level III springs are plotted. It can be seen that the three-level model with a third-level stiffness of $0.1k_{III}$ has a 20–30% higher adhesion coefficient than the three-level model. The results also show that the trends in the number of contacts are similar to that of the adhesion force. The study also investigated the effect of σ on adhesion energy. It was determined that the adhesion energy decreases with an increase of σ. For the smooth surface with $\sigma = 0.01$ μm, the adhesion energies for the two- and three-level hierarchical models are 2 times and 2.4 times larger than that for the one-level model, respectively, but these values decrease rapidly at surfaces with σ greater than 0.05 μm; and in every model the adhesion energy finally decreases to zero at surfaces with σ greater than 10 μm. The adhesion energy for the three-level model with $0.1k_{III}$ is 2–3 times higher than that for three-level model.

In order to demonstrate the effect of the hierarchical structure on adhesion enhancement, Kim and Bhushan (2007a) calculated the increases in the adhesion coefficient, the number of contacts, and the adhesion energy of the two-, three-, and three-level (with $0.1k_{III}$) models relative to the one-level model. These results are shown on the right side of Fig. 27.14b. It was found for the two- and three-level models, the relative increase of the adhesion coefficient increases slowly with an increase of σ and has the maximum values of about 70 and 80% at $\sigma = 1$ μm, respectively, and then decreases for surfaces with σ greater than 3 μm. On the whole, at the applied load of 1.6 μN, the effect of the variation of σ on the adhesion enhancement for both two- and three-level models is not so large. However, the relative increase of the adhesion coefficient for the three-level model with $0.1k_{III}$ has the maximum value of about 170% at $\sigma = 1$ μm, which shows significant adhesion enhancement. Owing to the relative increase of adhesion energy, the three-level model with $0.1k_{III}$ shows significant adhesion enhancement.

Figure 27.15 shows the variation of adhesion force and adhesion energy as a function of applied load for both one- and three-level models contacting with a surface with $\sigma = 1$ μm. It is shown that as the applied load increases, the adhesion force increases up to a certain applied load and then has a constant value, whereas adhesion energy continues to increase with an increase in the applied load. The one-

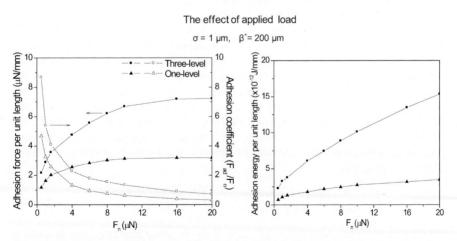

Fig. 27.15. The variation of adhesion force, adhesion coefficient, and adhesion energy as a function of applied loads for both one- and three-level models contacting with surface with $\sigma = 1$ μm. The value of k_{III} in the analysis is 2.908 N/m (Kim and Bhushan 2007a)

level model has a maximum value of adhesive force per unit length of about 3 μN/mm at an applied load of 10 μN, and the three-level model has a maximum value of about 7 μN/mm at an applied load of 16 μN. However, the adhesion coefficient continues to decrease at higher applied loads because adhesion force is constant even if the applied load increases.

The simulation results for the three-level model, which closely models gecko setae, presented in Fig. 27.14 show that roughness reduces the adhesion force. At a surface with σ greater than 10 μm, the adhesion force cannot support the gecko's weight. However, in practice, a gecko can cling or crawl on the surface of a ceiling with higher roughness. Kim and Bhushan (2007a) did not consider the effect of lamellae in their study. The authors state that the lamellae can adapt to the waviness of the surface, while the setae allow for the adaptation to microroughness or nanoroughness and expect that adding the lamellae of gecko skin to the model would lead to higher adhesion over a wider range of roughness. In addition, their hierarchical model considers only normal-to-surface deformation and motion of the seta. It should be noted that measurements of adhesion force of a single gecko seta made by Autumn et al. (2000) demonstrated that a load applied normal to the surface was insufficient for effective attachment. The lateral force required to pull parallel to the surface was observed by sliding the seta approximately 5 μm laterally along the surface under a preload.

27.5.5
Capillarity Effects

Kim and Bhushan (2007d) investigated the effects of capillarity on gecko adhesion by considering capillary force as well as the solid-to-solid interaction. The Laplace and surface tension components of the capillary force are treated according to Sect. 27.3.2. The solid-to-solid adhesion force was calculated by DMT theory according to (27.19) and will be denoted as F_{DMT}.

The work of adhesion was then calculated by (27.20). Kim and Bhushan (2007d) assumed typical values of the Hamaker constant to be 10^{-19} J in air and 6.7×10^{19} J in water (Israelachvili 1992). The work of adhesion of two surfaces in contact separated by an atomic distance $D \approx 0.2$ nm (Israelachvili 1992) is approximately -66 mJ/m^2 in air and -44 mJ/m^2 in water. Assuming the tip radius R is 50 nm, the DMT adhesion forces F_{DMT} of a single contact in air and water are $F_{\mathrm{DMT}}^{\mathrm{air}} = 11$ nN and $F_{\mathrm{DMT}}^{\mathrm{water}} = 7.3$ nN, respectively. As the humidity increases from 0 to 100%, the DMT adhesion force will take a value between $F_{\mathrm{DMT}}^{\mathrm{air}}$ and $F_{\mathrm{DMT}}^{\mathrm{water}}$. To calculate the DMT adhesion force for the intermediate humidity, an approximation method by Wan et al. (1992) was used. The work of adhesion W_{ad} for the intermediate humidity can be expressed as

$$W_{\mathrm{ad}} = \int_D^\infty \frac{H}{6\pi h^3} \, dh = \int_D^{h_{\mathrm{f}}} \frac{H_{\mathrm{water}}}{6\pi h^3} \, dh + \int_{h_{\mathrm{f}}}^\infty \frac{H_{\mathrm{air}}}{6\pi h^3} \, dh \,, \tag{27.26}$$

where h is the separation along the plane. h_{f} is the water film thickness at a filling angle ϕ, which can be calculated as

$$h_{\mathrm{f}} = D + R(1 - \cos\phi) \,. \tag{27.27}$$

Therefore, using (27.19), (27.26), and (27.27), the DMT adhesion force for intermediate humidity is

$$F_{\text{DMT}} = F_{\text{DMT}}^{\text{water}} \left(1 - \frac{1}{[1 + R(1 - \cos\phi)/D]^2}\right)$$
$$+ F_{\text{DMT}}^{\text{air}} \left(\frac{1}{[1 + R(1 - \cos\phi)/D]^2}\right). \qquad (27.28)$$

Finally, Kim and Bhushan (2007d) calculated the total adhesion force F_{ad} as the sum of (27.15) and (27.28):

$$F_{\text{ad}} = F_{\text{c}} + F_{\text{DMT}}. \qquad (27.29)$$

Kim and Bhushan (2007d) then used the total adhesion force as a critical force in the three-level hierarchical spring model discussed previously. In the spring model of a gecko seta the spring is regarded as having been detached if the force applied upon spring deformation is greater than the adhesion force.

27.5.6
Adhesion Results that Account for Capillarity Effects

To simulate the capillarity contribution to the adhesion force of a gecko spatula, Kim and Bhushan (2007d) set the contact angle on the gecko spatula tip θ_1 to 128° (Huber et al. 2005b). It was assumed that the spatula tip radius $R = 50$ nm, the ambient temperature $T = 25\,°\text{C}$, the surface tension of water $\gamma = 73$ mJ/m², and the molecular volume of water $V = 0.03$ nm³ (Israelachvili 1992).

Figure 27.16a shows the total adhesion force as a function of relative humidity for a single spatula in contact with surfaces with different contact angles. The total adhesion force decreases with an increase in the contact angle on the substrate, and the difference of the total adhesion force among different contact angles is larger in the intermediate-humidity regime. As the relative humidity increases, the total adhesion force for the surfaces with contact angle less than 60° has a higher value than the DMT adhesive force not considering wet contact, whereas for values above 60°, the total adhesion force has lower values at most relative humidities.

The simulation results of Kim and Bhushan (2007d) are compared with the experimental data of Huber et al. (2005b) in Fig. 27.16b. Huber et al. (2005b) measured the pull-off force of a single spatula in contact with four different types of Si wafer and glass at the ambient temperature of 25 °C and a relative humidity of 52%. According to their description, wafer families "N" and "T" in Fig. 27.16b differ by the thickness of the top amorphous silicon oxide layer. The "Phil" type is the cleaned silicon oxide surface which is hydrophilic with a water contact angle of approximately 10°, whereas the "Phob" type is a silicon wafer covered hydrophobic monolayer causing a water contact angle of more than 100°. The glass has a water contact angle of 58°. Huber et al. (2005b) showed that the adhesive force of a gecko spatula rises significantly for substrates with increasing hydrophilicity (adhesion force increases by a factor of 2 as mating surfaces go from hydrophobic to hydrophylic). As shown in Fig. 27.16b, the simulation results of Kim and Bhushan (2007d) closely match the experimental data of Huber et al. (2005b).

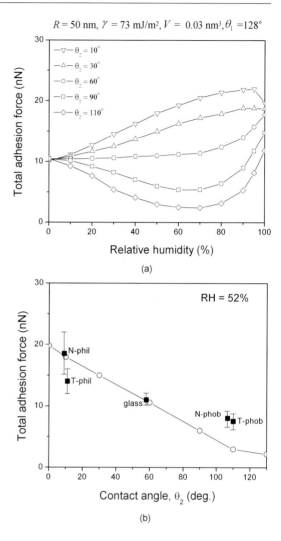

Fig. 27.16. a Total adhesion force as a function of relative humidity for a single spatula in contact with surfaces with different contact angles. **b** Comparison of the simulation results of Kim and Bhushan (2007d) with the measured data obtained by Huber et al. (2005b) for a single spatula in contact with hydrophilic and hydrophobic surfaces (Kim and Bhushan 2007d). *RH* relative humidity

Kim and Bhushan (2007d) performed an adhesion analysis for the three-level hierarchical model for gecko seta. Figure 27.17 shows the adhesion coefficient and number of contacts per unit length for the three-level hierarchical model in contact with rough surfaces with different values of RMS amplitude σ ranging from $\sigma = 0.01\,\mu\text{m}$ to $\sigma = 30\,\mu\text{m}$ for different relative humidities and contact angles of the surface. It can be seen that for the surface with contact angle $\theta_2 = 10°$ the adhesion coefficient is greatly influenced by relative humidity. At 0% relative humidity the maximum adhesion coefficient is about 36 at a value of σ smaller than 0.01 µm compared with 78 for 90% relative humidity with the same surface roughness. As expected, the effect of relative humidity on increasing the adhesion coefficient decreases as the contact angle becomes larger. For hydrophobic surfaces, relative humidity decreases the adhesion coefficient. Similar trends can be noticed in

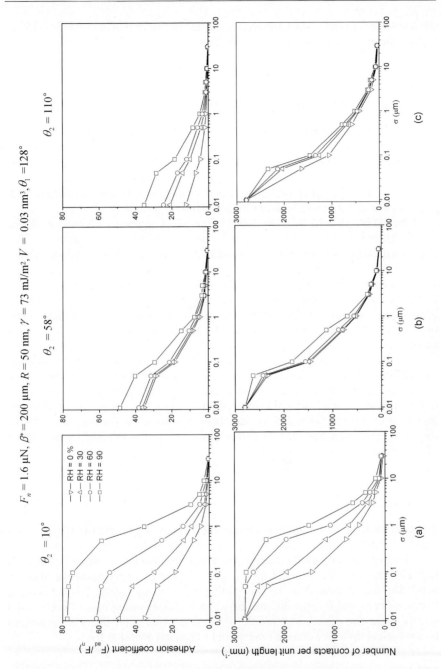

Fig. 27.17a–c. The adhesion coefficient and number of contacts per unit length for the three-level hierarchical model in contact with rough surfaces with different values of root mean square amplitudes σ and contact angles for different relative humidities (Kim and Bhushan 2007d)

the number of contacts. Thus, the conclusion can be drawn that hydrophilic surfaces are beneficial to gecko adhesion enhancement.

27.6
Modeling of Biomimetic Fibrillar Structures

The mechanics of adhesion between a fibrillar structure and a rough surface as it relates to the design of biomimetic structures has been a topic of investigation by many researchers (Jagota and Bennison 2002; Persson 2003; Sitti and Fearing 2003a; Glassmaker et al. 2004, 2005; Gao et al. 2005; Yao and Gao 2006; Kim and Bhushan 2007b,c). In order to better understand the mechanics of adhesion related to design, the approach of Kim and Bhushan (2007c) will be described.

Kim and Bhushan (2007c) developed a convenient, general, and useful guideline for understanding biological systems and for improving biomimetic attachment. This adhesion database was constructed by modeling the fibers as oriented cylindrical cantilever beams with spherical tips. The authors then carried out numerical simulation of the attachment system in contact with random rough surfaces considering three constraint conditions—buckling, fracture, and sticking of fiber structure. For a given applied load, roughness of contacting surface, and fiber material, a procedure to find the optimal fiber radius and aspect ratio for the desired adhesion coefficient was developed.

The model of Kim and Bhushan (2007c) is used to find the design parameters for fibers of a single-level attachment system capable of achieving the desired properties—high adhesion coefficient and durability. The design variables for an attachment system are as follows: fiber geometry (radius and aspect ratio of fibers, tip radius), fiber material, fiber density, and fiber orientation. The optimal values for the design variables to achieve the desired properties should be selected for fabrication of a biomimetic attachment system.

27.6.1
Fiber Model

The fiber model of Kim and Bhushan (2007c) consists of a simple idealized fibrillar structure consisting of a single-level array of microbeams/nanobeams protruding from a backing as shown in Fig. 27.18. The fibers are modeled as oriented cylindrical cantilever beams with spherical tips. In Fig. 27.18, l is the length of the fibers, θ is the fiber orientation, R is the fiber radius, R_t is the tip radius, S is the spacing between fibers, and h is the distance between the upper-spring base of each model and the mean line of the rough profile. The end terminal of the fibers is assumed to be a spherical tip with a constant radius and a constant adhesion force.

27.6.2
Single Fiber Contact Analysis

Kim and Bhushan (2007c) modeled an individual fiber as a beam oriented at an angle θ to the substrate and the contact load F is aligned normal to the substrate.

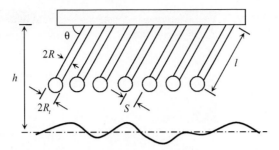

Fig. 27.18. Single-level attachment system with oriented cylindrical cantilever beams with spherical tip. l is the length of the fibers, θ is the fiber orientation, R is the fiber radius, R_t is the tip radius, S is the spacing between fibers, and h is the distance between the base of the model and the mean line of the rough profile (Kim and Bhushan 2007c)

The net displacement normal to the substrate can be calculated according to (27.16) and (27.17). The fiber stiffness ($k = F/\delta_\perp$) is given by (Glassmaker et al. 2004)

$$k = \frac{\pi R^2 E}{l \sin^2\theta \left(1 + \frac{4l^2 \cot^2\theta}{3R^2}\right)} = \frac{\pi R E}{2\lambda \sin^2\theta \left(1 + \frac{16\lambda^2 \cot^2\theta}{3}\right)}, \quad (27.30)$$

where $\lambda = l/2R$ is the aspect ratio of the fiber and θ is fixed at 30°.

Two alternative models dominate the world of contact mechanics—the JKR theory (Johnson et al. 1971) for compliant solids and the DMT theory (Derjaguin et al. 1975) for stiff solids. Although gecko setae are composed of β-keratin with a high elastic modulus (Russell 1986; Bertram and Gosline 1987) which is close to the DMT model, in general the JKR theory prevails for biological or artificial attachment systems. Therefore, the JKR theory was applied in the subsequent analysis of Kim and Bhushan (2007c) to compare the materials with wide ranges of elastic modulus. The adhesion force between a spherical tip and a rigid flat surface is thus calculated using the JKR theory as (Johnson et al. 1971)

$$F_{\text{ad}} = \frac{3}{2}\pi R_t W_{\text{ad}}, \quad (27.31)$$

where R_t is the radius of the spherical tip and W_{ad} is the work of adhesion (calculated according to (27.24)). Kim and Bhushan (2007c) used this adhesion force as a critical force. If the elastic force of a single spring is less than the adhesion force, they regarded the spring as having been detached.

27.6.3
Constraints

In the design of fibrillar structures a trade-off exists between the aspect ratio of the fibers and their adaptability to a rough surface. If the aspect ratio of the fibers is too large, they can adhere to each other or even collapse under their own weight as shown in Fig. 27.19a. If the aspect ratio is too small (Fig. 27.19b), the structures will lack the necessary compliance to conform to a rough surface. The spacing between the individual fibers is also important. If the spacing is too small, adjacent

Fig. 27.19. SEM micrographs of **a** high-aspect-ratio polymer fibrils that have collapsed under their own weight and **b** low-aspect-ratio polymer fibrils that are incapable of adapting to rough surfaces (Sitti and Fearing 2003a)

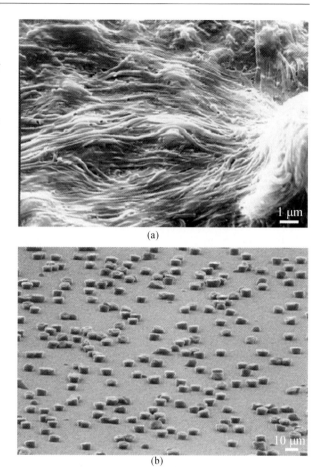

(a)

(b)

fibers can attract each other through intermolecular forces, which will lead to bunching. Therefore, Kim and Bhushan (2007c) considered three necessary conditions in their analysis, buckling, fracture, and sticking of fiber structure, which constrain the allowed geometry.

27.6.3.1
Nonbuckling Condition

A fibrillar interface can deliver a compliant response while still employing stiff materials because of bending and microbuckling of fibers. Based on classical Euler buckling, Glassmaker et al. (2004) established a stress–strain relationship and a critical compressive strain for buckling ε_{cr} for the fiber oriented at an angle θ to the substrate,

$$\varepsilon_{cr} = -\frac{b_c \pi^2}{3(Al^2/3I)}\left(1 + \frac{A_C l^2}{3I}\cot^2\theta\right), \quad (27.32)$$

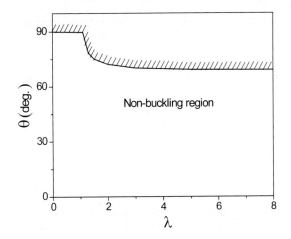

Fig. 27.20. Critical fiber orientation as a function of aspect ratio λ for the nonbuckling condition for pinned-clamped microbeams ($b_c = 2$) (Kim and Bhushan 2007c)

where A_c is the cross-sectional area of the fibril and b_c is a factor that depends on boundary conditions. The factor b_c has a value of 2 for pinned-clamped microbeams. For fibers having a circular cross section, ε_{cr} is calculated as

$$\varepsilon_{cr} = -\frac{b_c \pi^2}{3(4l^2/3R^2)} \left(1 + \frac{4l^2}{3R^2} \cot^2 \theta\right) = -b_c \pi^2 \left(\frac{1}{16\lambda^2} + \frac{\cot^2 \theta}{3}\right). \qquad (27.33)$$

In (27.33), ε_{cr} depends on both the aspect ratio λ and the orientation θ of the fibers. If $\varepsilon_{cr} = 1$, which means the fiber deforms up to the backing, buckling does not occur. Figure 27.20 plots the critical orientation θ as a function of aspect ratio for the case of $\varepsilon_{cr} = 1$. The critical fiber orientation for buckling is 90° at λ less than 1.1. This means that the buckling does not occur regardless of the orientation of the fiber at λ less than 1.1. For λ greater than 1.1, the critical fiber orientation for buckling decreases with an increase in λ, and has a constant value of 69° at λ greater than 3. Kim and Bhushan (2007c) used a fixed value at 30° for θ, because as stated earlier the maximum adhesive force is achieved at this orientation and buckling is not expected to occur.

27.6.3.2
Nonfiber Fracture Condition

For small contacts, the strength of the system will eventually be determined by fracture of the fibers. Spolenak et al. (2005) suggested the limit of fiber fracture as a function of the adhesion force. The axial stress σ_f in a fiber is limited by its theoretical fracture strength σ_{th}^f as

$$\sigma_f = \frac{F_{ad}}{R^2 \pi} \leq \sigma_{th}^f. \qquad (27.34)$$

Using (27.31), one calculates a lower limit for the useful fiber radius R as

$$R \geq \sqrt{\frac{3R_t W_{ad}}{2\sigma_{th}^f}} \approx \sqrt{\frac{15 R_t W_{ad}}{E}}, \qquad (27.35)$$

where the theoretical fracture strength is approximated by $E/10$ (Dieter 1988). The lower limit of the fiber radius for fiber fracture by adhesion force depends on the elastic modulus. By assuming $W_{ad} = 66\,\mathrm{mJ/m^2}$ stated earlier, Kim and Bhushan (2007c) calculated the lower limits of the fiber radius for $E = 1\,\mathrm{MPa}$, $0.1\,\mathrm{GPa}$, and $10\,\mathrm{GPa}$ to be 0.32, 0.032, and 0.0032 μm, respectively.

The contact stress cannot exceed the ideal contact strength transmitted through the actual contact area at the instant of tensile instability (Spolenak et al. 2005). Kim and Bhushan (2007c) used this condition (27.34) to extract the limit of the tip radius, R_t:

$$\sigma_c = \frac{F_{ad}}{a_c^2 \pi} \leq \sigma_{th}, \qquad (27.36)$$

where σ_c is the contact stress and σ_{th} is the ideal strength of van der Waals bonds, which is approximately W_{ad}/b, b is the characteristic length of the surface interaction, and a_c is the contact radius. Based on the JKR theory, for the rigid contacting surface, a_c at the instant of pull-off is calculated as

$$a_c = \left(\frac{9\pi W_{ad} R_t^2 (1 - \nu^2)}{8E} \right)^{1/3}, \qquad (27.37)$$

where ν is the Poisson ratio. The tip radius can then be calculated by combining (27.36) and (27.37) as

$$R_t \geq \frac{8 b^3 E^2}{3 \pi^2 (1 - \nu^2)^2 W_{ad}^2}. \qquad (27.38)$$

The lower limit of the tip radius also depends on the elastic modulus. Assuming $W_{ad} = 66\,\mathrm{mJ/m^2}$ and $b = 2 \times 10^{-10}\,\mathrm{m}$ (Dieter 1988), one calculates the lower limits of the tip radius for $E = 1\,\mathrm{MPa}$, $0.1\,\mathrm{GPa}$, and $10\,\mathrm{GPa}$ as 6×10^{-7}, 6×10^{-3}, and $60\,\mathrm{nm}$, respectively. In this study, Kim and Bhushan (2007c) fixed the tip radius at 100 nm, which satisfies the tip radius condition throughout a wide range of elastic modulus up to 10 GPa.

27.6.3.3
Nonsticking Condition

A high density of fibers is also important for high adhesion. However, if the space S between neighboring fibers is too small, the adhesion forces between them become stronger than the forces required to bend the fibers. Then, fibers might stick to each other and get entangled. Therefore, to prevent fibers from sticking to each other, they must be spaced apart and be stiff enough to prevent sticking or bunching. Several authors (e.g., Sitti and Fearing 2003a) have formulated a nonsticking criterion. Kim and Bhushan (2007c) adopted the approach of Sitti and Fearing (2003a). Both adhesion and elastic forces will act on bent structures. The adhesion force between neighboring two round tips is calculated as

$$F_{ad} = \frac{3}{2} \pi R'_t W_{ad}, \qquad (27.39)$$

where R'_t is the reduced radius of contact, which is calculated as $R'_t = 1/(1/R_{t1} + 1/R_{t2})$; R_{t1} and R_{t2} are the radii of the contacting tips; for the case of similar tips, $R_{t1} = R_{t2}$, $R'_t = 2/R_t$.

The elastic force of a bent structure can be calculated by multiplying the bending stiffness ($k_b = 3\pi R^4 E/4l^3$) by a given bending displacement δ as

$$F_{el} = \frac{3}{4} \frac{\pi R^4 E \delta}{l^3}. \tag{27.40}$$

The condition for the prevention of sticking is $F_{el} > F_{ad}$. By combining (27.39) and (27.40), a requirement for the minimum distance S between structures which will prevent sticking of the structures is (Kim and Bhushan 2007c)

$$S > 2\delta = 2\left(\frac{4}{3} \frac{W_{ad} l^3}{ER^3}\right) = 2\left(\frac{32}{3} \frac{W_{ad} \lambda^3}{E}\right). \tag{27.41}$$

The constant 2 takes into account two nearest structures. Using the distance S, the fiber density ρ is

$$\rho = \frac{1}{(S+2R)^2}. \tag{27.42}$$

Equation (27.42) was then used to calculate the allowed minimum density of fibers without sticking or bunching. In (27.41), it is shown that the minimum distance S depends on both the aspect ratio λ and the elastic modulus E. A smaller aspect ratio and a higher elastic modulus allow for greater packing density. However, fibers with a low aspect ratio and a high modulus are not desirable for adhering to rough surfaces owing to lack of compliance.

27.6.4
Numerical Simulation

The simulation of adhesion of an attachment system in contact with random rough surfaces was carried out numerically. In order to conduct two-dimensional simulations it is necessary to calculate the applied load F_n as a function of applied pressure P_n as an input condition. Using ρ calculated for the nonsticking condition, Kim and Bhushan (2007c) calculated F_n as

$$F_n = \frac{P_n p}{\rho}, \tag{27.43}$$

where p is the number of springs in scan length L, which equals $L/(S+2R)$.

Fibers of the attachment system are modeled as one-level-hierarchy elastic springs (Fig. 27.12) (Kim and Bhushan 2007c). The deflection of each spring and the elastic force arising in the springs are calculated according to (27.21) and (27.22), respectively. The adhesion force is the lowest value of elastic force F_{el} when the fiber has detached from the contacting surface. Kim and Bhushan (2007c) used an iterative process to obtain the optimal fiber geometry—fiber radius and aspect ratio. If the applied load, the roughness of the contacting surface and the fiber material are given, the procedure for calculating the adhesion force is repeated iteratively until the desired adhesion force is satisfied. In order to simplify the design problem, the fiber material is regarded as a known variable. The next step is constructing the design database. The left side of Fig. 27.21 shows the flow chart for the construction

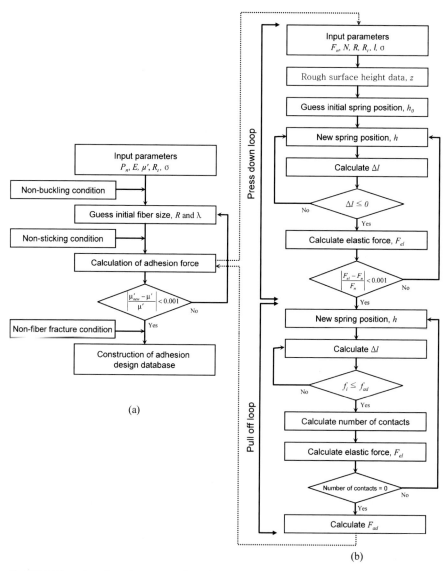

Fig. 27.21. Flow chart for **a** the construction of the adhesion design database and **b** the calculation of the adhesion force. P_n is the applied pressure, E is the elastic modulus, μ' is the adhesion coefficient, R_t is the tip radius, σ is the root mean square amplitude, R is the fiber radius, λ is the aspect ratio of the fiber, F_n is the applied load, N is the number of springs, k and l are the stiffness and the length of the structures, Δl is the spring deformation, f_i is the elastic force of a single spring, and f_{ad} is the adhesion force of a single contact (Kim and Bhushan 2007c)

of the adhesion design database and the right side of Fig. 27.21 shows the calculation of the adhesion force that is a part of the procedure to construct the adhesion design database.

27.6.5
Results and Discussion

Figure 27.22 shows an example of the adhesion design database for biomimetic attachment systems consisting of single-level cylindrical fibers with an orientation angle of 30° and spherical tips of $R_t = 100$ nm constructed by Kim and Bhushan (2007c). The minimum fiber radius calculated for the nonfiber fracture condition, which plays a role in the lower limit of the optimized fiber radius, is also added on the plot. The plots in Fig. 27.22 cover all applicable fiber materials from soft elastomer material such as poly(dimethylsiloxane) to stiffer polymers such as polyimide and β-keratin. The dashed lines in each plot represent the limits of fiber fracture due to the adhesion force. For a soft material with $E = 1$ MPa in Fig. 27.22a, the range of the desirable fiber radius is more than 0.3 μm and that of the aspect ratio is approximately less than 1. As the elastic modulus increases, the feasible ranges of both the fiber radius and the aspect ratio also increase as shown in Fig. 27.22b and c. In Fig. 27.22, the fiber radius has a linear relation with the surface roughness on a logarithmic scale.

If the applied load, the roughness of the contacting surface, and the elastic modulus of a fiber material are specified, the optimal fiber radius and aspect ratio for the desired adhesion coefficient can be selected from this design database. The adhesion databases are useful for understanding biological systems and for guiding the fabrication of biomimetic attachment systems. Two case studies (Kim and Bhushan 2007c) are calculated below.

Case study 1: Select the optimal size of fibrillar adhesive for a wall-climbing robot with the following requirements:

- Material: polymer with $E \approx 100$ MPa
- Applied pressure by weight less than 10 kPa
- Adhesion coefficient less than 5
- Surface roughness $\sigma < 1$ μm

The subplot of the adhesion database that satisfies the requirement is in the second column and second row in Fig. 27.22b. From this subplot, any values on the marked line can be selected to meet the requirements. For example, a fiber radius of 0.4 μm with an aspect ratio of 1 or a fiber radius of 10 μm an with aspect ratio of 0.8 satisfies the specified requirements.

Case study 2: Compare with the adhesion test for a single gecko seta (Autumn et al. 2000, 2002)

- Material: β-keratin with $E \approx 10$ GPa
- Applied pressure 57 kPa (2.5 μN on an area of 43.6 μm^2)
- Adhesion coefficient 8–16
- Surface roughness $\sigma < 0.01$ μm

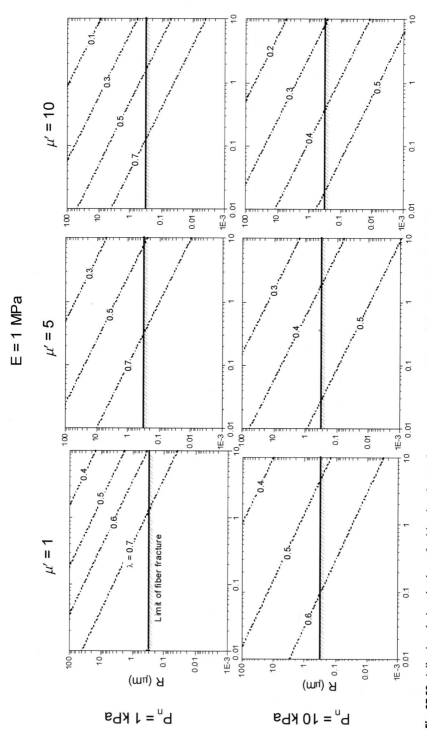

Fig. 27.22. Adhesion design database for biomimetic attachment system consisting of single-level cylindrical fibers with an orientation angle of 30° and spherical tips of 100 nm for an elastic modulus of **a** 1 MPa, **b** 100 MPa, and **c** 10 GPa (Kim and Bhushan 2007c). The *solid lines* shown in **b** and **c** correspond to the case studies I and II, respectively, which satisfy the specified requirements

27 Gecko Feet: Natural Attachment Systems

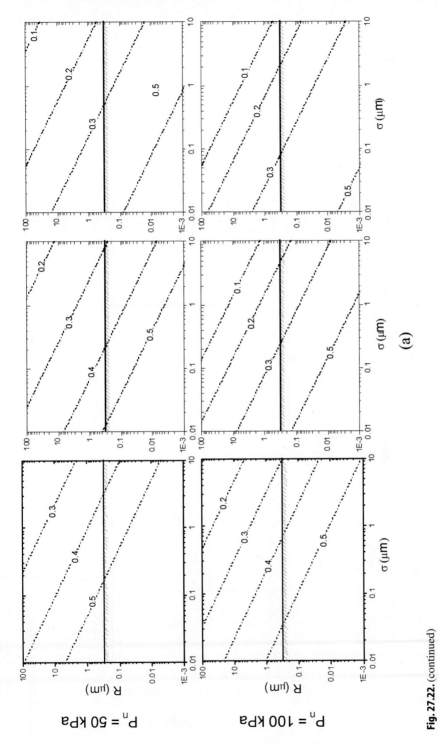

Fig. 27.22. (continued)

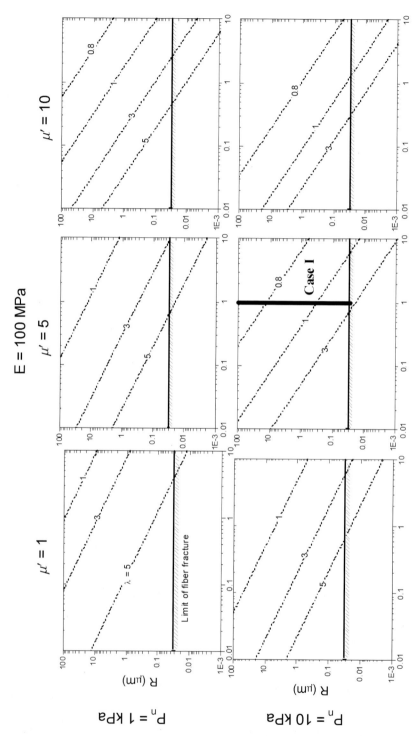

Fig. 27.22. (continued)

27 Gecko Feet: Natural Attachment Systems

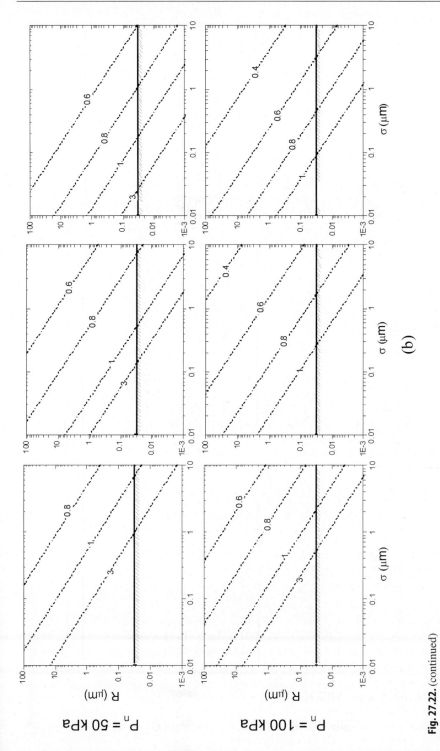

Fig. 27.22. (continued)

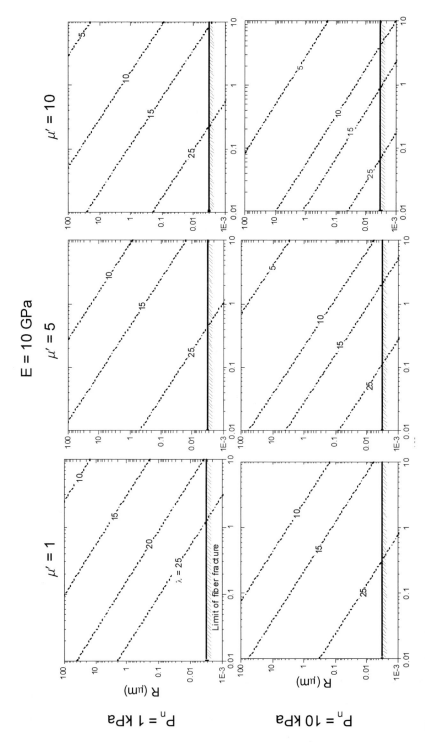

Fig. 27.22. (continued)

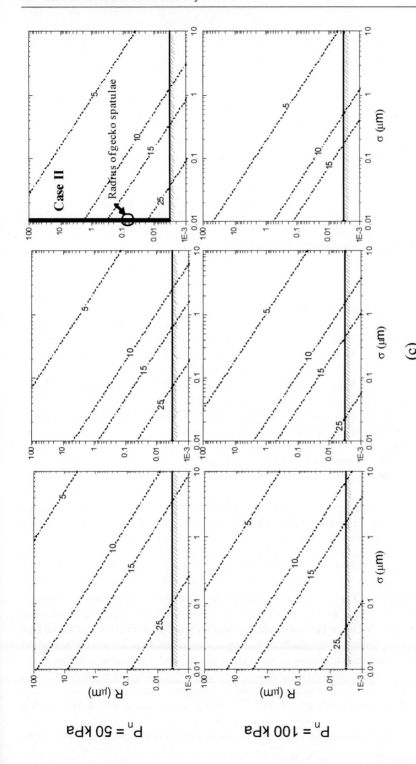

Fig. 27.22. (continued)

Autumn et al. (2000, 2002) showed that in isolated gecko setae contacting with the surface of a single crystalline silicon wafer, a 2.5-μN preload yielded adhesion of 20–40 μN and thus a value of the adhesion coefficient of 8–16. The region that satisfies the above requirements is marked in Fig. 27.22c. The spatulae have an approximate radius of 0.05 μm and an aspect ratio of 25. However, the radius corresponding to $\lambda = 25$ in the marked line is about 0.015 μm. This discrepancy is due to the difference between the simulated fiber model and real gecko setae model. Gecko setae are composed of a three-level hierarchical structure in practice, so higher adhesion can be generated than with a single-level model (Bhushan et al 2006; Kim and Bhushan 2007a, b). Given the simplification in the fiber model, this simulation result is very close to the experimental result.

27.7
Fabrication of Biomimetric Gecko Skin

On the basis of studies found in the literature, the dominant adhesion mechanism utilized by geckos and other spider attachment systems appears to be van der Waals forces. The complex divisions of the gecko skin (lamellae–setae–branches–spatulae) enable a large real area of contact between the gecko skin and the mating surface. Hence, a hierarchical fibrillar microstructure/nanostructure is desirable for dry, superadhesive tapes. The development of nanofabricated surfaces capable of replicating this adhesion force developed in nature is limited by current fabrication methods. Many different techniques have been used in an attempt to create (Geim et al. 2003; Sitti and Fearing 2003a; Northen and Turner 2005, 2006; Yurdumkan et al. 2005; Zhao et al. 2006) and characterize (Peressadko and Gorb 2004; Gorb et al. 2007; Bhushan and Sayer 2007) bio-inspired adhesive tapes.

Gorb et al. (2007) and Bhushan and Sayer (2007) characterized two poly(vinylsiloxane) PVS samples from the Max Planck Institute for Metals Research, Stuttgart, Germany: one consisting of mushroom-shaped pillars (Fig. 27.23a) and the other sample was an unstructured control surface (Fig. 27.23b). The structured sample is composed of pillars that are arranged in a hexagonal order to allow maximum packing density. They are approximately 100 μm in height, 60 μm in base diameter, 35 μm in middle diameter, and 25 μm in diameter at the narrowed region just below the terminal contact plates. These plates were about 40 μm in diameter and 2 μm in thickness at the lip edges. The adhesion force of the two samples in contact with a smooth flat glass substrate was measured by Gorb et al. (2007) using a homemade microtribometer. The results revealed that the structured specimens featured an adhesion force more than twice that of the unstructured specimens. The adhesion force was also found to be independent of the preload. Moreover, it was found that the adhesion force of the structured sample was more tolerant to contamination compared with the control and it could be easily cleaned with a soap solution.

Bhushan and Sayer (2007) characterized the surface roughness, friction force, and contact angle of the structured sample and compared the results with those of an unstructured control. As shown in Fig. 27.24a, the macroscale coefficient of kinetic friction of the structured sample was found to be almost 4 times greater than that of the unstructured sample. This increase was determined to be a result of the structured

Fig. 27.23. SEM micrographs of the **a** structured and **b** unstructured polyvinylsiloxane samples. *SH* shaft, *NR* neck region, *LP* lip (Bhushan and Sayer 2007)

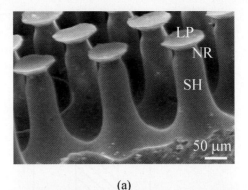

(a)

(b)

roughness of the sample and not the random nanoroughness. It is also noteworthy that the static and kinetic coefficients of friction are approximately equal for the structured sample. It is believed that the divided contacts allow the broken contacts of the structured sample to constantly recreate contact. As seen in Fig. 27.24b, the pillars also increased the hydrophobicity of the structured sample in comparison with the unstructured sample as expectedowing to increased surface roughness (Wenzel 1936; Burton and Bhushan 2005). A large contact angle is important for self-cleaning (Barthlott and Neinhius 1997), which agrees with the findings of Gorb et al. (2007) that the structured sample is more tolerant of contamination than the unstructured sample.

27.7.1
Single-Level Hierarchical Structures

One of the simplest approaches to create a single-level hierarchical surface employed an AFM tip to create a set of dimples on a wax surface. These dimples served as a mold for creating the polymer nanopyramids shown in Fig. 27.25a (Sitti and Fearing 2003a). The adhesive force to an individual pyramid was measured using another AFM cantilever. The force was found to be about 200 μN. Although each pyramid of the material is capable of producing large adhesion forces, the surface failed to replicate gecko adhesion on a macroscale. This was due to the lack of flexibility in the

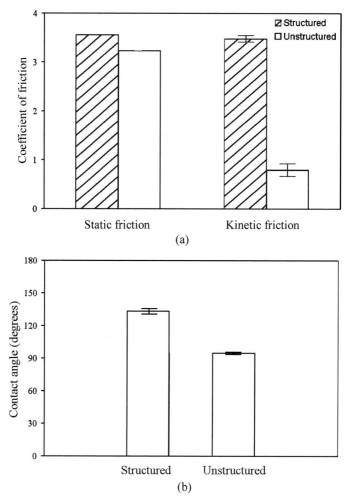

Fig. 27.24. a Coefficients of static and kinetic friction for the structured and unstructured samples slid against magnetic tape with a normal load of 130 mN. **b** Water contact angle for the structured and unstructured samples (Bhushan and Sayer 2007)

pyramids. In order to ensure that the largest possible area of contact occurs between the tape and the mating surface, a soft, compliant fibrillar structure would be desired (Jagota and Bennison 2002). As shown in previous calculations, the van der Waals adhesion force for two parallel surfaces is inversely proportional to the cube of the distance between two surfaces. Compliant fibrillar structures enable more fibrils to be in close proximity of a mating surface, thus increasing van der Waals forces.

Geim et al. (2003) created arrays of nanohairs using electron-beam lithography and dry etching in oxygen plasma (Fig. 27.25b, left). The original arrays were created on a rigid silicon wafer. This design was only capable of creating 0.01 N of adhesion force for a 1-cm^2 patch. The nanohairs were then transferred from the silicon wafer

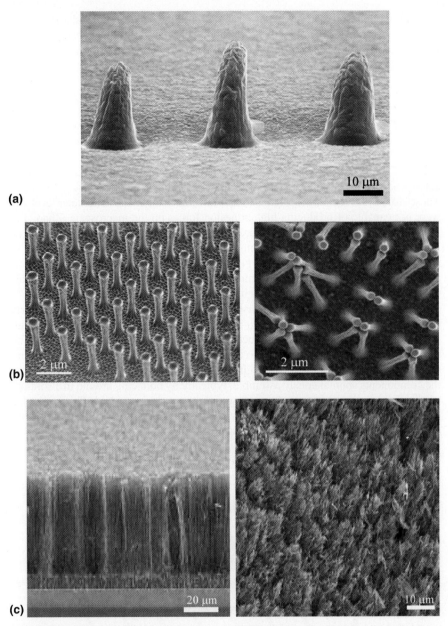

Fig. 27.25. SEM micrographs of **a** three pillars created by nanotip indentation (Sitti 2003), **b** *left* an array of polyimide nanohairs and *right* bunching of the nanohairs, which leads to a reduction in adhesive force (Geim et al. 2003), and **c** multiwalled carbon nanotube structures *left* grown on silicon by vapor deposition and *right* transferred into a poly(methyl methacrylate) matrix and then exposed on the surface after solvent etching (Yurdumakan et al. 2005)

to a soft bonding substrate. A 1-cm^2 sample was able to create 3 N of adhesion force under the new arrangement. This is approximately one third the adhesion strength of a gecko. Bunching (as described earlier) was determined to greatly reduce both the adhesion strength and the durability of the polymer tape. The bunching can be clearly seen in Fig. 27.25b (right).

Multiwalled carbon nanotube (MWCNT) hairs have been used to create superadhesive tapes. Yurdumakan et al. (2005) used chemical vapor deposition to grow MWCNT that are 50–100 µm in length on quartz or silicon substrates. Patterns were then created using a combination of photolithography and a wet and/or a dry etching. Scanning electron microscope images of the nanotube surfaces can be seen in Fig. 27.25c. On a small scale (nanometer level), the MWCNT surface was able to achieve adhesive forces 2 orders of magnitude greater than those of gecko foot-hairs. These structures were only designed to increase adhesion on the nanometer level and were not capable of producing high adhesion forces on the macroscale. Zhao et al (2006) created MWCNT arrays that are much more capable of macroscale adhesion. MWCNT of 5–10 µm were grown on a Si substrate. This arrangement was able to create an adhesion pressure of 11.7 N/cm^2. The durability of the adhesive tape is an issue as some of the nanotubes detach from the substrate with repeated use. This work shows promise for MWCNT being implemented in bio-inspired tapes.

Directed self-assembly has been proposed as a method to produce regularly spaced fibers (Schäffer et al. 2000; Sitti 2003). In this technique, a thin liquid polymer film is coated on a flat conductive substrate. As demonstrated in Fig. 27.26, a closely spaced metal plate is used to apply a DC electric field on the polymer film. Owing to instabilities, pillars will begin to grow. Self-assembly is desirable because the components spontaneously assemble, typically by bouncing around in a solution or gas phase until a stable structure of minimum energy is reached. This method is crucial in biomolecular nanotechnology, and has the potential to be used in precise devices (Anonymous 2002). These surface coatings have been demonstrated to be both durable and capable of creating superhydrophobic conditions and have been used to form clusters on the nanoscale (Pan et al. 2005).

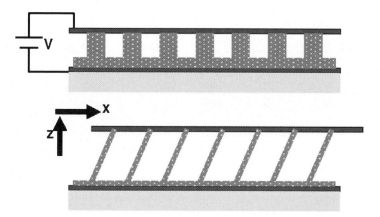

Fig. 27.26. Directed self-assembly based method of producing high-aspect-ratio microhairs/nanohairs (Sitti 2003)

27.7.2
Multilevel Hierarchical Structures

The aforementioned fabricated surfaces only have one level of hierarchy. Although these surfaces are capable of producing high adhesion on the microscale/nanoscale, all have failed in producing large-scale adhesion owing to a lack of compliance and bunching. In order to overcome these problems, Northen and Turner (2005, 2006) created a multilevel compliant system by employing a microelectromechanical-based approach. They created a layer of nanorods which they deemed "organorods" (Fig. 27.27a). These organorods are comparable in size to that of gecko spatulae (50–200 nm in diameter and 2 μm tall). They sit atop a silicon dioxide chip (approximately 2 μm thick and 100–150 μm across a side), which was created using photolithography (Fig. 27.27b). Each chip is supported on top of a pillar (1 μm in diameter and 50 μm tall) that attaches to a silicon wafer (Fig. 27.27c). The multilevel structures have been created across a 100-mm wafer (Fig. 27.27d).

Adhesion testing was performed using a nanorod surface on a solid substrate and on the multilevel structures. As seen in Fig. 27.28, adhesive pressure of the multilevel structures was several times higher than that of the surfaces with only one level of hierarchy. The durability of the multilevel structure was also much greater than that of the single-level structure. The adhesion of the multilevel structure did not change between iterations one and five. During the same number of iterations, the adhesive pressure of the single level structure decreased to zero.

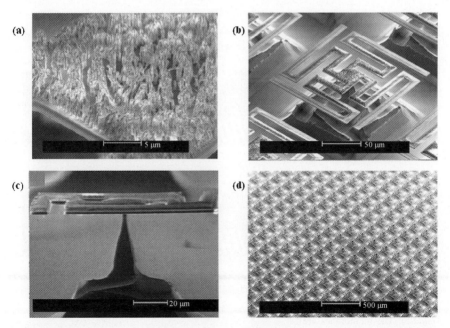

Fig. 27.27. Multilevel fabricated adhesive structure composed of **a** organorods, **b** silicon dioxide chips, and **c** support pillars. **d** This structure was repeated multiple times over a silicon wafer (Northen and Turner 2005 2006)

Sitti (2003) proposed a nanomolding technique for creating structures with two levels of structures. In this method two different molds are created—one with pores on the order of magnitude of microns in diameter and a second with pores of nanometer-scale diameter. One potential mold material is porous anodic alumina, which has been demonstrated to produce ordered pores on the nanometer scale of equal size (Maschmann et al. 2006). Pore-widening techniques could be used to cre-

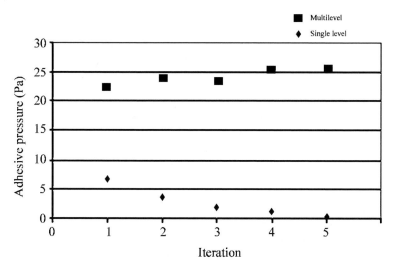

Fig. 27.28. Adhesion test results of a multilevel hierarchical structure and a single-level hierarchical structure repeated for five iterations (Northen and Turner 2006)

Fig. 27.29. Proposed process of creating multilevel synthetic gecko foot-hairs using nanomolding. Micron- and nanometer-sized pore membranes are bonded together (*top*) and filled with liquid polymer (*middle*). The membranes are then etched away leaving the polymer surface (*bottom*) (Sitti 2003)

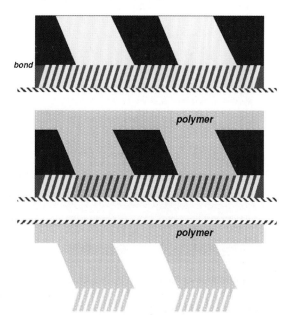

ate micron-scale pores. As seen in Fig. 27.29, the two molds would be bonded to each other and then filled with a liquid polymer. According to Sitti (2003), the method would enable the manufacturing of a high volume of synthetic gecko foot-hairs at low cost.

The literature clearly indicates that in order to create a dry superadhesive, a fibrillar surface construction is necessary to maximize the van der Waals forces by decreasing the distance between the two surfaces. It is also desirable to have a superhydrophobic surface in order to utilize self-cleaning. A material must be soft enough to conform to rough surfaces yet hard enough to avoid bunching, which will decrease the adhesion force.

27.8
Closure

The adhesive properties of geckos and other creatures such as flies, beetles, and spiders are due to the hierarchical structures present on each creature's attachment pads. Geckos have developed the most intricate adhesive structures of any of the aforementioned creatures. The attachment system consists of ridges called lamellae that are covered in microscale setae that branch off into nanoscale spatulae. Each structure plays an important role in adapting to surface roughness, bringing the spatulae in close proximity with the mating surface. These structures as well as material properties allow the gecko to obtain a much larger real area of contact between its feet and a mating surface than is possible with a nonfibrillar material. Two feet of a Tokay gecko have about 220 mm^2 of attachment pad area on which the gecko is able to generate approximately 20 N of adhesion force. Although capable of generating high adhesion forces, a gecko is able to detach from a surface at will—an ability known as smart adhesion. Detachment is achieved by a peeling motion of the gecko's feet from a surface.

Experimental results have supported the adhesion theories of intermolecular forces (van der Waals) as a primary adhesion mechanism and capillary forces as a secondary mechanism, and have been used to rule out several other mechanisms of adhesion, including the secretion of sticky fluids, suction, and increased frictional forces. Atomic force microscopy has been employed by several investigators to determine the adhesive strength of gecko foot-hairs. The measured values of the lateral force required to pull parallel to the surface of a single seta (194 μN) and the adhesion force (normal to the surface) of a single spatula (11 nN) are comparable to the van der Waals prediction of 270 and 11 nN for a seta and a spatula, respectively. The adhesion force generated by a seta increases with preload and reaches a maximum when both perpendicular and parallel preloads are applied. Although gecko feet are strong adhesives, they remain free of contaminant particles through self-cleaning. Spatular size along with material properties enable geckos to easily expel any dust particles that come into contact with their feet.

Recent creation of a three-level hierarchical model for a gecko lamella consisting of setae, branches, and spatulae has brought more insight into adhesion of biological attachment systems. One-, two-, and three-level hierarchically structured spring models for simulation of a seta contacting with random rough surfaces were considered. The simulation results show that the multilevel hierarchical structure has a higher adhesion force as well as adhesion energy than the one-level structure for a given applied load, owing to better adaptation and attachment ability. It is concluded that the multilevel hierarchical structure produces adhesion enhancement, and this enhancement increases with an increase in the applied load and a decrease in the stiffness of springs. The condition at which a significant adhesion enhancement occurs appears to be related to the maximum spring deformation. The result shows that significant adhesion enhancement occurs when the maximum spring deformation is greater than 2–3 times larger than the σ value of surface roughness. For the effect of applied load, as the applied load increases, adhesion force increases up to a certain applied load and then has a constant value, whereas adhesion energy continues to increase with an increase in the applied load. Inclusion of capillary forces in the spring model shows that the total adhesion force decreases with an increase in the contact angle of water on the substrate, and the difference of the total adhesion force among contact angles is larger in the intermediate-humidity regime. In addition, the simulation results match measured data for a single spatula in contact with a hydrophilic and a hydrophobic surface, which further supports van der Waals forces as the dominant mechanism of adhesion and capillary forces as a secondary mechanism.

There is great interest among the scientific community to create surfaces that replicate the adhesion strength of gecko feet. These surfaces would be capable of reusable dry adhesion and would have uses in a wide range of applications from everyday objects such as tapes, fasteners, and toys to microelectric and space applications and even wall-climbing robots. In the design of fibrillar structures, it is necessary to ensure that the fibrils are compliant enough to easily deform to the mating surface's roughness profile, yet rigid enough to not collapse under their own weight. The spacing between the individual fibrils is also important. If the spacing is too small, adjacent fibrils can attract each other through intermolecular forces, which will lead to bunching. The adhesion design database developed by Kim and Bhushan (2007c) serves as a reference for choosing design parameters.

Nanoindentation, electron-beam lithography, and growing of carbon nanotube arrays are all methods that have been used to create fibrillar structures. The limitations of current machining methods on the microscale/nanoscale have resulted in the majority of fabricated surfaces consisting of only one level of hierarchy. Although typically capable of producing high adhesion force with an individual fibril, these surfaces have failed to generate high adhesion forces on the macroscale. Bunching, lack of compliance, and lack of durability are all problems that have arisen with the aforementioned structures. Recently, a multilayered compliant system was created using a microelectromechanical-based approach in combination with nanorods. This method as well as other proposed methods of multilevel nanomolding and directed self-assembly show great promise in the creation of adhesive structures with multiple levels of hierarchy, much like those of gecko feet.

Appendix

Several natural (sycamore tree bark and siltstone) and artificial surfaces (dry wall, wood laminate, steel, aluminum, and glass) were chosen to determine the surface parameters of typical rough surfaces that a gecko might encounter. An Alpha-step® 200 (Tencor Instruments, Mountain View, CA, USA) was used to obtain surface profiles for three different scan lengths: 80 μm, which is approximately the size of a single gecko seta, 2000 μm, which is close to the size of a gecko lamella; and an intermediate scan length of 400 μm. The radius of the stylus tip was 1.5–2.5 μm and the applied normal load was 30 μN. The surface profiles were then analyzed using a specialized computer program to determine the RMS amplitude σ, correlation length β^*, peak-to-valley distance P-V, skewness Sk, and kurtosis K.

Samples of surface profiles and their corresponding parameters at a scan length of 2000 μm can be seen in Fig. 27.30a. The roughness amplitude σ, varies from as little as 0.01 μm in glass to as much as 30 μm in tree bark. Similarly, the correlation

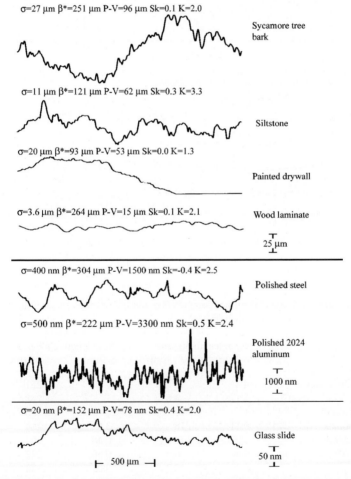

Fig. 27.30. a Surface height profiles of various random rough surfaces of interest at a scan length of 2000 μm and **b** a comparison of the profiles of two surfaces at scan lengths of 80, 400, and 2000 μm. (From Bhushan et al. 2006)

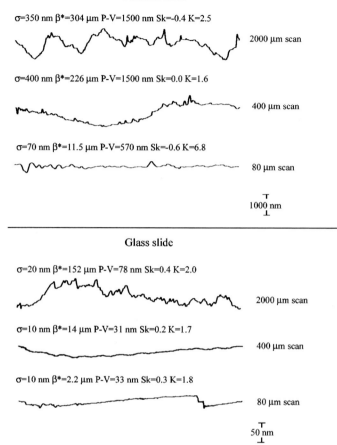

Fig. 27.30. (continued)

Table 27.3. Scale dependence of surface parameters σ and β^* for rough surfaces at scan lengths of 80 and 2000 μm (Bhushan et al. 2006)

Surface	Scan length 80 μm		Scan length 2000 μm	
	σ (μm)	β^* (μm)	σ (μm)	β^* (μm)
Sycamore tree bark	4.4	17	27	251
Siltstone	1.1	4.8	11	268
Painted drywall	1	11	20	93
Wood laminate	0.11	18	3.6	264
Polished steel	0.07	12	0.40	304
Polished 2024 aluminum	0.40	6.5	0.50	222
Glass	0.01	2.2	0.02	152

length varies from 2 to 300 μm. The scale dependency of the surface parameters is illustrated in Fig. 27.30b. As the scan length of the profile increases, so too does the roughness amplitude and the correlation length. Table 27.3 summarizes the scan length dependent factors σ and β^* for all seven sampled surfaces. At a scan length of 80 μm (size of seta), the roughness amplitude does not exceed 5 μm, while at a scan length of 2000 μm (size of lamella), the roughness amplitude is as high as 30 μm. This suggests that setae are responsible for adapting to surfaces with roughness on the order of several microns, while lamellae must adapt to roughness on the order of tens of microns. Larger roughness values would be adapted to by the skin of the gecko. The spring model of Bhushan et al. (2006) verifies that setae are only capable of adapting to roughnesses of a few microns and suggests that lamellae are responsible for adaptation to rougher surfaces.

References

1. Anonymous (2002) Merriam-Webster's medical dictionary, Merriam-Webster, Springfield
2. Aristotle (1918) Historia animalium, book IX, part 9. Translated by Thompson DAW. http://classics.mit.edu/Aristotle/history_anim.html
3. Arzt E, Gorb S, Spolenak R (2003) Proc Natl Acad Sci USA 100:10603–10606
4. Autumn K (2006) Am Sci 94:124–132
5. Autumn K, Peattie AM (2002) Integr Comp Biol 42:1081–1090
6. Autumn K, Liang YA, Hsieh ST, Zesch W, Chan WP, Kenny TW, Fearing R, Full RJ (2000) Nature 405:681–685
7. Autumn K, Sitti M, Liang YA, Peattie AM, Hansen WR, Sponberg S, Kenny TW, Fearing R, Israelachvili JN, Full RJ (2002) Proc Natl Acad Sci USA 99:12252–12256
8. Autumn K, Buehler M, Cutkosky M, Fearing R, Full RJ, Goldman D, Groff R, Provancher W, Rizzi AA, Sranli U, Saunders A, Koditschek DE (2005) Proc SPIE 5804:291–302
9. Autumn K, Majidi C, Groff RE, Dittmore A, Fearing R (2006) J Exp Biol 209:3558–3568
10. Barthlott W, Neinhuis C (1997) Planta 202:1–8
11. Bergmann PJ, Irschick DJ (2005) J Exp Zool 303A:785–791
12. Bertram JEA, Gosline JM (1987) J Exp Biol 130:121–136
13. Bhushan B (1996) Tribology and mechanics of magnetic storage devices, 2nd edn. Springer, New York
14. Bhushan B (1999) Principles and applications of tribology. Wiley, New York
15. Bhushan B (2002) Introduction to tribology. Wiley, New York
16. Bhushan B (ed) (2005) Nanotribology and nanomechanics—an introduction. Springer, Heidelberg
17. Bhushan B (ed) (2007a) Springer handbook of nanotechnology, 2nd edn. Springer, Heidelberg
18. Bhushan B (2007b) J Adhes Sci Technol (in press)
19. Bhushan B, Sayer RA (2007) Microsyst Technol 13:71–78
20. Bhushan B, Peressadko AG, Kim TW (2006) J Adhes Sci Technol 20:1475–1491
21. Bikerman JJ (1961) The science of adhesive joints. Academic, New York
22. Burton Z, Bhushan B (2005) Nano Lett 5:1607–1613
23. Chui BW, Kenny TW, Mamin HJ, Terris BD, Rugar D (1998) Appl Phys Lett 72:1388–1390
24. Dellit WD (1934) Jena Z Naturwissen 68:613–658
25. Derjaguin BV, Muller VM, Toporov YP (1975) J Colloid Interface Sci 53:314–326
26. Dieter GE (1988) Mechanical metallurgy. McGraw-Hill, London
27. Fan PL, O'Brien MJ (1975) In: Lee LH (ed) Adhesion science and technology, vol 9A. Plenum, New York, p 635

28. Federle W (2006) J Exp Biol 209:2611–2621
29. Federle W, Riehle M, Curtis ASG, Full RJ (2002) Integr Comp Biol 42:1100–1106
30. Gao H, Wang X, Yao H, Gorb S, Arzt E (2005) Mech Mater 37:275–285
31. Geim AK, Dubonos SV, Grigorieva IV, Novoselov KS, Zhukov AA, Shapoval SY (2003) Nat Mater 2:461–463
32. Gennaro JGJ (1969) Nat Hist 78:36–43
33. Glassmaker NJ, Jagota A, Hui CY, Kim J (2004) J R Soc Interface 1:23–33
34. Glassmaker NJ, Jagota A, Hui CY (2005) Acta Biomater 1:367–375
35. Gorb S, Varenberg M, Peressadko A, Tuma J (2007) J Royal Soc Interface 4:271–275
36. Hamaker HC (1937) Physica 4:1058
37. Han D, Zhou K, Bauer AM (2004) Biol J Linn Soc 83:353–368
38. Hanna G, Barnes WJP (1991) J Exp Biol 155:103–125
39. Hansen WR, Autumn K (2005) Proc Natl Acad Sci USA 102:385–389
40. Hiller U (1968) Z Morphol Tiere 62:307–362
41. Hinds WC (1982) Aerosol technology: properties, behavior, and measurement of airborne particles. Wiley, New York
42. Hora SL (1923) J Asiat Soc Beng 9:137–145
43. Houwink R, Salomon G (1967) J Appl Phys 38:1896–1904
44. Huber G, Gorb SN, Spolenak R, Arzt E (2005a) Biol Lett 1:2–4
45. Huber G, Mantz H, Spolenak R, Mecke K, Jacobs K, Gorb SN, Arzt E (2005b) Proc Natl Acad Sci USA 102:16293–16296
46. Irschick DJ, Austin CC, Petren K, Fisher RN, Losos JB, Ellers O (1996) Biol J Linn Soc 59:21–35
47. Israelachvili JN, Tabor D (1972) Proc R Soc Lond Ser A 331:19–38
48. Israelachvili JN (1992) Intermolecular and surface Forces, 2nd edn. Academic, San Diego
49. Jaenicke R (1998) In: Harrison RM, van Grieken R (eds) Atmospheric particles. Wiley, New York, pp 1–29
50. Jagota A, Bennison SJ (2002) Integr Comp Biol 42:1140–1145
51. Johnson KL, Kendall K, Roberts AD (1971) Proc R Soc Lond Ser A 324:301–313
52. Kesel AB, Martin A, Seidl T (2003) J Exp Biol 206:2733–2738
53. Kim TW, Bhushan B (2007a) J Adhes Sci Technol 21:1–20
54. Kim TW, Bhushan B (2007b) Ultramicroscopy 107:902–912
55. Kim TW, Bhushan B (2007c) J Vac Sci Technol A 25:1003–1012
56. Kim TW, Bhushan B (2007d) J Royal Soc Interface (in press)
57. Kluge AG (2001) Hamadryad 26:1–209
58. Losos JB (1990) Asiat Herp Res 3:54–59
59. Maderson PFA (1964) Nature 2003:780–781
60. Maschmann MR, Franklin AD, Amama PB, Zakharov DN, Stach EA, Sands TD, Fisher TS (2006) Nanotechnology 17:3925–3929
61. Menon C, Murphy M, Sitti M (2004) In: Proceedings of the IEEE international conference on robotics and biomimetics, 22–26 August, pp 431–436
62. Northen MT, Turner KL (2005) Nanotechnology 16:1159–1166
63. Northen MT, Turner KL (2006) Sens Actuators A 130–131:583–587
64. Orr FM, Scriven LE, Rivas AP (1975) J Fluid Mech 67:723–742
65. Pan B, Gao F, Ao L, Tian H, He R, Cui D (2005) Colloids Surf A 259:89–94
66. Peressadko A, Gorb SN (2004) J Adhes 80:247–261
67. Persson BNJ (2003) J Chem Phys 118:7614–7621
68. Persson BNJ, Gorb S (2003) J Chem Phys 119:11437–11444
69. Phipps PB, Rice DW (1979) In: Brubaker GR, Phipps PB (eds) ACS symposium series no 89. American Chemical Society, Washington

70. Rizzo N, Gardner K, Walls D, Keiper-Hrynko N, Hallahan D (2006) J Royal Soc Interface 3:441–451
71. Ruibal R, Ernst V (1965) J Morphol 117:271–294
72. Russell AP (1975) J Zool Lond 176:437–476
73. Russell AP (1986) Can J Zool 64:948–955
74. Schäffer E, Thurn-Albrecht T, Russell TP, Steiner U (2000) Nature 403:874–877
75. Schleich HH, Kästle W (1986) Amphib Reptil 7:141–166
76. Schmidt HR (1904) Jena Z Naturwissen 39:551
77. Shah GJ, Sitti M (2004) In: Proceedings of the IEEE international conference on robotics and biomimetics, pp 873–878
78. Simmermacher G (1884) Zeitschr Wissen Zool 40:481–556
79. Sitti M (2003) In: Proceedings of the IEEE/ASME advanced mechatronics conference, vol 2, pp 886–890
80. Sitti M, Fearing RS (2003a) J Adhes Sci Technol 17:1055–1073
81. Sitti M, Fearing RS (2003b) In: Proceedings of the IEEE international conference on robotics and automation, vol 1, pp 1164–1170
82. Spolenak R, Gorb S, Arzt E (2005) Acta Biomater 1:5–13
83. Stork NE (1980) J Exp Biol 88:91–107
84. Stork NE (1983) J Nat Hist 17:829–835
85. Tian Y, Pesika N, Zeng H, Rosenberg K, Zhao B, McGuiggan P, Autumn K, Israelachvili J (2006) Proc Natl Acad Sci USA 103:19320–19325
86. Tinkle DW (1992) Encyl Am 12:359
87. Van der Kloot WG (1992) Encyl Am 19:336–337
88. Wagler J (1830) Natürliches System der Amphibien. Cotta'schen, Munich
89. Wan KT, Smith DT, Lawn BR (1992) J Am Ceram Soc 75:667–676
90. Wenzel RN (1936) Ind Eng Chem 28:988–994
91. Williams EE, Peterson JA (1982) designs in the digital adhesive pads of scincid lizards. Science 215:1509–1511
92. Yao H, Gao H (2006) J Mech Phys Solids 54:1120–1146
93. Young WC, Budynas R (2001) Roark's formulas for stress and strain, 7th edn. McGraw-Hill, New York
94. Yurdumakan B, Raravikar NR, Ajayan PM, Dhinojwala A (2005) Chem Commun 3799–3801
95. Zhao Y, Tong T, Delzeit L, Kashani A, Meyyappan M, Majumdar A (2006) J Vac Sci Technol B 24:331–335
96. Zimon AD (1969) Adhesion of dust and powder. Translated from Russian by M Corn. Plenum, New York
97. Zisman WA (1963) Ind Eng Chem 55(10):18–38

28 Carrier Transport in Advanced Semiconductor Materials

Filippo Giannazzo · Patrick Fiorenza · Vito Raineri

Abstract. In this chapter, the main scanning probe microscopy based methods to measure transport properties in advanced semiconductor materials are presented. The two major approaches to determine the majority carrier distribution, i.e., scanning capacitance microscopy (SCM) and scanning spreading resistance microscopy (SSRM), are illustrated, starting from their basic principles. The imaging capabilities for critical structures and the quantification of SCM and SSRM raw data to carrier concentration profiles are described and discussed. The determination of drift mobility in semiconductors by combined application of SCM and SSRM is illustrated considering quantum wells. The carrier transport through metal–semiconductor barriers by conductive atomic force microscopy (CAFM) is reviewed. Finally, the charge transport in dielectrics is studied locally by CAFM, and a method for the direct determination of dielectric breakdown and Weibull statistics is illustrated.

Keywords: Transport in semiconductors, Carrier profiling, Scanning probe microscopy, Scanning capacitance microscopy, Conductive atomic force microscopy, Scanning spreading resistance microscopy

Abbreviations

AFM	Atomic force microscopy
BEEM	Ballistic electron emission microscopy
C-AFM	Conducive atomic force microscopy
CF	Correction factor
CVS	Constant voltage stress
EJ	Electrical junction
MBE	Molecular beam epitaxy
MJ	Metallurgical junction
MOS	Metal oxide semiconductor
MOSFET	Metal oxide semiconductor field-effect transistor
nanoMIS	Nanometric metal–insulator–semiconductor
QW	Quantum well
RVS	Ramped voltage stress
SBH	Schottky barrier height
SCS	Scanning capacitance spectroscopy
SCM	Scanning capacitance microscopy
SIMS	Secondary ion mass spectrometry
SPM	Scanning probe microscopy
SSRM	Scanning spreading resistance microscopy

STM Scanning tunneling microscopy
TSCM Transient scanning capacitance microscopy

28.1
Majority Carrier Distribution in Semiconductors: Imaging and Quantification

The methods based on scanning probe microscopy (SPM) for electrical characterization at the nanoscale emerged in the few last years as the most promising candidates to meet the requirements of the International Technology Roadmap for Semiconductors [1] for 2D carrier profiling. In particular, the spatial resolution, the carrier concentration dynamic range and the concentration gradient sensitivity are the most important targets. Scanning capacitance microscopy (SCM) and scanning spreading resistance microscopy (SSRM) are the most attractive methods for the wide dynamic range (from 10^{15} to 10^{20} cm^{-3}) and high spatial resolution. To date, both approaches have been applied independently for carrier profiling in advanced Si devices. Furthermore, different applications have been demonstrated, depending on the specific working principles of the two methods, i.e., capacitance variations and resistance measurements. As an example, the domain structure of ferroelectric materials [2] was imaged by SCM. Variable-temperature SCM was used to probe charge traps in dielectrics and semiconductors [3], recently leading to some demonstrations of transient SCM (TSCM), to probe deep levels in semiconductors [4]. The resistivity from SSRM has been coupled to the carrier concentration from SCM to determine the carrier mobility in semiconductors [5]. In the following, the application of SCM and SSRM to carrier profiling in semiconductors is described in detail.

28.1.1
Basic Principles of SCM

SCM is based on the use of an atomic force microscopy (AFM) conductive tip, connected to a high-sensitivity capacitance sensor. In the typical application of SCM [6], the probe is scanned in contact mode on the surface of a semiconductor coated by an ultrathin (a few nanometers) insulating film (usually SiO$_2$, but also high-k dielectrics [7]), thus forming a nanometric metal–insulator–semiconductor (nanoMIS) device. The capacitance variations of the nanoMIS are mainly related to the local carrier concentration underneath the tip, according to the MIS theory [8]. In some cases, the thin insulator is not present and a tip–semiconductor nano-Schottky contact is formed. Carrier concentration profiling in the semiconductor has also been demonstrated by Schottky SCM [9].

Owing to its nanoscale dimensions, the capacitance of the tip–insulator–semiconductor nanoMIS structure (Fig. 28.1, panel b, insert) is on the order of the attofarads (10^{-18} F) or sub-attofarads. This is illustrated in Fig. 28.1, panel a, where the calculated C–V_{tip} curves are reported for the case of an AFM probe (with 20-nm typical curvature radius) on p- and n-type Si (10^{15}, 10^{17} cm^{-3} concentrations) coated by a 2-nm-thick oxide layer.

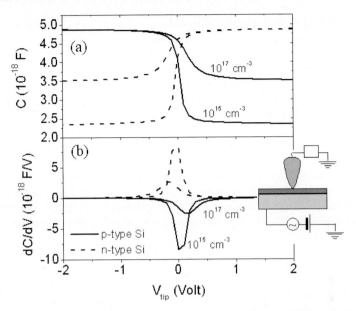

Fig. 28.1. Calculated C–V_{tip} (*a*) and dC/dV–V_{tip} (*b*) curves for a nanometric tip–insulator–semiconductor (*insert*) formed by an atomic force microscopy (AFM) probe (20-nm curvature radius) on p- and n-type Si (10^{15}, 10^{17} cm^{-3} concentrations) coated by a 2-nm-thick oxide layer

During the measurement, the sample is placed on a conductive chuck, with a DC bias and a high frequency (approximately 100 kHz) small-amplitude AC bias. A high-sensitivity capacitance sensor connected to the probe measures the capacitance variations induced by the modulating AC bias in the nanoMIS structure. At the same time, the morphology of the scanned surface can be acquired by contact-mode AFM. Since the first implementation of SCM by Matey and Blanc [10] and its first application to carrier profiling in Si by Williams et al. [11], the capacitance sensor was based on the RCA video disc capacitive-pickup circuitry [12], and this is still the basic principle of state-of-the-art equipment. This kind of sensor, which works at 900–1000 MHz with a sensitivity from approximately 10^{-19} F Hz$^{-1/2}$ [13] to approximately 10^{-21} F Hz$^{-1/2}$ [14], is the only one able to measure the small capacitance variations between the scanning probe and the sample. Moreover, since the sensor has large low-frequency noise and the stray capacitance is several orders of magnitude larger (on the order of picofarads) than the tip/sample capacitance (on the order of attofarads), the absolute value of the capacitance cannot be easily related to the MIS structure capacitance value. Therefore, the dC/dV signal is normally measured in arbitrary units by a lock-in amplifier (locked at the modulating AC bias frequency).

Hence, a SCM map is formed by a 2D array of values proportional (through the capacitance sensor gain G) to the derivative dC/dV at the fixed DC bias applied to the sample. The DC bias is usually fixed on the peak of the dC/dV curve, corresponding to the flat-band voltage condition [8], as illustrated in Fig. 28.1, panel b. Obviously, the sign of dC/dV is negative for p-type doped semiconductors and is positive for n-type doped semiconductors, thus enabling us to distinguish between different types of doped areas. Moreover, the absolute value of the of the dC/dV curve decreases monotonically with increasing carrier concentration.

In the case of Si, a sensitivity extending to a wide carrier concentration range (from approximately 10^{15} to approximately 10^{20} cm^{-3}) has been demonstrated on dedicated p- and n-type calibration samples [15]. The dynamic range is limited at the highest concentrations (above 10^{20} cm^{-3}), owing to the very small capacitance variations, i.e., no depletion region is formed in the semiconductor [8]. For the lowest concentrations (below 10^{15} cm^{-3}) the sensitivity is limited by oxide fixed and trapped charges and oxide–semiconductor interface states [16].

The SCM lateral resolution is concentration-dependent too, since the probed semiconductor volume underneath the tip is ultimately related to the maximum depletion extension W [8]. Simulations of the SCM nanometric metal oxide semi-

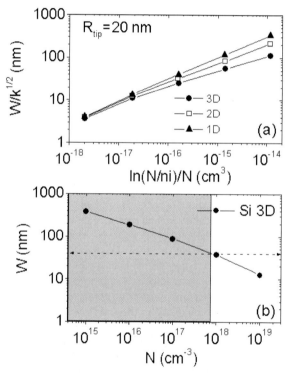

Fig. 28.2. a $W/k^{1/2}$ (W is the maximum depletion extension at $V_{DC} = 0$, k is the semiconductor dielectric constant) versus $\ln(N/n_i)/N$ (N is the dopant concentration and n_i is the intrinsic carrier density in a semiconductor) for a 1D, 2D and 3D nanometric metal oxide semiconductor model. A tip radius $R_{tip} = 20$ nm was assumed. The plot is independent of the material. **b** W versus N for Si. For carrier concentrations higher than 10^{18} cm^{-3}, the lateral resolution on the uniformly doped sample is set by the tip dimension itself (being $W < 2R_{tip}$). For carrier concentrations lower than 10^{18} cm^{-3}, $W < 2R_{tip}$, and the lateral resolution is set by the depleted semiconductor volume. (Simulations from [17])

conductor (nanoMOS) [17] demonstrated that W is critically dependent on the 3D distribution of the electric field around the probe, and, hence, on the probe geometry itself. This is shown in Fig. 28.2a, where $W/k^{1/2}$ is reported as a function of $\ln(N/n_i)/N$ for a 1D, 2D and 3D nanoMOS model. k is the dielectric constant and n_i is the intrinsic carrier density in the semiconductor. The plot in Fig. 28.2a is independent on the kind of semiconductor. A tip radius $R_{tip} = 20$ nm was assumed in the calculations, i.e., the typical radius of commercial tips for SCM. The plot of W versus N in the specific case of Si is reported in Fig. 28.2b. For $R_{tip} = 20$ nm and for carrier concentrations higher than approximately 10^{18} cm^{-3}, the lateral resolution on a uniformly doped sample is set by the tip dimension itself (being $W < 2R_{tip}$). For carrier concentrations lower than approximately 10^{18} cm^{-3}, $W > 2R_{tip}$, and the lateral resolution is set by the depleted semiconductor volume [17].

Recently, sub-10-nm lateral resolution has been demonstrated using special ultrasharp ($R_{tip} < 10$ nm) solid Pt probes [18] on the cross section of different critical structures: (1) the abrupt junction between a silicon-on-insulator oxide layer and a neighboring Si region; (2) a shallow implant (n$^+$/p, 24-nm junction depth); (3) an epitaxial staircase (75-nm steps).

Several concerns are related to the use of ultrasharp tips, like their duration and the reduced signal-to-noise ratio due to the small contact area and the very small depleted volume including few carriers (one to ten). A different approach to achieve an improved lateral resolution relies on the geometrical magnification of the sample region under investigation by angle beveling [19]. Quantitative carrier profiles with nanometric resolution (along the bevel direction) have been obtained by this method on unipolar doped semiconductors [20]. This approach proved to be extremely useful in the study of diffusion and electrical activation of dopant species implanted at ultralow energy in laterally confined windows [21]. However, some caution must be observed in the interpretation of SCM measurements on beveled bipolar devices, owing to the occurrence of the carrier spilling effect [22], determining a shift of the electrical junction (EJ) position towards the surface.

28.1.2
Carrier Imaging Capability by SCM

SCM has been extensively applied to image the critical features in advanced Si devices [18], in SiGe [23], SiC [24] and III–V semiconductors [25].

A unique capability of this method is the possibility to image the 2D majority carrier distribution even in operating devices [26–28]. As an example, independent voltages can be applied to the source/gate/drain/well regions of a p-channel MOS field-effect transistor (MOSFET) (see cross-section and top-view schematics in Fig. 28.3a, b), when the biased SCM tip is scanned on the device cross section. In Fig. 28.3c, the I_{drain}–V_{gate} characteristics on the cross-sectioned device are reported, showing a nanoampere leakage current (for $V_{gate} > 0$), provided a good quality passivating oxide is grown on that polished surface [26]. Moreover, a sequence of SCM images for different V_{gate} values below and beyond the channel inversion threshold voltage is reported [26]. The formation of the conductive channel after inversion (-1.75 and -3 V V_{gate}) is clearly imaged.

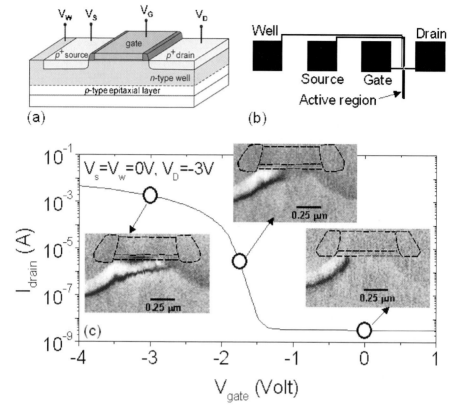

Fig. 28.3. Cross section (**a**) and top view (**b**) of a p-channel metal oxide semiconductor field-effect transistor. I_{drain}–V_{gate} characteristic (**c**) on the cross-sectioned device shows a nanoampere leakage current (for $V_{gate} > 0$). Sequence of three scanning capacitance microscopy (SCM) images for different V_{gate} values below and above the channel inversion threshold voltage. The formation of the conductive channel after inversion (−1.75 and −3 V V_{gate}) is clearly imaged. (Data from [26])

A crucial issue for device engineers is the measurement of critical features such as the effective channel length in MOSFET devices or the net base width in bipolar devices. These measures involve the capability to exactly locate the EJ in 2D section maps. The EJ in a p^+–n (n^+–p) junction is located within the space charge region and corresponds to the isoconcentration surface where the concentration of the holes (electrons) is equal to the concentration of the electrons (holes). Owing to the built-in electric field, the EJ is deeper with respect to the metallurgical junction (MJ), defined as the isoconcentration surface where the "chemical" concentration of the p-type (n-type) dopant atoms is equal to the concentration of the n-type (p-type) dopant atoms.

It has been shown that different electrical profiling methods give different EJ position values [22]. In the case of SCM, since p-type and n-type majority carriers yield dC/dV values with opposite sign (negative and positive, respectively), a straightforward way to locate the EJ in a 2D SCM map could be to take the $dC/dV = 0$ isocontour line. Although this criterion could seem univocal, it has been observed that the $dC/dV = 0$ position can be substantially shifted depending on the DC bias applied to the tip during measurements [29]. This is schematically illustrated in Fig. 28.4, showing a tip moving across an n–p junction, under different bias conditions. If a negative bias is applied to the tip, when it is on the depletion region, holes are attracted from the neighboring p-type region. As a consequence, a deformation is produced in the carrier distribution inside the depletion region and the EJ position is shifted towards the n-type Si side. Similarly, if a positive bias is applied to the tip, the EJ position is shifted towards the p-type Si side. The behavior of the C–V_{tip} characteristics when the tip is located on the n-type, and p-type sides and on the depletion region is also shown in Fig. 28.4. In particular, a typical symmetrical U-shaped curve is associated with the tip located on the EJ (Fig. 28.4e).

The DC bias dependence of the $dC/dV = 0$ position has been studied in details on ad hoc n^+–p and p^+–n source–drain implants [30]. In Fig. 28.5 the EJ in the n^+–p sample versus the applied tip DC bias, both on cross section (EJ_{cs}) and on bevel (EJ_{bev}), is reported. The MJ (650 nm, as determined by secondary ions mass spectrometry, SIMS) is indicated by the horizontal dashed line. It is evident that the EJ_{cs} depends more strongly than the EJ_{bev} on the DC bias. In particular, the EJ_{cs} shifts towards the surface on decreasing the tip DC bias, according to the model in Fig. 28.4. Moreover, the EJ_{bev} is always shallower than the MJ owing to the carrier spilling [30].

The determination of the EJ position is critically dependent on accurate sample preparation, in particular on the quality of polishing and the surface passivation. In fact a high surface state density can induce an additional shift in the EJ position [30]. Accurate 3D simulations of the EJ both on cross section and on bevel have been recently carried out by Stangoni et al. [31].

Scanning capacitance spectroscopy (SCS) [32], a variant of SCM, was proposed as the approach to unambiguously localize the EJ inside a narrow and well-defined

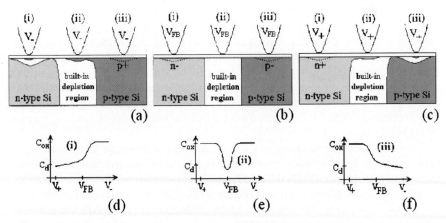

Fig. 28.4. A tip moving across an n–p junction, under different bias conditions, i.e., a negative bias (V_-) (**a**), the flat-band voltage (V_{FB}) (**b**) and a positive bias (V_+) (**c**). The behavior of the C–V_{tip} characteristics when the tip is located on the n-type (**a**) and p-type (**c**) sides and on the depletion region (**b**) is also shown

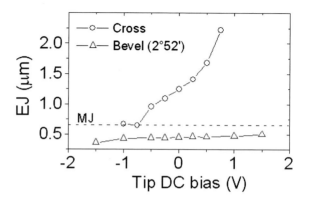

Fig. 28.5. The electrical junction (EJ) versus the applied tip DC bias, both on cross section and on bevel of an n^+–p source–drain implant. The metallurgical junction (MJ) is indicated by the *horizontal dashed line*. (Data from [30])

region. SCS differs from conventional single-voltage SCM, since each pixel in a SCS image contains a complete C–V curve, obtained by cycling the applied DC bias for each fixed probe position. The EJ can be determined from the pixels with the typical symmetrical (U-shaped) C–V curve, as already illustrated in Fig. 28.4e.

28.1.3
Quantification of SCM Raw Data

To date, the quantification of the SCM raw data to majority carrier concentration profiles has been addressed by different authors [33–35]. The most straightforward approach to quantification is the "direct inversion method" or "calibration curve method", which relies on the use of a dC/dV versus N calibration curve, where N is the majority carrier concentration. This curve is calculated starting from the solution of Poisson–Fermi equation coupled with the drift-diffusion equations for electrons and holes in the nanoMIS system with realistic 3D geometry. The fastest way to implement this approach is the "offline" calculation of an extended C–V curve database from those basic equations, for different uniform semiconductor concentrations, tip geometries and oxide properties. The $dC/dV - N$ calibration curve is extracted from that database for a fixed set of geometrical (tip radius R_{tip}, oxide thickness t_{ox}) and electrical (DC bias V_{DC}, AC bias V_{AC} and ultra-high-frequency capacitance sensor bias V_{hf}) parameters. Since dC/dV in the calculated curve is expressed in attofarads per volt, while the measured SCM signal is in arbitrary units, a normalization to the capacitance sensor gain is carried out [35]. The theoretical validity of the method has been discussed for unipolar doping profiles in Si [34].

From the experimental point of view, the reproducibility of SCM measurements is the key point for quantification. Critical studies on this subject have been carried out recently, using specifically designed p- and n-type doped Si calibration standards, containing a set of different uniform doping levels with concentrations ranging from 10^{15} to 10^{20} cm^{-3} [35]. The reliability of the calibration curve approach is critically dependent on the quality of the nanoMIS structure, and on the possibility to describe its properties by the "ideal" nanoMIS model. In Fig. 28.6 the concentration and resistivity profiles measured by spreading resistance profiling on special multilayer Si structures proposed by IMEC [21] are shown.

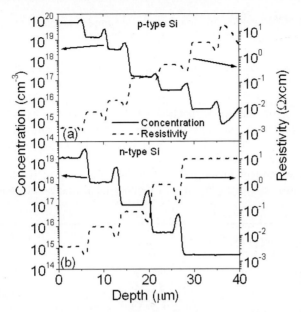

Fig. 28.6. Concentration and and resistivity profiles measured by spreading resistance profiling on special multilayer p-type (**a**) and n-type (**b**) Si structures. (Data from [21])

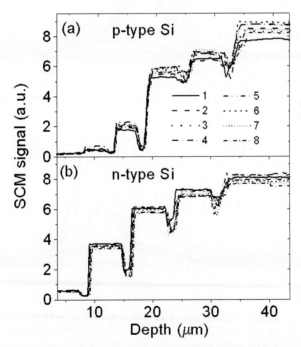

Fig. 28.7. SCM versus depth profiles measured on p-type (**a**) and n-type (**b**) Si samples prepared at different times to illustrate the reproducibility of the technique. (Data from [35])

Different SCM profiles measured on those samples prepared at different times (different days) are reported in Fig. 28.7 to illustrate the reproducibility of the technique.

It was shown that the conductive tip coating (metal, i.e., Pt/Ir and Co/Cr, and conductive diamond), low-temperature oxidation method (wet and UV/ozone) and Si–SiO$_2$ interface microroughness have an impact on the flat-band voltage (depending on the work function of the tip), on the hysteresis (depending on the density of the trapped charges in the oxide) and on the distortion of the $dC/dV - V$ characteristics (depending on the surface state density), respectively [35]. Some criteria to evaluate the quality of the tip, of the oxide and of the semiconductor polished surface for quantitative SCM measurements were also provided [35]. It was demonstrated that an accurate control of

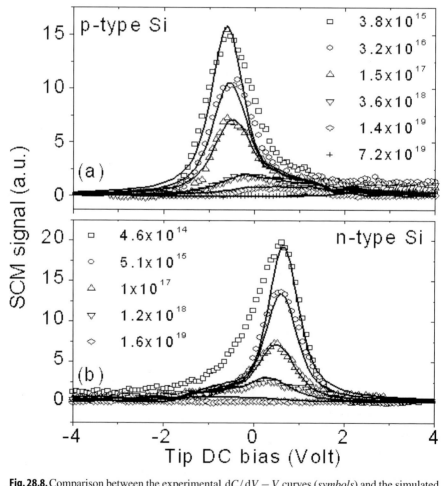

Fig. 28.8. Comparison between the experimental $dC/dV - V$ curves (*symbols*) and the simulated characteristics (*lines*) for a set of different concentration levels in p-type Si (**a**) and n-type Si (**b**) calibration standards. (Data from [35])

all of these issues allows measurements of $dC/dV - V$ to be performed on a set of different concentration levels that can be very well reproduced by the simulation of an ideal nanoMOS, for an optimized set of geometrical parameters (R_{tip}, t_{ox}). This is illustrated in Fig. 28.8, where the symbols represent the experimental SCM curves, while the lines represent the simulated characteristics [35]. From those characteristics, the calibration curve is extracted for fixed V_{DC}.

In Fig. 28.9, a comparison between the calculated calibration curves (lines) and the experimental ones (symbols) is reported. The error bars represent the maximum errors on the measured SCM signal for the different concentration levels, obtained from the reproducibility study reported in Fig. 28.7. The errors affecting the different concentration values obtained by the calibration have been subsequently estimated [35]. In particular, in the case of p-type Si, the maximum percentage errors are lower than 20% for concentrations higher than 5×10^{16} cm^{-3}. In the case of n-type Si, the maximum percentage errors are lower than 35% for concentrations higher than 7×10^{16} cm^{-3}.

The data in Figs. 28.8 and 28.9 show that, provided there are two known concentration levels in a sample with unknown carrier profile ("internal calibration"), the SCM map on that sample can be converted into a carrier concentration profile. The highest concentration level (normalization level N_n) is used to calculate the SCM sensor gain G, defined as the ratio between the measured SCM signal (N_n) and the calculated $dC/dV(N_n)$. The lowest concentration level (control level N_c) is used to

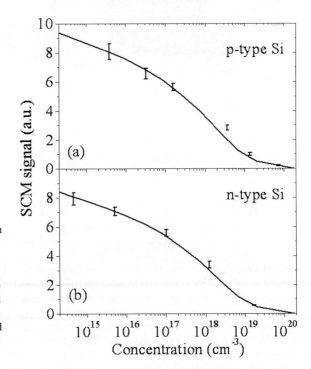

Fig. 28.9. Comparison between the calculated calibration curves (*lines*) and the experimental ones (*symbols*). The *error bars* represent the maximum errors on the measured SCM signal for the different concentration levels, obtained from the data in Fig. 28.7. (Data from [35])

verify that the calculated $dC/dV(N_c)$ for the fixed set of parameters is coincident with the experimental SCM value (N_c).

Recently, calibrated SCM has been presented as a method of choice to study the electrical activation of low-energy implanted B within deep submicron lateral extension. In fact, capacitance measurements directly provide the concentration of electrically active B , without being affected by phenomena such as the reduction of free carrier mobility due to the presence of B-interstitial clusters in the peak of

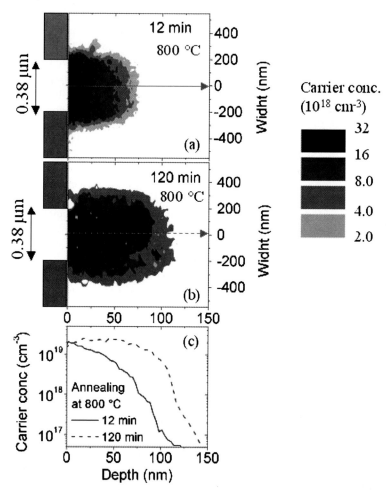

Fig. 28.10. *Gray scale* map of 2D carrier concentration profiles on a low-energy (3-keV) B implant (dose 2×10^{14} cm^{-2}) in a p-type Si substrate (through a patterned oxide mask with 0.38-μm windows) after 12 min (**a**) and 120 min (**b**) of annealing at 800 °C in N_2. The samples were angle-beveled to obtain ×10 magnification in the depth direction (note that in the lateral direction the resolution is the same as in the cross section, since no magnification in that direction is carried out). The concentration versus depth profiles at the center of the implantation window are reported (**c**). The electrically active percentage of the implanted dose for the two annealing times (38% after 12 min, 98% after 120 min) is also reported. (Data from [30])

the implanted region. In Fig. 28.10, results on the electrical activation and diffusion of B implanted at low energy (3 keV), with a dose of $2 \times 10^{14}\,\text{cm}^{-2}$ in a p-type Si substrate, through a patterned oxide mask with 0.38-μm window openings are reported [30]. The electrically active B profiles were measured by calibrated SCM after different annealing processes at 800 °C in N_2 for different times ranging from 12 to 120 min. The 2D carrier concentration profiles after 12 and 120 min of annealing are reported in gray scale in Fig. 28.10, panels a and b, respectively. One can observe the combined effect of electrical activation and diffusion of B during the annealing, causing an increase in both the peak concentration and the spatial extension of the high-concentration region (the black region in the map). In Fig. 28.10, panel c, the concentration versus depth profiles at the center of the implantation window mask are reported. By calculating the area under these profiles, one can obtain the electrically active percentage of the implanted dose for the two annealing times (38% after 12 min, 98% after 120 min).

Quantitative SCM has been applied to study the electrical activation of N- or Al-implanted SiC upon thermal annealing [24]. This is illustrated in Fig. 28.11, where the SIMS profiles of implanted N concentration ($1 \times 10^{14}\,\text{cm}^{-2}$, Fig. 28.11, panel a, and $5 \times 10^{14}\,\text{cm}^{-2}$, Fig. 28.11, panel b, implanted doses) are compared with the substitutional N concentration (by SCM) after a thermal annealing at 1300 °C for 1 h in Ar. The electrically active fraction (approximately 10% for both doses) is obtained from the ratio between the chemical and the electrical profiles.

Quantitative carrier profiling has been extended also to ultranarrow doping profiles in Si and in SiGe. Recently, a set of dedicated test samples to determine the

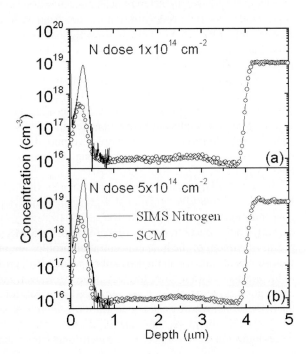

Fig. 28.11. Comparison between the secondary ion mass spectrometry (*SIMS*) and the SCM substitutional dopant concentration profiles on N implanted in 6H–SiC **a** $1 \times 10^{14}\,\text{cm}^{-2}$ and **b** $5 \times 10^{14}\,\text{cm}^{-2}$ implanted doses), after a thermal annealing at 1300 °C for 1 h in Ar. The electrically active fraction (approximately 10% for both doses) is obtained from the ratio between the chemical and electrical profiles. (Data from [24])

Fig. 28.12. Nominal B concentration profile (**a**) and hole concentration profiles from SCM (**b**) on a test Si sample (grown by molecular beam epitaxy), containing five ultranarrow (3-nm-thick) B spikes with different peak concentrations in the range 10^{19} –10^{17} cm^{-3}. The sample was angle-beveled ($\times 100$ magnification). *FWHM* full width at half maximum. (Data from [20])

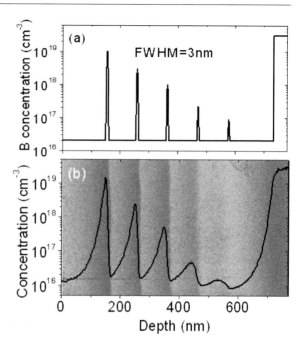

ultimate depth resolution by SCM were proposed [20]. In Fig. 28.12, panel a, a test Si sample (grown by molecular beam epitaxy, MBE), consisting of five ultranarrow (3-nm-thick) B spikes with different peak concentrations in the range 10^{19} – 10^{17} cm^{-3} is presented. The hole concentration profiles from SCM data on the beveled sample are reported in Fig. 28.12, panel b, showing that the carrier profiles are more broadened than the nominal doping profile, because of the free carrier redistribution in the presence of the biased tip.

MBE-grown p-type doped samples with narrow B doped layers (B spikes) embedded in Si/Si$_{0.75}$Ge$_{0.25}$/Si heterostructures (very efficient quantum wells, QWs, for holes) have also been proposed as test samples [23]. In Fig. 28.13, panel a, a sample with B spikes embedded and not embedded in the QWs is illustrated. In Fig. 28.13, panel b the SCM profile for that sample is reported, showing the strong confinement of the carriers within the QWs and the huge broadening of the holes not confined in the QWs.

In Fig. 28.14, panel a, the SCM profile of a Si/SiGe/Si heterostructure is reported. It consists of five QWs (identical widths of 5 nm), doped with different B peak concentrations in the range 7×10^{18} – 2×10^{16} cm^{-3}. In Fig. 28.14, panel b the converted hole concentration is compared with the SIMS profile of B. The full widths at half maximum of the hole concentration profiles are included inside the nominal well widths of 5 nm for all the five spikes, while SIMS profiles are much larger owing to the intrinsic resolution of the technique. Moreover, the hole peak concentrations are in good agreement with those measured by SIMS for all the B spikes, except for the highest one, whose peak concentration is more than a factor of 2 lower than the SIMS one. Furthermore, the same SCM spike exhibits very long tails starting from a concentration of 8×10^{17} cm^{-3}, much longer than the tails exhibited by all the

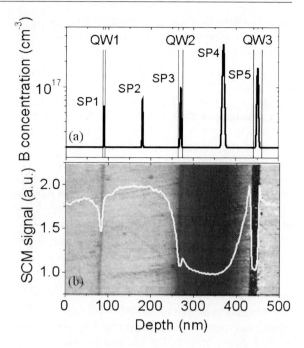

Fig. 28.13. Nominal B concentration profile (**a**) on a molecular beam epitaxy grown sample with B spikes embedded and not embedded in Si/Si$_{0.75}$Ge$_{0.25}$/Si quantum wells (*QWs*). The SCM profile for that sample is reported in **b**, showing the strong confinement of the carriers within the QWs and the huge broadening of the holes not confined in the QWs. (Data from [23])

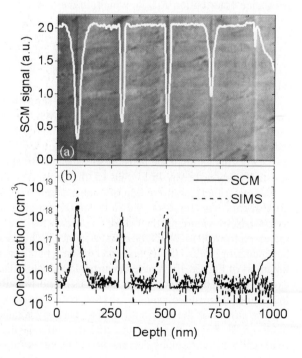

Fig. 28.14. A 2D SCM map with superimposed SCM versus depth profile (**a**) for a Si/SiGe/Si heterostructure consisting of five QWs (identical widths of 5 nm), doped with different B peak concentrations in the range $7 \times 10^{18} - 2 \times 10^{16}$ cm^{-3}. The hole concentration profile by quantification of the SCM raw data is compared with SIMS profiles of B in **b**. (Data from [23])

other spikes, which can be attributed to the quantum hole distribution beyond the potential well. This behavior for the highest concentration spike indicates that a hole concentration of 2×10^{18} cm^{-3} completely fills all the energy states available in the QW, and holes in excess of this concentration are distributed outside the well.

28.1.4
Basic Principles of SSRM

In SSRM a hard conductive AFM (C-AFM) probe is scanned in contact mode across the sample, while a DC bias is applied between an ohmic contact on the backside of the sample and the tip. The resulting current is measured using a logarithmic current amplifier providing a typical range from 1 pA to 0.1 mA.

The total measured resistance R in SSRM experiments includes the resistance of the probe R_{tip}, the tip–semiconductor nanocontact resistance $R_{contact}$, the spreading resistance R_{spread} encountered by the current under the tip, the sample series resistance R_{series} and the back-contact resistance R_{back}:

$$R = R_{tip} + R_{contact} + R_{spread} + R_{series} + R_{back} . \tag{28.1}$$

The working principle of SSRM relies on the assumption that, using a high force (above micronewtons) to realize an intimate contact between the probe and the semiconductor sample, the measured resistance R is dominated by the spreading resistance contribution R_{spread}, which, in the first approximation, is linearly related to the local semiconductor resistivity ϱ by

$$R_{spread} = \frac{\varrho}{4a} , \tag{28.2}$$

where a is the effective contact radius.

SSRM has been applied to Si [37], III–V semiconductors (GaAs, InP) [38–40] and SiC [41]. The first implementation of SSRM on Si was carried out at IMEC by De Wolf et al. [42], who demonstrated that, on increasing the force above a threshold value, a significant increase in the current flowing between the tip and the back-contact occurs and that the resistance is strongly related to the local semiconductor resistivity under the tip. In Fig. 28.15, the current–voltage (I–V_{tip}) curves acquired in that experiment as a function of load are reported for an n-type (100) Si sample (1.03 Ωcm). The force on the indenter (a diamond tip) was increased from 50 to 260 μN by changing the AFM feedback set-point voltage. When the force is increased from 50 to 70 μN the contact suddenly changes from being nearly isolating to conducting. Below this critical load the current never rises above 5 nA, as shown in the insert of Fig. 28.15. Above the critical load the I–V_{tip} curves are nearly identical except for a scaling factor which increases proportionally with the square root of the load minus a threshold value.

The origin of the threshold load is partially related to the presence of native oxide on Si. Below this load the current is mainly the tunnel current through the oxide and it is only weakly dependent on the underlying sample conductivity. Above a certain minimum load the indenter pushes partially through this layer, and a behavior specific for the sample under investigation appears. Furthermore, the large pressure involved in SSRM measurement is responsible for elastoplastic or even fully plastic deformations of the indented sample. In particular, in the case of Si, the occurrence of a temporary phase transformation in the so-called β-tin phase has been demonstrated [43]. The β-tin structure, also named Si(II), is a denser structure than the cubic structure Si(I) (22% volume reduction) obtained at moderately elevated pressures [44]. Under pure hydrostatic conditions, the transformation takes place at about 12 GPa, but this is reduced to values as low as 8 GPa when shear stresses are present (that is the case in the scanning contact mode SSRM). The Si(II) phase has material properties which are substantially different from those of Si(I). The most apparent property is represented by its metallic behavior [45]. The contact between the probe and the Si(II) pocket may be considered as ohmic, while the dominant SSRM nanocontact (linked to the silicon resistivity) is in reality the contact between the Si(II) pocket and the Si(I) (Fig. 28.16).

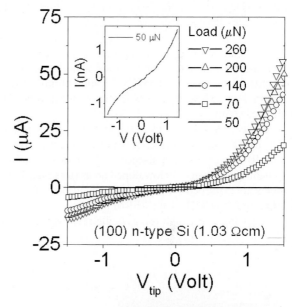

Fig. 28.15. Variation of the current–voltage curves with load (from 50 to 260 μN) for a diamond tip on (100) n-type Si (1.03 Ωcm). (Data from [42])

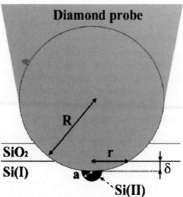

Fig. 28.16. A spherical scanning spreading resistance microscopy (SSRM) indentation in a flat silicon surface with elastoplastic deformation and β-tin Si(II) formation

The resolution of the SSRM technique is set by the size of the β-tin pocket created under the probe rather than by the probe itself, giving an explanation for the nanometer spatial resolution of the SSRM technique [46]. The β-tin pocket can thus be considered as a "virtual SSRM probe." Nevertheless, the shape of the probe–sample contact is of crucial importance, as it will determine the stress distribution in the silicon sample. Above the threshold load, the resistance is strongly related to the local semiconductor resistivity under the tip. In particular, at a constant load of 200 µN, an evolution in the I–V_{tip} curves with increasing resistivity from nearly ohmic behavior to a typical rectifying contact was shown [42].

In Fig. 28.17, a plot of the measured resistance versus resistivity is shown for (100) n-type and p-type Si for a 50-, 70- and 200-µN load. The resistance represented in these curves corresponds to the reciprocal slope of the I–V_{tip} curves at zero bias. The saturation exhibited by the n-type and p-type calibration curves for highly conductive samples was due to the high resistance (6.5 kΩ) of the implanted diamond tips used in this first experiment. After subtraction of this series resistance one obtains the resistance values characteristic of the different substrates. Further improvements in tip construction [47] reduced the tip resistance to a smaller value (approximately 1 kΩ). Indeed, progress in the SSRM technique is intimately linked with the use of silicon probes coated with doped diamond. Those probes mounted on stiff cantilevers (10–100 N m^{-1}) are the only ones that could survive the forces required for SSRM. In the few last years, the advent of full diamond probes [48] has allowed significant progress in SSRM.

Hence, SSRM benefits in theory from an extremely large dynamic range (from 10^{14} to 10^{21} cm^{-3}). In practice, this range is, however, limited at high doping (more than 10^{20} cm^{-3}) by the presence of the diamond probe resistance in series with the spreading resistance. In low doped areas (less than 10^{17} cm^{-3}), the SSRM measurements are disturbed by the presence of surface charges, as demonstrated in [49]. An important drawback of the SSRM technique is the damage to the sample and to the probe owing to the high forces while scanning (in particular owing to the presence of an important shear force).

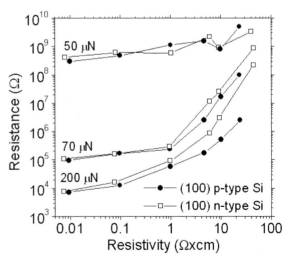

Fig. 28.17. Variation of the resistance (slope at zero bias) with resistivity at a load of 50, 70 and 200 µN for n- and p-type doped Si. Below the transition load the resistance is insensitive to sample resistivity. (Data from [42])

28.1.5
Carrier Imaging Capability by SSRM

SSRM demonstrated high-performance imaging capabilities for ultrascaled devices in Si, but also for III–V semiconductor devices. In Fig. 28.18, SSRM images are reported for the cross section of a 0.18 μm p-type MOS transistor (illustrating the sensitivity of SSRM also to threshold voltage-adjust and halo implants) [37] and for the cross section of an n–p–n bipolar transistor [37]. For both devices, the critical structures in the 2D doping profiles are clearly imaged with high lateral resolution.

A sub-5-nm spatial resolution in cross section, using special ultrasharp (approximately 3.5 nm contact radius from scanning electron microscopy) full diamond tips, has been estimated and reported [50], when characterizing a fully depleted MOSFET device fabricated on SOI (Fig. 28.19b). In Fig. 28.19a, a SSRM image of the device is reported, while in Fig. 28.19b, the resistance versus depth profile along the dashed white line in Fig. 28.19a is reported. A feature associated with the 3-nm gate oxide is evident in the resistance profile and, from consideration on the peak resistance measured for this feature [51], a sub-5-nm "effective contact radius" was estimated by the authors [50].

28.1.6
Quantification of SSRM Raw Data

The simple linear relation (28.2) linking the spreading resistance to the resistivity of the sample under the probe is not fully satisfactory and it presents various inaccuracies and deficiencies. Electrical properties of the probe–semiconductor contact depend on many different parameters, such as the concentration and energy distribution of surface states, the precise probe shape or the pressure. It is almost impossible to express mathematically all those nonidealities. Comparisons with calibration samples should, therefore, be performed and the deviation from the ideal linear relation (28.2) is expressed with a resistivity-dependent barrier resistance component R_{barrier} [52]:

$$R = \frac{\varrho}{4a} + R_{\text{barrier}}(\varrho) \, . \tag{28.3}$$

Several dedicated calibration samples have been proposed both for Si [15] and for III–V semiconductors (GaAs [38, 39] and InP [38–40]).

In Fig. 28.20 the SSRM raw data (output of the SSRM logarithmic current amplifier [53]) and the resistance profiles obtained through the current amplifier transfer functions (inserts in Fig. 28.20) are reported for the p- and n-type Si calibration samples proposed by IMEC (Fig. 28.7) [37].

Finally, in Fig. 28.21 the resistance versus resistivity curves obtained for the calibration samples are reported [37].

In a first approximation, the SSRM profile measured on an arbitrarily doped Si sample can be converted into the resistivity profile by applying the calibration curves of Fig. 28.21.

However, for nonuniformly doped samples, the resistance at every position is no longer exclusively determined by the resistivity at this same place as the spreading of

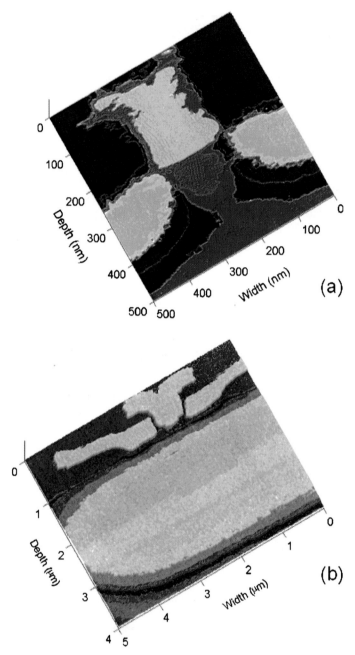

Fig. 28.18. SSRM images of the cross section of a 0.18 µm *p*-type metal oxide semiconductor transistor (**a**) and of the cross-section of an n–p–n bipolar transistor (**b**). (Data from [37])

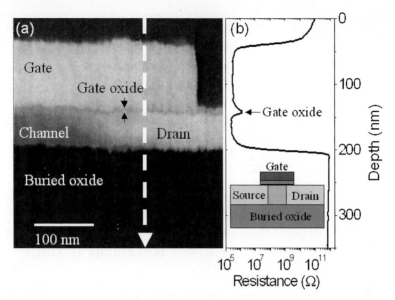

Fig. 28.19. a SSRM image of a fully depleted metal oxide semiconductor field-effect transistor device fabricated on SOI (see the *insert* in **b**) and **b** resistance versus depth profile along the *dashed white line* in **a**. (Data from [50])

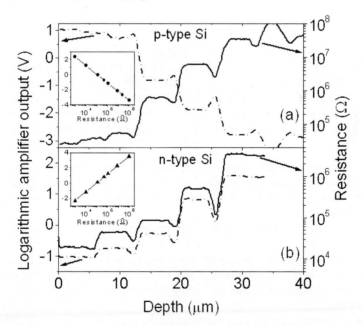

Fig. 28.20. SSRM raw data (output of the SSRM logarithmic current amplifier) and resistance profiles obtained through the current amplifier transfer functions (*inserts* in **a** and **b**) for the p-type (**a**) and n-type (**b**) Si calibration samples. (Data from [37])

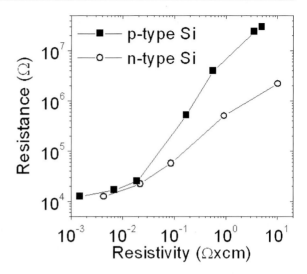

Fig. 28.21. Resistance versus resistivity curves obtained for the p-type and n-type Si calibration samples. (Data from [37])

the current might be influenced by neighboring conducting or insulating regions [52]. Hence, to obtain a better estimation of the actual resistivity profile, a correction factor (CF) ntroduced [52] in (28.3):

$$R = \text{CF}(\varrho, a) \times \frac{\varrho}{4a} + R_{\text{barrier}}(\varrho) \,. \tag{28.4}$$

CF is a function of the contact radius a, of the resistivity values around the contact and of the resistivity gradient. A correct evaluation of this current spreading effect requires detailed 3D calculations of the current and potential distribution around the spreading resistance point contact. A CF database was calculated from those simulations [52] for different tip radii and resistance profiles around the point contact. A conversion algorithm based upon this database to calculate the exact resistivity profile from the measured resistance profile includes the following steps:

1. Since a minor change of the input resistance values can be the indication of a large resistivity change, an adequate smoothing step is included. This eliminates measurement noise, removing as little information as possible from the underlying physical profile.
   In the subsequent steps, the smoothed resistance data are transformed into resistivity values.
2. A starting value ϱ_i ($i = 1$) for the resistivity profile is calculated by using the calibration curve (28.3) assuming there are no corrections (CF = 1).
3. The corresponding CF profile CF(ϱ_i) is calculated by interpolation of the database.
4. A new resistivity profile ϱ_{i+1} is calculated by applying (28.3) in the following way:

$$\varrho_{i+1} = \frac{4a\left[R - R_{\text{barrier}}(\varrho_i)\right]}{\text{CF}(\varrho_i, a)} \,. \tag{28.5}$$

5. Steps 1–4 are repeated until a stopping criterion is fulfilled. One possible stopping criterion is to compare ϱ_i and ϱ_{i+1} and stop the repetition when sufficient agreement is observed, for example, expressed in terms of the standard deviation:

$$\sum_{k=1}^{N}(\varrho_{i+1,k} - \varrho_{i,k})^2 \leq \varepsilon ,\qquad(28.6)$$

where ϱ_i represents the complete resistivity profile at step i in matrix notation.

6. Finally, the carrier profile (p, n) and the dopant profile (N_A, N_D) can be calculated from the resistivity profile by solving the mobility equation (28.7) together with the Poisson equation (28.8):

$$\varrho = \frac{1}{q(n\mu_e + p\mu_h)} \qquad(28.7)$$

$$\nabla^2 V = -\frac{q}{\varepsilon\varepsilon_0}(N_D - N_A + p - n) ,\qquad(28.8)$$

where the electron μ_e and hole μ_h mobilities are functions of the total dopant concentration $N_A + N_D$.

28.1.7
Drift Mobility by SCM and SSRM

SCM and SSRM methods present some peculiar differences, owing to the different physical working principles. SSRM is strongly affected by changes in the carrier mobility and in the tip–semiconductor Schottky contact properties. Therefore, the quantification of SSRM raw data in carrier profiles is hampered, unless the mobility is independently known or assumptions for the mobility are made. If the mobility versus doping concentration relation is well known for the most common crystalline bulk semiconductors, this relation is not obvious in the case of semiconductor crystals with high level of damage (e.g., ion-implanted samples), and it is sometimes unknown in the case of semiconductor nanostructures. In contrast, capacitance measurements on MOS structures are directly connected to carrier concentration. Recently, it was demonstrated [5] that the two methods (SCM and SSRM) can be used as two complementary techniques to determine the local drift mobility of carriers in semiconductor heterostructures from the carrier concentration (obtained by SCM [23]) and the resistivity (obtained by SSRM) depth profiles. This approach was applied to compressively strained $Si_{0.75}Ge_{0.25}$ layers with different widths (from 20 to 1 nm) and doped with different B concentrations (1×10^{18} and 3×10^{18} cm^{-3}). The strain-induced modification of the band structure is expected to cause an enhancement of the in-plane mobility (i.e., the mobility in the direction parallel to the heterointerface) for the 2D confined hole gas. On the other hand, in very narrow QWs, additional effects can deteriorate the carrier mobility, like the Si/SiGe interface roughness scattering [54], the scattering due to the interface charge and the local strain fluctuations induced by alloy composition variations. The Hall effect is the usual method to measure the mobility of a 2D hole gas. However, those measurements provide average information on the mobility of the 2D hole gas and they are limited to lower temperatures, since the parallel conduction due to the holes that are not confined in the QW, for temperatures higher than −123 °C, is comparable to that of the confined holes, making the data interpretation questionable. Moreover the Hall scattering factors must be known to calculate the drift mobility from the Hall mobility. From this point of view, the nanoscale drift mobility determination reported in [5] presents several advantages, since it is a truly local measurement and it can be carried out at room temperature.

In Fig. 28.22, panel a, the 2D SCM map on a multi-QW sample (QW widths from 20 to 1 nm, nominal QW doping 3×10^{18} cm^{-3}) is reported [5], together with the hole concentration profile obtained from the quantification of the SCM signal versus depth profiles [35]. In Fig. 28.22, panel b, the 2D current map measured on the same sample for a fixed DC bias (1 V) is reported, with the current versus depth profile obtained superimposed on it. The box-shaped current profile, particularly evident for the thicker QWs (20 and 10 nm), demonstrates that the current flow is confined inside the width of the QW.

The I–V characteristics measured by placing the tip on the individual QWs were ohmic for a DC bias from −1 to 1 V. From the linear fit of the I–V characteristics, the resistance R is accurately determined, showing the increase of R on decreasing the

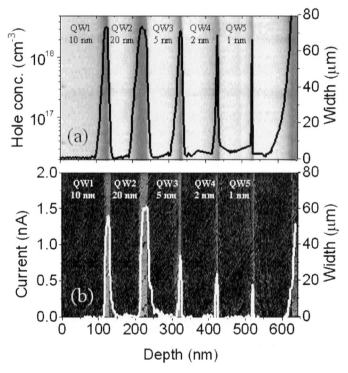

Fig. 28.22. a 2D SCM map on a multi-QW sample (QW widths from 20 to 1 nm, nominal QW doping 3×10^{18} cm^{-3}), together with the hole concentration profile obtained by the SCM signal. **b** 2D current map at fixed DC bias (1 V), with the current versus depth profile superimposed. (Data from [5])

width of the QW from 20 to 1 nm. The measured resistances result from the different contributions in (28.1). R_{tip} and $R_{contact}$ are identical for all the different QWs, since exactly the same probe is used and the same pressure is applied by the tip on the SiGe. Both R_{spread} and R_{series} depend on the resistivity ϱ in the semiconductor. Therefore, the increasing R values with decreasing QW width can be explained in terms of increasing ϱ values in the QWs. R versus ϱ calibration curve were determined for a set of calibration samples, consisting of compressively strained $Si_{0.75}Ge_{0.25}$ layers with uniform B doping (from 10^{18} to 10^{19} cm^{-3}) grown on n-type doped Si. The resistivity ϱ of the $Si_{0.75}Ge_{0.25}$ thin films was independently determined by four point probe measurements, while the resistance R was obtained by measuring the I–V characteristics, with the current I flowing between a fixed gold contact on the $Si_{0.75}Ge_{0.25}$ top film and the tip.

Application of those R versus ϱ calibration curves allowed the resistivity ϱ or all the QW widths (with 1×10^{18} and 3×10^{18} cm^{-3}) to be determined (Fig. 28.23, panel a). In Fig. 28.23, panel b the hole peak concentration values p as determined from the hole concentration versus depth profiles of the same samples are reported [5]. The values of the hole mobility μ obtained by combining the data in Fig. 28.23, panels a and b, according to the relation $\mu = 1/(q \times p \times \varrho)$, where q is the electron charge, are reported in Fig. 28.23, panel c.

The mobilities in the sample with 3×10^{18} cm^{-3} doped QWs are lower than those in the sample with 1×10^{18} cm^{-3} doped QWs, owing to the 3 times higher B concentration. Looking at the sample with 3×10^{18} cm^{-3} doped QWs, the mobility decreases by only 9% from the 20-nm-thick to the 10-nm-thick QWs, while the

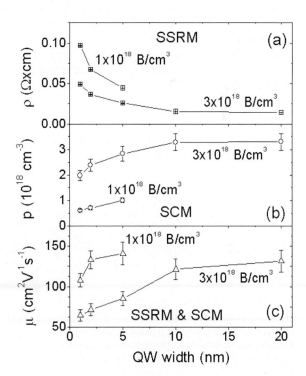

Fig. 28.23. Resistivity (**a**), hole peak concentration (**b**) and hole mobility (**c**) versus QW width for QWs doped with 1×10^{18} and 3×10^{18} B cm^{-3}. (Data from [5])

decrease becomes more relevant (approximately 87%) from the 10-nm-thick to the 1-nm-thick QWs. In the case of the sample with $3 \times 10^{18}\,\text{cm}^{-3}$ doped QWs, the decrease in mobility is approximately 23.5% on decreasing the QW width from 5 to 1 nm. The agreement between the value of the hole mobility for a 20-nm-thick surface SiGe layer, presenting a single Si/SiGe interface, and the value for the buried SiGe layer with the same thickness, presenting two heterointerfaces, indicates that the hole scattering mechanism is a minor contribution to the mobility for this thickness. These effects become very relevant when decreasing the QW width below 10 nm.

28.2
Carrier Transport Through Metal–Semiconductor Barriers by C-AFM

In 1988, Bell and Kaiser [55] demonstrated ballistic electron emission microscopy (BEEM), i.e., a method based on scanning tunneling microscopy (STM), enabling spatially resolved carrier-transport spectroscopy at interfaces. The energy-band diagram for the BEEM setup in the case of a metal–n-type semiconductor Schottky barrier is represented in Fig. 28.24. The method is based on the injection of hot electrons by tunneling from the reverse biased STM tip to the metal film. In the case of ultrathin films (compared with the electron mean free path in the metal), a not negligible fraction of the hot electrons propagate ballistically in the metal; the electrons are able to overcome the barrier and can be collected at the substrate terminal (collector), when a bias higher than the Schottky barrier height (SBH) is applied to the tip.

In its first demonstration, BEEM was applied to two interface systems having very different band structure, i.e., the Au/Si and Au/GaAs Schottky contacts. The collector current (I_c) versus tip bias curves for the two systems are reported in

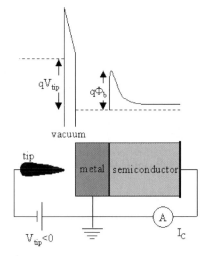

Fig. 28.24. Energy-band diagram for the ballistic electron emission microscopy setup in the case of a metal–n-type semiconductor Schottky barrier

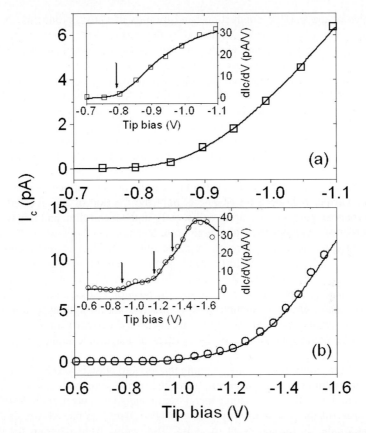

Fig. 28.25. Ballistic electron emission microscopy collector current (I_c) versus tip bias curves (*symbols*) and fits with the Bell–Kaiser model (*lines*) for the Au/Si (**a**) and Au/GaAs (**b**) Schottky contacts. The *inserts* in a and b show the experimental dI_c/dV and the derivatives of the fitting law. The Schottky barrier heights correspond to the step in the $dI_c/dV - V$ curve. (Data from [55])

Fig. 28.25. Those curves were fitted by a model (Bell–Kaiser formalism [55]) taking into account the following: (1) the transmission probability for an electron to tunnel from the tip through the vacuum barrier; (2) the electron current attenuation due to scattering in the metal base; (3) the constraints on the electron kinetic energy, based on the conservation of the transverse momentum at the metal–semiconductor interface (in the absence of scattering); (4) the parabolic shape of the conduction-band minimum in the k space. An important consequence of point 4 is that the I_c–V curve can be fitted by a parabolic law [$I \propto (V - V_{th})^2$] around the current onset threshold $V_{th} = \Phi_B/q$, as shown in Fig. 28.25, panel a by the line. The inserts in Fig. 28.25 show the experimental dI_c/dV and the derivatives of the fitting law. The SBH corresponds to the step in the $dI_c/dV - V$ curve. In the case of Au/Si, a single step at 0.8 eV is observed, indicating a single conduction-band minimum in the interfacial semiconductor-band structure. For the Au–GaAs interface, BEEM clearly reveals properties of the satellite conduction-band minima of GaAs, producing the three steps at 0.89, 1.18 and 1.36 eV (indicated by the arrows in the insert in Fig. 28.25, panel b). Further, BEEM has yielded the first direct measurement of the lateral variations of the GaAs conduction-band structure [56].

More recently, BEEM has been applied to study biased GaAs/AlGaAs superlattices [57], or to demonstrate the QW behavior of 3C–SiC inclusions in 4H–SiC [58]. BEEM spatial resolution has been directly measured on metal–QW Schottky contact [59]

Notwithstanding its high spatial and energy resolutions [6], BEEM suffers some intrinsic limitations of STM. In fact, since the injected tunnel current intensity is used also in the tip-height control feedback, BEEM cannot be applied to samples containing both conductive and insulating regions on the surface. Moreover, the collector current measured by BEEM is only the fraction of the injected tunnel current (nanoamperes) propagating ballistically in the metal film, and it is typically on the order of picoamperes, with a high noise level.

Recently, an alternative approach for nanometer-scale SBH measurements based on C-AFM was demonstrated [60]. It is based on quite different physical models and it overcomes some BEEM limitations. In fact, since C-AFM operates in contact mode and the feedback for the tip-height position is based on the cantilever deflection, also samples containing both conductive and insulating regions on the surface can be characterized. Moreover, the measured current values are typically on the order of nanoamperes, i.e., 2–3 orders of magnitude higher than BEEM currents.

It has been shown [60] that by scanning the forward biased C-AFM tip in contact with an ultrathin (1–5-nm) metal film deposited on the semiconductor, only a very localized region in the macroscopic metal–semiconductor contact, with dimensions on the order of the nanometric tip diameter (10–20 nm), is biased, i.e., a "nano-Schottky diode" is formed point by point. The energy-band diagram for the C-AFM-based setup in the case of a metal–n-type semiconductor Schottky barrier is represented in Fig. 28.26. The biased C-AFM tip is in contact on a nanometric area with the ultrathin metal layer, which is connected to the ground potential. The backside ohmic contact of the semiconductor is connected to a current amplifier, enabling the measurement of the collector current (I_c) on the nanoampere scale with picoampere sensitivity.

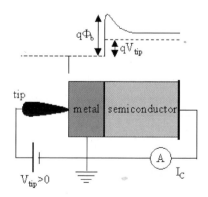

Fig. 28.26. Energy-band diagram for the conductive AFM based setup in the case of a metal–n-type semiconductor Schottky barrier

The typical I_c–V characteristic on a linear scale qualitatively exhibits in the forward DC bias condition a Schottky diode behavior (thermoionic emission law [8]), i.e., a negligible current for voltages below a threshold V_{th} and a current increase for biases higher than V_{th}. In order to determine V_{th} for each tip position, the region of the current onset was fitted by a parabolic law. The V_{th} bias represents the voltage corresponding to the minimum of the parabola. By Taylor series expansion of the thermoionic emission law for small values of the variable $q(V - n\Phi_B)/nk_BT$, the barrier height Φ_B is related to V_{th} by $\Phi_B = V_{th}/n + k_BT/q$, where n is the ideality factor of the diode, k_B the Boltzmann constant and T the temperature. Hence, by spectroscopic analysis of the forward bias I_c–V_{tip} characteristics for different tip positions, 2D maps of the metal–semiconductor SBH can be obtained with a 10–20-nm spatial resolution and with an energy resolution in the sub-0.1-eV range.

The nanometric localization of the current across the metal–semiconductor interface is obtained by the combined effects of the electric field localization at the apex of the biased conductive tip and of the high resistivity of ultrathin (1–5-nm) metal films [60]. This effect related to the film thickness has been investigated experimentally and modeled by a quantum mechanical description [61–63]. As an example, the data in Fig. 28.27 show in the case of polycrystalline Au films a resistivity ($\varrho \approx 10^{-3}$ Ωcm) 2 orders of magnitude higher than the ordinary Au bulk resistivity [60]. The collector current I_c is represented by the local thermoionic current flowing across the Schottky barrier in the nanometric region defined by the forward-biased tip in contact with the high-resistivity metal film.

The C-AFM-based approach described has been applied to map the lateral distribution of the barriers in a Au/4H–SiC Schottky system [60]. In particular, the effect of an ultrathin (approximately 2 nm) inhomogeneous SiO_2 layer between the 4H–SiC and the Au film was studied. Macroscopic current–voltage measurements (Fig. 28.28) carried out as a reference on Au dots deposited on oxidized and on oxide-free sample areas, respectively, indicate that the average macroscopic effect of an inhomogeneous oxide layer at the interface is a lowering of the SBH. The microscopic investigation by C-AFM was carried out using conductive diamond-coated Si tips and the collector current I_c was amplified by a linear current amplifier (10^{10} V A^{-1} gain).

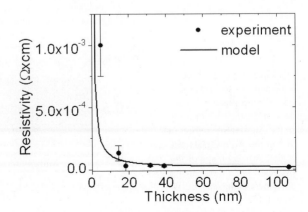

Fig. 28.27. Resistivity versus film thickness in the case of polycrystalline Au films. (Data from [60])

A set of I_c–V curves collected on an array of 25 × 25 tip positions in an area of 1 μm × 1 μm in the Au/4H–SiC region is reported in Fig. 28.29, panel a. An identical array of curves was collected in the Au/SiO$_2$/4H–SiC region, and is shown in Fig. 28.29, panel b. Each set of 625 curves exhibits a spread, which is wider in the case of the Au/SiO$_2$/4H–SiC system. Two arrays of Φ_B values for the different tip positions were obtained, and 2D maps of Φ_B were plotted from those arrays (Fig. 28.29, inserts).

A statistical analysis on the arrays of Φ_B values [60] resulted in two well-separated histograms being obtained (Fig. 28.30). In the case of the Au/4H–SiC system, the distribution is peaked at a barrier height of 1.8 eV and it extends from 1.6 to 2.1 eV. In the case of the Au/SiO$_2$/4H–SiC system, the distribution is peaked at 1.5 eV, but it extends from 1.1 to 2.1 eV. The high barrier energy tail is probably related to the interface regions locally free from the not uniform oxide.

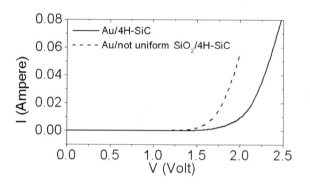

Fig. 28.28. Macroscopic current–voltage measurements carried out on Au dots deposited on oxidized approximately (2 nm inhomogeneous SiO$_2$ layer) and on oxide-free 4H–SiC. (Data from [60])

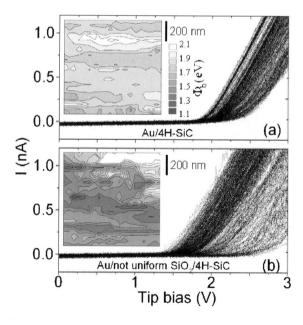

Fig. 28.29. Set of I_c–V curves collected on an array of 25 × 25 tip positions in an area of 1 μm × 1 μm area in the Au/4H–SiC region (**a**) and in the Au/SiO$_2$/4H–SiC region (**b**). 2D maps of Φ_B extracted from those arrays are shown in the *inserts*. (Data from [60])

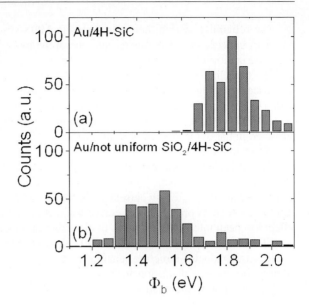

Fig. 28.30. Histograms of Φ_B for the Au/4H–SiC (**a**) and Au/SiO$_2$/4H–SiC (**b**) samples obtained from a statistical analysis on the arrays in Fig. 28.29. (Data from [60])

28.3
Charge Transport in Dielectrics by C-AFM

When the gate insulator of a MOS structure is subjected to electrical stress, traps or defects are progressively generated inside the oxide and eventually lead to the formation of a low-resistance conducting path between the electrodes. The occurrence of such an event can be detected either as a gradual or as a sudden change in the conductance of the system and it is associated with the appearance of a localized conduction mechanism in parallel with the area-distributed tunneling current. According to the magnitude and shape of the resulting current–voltage characteristics, the failure mode is usually referred to as soft or hard breakdown. Several models have been proposed for current transport, and can be classified according to the underlying mechanism: junctionlike, hopping, percolation and tunneling conduction. Within the last of these mechanisms direct tunnel, Fowler–Nordheim, trap-assisted, resonant, inelastic tunneling and point contact conduction can be mentioned.

The properties of SiO$_2$ are well known from many macroscopic measurements [64], including I–V and C–V measurements and internal photoemission [65]. Investigations by SPM-based techniques have contributed to the determination of transport properties at the nanoscale. STM [66, 67], BEEM [68, 69] and C-AFM [70–74] have been used to electrically characterize SiO$_2$ films. Although BEEM and STM may be more spatially resolved than atmospheric AFM-based techniques, from the experimental point of view they show some drawbacks, such as tedious sample preparation and/or ultrahigh vacuum conditions. In contrast to STM-based techniques, by C-AFM, electrical and topographical information can be obtained simultaneously and they can be measured independently on an insulating surface using a typical lateral resolution of approximately 10 nm with a quite simple experimental setup.

The reliability of thin dielectric layers in MOS structures is one of the issues of major concern [75]. For this reason, one of the most discussed arguments on the characterization of thin dielectric layers is the charge trapping due to injection of carriers from the substrate in the dielectric during stress induced by an applied bias. Electrical stress tests, namely, constant voltage stress (CVS) and ramped voltage stress (RVS), applied to ultrathin SiO_2 films using C-AFM demonstrated the ability of C-AFM to discern the local (with nanometer spatial sensitivity) mechanisms involved in the oxide degradation till breakdown [72].

An example of charging effects in SiO_2 layers monitored by C-AFM is given in Fig. 28.31, where the current maps after CVS applied for different times are reported. The current maps present dark and bright zones. The dark zones show the typical behavior, i.e., a high current value for a 2-nm-thick SiO_2 layer biased at 4 V. The bright regions present lower currents and coincide with the stressed areas. The effect has been interpreted in terms of electron trapping during the CVS [72].

Electrons trapped in the SiO_2 modify the flat-band voltage of the metal (tip) oxide semiconductor structure, changing the conduction mechanism. The charge trapped produces a distortion in both the oxide barrier and the flat-band voltage (V_{FB}), according to the relation

$$V_{FB} = \Phi_{ms} \pm \frac{Q_t}{C} \pm \dots , \qquad (28.9)$$

where Φ_{ms} is the theoretical difference between the metal work function and the distance between the Fermi and the vacuum levels in the semiconductor, C is the capacitance of the device, Q_t is the amount of charge trapped, and the sign $\pm$ depends on the negative or positive charge trapped. A negative (positive) charge produces an increase (decrease) in the barrier height.

According to (28.9), it is possible to extrapolate the density of the charge trapped (N_t) within the dielectric layer by

$$N_t = \frac{\varepsilon_0 \kappa \Delta V_{FB}}{et} , \qquad (28.10)$$

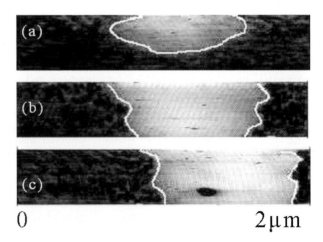

Fig. 28.31. Current maps at 4 V of 2 μm × 0.5 μm after constant voltage stress at 4.5 V for **a** 30 s, **b** 60 s and **c** 120 s. (Data from [72])

where ε_0 is the vacuum permittivity, κ is the relative dielectric constant of the insulator layer, ΔV_{FB} is the shift produced in the flat-band voltage during the stress process, e is the electron charge, and t is the dielectric thickness. Figure 28.31 shows the area influenced by the stress (represented by the bright region) changing at the different stress times: shorter stress times correspond to less-stressed regions (Fig. 28.31, panels a and b). If the stress time is long enough, some breakdown processes could be triggered in the stressed region as is shown by the big black spot (high breakdown current) inside the bright/stressed area in Fig. 28.31, panel c.

RVS cycles on SiO_2 layers allow the charge "spreading" in dielectrics to be imaged (Fig. 28.32). In this case, the stress parameter considered is the number of ramps performed on a single point. The stressed area is the contact region between the SiO_2 surface and the tip, and it can be estimated to be a few hundred square nanometers. Figure 28.32 shows current maps after several RVS. From top to bottom, the current maps collected after 10, 20, 40 and 60 voltage ramps are depicted. The more the stressing by ramps, the larger is the area affected by the stress. The dimension of the stressed region imaged is always larger than the contact area of the metallic tip on the surface. This effect can be partially due to a little hysteresis of the piezoelectric material, responsible for the motion of the tip, which causes a drift movement on the surface and an increase of the stressed area. This phenomenon is, however, assumed to be negligible compared with the lateral spread of trapped charge and anyway can be quite significantly reduced in modern microscopes. Moreover, drift would not impose a circular stress zone as the drift is always in one particular direction. In fact, after several numbers of ramps (40–60 RVS) there is no futher increase in the dimension of the stressed region. During the RVS process each $I-V$ curve was saved. Figure 28.33a shows the characteristic $I-V$ curves after several ramps.

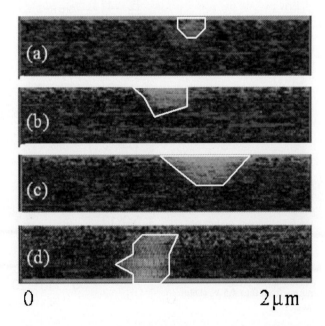

Fig. 28.32. Current maps at 4 V of 2 μm × 0.5 μm after ramped voltage stress (RVS) from 0 to 4.5 V for **a** 10 ramps, **b** 20 ramps, **c** 40 ramps and **d** 60 ramps. (Data from [72])

The threshold voltage shifts at higher voltages especially for the first three curves. This observation, corroborated by the growth of the stressed regions after subsequent ramps, indicates a direct relation between the voltage shifts and the number of the trapped charges. After the first ramps, the threshold voltage tends to reach a fixed value till the difference is practically zero. The slopes of all the curves are the same, suggesting that Fowler–Nordheim injection [76] is still occurring. At the same time, the trapped charges produce a shift in the flat-band voltage according to (28.9).

In this way, the density of trapped charges can be estimated using (28.10). Consideration of the the voltage shift between the first and the 60th ramp resulted in a value of $N_t = 1.3 \times 10^{13}$ cm^{-2} has being calculated. It might be pointed out that the contact area of the tip is not required, because its dimension does not during measurements. In fact, sequential RVS was reproducible using the same tip at different points. After several ramps the dimension of the stressed region stops its lateral growth, (Fig. 28.32, panels c and d). Simultaneously, the I–V characteristics do not present any threshold voltage shift. This result indicates a kind of saturation in the trapping effect. Figure 28.33b shows the density of the traps created versus the number of ramps (from 0 to 4.5 V). This result points out that the maximum density of traps created during the stress is close to the critical defect density to trigger a breakdown event. In fact, breakdown events are triggered when, according to the percolation theory, a critical trap density is reached [77]. The saturation density, or the critical density, has been found.

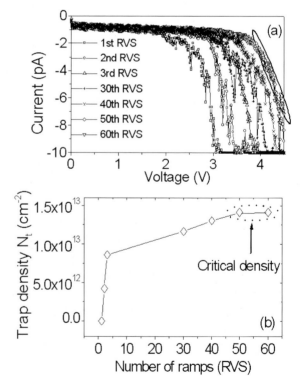

Fig. 28.33. a I–V characteristics after several voltage ramps. Between the first and the 60th there is a shift of about 1 V. In the *circle* the 50th and the 60th ramps are depicted. **b** Trend of the trap density versus the number of ramps from 0 to 4.5 V. (Data from [72])

28.3.1
Direct Determination of Breakdown

The breakdown spots have been observed in the C-AFM images under the application of a constant voltage. Their density was determined for different electrical stresses (applied voltages). According to previous studies, the intrinsic dielectric breakdown can be considered a three-stage process [78, 79]: the wear-out phase, the breakdown event itself, eventually followed by a damage phase. The third phase can be classified as either thermal, when it is due to the dissipation of the energy stored in the capacitor, or progressive, when it can be related to the gradual accumulation damage occurring at the onset of the breakdown event. This ideal percolation mechanism happens when only intrinsic phenomena are involved [80]. The presence of a large number of defects at the interface or in the bulk already in the unstressed device could induce extrinsic phenomena and the breakdown kinetics would not follow this percolation mechanism.

In the following we will show how C-AFM can be used to study the breakdown kinetics, i.e., the evolution of breakdown spots, as a function of time for different applied voltages. The method has been applied successfully to many dielectrics [81, 82].

As an example, the average I–V characteristics at room temperature recorded on large-area (25 μm in radius) MOS devices on 4H–SiC are reported in Fig. 28.34. Different oxide thicknesses, 5 and 7 nm, are considered.

A breakdown field of 12.7 MV cm^{-1} was measured in both cases, in agreement with the theoretical prediction. Moreover, from the fitting of the I–V curves in Fig. 28.34, the direct and Fowler–Nordheim tunnel mechanisms for the 5-nm-thick SiO$_2$ layer, and the Fowler–Nordheim mechanism in the case of the 7-nm-thick insulator layer were highlighted, in good agreement with the ideal behavior.

On the other hand at the nanoscale, the current recorded under the AFM tip area is so small that only the breakdown events can be distinguished (in the appropriate current scale). The color scale reported in Fig. 28.35 (difference between black and

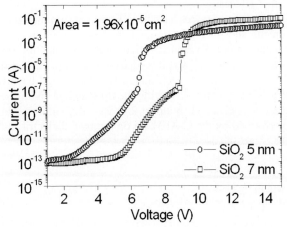

Fig. 28.34. Current density versus voltage measured on 5- and 7-nm-thick SiO$_2$ on SiC metal oxide semiconductor devices. The current signal increases rapidly at the beginning of the breakdown process at 6.3 and 9.0 V, respectively. (Data from [81])

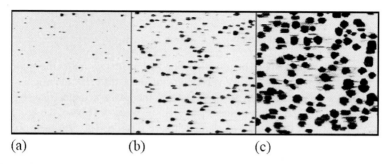

Fig. 28.35. Conductive AFM current maps recorded on different areas of the SiO$_2$ dielectric at 6.3 V negative bias applied to the sample on a 5-nm-thick SiO$_2$ layer. **a** 16 × 16 µm^2 **b** 8 × 8 µm^2 **c** 4 × 4 µm^2. A larger stress time is applied, decreasing the scan side. The breakdown spot concentration increases upon increasing the stress time. (Data from [81])

white) is 100 pA. Attention has been focussed on the weak breakdown spots in the C-AFM images.

Considering such maps, it is possible to discriminate between unbroken and broken nanodevices. In fact, when the voltage is large enough, the device breaks down and the current signal increases by several orders of magnitude, changing from white to black in the related current maps. It is possible to distinguish between breakdown events and intrinsic defects by comparing the local and the large-area I–V curves. In fact, C-AFM breakdown current values (Fig. 28.35) are comparable to those obtained in the microscopic measurements during the breakdown transient (Fig. 28.34), while intrinsic defects cause current density whose values are smaller than those associated with microscopic breakdown events. Thus, it is possible to conclude that the weak spots observed in the current maps are related only to breakdown events.

The breakdown spots obtained under different stress times (Fig. 28.35) have been evaluated at 6.3 V constant stress voltage on a 5-nm-thick SiO$_2$ layer. This voltage was selected because of the breakdown voltage (6.3 V) found in the microscopic measurements and to perform an accelerated testing of the nanoMOS "tip–dielectric–SiC" devices.

The nanodevices have been kept in accumulation by applying a negative voltage on the substrate with respect to the C-AFM tip. In these conditions, the anodic oxidation of the scanned surface is avoided and the stress was at constant voltages. Each current map was recorded by keeping constant the scan frequency at 0.3 Hz. The stress time on each nanoMOS was varied by investigation of different areas and consequently by varying the tip scan rate. With use of this procedure, the stress time was changed by up to 2 orders of magnitude. In fact, the stress time (t_{stress}) performed by the C-AFM tip on the bare dielectric surface on each nanoMOS can be expressed as

$$t_{\text{stress}} = T/(L/a) , \qquad (28.11)$$

where

$$1/T = f , \qquad (28.12)$$

where T is the scan time per line, L is the scan length, f is the scan rate and a is the tip diameter. The choice of a suffers from a certain degree of arbitrariness and the t_{stress} values reported here are considered estimates of the real stress times. Diamond-coated tips have also been used because of their hardness, in order to keep constant a.

28.3.2
Weibull Statistics by C-AFM

The C-AFM images of different scanned areas varying from 16×16 to $4 \times 4\,\mu m^2$ are shown in Fig. 28.35. The breakdown dots are about 100 nm in diameter as measured in the higher-magnification image. Thus, the maps in Fig. 28.35 were acquired at different stress times, from 2.5×10^{-3} to 1×10^{-1} s. The density of breakdown spots increases upon increasing the stress time per nanodevice.

Figure 28.36 shows the breakdown kinetics obtained in SiO_2/SiC. Each scanned region can be considered as an array of nanoMOS capacitors, and the total area of the weak spots is proportional to the number of broken nanoMOS devices. The dimension of the single MOS device (A_{dev}) is the electrical contact region between the tip and the dielectric. The number of failed devices (N_f) has been normalized with respect to the different total number of devices (N_i) in each region investigated. Considering the ratio between the number of failed devices and the total number of stressed devices in each scanned region, one obtains

$$\frac{N_f}{N_i} = \frac{A_{tf}}{A_{dev}} \cdot \frac{A_{dev}}{A_i}, \quad (28.13)$$

where A_{tf} is the total area covered by the failed devices and A_i is the total area of the stressed region. The contact area (A_{dev}) is constant in the scanned region, both in the failed and in the stressed area, and can be neglected.

The Weibull plot showing the breakdown kinetics for both thicknesses of our dielectric is reported in Fig. 28.36. The continuous and the dashed lines represent the fits of the experimental points using the Weibull distribution for 5 and 7 nm,

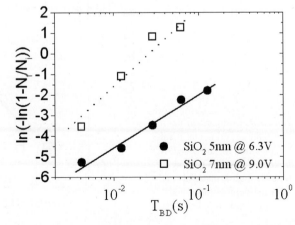

Fig. 28.36. Breakdown kinetics for different oxide thicknesses and at several stress voltages. The fits to the experimental points (concentration of failed nanodevices) using Weibull statistics are reported as a *continuous line* for the 5-nm-thick SiO_2 and as a *dashed line* for the 7-nm-thick oxide. (Data from [81])

respectively. The squares are the experimental data obtained by stressing the 7-nm-thick dielectric layer at 9.0 V, while the circles refer to the 5-nm-thick dielectric stressed at 6.3 V. The Weibull distribution considered is [83]

$$F(x) = 1 - \exp\left[-(x/\alpha)^\beta\right], \tag{28.14}$$

where F is the cumulative failure probability (representing in our case the ratio N_f/N_i expressed in (28.13)), x is the time of stress that induces the breakdown phenomena (28.11), α is the characteristic lifetime of the dielectric and β is the shape factor, also called the Weibull slope. It represents an important value distribution in $\ln x$ and it is appropriate for a weakest-link type of problem. In particular, $\ln[-\ln(1-F)]$ of the nanoMOS capacitors is plotted versus the natural logarithm of the stress times. In this way, the breakdown kinetics at nanometer scale can be determined before the ultimate fabrication of the device.

A fundamental parameter such β of the Weibull distribution is usually obtained by indirect methods measuring large-area devices. In contrast, this method allows its determination directly at nanometer scale. The values determined for the SiO_2/SiC system are $\beta = 2.35 \pm 0.18$ and 3.9 ± 0.5 for the 5- and 7-nm-thick dielectric layers, respectively. These results agree with data reported for the percolation theory [84].

Figure 28.37 shows the experimental data obtained for SiO_2 thermally grown on 4H-SiC and Si, respectively, and for different high-k dielectrics (Al_2O_3, Pr_2O_3) deposited on Si. They are compared with the theoretical prediction from the percolation mechanism. These results show an intrinsic dielectric breakdown behavior in the case of SiO_2 thermally grown on 4H–SiC, as well as on Si. This means that probably the presence of CO components in the SiO_2 films grown on the 4H–SiC layer does not play a relevant role in the breakdown mechanism till a critical thickness is reached [85]. In contrast, in the case of the high-κ dielectrics, the β values obtained demonstrate the extrinsic character of the breakdown process.

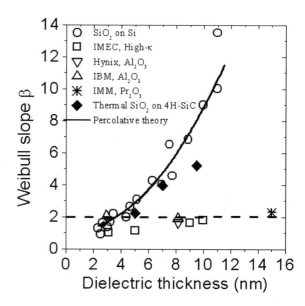

Fig. 28.37. Comparison between the experimental β values obtained for the SiO_2/SiC system, for the SiO_2/Si system, for high-κ materials and with the percolation mode (*curve*). (Data from [81,82])

28.4 Conclusion

In conclusion, we have shown that SPM-based electrical characterization methods (SCM, SSRM and C-AFM) represent powerful tools for the nanoscale determination of critical properties in advanced semiconductor materials. In particular, SCM and SSRM demonstrated high resolution and sensitivity in the determination of 2D majority carrier concentration profiles in semiconductors, while C-AFM allowed the nanoscale determination of the SBH distribution in laterally inhomogeneous metal–semiconductor Schottky contacts, and the study of degradation phenomena connected with charge transport in dielectrics.

References

1. International Technology Roadmap for Semiconductors (2007) Welcome to the ITRS! http://public.itrs.net
2. Leu C-C, Chen C-Y, Chien C-H, Chang M-N, Hsu F-Y, Hu C-T (2003) Appl Phys Lett 82:3493
3. Kim CK, Yoon IT, Kuk Y, Lim H (2001) Appl Phys Lett 78:613
4. Toth AL, Dozsa L, Gyulai J, Giannazzo F, Raineri V (2001) Mater Sci Semicond Process 4:89
5. Giannazzo F, Raineri V, Mirabella S, Impellizzeri G, Priolo F (2006) Appl Phys Lett 88:043117
6. Williams CC (1999) Annu Rev Mater Sci 29:471
7. Brezna W, Harasek S, Bertagnolli E, Gornik E, Smoliner J, Enichlmair H (2002) J Appl Phys 92:2144
8. Sze SM (1981) Physics of semiconductor devices, 2nd edn. Wiley, New York
9. Tran T, Nxumalo JN, Li Y, Thomson DJ, Bridges GE, Oliver DR (2000) J Appl Phys 88:6752
10. Matey JR, Blanc J (1985) J Appl Phys 57:1437
11. Williams CC, Slinkman J, Hogh WP, Wickramasinghe HK (1989) Appl Phys Lett 55:1662
12. Palmer RC, Denlinger EJ, Kawamoto H (1982) RCA Rev 43:194
13. De Wolf P, Stephenson R, Trenkler T, Clarysse T, Hantschel T, Vandervorst W (2000) J Vac Sci Technol B 18:361
14. Tran T, Oliver DR, Thomson DJ, Bridges GE (2001) Rev Sci Instrum 72:2618
15. Clarysse T, Caymax M, De Wolf P, Trenkler T, Vandervorst W, McMurray JS, Kim J, Williams CC, Clark JG, Neubauer G (1998) J Vac Sci Technol B 16:394
16. Goghero D, Raineri V, Giannazzo F (2002) Appl Phys Lett 81:1824
17. Stangoni M (2005) PhD thesis, ETH Zurich
18. Bussmann E, Williams CC (2004) Rev Sci Instrum 75:422
19. Giannazzo F, Priolo F, Raineri V, Privitera V (2000) Appl Phys Lett 76:2565
20. Giannazzo F, Goghero D, Raineri V, Mirabella S, Priolo F (2003) Appl Phys Lett 83:2659
21. Bruno E, Mirabella S, Impellizzeri G, Priolo F, Giannazzo F, Raineri V, Napolitani E (2005) Appl Phys Lett 87:133110
22. Clarysse T, Eyben P, Duhayon N, Xu MW, Vandervorst W (2003) J Vac Sci Technol B 21:729
23. Giannazzo F, Raineri V, La Magna A, Mirabella S, Impellizzeri G, Piro AM, Priolo F, Napolitani E, Liotta SF (2005) J Appl Phys 97:014302
24. Giannazzo F, Calcagno L, Raineri V, Ciampolini L, Ciappa M, Napolitani E (2001) Appl Phys Lett 79:1211

25. Maknys K, Douheret O, Anand S (2003) Appl Phys Lett 83:4205
26. Nakakura CY, Hetherington DL, Shaneyfelt MR, Shea PJ, Erickson AN (1999) Appl Phys Lett 75:2319
27. Nakakura CY, Tangyunyong P, Hetherington DL, Shaneyfelt MR (2003) Rev Sci Instrum 74:127
28. Kimura K, Kobayashi K, Yamada H, Matsushige K, Usuda K (2006) J Vac Sci Technol B 24:1371
29. O'Malley ML, Timp GL, Moccio SV, Garno JP, Kleiman RN (1999) Appl Phys Lett 74:272
30. Giannazzo F, Raineri V, Mirabella S, Bruno E, Impellizzeri G, Priolo F (2005) Mater Sci Eng B 124:54
31. Stangoni M, Ciappa M, Fichtner W (2004) J Vac Sci Technol B 22:406
32. Edwards H, McGlothlin R, San Martin R, U E, Gribelyuk M, Mahaffy R, Shih CK, List RS, Ukraintsev VA (1998) Appl Phys Lett 72:698
33. Ruda HE, Shik A (2003) Phys Rev B 67:235309
34. Marchiando JF, Kopanski JJ (2004) J Vac Sci Technol B 22:411
35. Giannazzo F, Goghero D, Raineri V (2004) J Vac Sci Technol B 22:2391
36. Duhayon N, Eyben P, Fouchier M, Clarysse T, Vandervorst W, Alvarez D, Schoemann S, Ciappa M, Stangoni M, Fichtner W, Formanek P, Kittler M, Raineri V, Giannazzo F, Goghero D, Rosenwaks Y, Shikler R, Saraf S, Sadewasser S, Barreau N, Glatzel T, Verheijen M, Mentink SAM, von Sprekelsen M, Maltezopoulos T, Wiesendanger R, Hellemans L (2004) J Vac Sci Technol B 22:385
37. Eyben P, Xu M, Duhayon N, Clarysse T, Callewaert S, Vandervorst W (2002) J Vac Sci Technol B 20:471
38. Lu RP, Kavanagh KL, Dixon-Warren St J, Kuhl A, SpringThorpe AJ, Griswold E, Hillier G, Calder I, Ares R, Streater R, (2001) J Vac Sci Technol B 19:1662
39. Lu RP, Kavanagh KL, Dixon-Warren St J, Spring-Thorpe AJ, Streater R, Calder I (2002) J Vac Sci Technol B 20:1682
40. De Wolf P, Geva M, Reynolds CL, Hantschel T, Vandervorst W, Bylsma RB (1999) J Vac Sci Technol A 17:1285
41. Osterman J, Abtin L, Zimmermann U, Janson MS, Anand S, Hallin C, Hallen A (2003) Mater Sci Eng B 102:128
42. De Wolf P, Snauwaert J, Clarysse T, Vandervorst W, Hellemans L (1995) Appl Phys Lett 66:1530
43. Clarysse T, De Wolf P, Bender H, Vandervorst W (1996) J Vac Sci Technol B 14:358
44. Hu JZ, Merkle LD, Menoni CS, Spain IL (1986) Phys Rev B 34:4679
45. Gupta MC, Ruo AL (1980) J Appl Phys 51:1072
46. Eyben P (2004) PhD thesis, University of Leuven
47. Hantschel T, Slesazeck S, Niedermann P, Eyben P, Vandervorst W (2001) Microelectron Eng 57:749
48. Fouchier M, Eyben P, Alvarez D, Duhayon N, Xu M, Lisoni J, Brongersma S, Vandervorst W (2003) Proc SPIE 5116:607
49. Eyben P, Denis S, Clarysse T, W Vandervorst (2003) Mater Sci Eng B 102:132
50. Alvarez D, Hartwich J, Fouchier M, Eyben P, Vandervorst W (2003) Appl Phys Lett 82:1724
51. De Wolf P (1998) PhD thesis, University of Leuven
52. De Wolf P, Clarysse T, Vandervorst W (1997) J Vac Sci Technol B 16:320
53. Dürig U, Novotny L, Michel B, Stalder A (1997) Rev Sci Instrum 68:3814
54. Tsujino S, Falub CV, Müller E, Scheinert M, Diehl L, Gennser U, Fromherz T, Borak A, Sigg H, Grützmacher D, Campidelli Y, Kermarrec O, Bensahel D (2004) Appl Phys Lett 84:2829
55. Bell LD, Kaiser WJ (1988) Phys Rev Lett 61:2368
56. Bell LD, Kaiser WJ (1988) Phys Rev Lett 60:1406

57. Heer R, Smoliner J, Strasser G, Gornik E (1998) Appl Phys Lett 73:3138
58. Park K-B, Pelz JP, Grim J, Skowronski M (2005) Appl Phys Lett 87:232103
59. .Tivarus C, Pelz JP, Hudait MK, Ringel SA (2005) Appl Phys Lett, 87:182105
60. Giannazzo F, Roccaforte F, Raineri V, Liotta SF (2006) Europhys Lett 74:686
61. Trivedi N, Ashcroft NW (1988) Phys Rev B 38:012298
62. Tesanovic Z, Jaric MV, Maekawa S (1986) Phys Rev Lett 57:2760
63. Mayadas AF, Shatzkes M (1970) Phys Rev B 1:001382
64. Miranda E, Sune J (2004) Microelectron Rel 44:1
65. Nicollian EH, Brews JR (1982) MOS physics and technology. Wiley, New York
66. Watanabe H, Fujita K, Ichikawa M (1998) Appl Phys Lett 72:1987
67. Daniel ES, Jones JT, Marsh OJ, McGill TC (1997) J Vac Sci Technol B 13:1945
68. Wen HJ, Ludeke R (1998) J Vac Sci Technol A 16:1735
69. Kaczer B, Pelz JP (1996) J. Vac Sci Technol. B 14 :2864
70. Porti M, Nafria N, Aymerich X, Olbrich A, Ebersberger B (2002) J Appl Phys 91:2071
71. Kyuno K, Kita K, Toriumi A (2005) Appl Phys Lett 86:063510
72. Fiorenza P, Polspoel W, Vandervorst W (2006) Appl Phys Lett 88:222104
73. Watanabe Y, Seko A, Kondo H, Sakai A, Zaima S, Yasuda J (2004) J Appl Phys 43:4679
74. Petry J, Vandervorst W, Pantisano L, Degraeve R (2005) Microelectron Rel 45:815
75. Lombardo S, Stathis JH, Linder BP,Pey KL, Palumbo F, Tung CH (2006) J Appl Phys 98:121301
76. Maserjian J, Zamani N (1982) J Appl Phys 53:559
77. Degraeve R, Groeseneken G, Bellens R, Ogier JL, Depas M, Roussel PJ, Maes HE (1998) IEEE Trans Electron Device 45:904
78. Nafria M, Surie J, Aymerich X (1993) J Appl Phys 73:205
79. Depas H, Nigam T, Heyns MM (1996) IEEE Trans Electron Device 43:1499
80. Degraeve R, Groeseneken G, Bellens R, Ogier JL, Depas M, Roussel PJ, Maes HE (1998) IEEE Trans Electron Device 45:904
81. Fiorenza P, Raineri V (2006) Appl Phys Lett 88:212112
82. Fiorenza P, Lo Nigro R, Raineri V, Lombardo S, Toro RG, Malandrino G, Fragalà IL (2005) Appl Phys Lett 87:231913
83. Degraeve R, Groeseneken G, Bellens R, Depas M, Maes HE (1995) IEDM Tech Dig 863
84. Stathis JH (1999) J Appl Phys 86:5757
85. Fiorenza P, Lo Nigro R, Raineri V, Salinas D (2007) Microelectron Eng 84:441

29 Visualization of Fixed Charges Stored in Condensed Matter and Its Application to Memory Technology

Yasuo Cho

Abstract. Scanning nonlinear dielectric microscopy (SNDM) with super-high resolution is described. Experimental results for the ferroelectric domain and the visualization of charge stored in flash memories are given, following the description of the theory and principle of SNDM. Next, a higher-order nonlinear dielectric imaging method and noncontact SNDM (NC-SNDM) are proposed. The first achievement of atomic resolution in capacitance measurement is successfully demonstrated using this NC-SNDM technique. In addition to these techniques, a new 3D-type of SNDM to measure the 3D distribution of ferroelectric polarization is developed. Finally, a very high density next-generation ferroelectric data storage device based on SNDM is demonstrated.

Key words: Scanning nonlinear dielectric microscopy, Ferroelectric domain, Electric dipole moment, Ferroelectric data storage, Flash memory

29.1
Introduction

In recent years, we have developed and reported on the use of scanning nonlinear dielectric microscopy (SNDM) for the measurement of the microscopic distribution of dielectric properties on a dielectric material surface [1–5]. This is the first successful purely electrical method for observing the ferroelectric polarization distribution without the influence of the screening effect from free charges. As this microscopic technique has quite a high sensitivity to variation in capacitance, on the order of 10^{-22}–10^{-23} F, we have also succeeded in visualizing the electron and hole distributions stored in semiconductor flash memory devices using SNDM [6–9]. Therefore, this technique is expected to apply in the development of new semiconductor devices.

New techniques have been developed in order to increase the resolution of SNDM. One technique is the higher-order SNDM (HO-SNDM) technique, which detects the higher-order nonlinear dielectric constants, $\varepsilon(4)$ and $\varepsilon(5)$, in addition to the lowest-order nonlinear dielectric constant, $\varepsilon(3)$ [10]. The other technique is noncontact SNDM (NC-SNDM) [11]. NC-SNDM utilizes the higher-order nonlinear dielectric signal $\varepsilon(4)$ for control of the noncontact state (tip–sample gap control), and simultaneous detection of the lowest-order nonlinear dielectric constant, $\varepsilon(3)$, and the linear dielectric constant, $\varepsilon(2)$. Thus, NC-SNDM can observe both topography and dielectric properties of materials, including insulators.

To date, the highest confirmed resolution of SNDM in contact mode operation has been in the subnanometer order of a ferroelectric domain measurement. However, evaluation of the single dipole moment level of a material is expected to be characterized to a greater extent. From calculated results, it is expected that NC-SNDM has higher sensitivity to the gap between a tip and a specimen than scanning tunneling microscopy (STM), and will also have lateral atomic resolution. However, among the researchers in this field, there is the thought that "as the polarization is defined at a macroscopic volume-averaged value, the achievement of atomic resolution by a dielectric (capacitance) measurement is impossible". There have even been theoretical reports that deny atomic resolution in dielectric microscopy techniques including scanning capacitance microscopy (SCM), which probably arise from this reasoning [12, 13].

Most recently, we have developed a NC-SNDM system operated in ultrahigh vacuum (UHV), in order to prove that NC-SNDM has a real atomic resolution. Using this microscopic technique, we have succeeded in observing the Si(111)7×7 atomic structure [14]. This is the first successful demonstration of the achievement of atomic resolution in a dielectric microscopic technique. Thus, SNDM is recognized as the fifth microscopy technology with an atomic resolution, following field ion microscopy (FIM), transmission electron microscopy (TEM), STM and atomic force microscopy (AFM).

In addition to the abovementioned SNDM techniques which can detect the polarization components perpendicular to the specimen surface only, a new 3D-type of SNDM (3D-SNDM) to measure the 3D distribution of ferroelectric polarization has been developed [15–17]. Using this 3D-SNDM, one can measure the polarization components both perpendicular (vertical) and parallel (lateral) to the specimen surface by selecting the direction of the applied electric field.

Moreover, we also have proposed next-generation ultra-high-density ferroelectric data storage based on SNDM and already have reported that 10 Tbit/in.^{2} in memory density was achieved and actual information composed of a data bit array of 128×82 was successfully stored at an areal data storage density of 1 Tbit/in.^{2} [18, 19]. These results confirm the potential of ferroelectric materials in storing information at high densities.

In this chapter, we review and summarize these SNDM techniques and related technologies.

29.2
Principle and Theory for SNDM

Figure 29.1 shows the system setup of SNDM using the LC lumped constant resonator probe. In the figure, C_s denotes the capacitance of the specimen under the center conductor (the tip) of the probe. C_s is a function of time because of the nonlinear dielectric response under an applied alternating electric field $E_{p3} (= E_p \cos \omega_p t, f_p = 5\,\text{kHz}^{-1}\,\text{MHz})$.

This LC resonator is connected to the oscillator tuned to the resonance frequency of the resonator. The abovementioned electrical parts, i.e., tip (needle or cantilever), ring, inductance and oscillator, are assembled into a small probe for SNDM. The

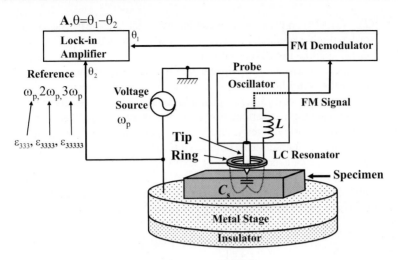

Fig. 29.1. Scanning nonlinear dielectric microscopy (SNDM)

oscillating frequency of the probe (or oscillator) (around 1–6 GHz) is modulated by the change of capacitance $\Delta C_s(t)$ due to the nonlinear dielectric response under the applied electric field. As a result, the probe (oscillator) produces a frequency-modulated (FM) signal. By detecting this FM signal using the FM demodulator and the lock-in amplifier, we obtain a voltage signal proportional to the capacitance variation. Thus, we can detect the nonlinear dielectric constant just under the tip and can obtain the fine resolution determined by the diameter of the pointed end of the tip and the linear dielectric constant of specimens. The capacitance variation caused by the nonlinear dielectric response is quite small [$\Delta C_s(t)/C_{s0}$ is in the range from 10^{-3} to 10^{-8}.]. Therefore, the sensitivity of the SNDM probe must be very high. The measured value of the sensitivity of the abovementioned lumped constant resonator probe is 10^{-22}–10^{-23} F, which is much higher than that of SCM, whose typical sensitivity is 10^{-18} F.

29.3
Microscopic Observation of Area Distribution of the Ferroelectric Domain Using SNDM

Originally, this microscope was developed for the purpose of measuring the ferroelectric polarization and the local anisotropy of dielectrics through the detection of the third-order nonlinear dielectric constant. In Fig. 29.1, the capacitance immediately beneath the tip C_s is modulated alternately owing to the nonlinear dielectric properties of the specimen caused by the application of an alternating electric field between the tip and the metal stage.

Equation (29.1) is a polynomial expansion of the electric displacement D_3 as a function of electric field E_3.

$$D_3 = P_{s3} + \varepsilon(2)E_3 + \frac{1}{2}\varepsilon(3)E_3^2 + \frac{1}{6}\varepsilon(4)E_3^3 + \frac{1}{24}\varepsilon(5)E_3^4 + \cdots, \quad (29.1)$$

where $\varepsilon(2)$, $\varepsilon(3)$, $\varepsilon(4)$ and $\varepsilon(5)$ are a linear, lowest-order nonlinear, one-order-higher and two-orders-higher nonlinear dielectric constants, respectively. (The numbers in the parentheses denote the rank of of tensor.) The even-rank tensors, including the linear dielectric constant $\varepsilon(2)$, are insensitive to the states of the spontaneous polarization. On the other hand, the lowest-order nonlinear dielectric constant $\varepsilon(3)$ and other higher-order odd-rank tensor nonlinear dielectric constants are very sensitive to spontaneous polarization. For example, there is negative $\varepsilon(3)$ in a material with a center of symmetry, the sign of $\varepsilon(3)$ changes in accordance with the inversion of the spontaneous polarization. Therefore, by detecting this $\varepsilon(3)$ microscopically, we can measure the of area distribution of the ferroelectric domain.

The ratio of this derivative capacitance variation ΔC_s to the static value to the capacitance C_{s0} is given by the following equation:

$$\frac{\Delta C_s}{C_{s0}} = \frac{\varepsilon(3)}{\varepsilon(2)} E_p \cos(\omega_p t) + \frac{1}{4} \frac{\varepsilon(4)}{\varepsilon(2)} E_p^2 \cos(2\omega_p t) + \frac{1}{24} \frac{\varepsilon(5)}{\varepsilon(2)} E_p^3 \cos(3\omega_p t) + \ldots \qquad (29.2)$$

This equation shows that the alternating capacitance of different frequencies corresponds to each order of the nonlinear dielectric constant. Signals corresponding to $\varepsilon(3)$, $\varepsilon(4)$ and $\varepsilon(5)$ were obtained by setting the reference signal of the lock-in amplifier in Fig. 29.1 to frequencies ω_p, $2\omega_p$ and $3\omega_p$ of the applied electric field, respectively.

As one example of domain measurement, here, we show the $\varepsilon(3)$ image in Fig. 29.2 which demonstrates the resolution of SNDM is really of subnanometer order. The tip used in this measurement was a metal-coated conductive cantilever with a tip radius of 25 nm.

These images were taken from epitaxial lead zirconate titanate (PZT) (400-nm) thin film/La–Sr–Co–O/SrTiO$_3$. The strip shape 90° a–c domain pattern is seen.

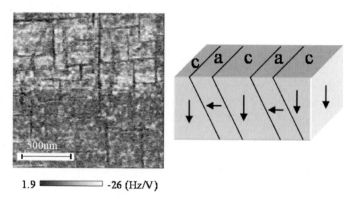

Fig. 29.2. Domain pattern taken from an epitaxial lead zirconate titanate (PZT) thin film on La–Sr–Co–O/SrTiO$_3$ [nonlinear dielectric $\varepsilon(3)$ image]

29.4
Visualization of Stored Charge in Semiconductor Flash Memories Using SNDM

Because SNDM is capable of detecting an ac capacitance change of 10^{-22}–10^{-23} F, which is extraordinarily small, the methods already described can be used not only to detect ferroelectric polarization, but also in any applications that involve samples containing dipoles. As an illustrative example of investigating systems other than ferroelectrics, in Fig. 29.3 we show the results of our experimental visualizations of electrons and holes stored in MONOS type flash memory and of electrons in floating gate type flash memory. In MONOS type flash memory (Fig. 29.3a), we were able to visualize the dipoles created when an electron (hole) stored in the ONO layer induces a hole (electron) at the silicon surface, which amounts to detecting the negatively charged electrons and the positively charged holes. And also, we can clearly visualize the electron distributions stored in the floating gate of floating gate type flash memory as shown in Fig. 29.3b. This is a first successful achievement of visualization of charges stored in semiconductor devices. As great numbers of flash memories have been used as a nonvolatile semiconductor memory for mobile communication devices, this result is quite important, not only from the viewpoint of semiconductor science but also because of the influence on the electronics industry.

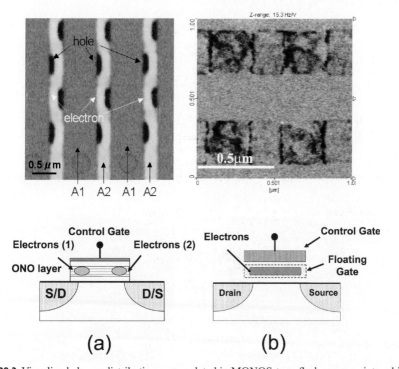

Fig. 29.3. Visualized charge distribution accumulated in MONOS-type flash memory into which both electrons and holes are injected, creating a checked pattern (**a**), and visualized electron distribution floating gate type flash memory (**b**)

29.5
Higher-Order SNDM

A higher-order nonlinear dielectric microscopy technique with higher lateral and depth resolution than conventional nonlinear dielectric imaging is investigated. From (29.1), the resolution of SNDM is found to be a function of the electric field E. We note that the distributions of E^2, E^3 and E^4 fields underneath the tip become much more concentrated in accordance with their power than that of the field E, as shown in Fig. 29.4. From this figure, we find that higher-order nonlinear dielectric imaging has both higher lateral and depth resolution than lower-order nonlinear dielectric imaging.

We experimentally confirmed that $\varepsilon(5)$ imaging has higher lateral resolution than $\varepsilon(3)$ imaging using an electroconductive cantilever as a tip with a radius of 25 nm. For example, Fig. 29.5 shows $\varepsilon(3)$ and $\varepsilon(5)$ images of the 2D distribution of PZT thin film. The two images can be correlated, and it is clear that the $\varepsilon(5)$ image resolves greater detail than the $\varepsilon(3)$ image owing to the higher lateral and depth resolution. Thus, this technique was demonstrated to be very useful for observing surface layers of the order of unit cell thickness on ferroelectric materials.

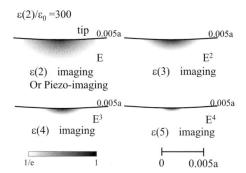

Fig. 29.4. Distribution of E, E^2, E^3 and E^4 fields under the needle tip. a denotes the tip radius

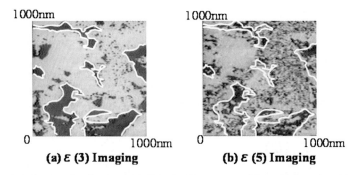

Fig. 29.5. Polarization distribution of PZT thin film measured by $\varepsilon(3)$ imaging (**a**) and $\varepsilon(5)$ imaging (**b**)

29.6
Noncontact SNDM

Recently, we developed NC-SNDM [11]. NC-SNDM utilizes the higher-order nonlinear dielectric signal $\varepsilon(4)$ for control of the noncontact state (tip–sample gap control), and simultaneous detection of the lowest-order nonlinear dielectric constant, $\varepsilon(3)$, and the linear dielectric constant, $\varepsilon(2)$. Thus, NC-SNDM can observe both topography and dielectric properties of materials, including insulators.

Most recently, we developed a NC-SNDM system operated in UHV in order to prove that NC-SNDM has a real atomic resolution.

All measurements were performed under UHV conditions. Equation (29.1), using the dielectric constant, is a somewhat macroscopic description. However, even if a very localized situation is considered, for example, with atomic resolution in mind, as the localized dipole moment also causes capacitance variation, (29.1) is still applicable by replacing the macroscopic dielectric constant, ε, with the local polarizability factor, ε_{local}, of a single dipole moment as a matter of form.

In the actual measurements, the ambient condition should be UHV, because adsorbed water, with a relatively high dielectric constant on the order of 80, on the specimen surface causes the electric field under the tip to broaden under such an atmospheric condition [20]. Therefore, an UHV-type SNDM was developed. In this study, the specimen and the probe were in a vacuum chamber in which the degree of vacuum was below 1.0×10^{-10} Torr.

In order to achieve atomic resolution, the sample surface should be cleaned at the atomic scale, and to prove the atomic resolution, a well-established standard sample is required; therefore, Si(111) was chosen as a representative specimen. The specimen was heat-treated by resistance heating at 600 °C for 12 h and 1000 °C for 1 min to form a clean surface. For this experiment, we used a Pt–Ir tip which was made sharper by electrochemical etching. The typical radius of the tip was approximately 10–50 nm. After fabrication of the probes, they were cleaned by Ar ion sputtering in the UHV preparation chamber.

Finally, the Si(111) clean surface was observed using NC-SNDM. In this experiment, an alternating voltage of 3 V_{p-p} at 30 kHz was applied. Figure 29.6 shows the simultaneously measured topographical image [the image of the height control signal applied to the Z-axis piezo scanner to maintain a constant $\varepsilon_{local}(4)$ signal (Fig. 29.6a), and the image of the $\varepsilon_{local}(3)$ signal distribution (Fig. 29.6b)]. In the topography, the Si(111)7×7 atomic structure is clearly recognized; therefore, it is confirmed that atomic resolution was achieved by NC-SNDM. Moreover, it is also confirmed that the distribution of the $\varepsilon_{local}(3)$ signal was obtained simultaneously. Because the $\varepsilon_{local}(3)$ image has a contrast that corresponds to the individual Si atoms, this image includes information of the dipole moment of the single Si atom. Of course, this is the first successful demonstration of the achievement of atomic resolution in a dielectric microscopic technique. Thus, SNDM is recognized as the fifth microscopy technology with an atomic resolution, following FIM, TEM, STM and AFM.

This technique, which is used to detect dielectric properties and topography, is applicable not only to semiconductors but also to both polar and nonpolar dielectric materials. Thus, it is expected that NC-SNDM will contribute to the study of electric dipole moment distribution in insulating materials at the atomic level.

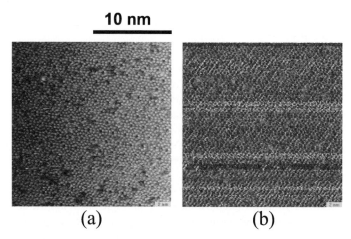

Fig. 29.6. a Topography of the Si(111)7×7 atomic structure observed by noncontact SNDM. **b** Simultaneously measured $\varepsilon_{local}(3)$

29.7
SNDM for 3D Observation of Nanoscale Ferroelectric Domains

The other advantage of SNDM in the investigation of ferroelectric materials is that it is possible to measure the polarization components both perpendicular (vertical) and parallel (lateral) to the specimen surface by selecting the direction of the applied electric field [15–17]. The 3D measurement of ferroelectric polarization is quite important, because many ferroelectric materials have 3D domain structures.

In ferroelectric materials, the sign of the lowest-order component of the nonlinear dielectric constant (third-rank tensor) changes with inversion of the polarization direction. Hence, by measuring the sign of the nonlinear dielectric constant we can obtain the polarization direction. In vertical measurement, SNDM detects the component of polarization perpendicular to the specimen surface (ε_{333} component) and in lateral measurement, it detects the parallel component (ε_{311} component).

Figure 29.7 shows a schematic diagram of a SNDM system designed to measure in both the vertical and the lateral polarization detections. In vertical measurement, the bottom electrode is used to apply an alternating electric field in the vertical direction. In lateral measurement, an alternating electric field is generated by applying an ac voltage to four electrodes denoted A, B, C and D on the electrode plate. By producing a 90° phase shift between the applied voltages V_{A-B} ($2V_1 \cos\theta$) and V_{C-D} ($2V_1 \sin\theta$), one generates a rotating electric field parallel to the specimen surface. With use of the rotating electric field, the phase output of the lock-in amplifier directly indicates the lateral azimuth angle.

In this experiment, lateral measurement was carried out in the cross-sectional plane (the x-cut plane) of the c–c domain structure of multidomain congruent LiTaO$_3$ (CLT) with a thickness of 200 μm, which has a lateral polarization component only.

Figure 29.8a shows a phase image obtained for lateral measurement in the x-plane of LiTaO$_3$. Antiparallel polarization orientations are clearly distinguished in this image. This image further reveals that the c–c domains do not penetrate along

29 Visualization of Fixed Charges by SNDM

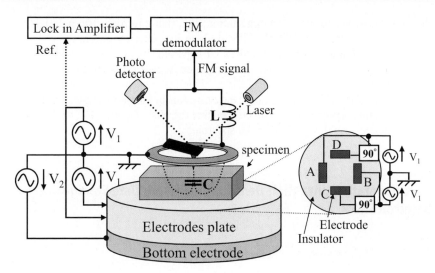

Fig. 29.7. The 3D-SNDM system

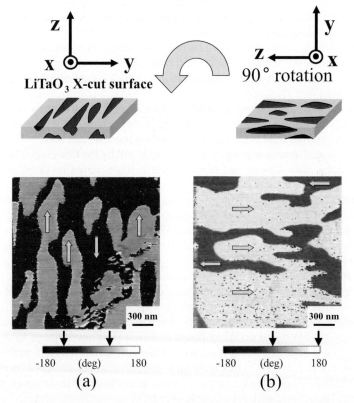

Fig. 29.8. a Phase image in the x-cut plane of multidomain LiTaO$_3$ by lateral measurement. **b** Phase image by lateral measurement after physical rotation of the specimen by 90°

the z-axis contrary to the usual expectation; i.e., that a ferroelectric polarization is expected to penetrate through the crystal along the polarization direction. In other words, the multidomain structure of the CLT generated by the depolarization process was found to be composed of numerous charged domains. Figure 29.8b also shows a phase image in the x-plane of LiTaO$_3$, in which the specimen has been physically rotated by 90°. In these images, where the azimuthal polarization angles are indicated by arrows on the color scale, it can be seen that the phase signals shift by around 90° in accordance with the rotation of the specimen.

29.8 Next-Generation Ultra-High-Density Ferroelectric Data Storage Based on SNDM

29.8.1 Overview of Ferroelectric Data Storage

With the advance of information processing technology, the importance of high-density data storage is increasing. Studies on thermal fluctuation predict that magnetic storage, which plays a major role in this field, will reach a theoretical limit in the near future, and thus a novel high-density storage method is required.

Ferroelectrics can hold bit information in the form of the polarization direction of individual domains. Moreover, the domain wall of typical ferroelectric materials is as thin as the order of a few lattice parameters [21], which is favorable for high-density data storage.

The basic idea of applying ferroelectrics to a recording system was suggested by Pulvari [22] in 1951. In ferroelectric data storage, the remanent polarization does not produce an external field even after the recording medium has been polarized. The reason is that the remanent polarization is neutralized by the free charge. Therefore, in ferroelectric data storage, the reproduction voltage cannot be detected without using some type of excitation.

In order to address this problem, several methods have been proposed. A destructive reading method applying an electric switching field to the medium was used in the system proposed by Anderson [23]. Application of ultrasonic vibration to the recording medium was proposed to obtain nondestructive reproduction by Crawford [24].

In 1969, Tanaka and Sato31-25 devised a reproduction method which used the pyroelectric effect and they experimentally demonstrated the feasibility of a ferroelectric recording system which used this concept [26].

Recently, with the progress in atomic force microscope technology, piezo force microscopy (PFM) based on AFM has been extensively investigated as a method of forming and detecting small inverted domain dots in ferroelectric material such as PZT thin film [27–31]. In this technique, domain dots are switched by applying a relatively large dc pulse to the probe, creating an electric field at the tip of the probe cantilever. These dots can then be detected via the ac surface displacement (vibration) of the ferroelectric material based on the piezoelectric response to an ac electric field applied at the same tip. This technology has clear implications for bit storage in ultra-high-density recording systems, with anticipated storage densities on the order of terabits per square inch. However, current techniques are capable of forming and detecting dots of around 100 nm (achieved memory density of 64 Gbit/in.^{2} in a BaTiO$_3$ single crystal [30] and of 39 Gbit/in.^{2} in PZT thin film [31]) and this falls far short of the 1 Tbit/in.^{2} thought to be possible and also short of really achieved memory density in ferromagnetic recording. The limiting factors involved are the resolution of the piezo-imaging method and the physical properties of the ferroelectric medium. For example, even if very small nanodots of less than 10 nm could be formed successfully, a detection method with resolution finer than 1 nm would be required to detect such dots with sufficient accuracy. Therefore, it is of primary importance to improve the resolution of the domain detection device such that the smallest formable domain sizes can be resolved.

The resolution of our SNDM is in the subnanometer range [32], much higher than that of the PFM method used for observing polarization distributions [33]. Moreover, our SNDM technique is a purely electrical method, allowing domain information to be read at much higher speed than by piezo-imaging, the read speed of which is limited by the mechanical resonance frequency of the atomic force microscope cantilever (typically around 100 kHz).

More recently, the present author also has studied the formation of small domain inverted dots in lead titanate [34] and PZT thin film [35] using SNDM, and has successfully produced and observed a domain inverted dot with a radius of 12.5 nm [35]. However, single-crystal material is expected to be more suitable for studying nanodomain engineering quantitatively with good reproducibility because current thin films still suffer from atomic-scale nonuniformities that prevent switching in the nanodomains in which they occur.

To date, barium titanate (BaTiO$_3$) is the only material that has been studied as a single-crystal material for nanodomain switching [30]. Although BaTiO$_3$ is a typical ferroelectric material, it is not suitable as a recording medium for a number of reasons. Most importantly, the BaTiO$_3$ single crystal belongs to the tetragonal system, with the result that it has two possible domains in which to store bits, a 90° a–c domain and a 180° c–c domain. Such a structure introduces a level of complexity not desired at present. Furthermore, the phase transition from the tetragonal phase to the orthorhombic phase occurs at 5 °C, which is dangerously close to room temperature for a storage medium, possibly resulting in data loss as a result of ambient temperature drift. Lastly, it is difficult to fabricate a large, high-quality, and practical usable BaTiO$_3$ single crystal at low cost. A ferroelectric nano-domain engineering material suitable for practical application as a storage medium should have a single 180° c–c domain, an adequately high Curie point without phase transitions below the Curie point, and should be producible as large single crystals with good homogeneity at low cost. The lithium tantalate (LiTaO$_3$) single crystal satisfies all these conditions, and has been used widely in optical and piezoelectric devices. The development of nanodomain engineering techniques based on LiTaO$_3$ single crystals will have applications in not only ultra-high-density data storage, but also in various electro-optical, integrated-optical and piezoelectric devices.

With these as a background, we have proposed an ultra-high-density ferroelectric data storage system using the scanning nonlinear dielectric microscope as a pickup device and a LiTaO$_3$ single crystal as a ferroelectric recording medium [18,36–43]. In this chapter, an investigation of ultra-high-density ferroelectric data storage based on SNDM is described. For the purpose of obtaining fundamental knowledge on high-density ferroelectric data storage, several experiments on nanodomain formation in a LiTaO$_3$ single crystal were conducted. Through domain engineering, a domain dot array with an areal density of 1.5 Tbit/in.^{2} was formed on CLT. Subnanosecond (500 ps) domain switching speed also has been achieved. Next, actual information storage with low bit error and high memory density was performed. A bit-error ratio (BER) of less than 1×10^{-4} was achieved at an areal density of 258 Gbit/in.^{2}. Moreover, actual information storage was demonstrated at a density of 1 Tbit/in.^{2}.

Finally, we examined the effect of a small dc offset voltage on the pulse amplitude and duration needed to switch and stabilize the nanodomains. Both the pulse amplitude and the application time required to achieve reversal of polarization vary significantly with even small variation in the dc offset voltage following the pulses. These results indicated that application of a very small dc offset voltage is very effective in accelerating the domain switching speed and in stabilizing the reversed nanodomain dots. Applying this offset application technique, we formed a smallest artificial nanodomain single dot of 5.1 nm in diameter and an artificial nanodomain dot-array with a memory density of 10.1 Tbit/in.^{2} and a bit spacing of 8.0 nm, representing the highest memory density for rewritable data storage reported to date.

29.8.2
SNDM Nanodomain Engineering System and Ferroelectric Recording Medium

The ferroelectric data storage system has been developed based on SNDM for fundamental read/write experiments. Figure 29.9 shows a schematic diagram of the system used in this study. When we read the nanodomain dots, the SNDM technique is used. On the other hand, writing is carried out by applying relatively large voltage pulses to the ferroelectric recording medium and locally switching the polarization direction. The pulse generator is connected to the bottom electrode of the medium; thus, the positive domains (which are observed on the surface of the probe side) are written by the positive voltage pulses and the negative domains are written by the negative voltage pulses. The positive domains are defined as data bits of "1" in this chapter. Piezoelectric scanners with displacement sensors are employed for highly accurate positioning. Two types of cantilevers were used as read/write probes. Conductive-diamond-coated cantilevers with a tip radius of 50 nm were used for the purpose of studying the basic recording technology, because this type of probe has excellent durability, and would be valid for practical use. Metal-coated cantilevers with sharp tip radius of 25 nm were also used as a probe in order to write the higher-density data.

We selected the LiTaO$_3$ single crystal as a recording medium because this material has suitable characteristics as follows: (1) there exist only 180° c–c domains; (2) it does not possess a transition point near room temperature; (3) high-quality and large single crystals can be fabricated at a low cost. When a reversed domain is formed by applying a voltage to a specimen using a sharp-pointed tip, the electric field is highly localized immediately beneath the tip. Therefore, the preparation of thin and homogeneous specimens is a matter of the highest priority. We have obtained very thin and uniform LiTaO$_3$ recording media with thicknesses of 10–250 nm

29 Visualization of Fixed Charges by SNDM

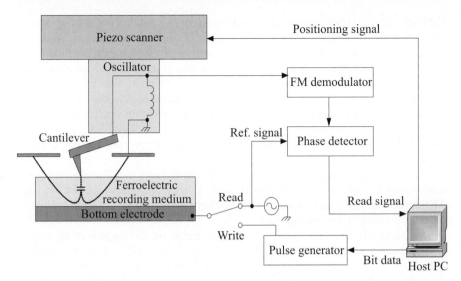

Fig. 29.9. The ferroelectric data storage system based on SNDM

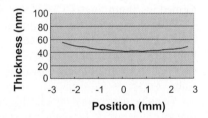

Fig. 29.10. A profile showing the thickness distribution of congruent LiTaO$_3$ (CLT) recording media (6 × 6 mm^2) at a thickness of 50 ± 7 nm

These specimens were fabricated by mechanically polishing a single-crystal wafer, which was mounted on a metal-coated (chromium) LiTaO$_3$ single-crystal wafer with a thickness of 400–500 μm, to a thickness of approximately 0.5 μm, followed by electron cyclotron resonance dry etching to the desired thickness. The same LiTaO$_3$ material was used for the substrate and the polished thin film to ensure matching of the thermal expansion coefficients. The thickness of the fabricated thin medium was measured using a spectrum reflectance thickness monitor (Otshuka Electronic FE-3000) with nanometer-scale precision. The area of the sample was 6 × 6 mm^2, all of which was usable for data storage. As an example, Fig. 29.10 shows the surface roughness of a recording medium with a thickness of 50 nm. The distribution in thickness is within ±7 nm. Thus, we now have a ferroelectric medium with sufficient homogeneity and space to perform practical information storage on the terabit per square inch density scale.

29.8.3
Nanodomain Formation in a LiTaO$_3$ Single Crystal

Presently, two types of LiTaO$_3$ single crystals are widely known; one is stoichiometric LiTaO$_3$ (SLT) [44] and the other is congruent LiTaO$_3$ (CLT). It was reported that the polarization reversal characteristics of these crystals are distinctly different from each other [45–47]. SLT has few pinning sites of domain switching derived from Li point defects; therefore, SLT has the characteristics that the coercive field is low and the switching time is short. This means that SLT is favorable for low-power and high-speed writing. On the other hand, CLT has many pinning sites because it is Ta-rich crystal, and the natural domain size of CLT is much smaller than that of SLT. So we expect that CLT is suitable for higher-density storage with smaller domain dots.

Figure 29.11 shows the SNDM images of typical nanosized inverted domains formed in a 100-nm-thick SLT medium by applying voltage pulses at amplitude of 15 V and durations of 500, 100 and 60 ns. Figure 29.11a–c shows $A \cos \theta$ images and their polarity images are shown in Fig. 29.11d–f. From these images, we found that the area of the domain decreased with decreasing voltage application time. The dependence of domain size on voltage application time was observed from the sidewise motion of the domain wall.

Subsequently, we conducted some experiments on the formation of domain shape by applying voltage pulses at multipoints while controlling the probe position. Figure 29.12 shows the domain characters "TOHOKU UNIV" written on SLT and CLT. The inverted domain shown in Fig. 29.12a was formed in 150-nm-thick SLT by applying 15-V, 100-ns pulses, and the inverted domain shown in Fig. 29.12b was formed in 70-nm-thick CLT by applying 14-V, 10-μs pulses for the left image and 14-V,5-μs pulses for the right image. From these images, we found that a small inverted domain pattern was successfully formed in CLT despite the application of long pulses on a thinner sample. This observation clearly demonstrated that pinning sites derived from lithium nonstoichiometry prevent the sidewise motion of domain walls.

Thereby, we verified the feasible storage density using CLT. Figure 29.13 shows images of the inverted domain patterns with densities of 0.62, 1.10 and 1.50 Tbit/in.^{2}. These domain patterns were formed by applying voltage pulses of 11, 12 and 12 V,

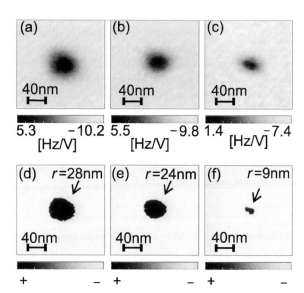

Fig. 29.11. Images of typical inverted domains formed in stoichiometric LiTaO$_3$ (SLT) by applying voltage pulses at an amplitude of 15 V and for durations of **a, d** 500 ns, **b, e** 100 ns and **c, f** 60 ns. **a–c** $A \cos \theta$ images. **d–f** Polarity images

Fig. 29.12. Nanodomain characters "TOHOKU UNIV." written on **a** SLT and **b** CLT

(a) SLT (Sample thickness: 150nm)

Applied pulse: 15V, 100nsec

(b) CLT (Sample thickness: 70nm)

Applied pulse: 14V, 10 μsec   Applied pulse: 14V, 5 μsec

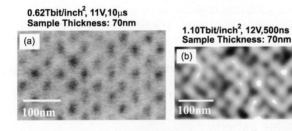

Fig. 29.13. Images of inverted domain patterns formed in CLT with densities of (**a**) 0.62 Tbit/inch^2, (**b**) 1.10 Tbit/inch^2 and (**c**) 1.50 Tbit/inch^2

and for durations of 10 μs, 500 ns and 80 ns, respectively. A close-packed dot array composed of positive and negative domains can be observed in these images.

We have also investigated long-term retention characteristic of a ferroelectric data storage system. CLT plates, 80 nm thick, with an inverted domain dot array composed of 100-nm-diameter dots were baked at high temperature. After heat treatment for several time intervals, the dots shrinked. From the variation data of dot size, the activation energy (E_a) and the frequency factor (α), a parameter of the Arrhenius equation, were determined Using these parameters, we estimated the rate of dot-radius variation in the temperature range from 24 to 80 °C in this temperature range, general memory devices are used. We conclude that it takes more than 16 years until the dot radius decreases by 20%. This time is long enough compared with lifetime of general memory devices, at least only under one condition of one thickness of specimens (CLT 80 nm) and one dot radius (50 nm).

29.8.4
High-Speed Switching of Nanoscale Ferroelectric Domains in Congruent Single-Crystal LiTaO$_3$

The recording medium for large-capacity memory devices also requires excellent high-speed writing characteristics in addition to a high memory density. The switching speed of polarization reversal immediately below the tip of the probe is directly related to writing time, and should be as fast as possible. Therefore, the investigation of the kinetics of polarization reversal below a probe tip is important in estimating the expected writing speed of any recording system, and may also lead to a better understanding of the polarization reversal process near such tips.

The sizes of the domain dots were measured as a function of pulse amplitude and duration for various sample thicknesses. Figure 29.14 shows the SNDM images of

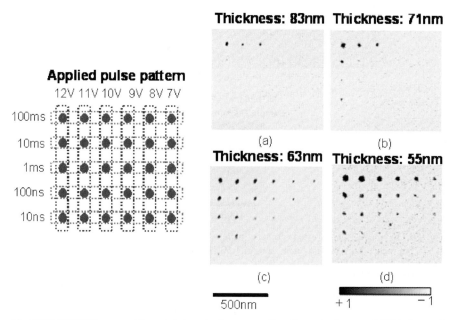

Fig. 29.14. SNDM images of inverted domain dot arrays (5 × 6) on single-crystal CLT plates of thicknesses **a** 83, **b** 71, **c** 63 and **d** 55 nm. The pulse amplitude was varied from 12 to 7 V in 1-V steps from *left* to *right*, and the pulse application time was varied from 100 μs to 10 ns in 10× steps from *top* to *bottom*

the inverted domain dot array formed on single-crystal CLT plates with thicknesses of 83, 71, 63 and 55 nm. An array of 5 × 6 pulses was applied to each sample. From left to right in the figure, the pulse amplitude was varied from 12 to 7 V in 1-V steps, and from top to bottom, the pulse application time was varied from 100 μs to 10 ns in 10× steps. Both the pulse amplitude and the application time required to achieve the reversal of polarization varied significantly with even small variations in sample thickness. A very fast domain switching of less than 100 ns was achieved for the single-crystal CLT plates with thicknesses of less than 63 nm. This strong thickness dependence originated from the inhomogeneous distribution of the electric field immediately beneath the tip. The domain switching speed markedly depended on the electric field strength, and it rapidly decreased along the depth direction because the field was highly concentrated immediately beneath the tip. Therefore, it is only when the thickness of the sample becomes less than the penetration depth of the field that high-speed switching possible. As a result, a small variation in sample thickness affects the nanodomain reversal characteristics materially.

Using a very thin medium with a thickness of 18 nm, we succeeded in switching the nanodomain at a speed of 500 ps without the dc offset voltage. Fig. 29.15 shows the dot shape and a cross-sectional image of the nanodomain dot formed by a 10-V, 500-ps pulse.

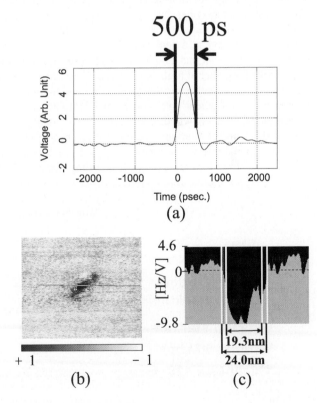

Fig. 29.15. Nanodomain dot formed by a 10-V, 500-ps pulse without dc offset voltage on an 18-nm-thick CLT thin plate: **a** applied pulse waveform; **b** 2D image; **c** cross-sectional image

29.8.5
Prototype of a High-Density Ferroelectric Data Storage System

The abovementioned SNDM domain engineering system, which is remodeled from a commercial AFM unit, is inadequate for establishing basic elemental technologies of high-density ferroelectric data storage. Moreover, a system equipped with all the components necessary for actual read/write functions is required in further studies aimed at practical applications. Therefore, we developed a prototype high-density ferroelectric data storage system.

Firstly, we evaluated the writing speed of this system. Figure 29.16 shows a bit array written on CLT with a linear recording density of 440 kbit/in. at a data transfer rate of 50 kbit/s. A discrete bit array was successfully written. Moreover, we demonstrated that rewritable storage could be realized using this system. Figure 29.17, panel a shows the SNDM image of a 5×5 inverted domain dot array formed on CLT by applying 12-V, 1-ms pulses, and these dots were sequentially erased by applying -12-V, 10-ms pulses as shown in Fig. 29.17, panels b–e. A longer "erase" pulse than the "write" pulse was used to ensure erasing of the dots even in the case where the tip position slightly misses the center of dot to be erased. Of course, an intrinsically needed "erase" pulse duration is the same as that of the "write" pulse as long as a quite precise tip positioning is possible. (In this figure, bright dots are found in some black dots; this phenomenon is known as ring-shaped domain formation [48]).

Next, real information was actually stored using this system. Although ferroelectric data storage at a density of 1 Tbit/in.^{2} was in principle possible, this does not necessarily mean that actual information storage, requiring an abundance of bits to be packed together at high density, is easy to achieve. To facilitate actual storage, a large medium of sufficient surface quality and homogeneity is needed. Therefore, demonstration of actual data storage at 1 Tbit/in.^{2} using ferroelectric materials is important for showing whether the technology is really achievable. At first, we considered the usage of diamond-coated cantilevers with much better durability for the storage of actual information data composed of numerous bits. Probes with sharper tips were required for increasing the recording density while keeping the BER low. Fine domain structures were obtained, written successfully using metal-coated can-

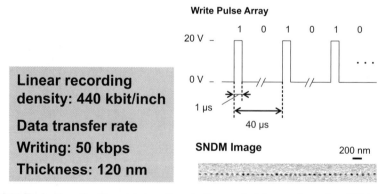

Fig. 29.16. Evaluation of writing data transfer rate. Image of a bit array formed on CLT at a data transfer rate of 50 kbit/s

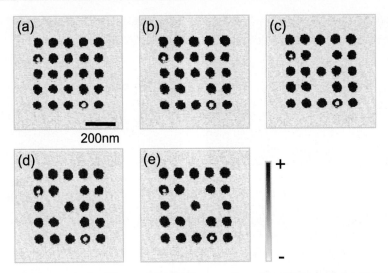

Fig. 29.17. a Image of a 5 × 5 inverted domain dot array formed on CLT by applying 12-V, 1-ms voltage pulses. **b–e** Inverted domain dots were sequentially erased by applying −12-V, 10-ms pulses

tilevers as read/write probes. The typical tip radius of a metal-coated cantilever was approximately 25 nm, whereas it was 100 nm for the diamond-coated cantilever used in this study. Therefore, a diamond-coated cantilever with a sharp tip was subsequently prepared using the ion milling method in order to write small domain dots. Figure 29.18 shows the tip of a diamond-coated cantilever observed by scanning electron microscopy after the ion milling process. The tip radius was reduced to less than 50 nm by the ion milling. Additionally, the thickness of the recording medium was reduced to 55 nm. With the improved cantilever, the information data was recorded at 50-nm bit spacing (areal density 258 Gbit/in.^{2}) as shown in Fig. 29.19. The pulse amplitude was kept constant at -16 V, and the pulse duration was changed in the range 5–50 µs. BERs were evaluated for each image. The longer-duration pulses create the too large domain dots that stick together. On the other hand, the shorter-duration pulses cannot form fully penetrating dots through the medium. At both ends of the pulse duration, the BER increases; therefore, an optimum pulse duration exists. In this case, a pulse duration of 10 µs resulted in the lowest BER, with a measured BER less than 1×10^{-4}.

Thereafter, the recording density was increased, since the tip abrasion was inhibited by improving the circuit of contacting load control and reducing the spring constant of the cantilevers. Data of 128 × 82 bits were written on a 40-nm-thick CLT plate. Figure 29.20a shows the SNDM image of the information data written at a bit spacing of 25.6 nm (areal recording density 0.98 Tbit/in.^{2}). The writing conditions were as follows: the data bits "0" were written by −14.1-V, 100-ns pulses, whereas the positive voltage pulses were not applied at the positions "1". The written data were read out with high signal-to-noise ratio, in spite of increasing the recording density. The BER was evaluated to be 7.2×10^{-2} from this image. There were some inhomogeneities of the BER depending on the bit arrangements, similarly to in the

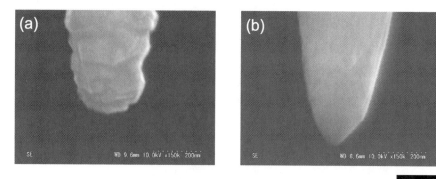

Fig. 29.18. Scanning electron microscopy images of the tip of a conductive-diamond-coated cantilever: **a** before ion milling; **b** after ion milling

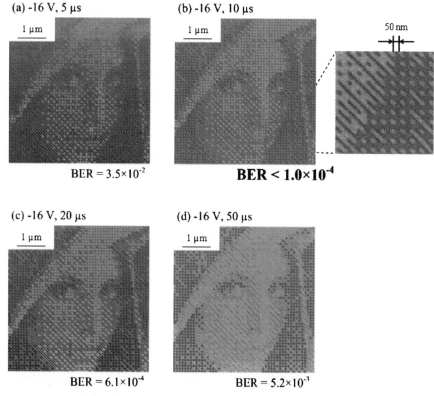

Fig. 29.19. Actual information data recorded on 55-nm-thick $LiTaO_3$ recording medium with a bit spacing of 50 nm (areal recording density: 258 Gbit/in.^{2}). The amplitudes of the writing pulses were kept constant at -16 V, and the pulse widths were changed in the range from 5 to 50 ms. The bit-error ratio (*BER*) is indicated below each image

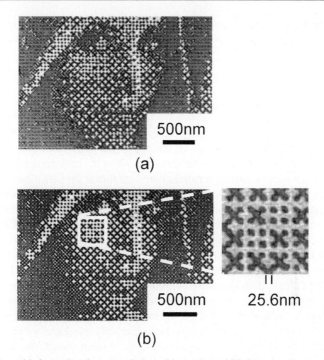

Fig. 29.20. Actual information data recorded at a bit spacing of 25.6 nm (areal recording density 0.98 Tbit/in.^{2}). Data of 128×82 bits written on a CLT single crystal formed by application of a 100-ns pulse without a dc offset voltage on a 40-nm-thick CLT plate. **a** All the negative domains were written under the fixed condition. (The bit error rate was 7.2×10^{-2}.) **b** The duration times of negative voltage were changed according to the bit arrangements. (The bit error rate was 1.8×10^{-2}.)

previous cases; thus, the writing parameters were modified. Some different pulse shapes were used for the different bit arrangements, since this was expected to be a direct approach and to be more effective for reducing the BER than bipolar pulse application. Specifically, the negative pulses for writing the data bits "0" were of the same amplitude as in Fig. 29.20 a when the bit just prior was "0" and were slightly larger when the bit just prior was "1", while the pulse widths were set at 100 ns in both cases. The BER of the information data written in this manner, which is shown in Fig. 29.20b, was 1.8×10^{-2}. A magnified image of the data bit array is also shown in Fig. 29.20b. The BER was lower by a factor of 4 compared with that in the case where all of the data bits "0" were written under fixed conditions. The number of bit errors where "0" was wrongly written as "1" was 124 out of 3889, whereas the number of errors of the opposite type was 24 out of 4251. The "0" bit errors, which were more than 5 times more numerous than "1" bit errors, were categorized in more detail according to the surrounding data bits. All "0" bit errors arose only when three or more bits out of the four nearest bits were "1". The BER was expected to be reduced by increasing the number of surrounding bits which were referred to in the categorization and setting the appropriate pulse voltage for each case, although this causes some complexities in deciding the pulse parameters such that estimations by means of simulations of domain expansion are required. Thus, we succeeded in actual information storage at a density of 1 Tbit/in.^{2} in a ferroelectric material.

29.8.6
Realization of 10 Tbit/in.^{2} Memory Density

Finally, we performed fundamental studies on highest memory density. We first examined the effect of a small dc offset voltage on the pulse amplitude and duration needed to switch and stabilize the nanodomains. All of the nanodomain dots obtained in this study were formed using a metal-coated cantilever tip with a radius of 25 nm. A dc offset voltage was applied to the sample following an initial writing pulse voltage. Figure 29.21 shows SNDM images of domain dot arrays formed on a 45-nm-thick single-crystal CLT plate with dc offset voltages of −0.5, 0, 0.5 and 1.0 V, respectively. An array of 6×6 pulses was applied by varying the offset voltage. From left to right in Fig. 29.21, the pulse amplitude was varied from 10 to 5 V in 1-V steps, and from top to bottom, the pulse duration was varied from 1 ms to 10 ns in logarithmic steps. Both the pulse amplitude and the application time required to achieve reversal of polarization vary significantly with even small variation in the dc offset voltage following the pulses. These results show that application of a very small dc offset voltage is very effective in accelerating the domain switching speed and in stabilizing the reversed nanodomain dots. In other words, the offset voltage application suppresses the domain backswitching effect known to occur in CLTs [47], and smaller pulse amplitudes and shorter pulse durations are needed to form the nanodomain dots under the offset voltage application. To demonstrate application of this acceleration effect of the offset voltage in high-density data storage, we formed our smallest artificial nanodomain single dot of 5.1-nm diameter and also formed an artificial nanodomain dot array with a memory density of 10.1 Tbit/in.^{2} and a bit spacing of 8.0 nm, as shown in Fig. 29.22. The single dot in Fig. 29.22a was formed by application of an 8.2-V, 100-ns pulse with a dc offset of 1.0 V on a CLT thin plate with a thickness of 40 nm and the dots in Fig. 29.22b were formed by a 7.8 V, 20 ns pulse application with a dc offset of 1.5 V on a CLT thin plate with a thickness of 35 nm. This reduction in film thickness from 45 nm (Fig. 29.21) to 35 nm (Fig. 29.22) in addition to the application of an offset voltage is essential to form much smaller nanodomains. As shown above, both the pulse amplitude and the application time required to achieve reversal of polarization vary significantly with even small variations in sample thickness. This strong thickness dependence originates from the inhomogeneous distribution of the electric field immediately beneath the tip. The domain switching speed and the domain size are heavily dependent on the electric field strength, which rapidly decreases along the depth direction because the field is highly concentrated just beneath the tip. As such, high-speed switching is only possible when the thickness of the sample is less than the penetration depth of the field. As a result, a small variation in sample thickness affects the nanodomain reversal characteristics materially. On the other hand, this highly concentrated field distribution enables the formation of domains as small as 8 nm when using a tip with a 25-nm radius. For example, in the case of $LiTaO_3$ with a linear dielectric constant of 42, the field is concentrated roughly within an area

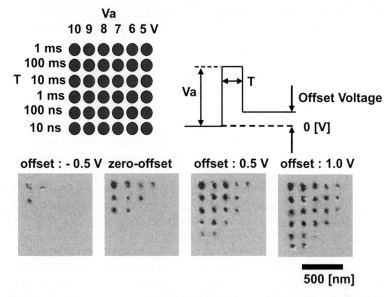

Fig. 29.21. The effect of dc offset voltage following the writing pulse. The sample thickness was 45 nm

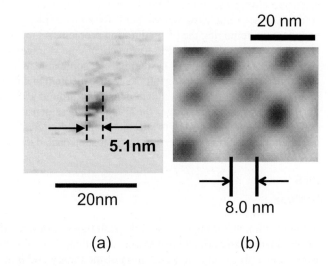

Fig. 29.22. a Smallest artificial nanodomain single dot of 5.1 nm in diameter. **b** Artificial nanodomain dot-array with a highest memory density of 10.1 Tbit/in.^{2} and a bit spacing of 8.0 nm

with a radius one tenth of the tip radius. This means that, in principle, a potential domain size of 2.5 nm in diameter is available for a relatively thin film; the small nanodomain can be stabilized by the offset voltage to suppress backswitching.

In addition to the above discussion, it is necessary to discuss the duration time of the offset voltage, as well as the advantage of applying the offset voltage. As application of the offset voltage is regarded as part of the write operation, it might be disadvantageous from the viewpoint of fast writing. Therefore, to evaluate the influence of the offset voltage on the writing speed, we performed point-to-point

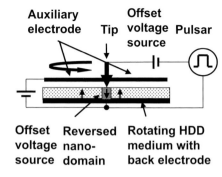

Fig. 29.23. A proposed example of a ferroelectric hard disk drive (*HDD*) system in which the auxiliary electrode covers the entire area of the recording medium to make the offset application method effective

scanning, varying the tip stopping time at each writing point under the constant offset voltage (1 V) condition. We subsequently found that even a 1-ms tip rest at each writing point was effective in stabilizing the domain. As this 1 ms was the minimum stopping duration available owing to the limits of the scanner speed employed in this study, the shortest duration is unknown but is expected to be not much faster than 1 ms. However, this does not necessarily mean that offset voltage is disadvantageous from the viewpoint of fast writing as long as a unipolar writing method is employed. For example, as shown in Fig. 29.23, if we fabricate the ferroelectric recording system for a hard disk drive such that the auxiliary electrode covers the entire area of the recording medium, we can apply a small constant offset voltage (or electric field) for an indefinitely long time to stabilize the nanodomain, removing the offset duration problem. As the dots in the 10.1 Tbit/in.^{2} array are sufficiently resolvable for practical data storage, it is fully expected that, with further refinements, this system will be applicable as a storage technology. We have thus demonstrated, using a ferroelectric medium and nanodomain engineering, that rewritable bit storage at a data density of around 10.1 Tbit/in.^{2} is achievable. To the best of our knowledge, this is the highest density reported for rewritable data storage, and is expected to stimulate renewed interest in this approach towards next-generation ultra-high-density rewritable electric data storage systems.

29.9
Outlook

In this chapter, SNDM with superhigh resolution was described. Experimental results for the ferroelectric domain and the visualization of charge stored in flash memories were shown, following the description of the theory and principle of SNDM. Next, a higher-order nonlinear dielectric imaging method (HO-SNDM) and NC-SNDM were proposed. With use of this NC-SNDM technique, the first achievement of atomic resolution in capacitance measurement was successfully demonstrated. In addition to these techniques, a new 3D-type of SNDM to measure the 3D distribution of ferroelectric polarization has developed. Finally, a very high density ferroelectric data storage based on SNDM with a density of 10 Tbits/in.^{2} was demonstrated.

Therefore, we conclude that the SNDM is very useful for observing the ferroelectric nanodomain and the local dipole moment of material with atomic resolution, and also has a quite high potential as a nanodomain engineering tool.

References

1. Cho Y, Kirihara A, Saeki T (1995) Denshi Joho Tsushin Gakkai Ronbunshi J78-C-1:593 (in Japanese)
2. Cho Y, Kirihara A, Saeki T (1996) Rev Sci Instrum 67:2297
3. Cho Y, Atsumi S, Nakamura K (1997) Jpn J Appl Phys 36:3152
4. Cho Y, Kazuta S, Matsuura K (1999) Appl Phys Lett 72:2833
5. Odagawa H, Cho Y (2000) Surf Sci 463:L621
6. Honda K, Cho Y (2005) Appl Phys Lett 86:013501
7. Honda K, Hashimoto S, Cho Y (2005) Appl Phys Lett 86:063515
8. Honda K, Hashimoto S, Cho Y (2005) Nanotechnology 16:90
9. Honda K, Hashimoto S, Cho Y (2006) Nanotechnology 17:S185
10. Cho Y, Ohara K (2001) Appl Phys Lett 79:3842
11. Ohara K, Cho Y (2005) Nanotechnology 16:54
12. Ruda HE, Shik A (2003) Phys Rev B 67:235309
13. Ruda HE, Shik A (2005) Phys Rev B 71:075316
14. Hirose R, Ohara K, Cho Y (2007) Nanotechnology 18:084014
15. Odagawa H, Cho Y (2002) Appl Phys Lett 80:2159
16. Sugihara T, Odagawa H, Cho Y (2005) Jpn J Appl Phys 44:4325
17. Sugihara T, Cho Y (2006) Nanotechnology 17:162
18. Cho Y, Hashimoto S, Odagawa N, Tanaka K, Hiranaga Y (2005) Appl Phys Lett 87:232907
19. Cho Y, Hashimoto S, Odagawa N, Tanaka K, Hiranaga Y (2006) Nanotechnology 17:S137.
20. Ohara K, Cho Y (2004) J Appl Phys 96:7460
21. Matsuura K, Cho Y, Ramesh R (2003) Appl Phys Lett 83:2650
22. Pulvari CF (1955) US Patent 2,698,928
23. Anderson JR (1952) Elect Eng 916
24. Crawford JC (1971) Trans IEEE ED-18:951
25. Tanaka H, Sato R (1969) Shingakuron 52-A:436 (in Japanese)
26. Niitsuma H, Sato R (1981)Ferroelectrics 34:37
27. Guthner P, Dransfeld K (1992) Appl Phys Lett 61:1137
28. Hidaka T, Maruyama T, Satoh M, Mikoshiba N, Shimizu M, Shiozaki T, Wills L, Hiskes R, Dicarolis SA, Amano J (1996) Appl Phys Lett 68:2358
29. Gruverman AL, Hatano J, Tokumoto H (1997) Jpn J Appl Phys 36:2207
30. Eng LM, Bammerlin M, Loppacher CH, Guggisberg M, Bennewitz R, Luthi R, Meyer E, Huser TH, Heinzelmann H, Guntherodt H-J (1999) Ferroelectrics 222:153
31. Paruch P, Tybell T, Triscone J-M (2001) Appl Phys Lett 79:530
32. Cho Y (2003) Advances in imaging. Electron physics, vol 127. Elsevier, Amsterdam, p 1
33. Matsuura K, Cho Y, Odagawa H (2001) Jpn J Appl Phys 40:3534
34. Cho Y, Kazuta S, Matsuura K (1999) Jpn J Appl Phys 38:5689
35. Matsuura K, Cho Y, Odagawa H (2001) Jpn J Appl Phys 40:4354
36. Cho Y, Fujimoto K, Hiranaga Y, Wagatsuma Y, Onoe A, Terabe K, Kitamura K (2002) Appl Phys Lett 81:4401
37. Cho Y, Fujimoto K, Hiranaga Y, Wagatsuma Y, Onoe A, Terabe K, Kitamura K (2003) Nanotechnology 14:637
38. Hiranaga Y, Cho Y, Fujimoto K, Wagatsuma Y, Onoe A (2003) Jpn J Appl Phys 42:6050
39. Fujimoto K, Cho Y (2003) Appl Phys Lett 83:5265
40. Hiranaga Y, Wagatsuma Y, Cho Y (2004) Jpn J Appl Phys 43:L569
41. Hiranaga Y, Cho Y (2004) Jpn J Appl Phys 43:6632
42. Cho Y, Odagawa H, Ohara K, Hiranaga Y (2005) IEICE Trans Electron C J88-C:1 (in Japanese)
43. Hiranaga Y, Cho Y (2005) Jpn J Appl Phys 44:6960

44. Furukawa Y, Kitamura K, Suzuki E, Niwa K (1999) J Cryst Growth 197:889
45. Kitamura K, Furukawa Y, Niwa K, Gopalan V, Mitchell TE (1998) Appl Phys Lett 73:3073
46. Gopalan V, Mitchell TE, Sicakfus KE (1999) Solid State Commun 109:111
47. Kim S, Gopalan V, Kitamura K, Furukawa Y (2001) J Appl Phys 90:2949
48. Katoh M, Morita T, Cho Y (2004) Integr Ferroelectr 68:207

30 Applications of Scanning Probe Methods in Chemical Mechanical Planarization

Toshi Kasai · Bharat Bhushan

Abstract. This chapter describes scanning probe methods (SPM) used in chemical mechanical planarization (CMP). The term "planarization" refers to the reduction in step height present on a surface. Planarization is a key step in the fabrication process of an integrated circuit (IC) on silicon wafers, which allows the construction of multilevel interconnect structures. To precisely characterize the surface morphology, including step height, SPM are especially useful because fabricated patterns on wafers are complicated and have dimensions down to the micrometer/nanometer scale. The topographic information obtained can be used for the improvement of CMP processes, such as the optimization of CMP slurry formulation and polishing conditions. SPM are also found to be beneficial for investigating the surface properties of IC wafers. The electrical, dynamic-mechanical and tribological properties can be characterized using their multifunctional capabilities.

For readers not familiar with CMP, the CMP process and its associated need for SPM are first outlined. Next, the two main SPM used in CMP, atomic force profilometry (AFP) and atomic force microscopy (AFM), are introduced and some representative data are presented for the characterization of line profile and surface roughness. Finally, techniques for understanding fundamental aspects of CMP mechanisms are discussed. The study of adhesion, friction, abrasion and associated material removal using an abrasive particle mounted on an atomic force microscope tip is presented to simulate microscale/nanoscale CMP processes. SPM provide insight into the microscopic CMP mechanisms to interpret the macroscopic CMP phenomena. Other related techniques, such as AFM phase imaging and scanning Kelvin probe microscopy (KPM) are also highlighted.

SPM have made critical contributions to the development of CMP processes. The further improvement of SPM will significantly enhance the understanding and development of CMP technology.

Key words: Scanning probe methods, Chemical mechanical planarization, Integrated circuit, Atomic force profilometry, Kelvin probe microscopy, Dishing, Erosion

30.1
Overview of CMP Technology and the Need for SPM

30.1.1
CMP Technology and Its Key Elements

Polishing technologies have favorably impacted industrial needs throughout history. For example, in the early period, the development of optical materials such as lenses used for eye glasses and microscopes was significantly enhanced through the use of glass polishing technology [1]. Since the invention of the transistor in

1947, silicon wafer lapping/polishing has been a crucial process in the IC industry [2, 3]. Perhaps the most important technological need in this regard was demonstrated in the case of large-scale integrated circuit (LSI) technology introduced in the 1970s [1]. Additional important/potential applications relying on CMP technology include the polishing of rigid disks for hard-disk drives [4–8] and microelectromechanical/nanoelectromechanical systems (MEMS/NEMS) [9, 10].

In the LSI fabrication process, highly planarized surfaces are required to construct microscale/nanoscale line patterns on silicon wafers. Metal lines connect IC elements to each other and often consist of a multilayered structure. Since topographical steps can be created by etching and deposition of conducting/insulating materials during the formation of each layer, if the surface is not planarized, the accumulated height nonuniformity will be significant. Height variations across the uppermost layer will exceed the acceptable range in depth of focus for the lithography process, resulting in a nonfunctioning device. The CMP technique has been demonstrated as the best commercial approach to achieve this critical global planarization over an entire wafer surface [11].

Figure 30.1 illustrates the key elements of a CMP-based polishing process for IC wafers. A wafer attached to a carrier is pressed against a polishing pad placed on a platen. The carrier and the platen rotate in the same direction with similar rotation speeds to provide a uniform polishing rate across the wafer [12]. Slurry is supplied to the polishing pad surface at a regulated rate. The polishing pad is usually made of polyurethane and has a porous structure. In order to maintain an optimal pad surface condition during CMP, in situ or ex-situ conditioning of the pad is generally conducted. The pad conditioner, usually made of a stainless steel plate with diamond abrasives embedded, continuously resurfaces the pad, thereby preventing the pores of the pad from being filed with slurry and polishing waste.

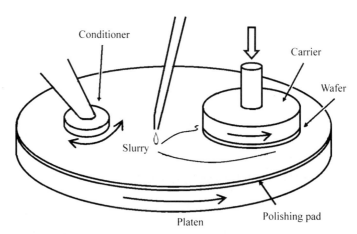

Fig. 30.1. A chemical mechanical planarization (CMP) based polishing process for integrated circuit (IC) wafers. A wafer attached to a carrier is polished with the aid of chemical reactions by slurry supplied and mechanical forces exerted at the wafer and polishing pad interface on the platen. An in situ or an ex-situ conditioning of the pad maintains an optimal pad surface condition during CMP

Figure 30.2 shows cross-sectional scanning electron microscopy (SEM) images of silicon dioxide deposited on aluminum/copper interconnect lines. A gaseous compound, tetraethyl orthosilicate [TEOS; $Si(OC_2H_5)_4$], is typically used in chemical vapor deposition of the silicon dioxide deposition process; therefore, the oxide is often called TEOS oxide. The topographical steps apparent in Fig. 30.2a are eliminated by CMP as shown in the image taken after the CMP process represented in Fig. 30.2b. The resulting silicon dioxide surface is uniformly planarized. To effectively achieve planarized surfaces, the process requires more aggressive material removal at topographically elevated regions than at the recessed regions.

One can consider the polishing slurry to be the most important component in a CMP process [13], which normally consists of deionized water, abrasive particles, an oxidizer, complexing agents and other chemical additives such as a pH adjuster and a corrosion inhibitor. Various factors such as particle type, particle size, percentage solid loading and pH affect polishing performance [11, 14–16].

Materials are removed by chemical and mechanical actions in CMP. The removal mechanisms are dependent on the chemical nature of the surface being polished. Silicon dioxide CMP requires more mechanical abrasive action by the slurry because oxides are relatively hard and chemically inactive. The silicon dioxide CMP slurry is typically maintained at higher pH, which allows the surface of silicon dioxide to be hydrolyzed and then to become softened. Abrasives driven by asperities of a pad primarily interact with topographically elevated regions of the wafer. In metal CMP, such as copper and tungsten polishing, chemical reactions at the wafer surface play a more dominant role. The surface is chemically modified by oxidants in the CMP slurry and transformed to more fragile compounds. For example, in tungsten CMP, Kaufman et al. [17] suggested passivation of tungsten in a slurry environment, forming softer tungsten oxide as suggested by the reaction:

$$W + 3H_2O \rightarrow WO_3 + 6H^+ + 6e^- \tag{30.1}$$

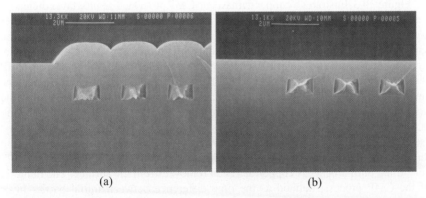

(a)   (b)

Fig. 30.2. Cross-sectional scanning electron microscopy (SEM) images of a silicon dioxide layer deposited on aluminum/copper interconnect lines **a** before and **b** after the CMP process. The topographical steps apparent in **a** are eliminated by CMP as shown in **b**, resulting in uniformly planarized silicon dioxide surface. (Courtesy of Duane Boning, MIT)

The newly formed oxide layer is removed with the mechanical force exerted by abrasives. The material removal is additionally enhanced by the use of complexing agents that produce water-soluble compounds from abraded materials [18]. It can be noted that the oxidation state of the surface depends on the chemical environment. For example, the passivation reaction proposed by Kaufman et al. is not applicable in the CMP process using slurry containing hydrogen peroxide as an oxidizer [11]. For relatively more chemically active metals, such as copper, a corrosion inhibitor, e. g., benzotriazole (BTA), is generally used to protect topographically recessed regions from being further etched and mechanically abraded [19].

30.1.2
Various CMP Processes and the Need for SPM

Applications of CMP can be readily found in the literature of IC fabrication processes [11,20]. These processes entail the interconnection, also called metallization, among transistor elements patterned on a silicon wafer, employing etching, deposition and planarization of materials to produce line patterns.

As the materials used in the fabrication of IC have advanced, so have the CMP processes used for their manufacture. For example, silicon dioxide and aluminum are typical materials used as interdielectric and interconnect metal, respectively. Recent advances have introduced low-k dielectrics and copper as the materials of IC construction owing to their improved resistance–capacitance delay performance. The reactive ion etching technique used for aluminum line pattern formation cannot be utilized for copper, because copper does not produce compounds with a vapor pressure high enough to be readily removed from the surface. In order to overcome this difficulty, a damascene CMP technique was introduced in 1975 [21] and later proposed by the IBM research group for IC wafer planarization [17]. The word, "damascene" was coined referring to artistic decoration with gold or silver line patterns on iron or steel initiated in the early period. This technique ushered in a new era of CMP technology.

Figure 30.3 illustrates a copper damascene CMP process. The interlayer dielectric is first etched for pattern formation, followed by the deposition of a thin barrier layer (approximately 250 Å) made typically of Ta or TaN and a copper seed layer. The barrier layer prevents copper from diffusing into silicon dioxide. Copper is then electroplated to fill the pattern with excess coverage of the whole surface. CMP is used to remove the excess copper and the barrier metal, leaving the copper line patterns designed. Copper CMP is usually performed using a two-step process. In the first step, called "bulk removal", the excess copper layer is removed at high polishing rate. After a certain amount of the layer has been removed, the second process, called "soft landing", is conducted using a slurry with a more moderate removal rate. During this step, the barrier layer starts to be revealed and the polishing is continuously performed until all excess copper has been removed from the surface. The duration of the CMP process, defined as the time difference between initial barrier layer appearance and total excess copper layer removal, is called the overpolish time [22]. The overpolishing is effective in the case where the thickness of the copper deposited varies across the wafer; however, as the overpolish time becomes longer, copper at some regions can suffer from being overpolished. This results in a nonplanarized

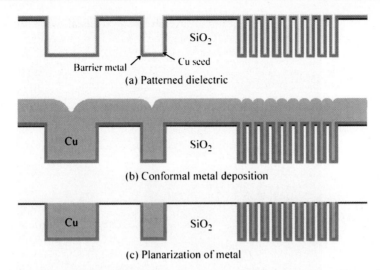

Fig. 30.3. A copper damascene CMP process. **a** Line patterns for interconnection are formed on the dielectric layer by etching silicon dioxide and depositing thin barrier and copper seed layers; **b** copper is electroplated; **c** excess copper and barrier metal are removed by CMP, leaving planarized copper lines as designed

surface defined as dishing and erosion. SPM play an important role in characterizing these topographical features. These issues are detailed in Sect. 30.2. After the second step copper CMP, the next step is a barrier layer removal. The polishing performance of a barrier material is very different from that of copper; therefore, a slurry is usually switched to the one formulated for barrier layer polishing. During this step, the surface is composed of mixed phases including barrier material, copper and silicon oxide. Therefore, the use of an optimum slurry with the desired selectivity in material removal is important in order to obtain a highly planarized surface. During the barrier CMP, copper and silicon oxide tend to become overpolished, leading to the dishing/erosion issues. Moreover, the barrier metal layer needs to be completely removed; otherwise an electrical short may occur. To characterize the metal layer residue, the use of a conductive AFM technique has been reported [23].

Another important application of metal CMP is the polishing of tungsten. In microelectronic devices, tungsten is used as a stud that connects the interlayer metal and the complementary metal oxide semiconductor (CMOS) field-effect transistor electrodes. As shown in Fig. 30.4, studs have a relatively deep and narrow geometry. Tungsten is preferably used to fill studs since it allows conformal deposition in the chemical vapor deposition process with the aid of fluorine gas. In a similar manner to copper removal, excess tungsten is removed using a damascene CMP process. SPM can be used to characterize the planarized tungsten stud geometry and alignment in spatially limited regions.

CMP has traditionally been restricted to the back end of line processes, i. e., fabrication processes that occur after the construction of CMOS structures (Fig. 30.4). In the front end of line (FEOL) processes, concentration of residual ions on a silicon substrate is highly restricted. Since CMP slurry contains various chemicals,

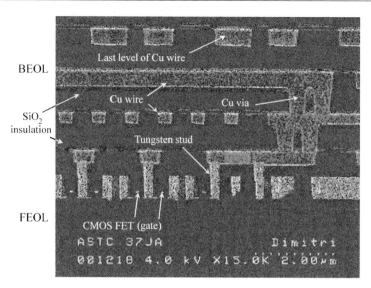

Fig. 30.4. A cross-sectional SEM image of an integrated circuit (IC) wafer. The fabrication process is categorized by front end of line (*FEOL*) and back end of line (*BEOL*) processes. The FEOL process contains the formation of shallow trench isolation (STI; see text) and transistor elements (*CMOS FET*, complementary metal oxide semiconductor field-effect transistor). The BEOL process allows multilayered intermetal connections consisting of tungsten studs, copper wires (lines) and vias. Studs connect the interlayer metal lines and the transistor electrodes, and vias bond interlayer lines. (Courtesy of Ernest Levine, University of Albany)

contamination issues were a concern for the application of CMP to the FEOL processes. Recent advances in the post-CMP cleaning process and related techniques allow CMP to be applied in the FEOL processes, such as the formation of shallow trench isolation (STI) [24, 25]. Isolation techniques are necessary for IC fabrication, which limit the electrical crosstalk between CMOS devices. The local oxidation of silicon (LOCOS) technique has been commonly used for this purpose; however, as the size of CMOS shrinks, alternative techniques need to be developed [11]. The STI technique utilizing CMP is a promising solution. Figure 30.5 illustrates pre- and post-STI CMP. STI structures consist of shallow trenches, filled with silicon dioxide (TEOS-based oxide). Thermal oxide is usually grown on a silicon substrate prior to the TEOS-oxide deposition. A silicon nitride layer is deposited on top of the silicon regions, which contributes as a material removal stopping layer. Similar to a copper damascene CMP, the excess silicon dioxide is polished until the nitride layer appears. A silica or ceria particle based slurry is used which demonstrates high selectivity in removal rate to silicon dioxide compared with silicon nitride; therefore, further material removal at silicon nitride regions is limited. During this process, dishing (in this case, overpolishing of silicon dioxide) is likely to occur. The nitride layer is chemically etched out after CMP and before CMOS structure formation on the silicon regions. SPM are found to be effective in the measurement of STI steps created at the border of silicon dioxide and the surrounding silicon substrate.

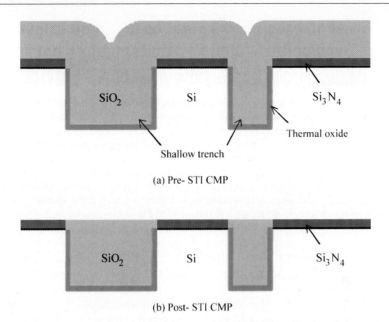

Fig. 30.5. STI damascene CMP process. **a** Shallow trenches filled with silicon dioxide are formed on a silicon wafer; **b** CMP eliminates the excess silicon dioxide so that STI is formed. Silicon nitride, used as a stopping layer of polishing, will be etched out in the next process and IC elements are constructed on the silicon area

30.2
AFP for the Evaluation of Dishing and Erosion

Metal loss caused by dishing and erosion is one of the most critical issues during metal CMP. The evaluation of dishing/erosion of a CMP process is often conducted using Sematech test pattern wafers [11]. These standardized wafers contain repeating dies with an individual size of about 20 mm square in which various metal line patterns are contained. The line patterns are specified with respect to line width, line space, line pitch and line density [11, 26, 27]. The line width refers to the width of metal lines and the line space is defined as the width between two adjacent metal lines, or the width of the dielectric. The sum of the line width and the line space is the line pitch. The line density can be calculated from the ratio of the line width to the line pitch, typically represented as a percentage.

As shown in Fig. 30.6, dishing is a metal line recess with respect to the surrounding dielectric. This occurs because the metal is more easily removed than the surrounding oxide materials. The amount of dishing can be measured as the vertical distance near the center of the line to the level of the oxide material. Dishing has a strong line width dependency [1]. The amount of dishing becomes larger as the line width increases. This can occur owing to more concentrated pressure exerted by a conformal pad. Therefore, dishing is typically evaluated in relatively wide line width regions, such as 50- or 100-μm lines.

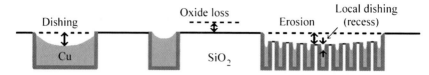

Fig. 30.6. Dishing, erosion and oxide loss in copper CMP. Dishing is a metal line recess with respect to the surrounding dielectric. This occurs when the metal is more easily removed than the surrounding oxide materials. The amount of dishing can be measured as the vertical distance near the center of the line to the level of the oxide material. Erosion is a material loss of dielectric occurring at mixed line patterns. At each line, local dishing (recess) may be expected to occur as illustrated. Dielectric loss is measured at regions mainly dominated by silicon dioxide. Total metal loss at a location is the sum of oxide loss, erosion and local dishing. The topographic variation caused by dishing/erosion is characterized using scanning probe methods. Note that the degree of topographical variations with respect to line thickness is not drawn to scale

Erosion is a material loss of dielectric occurring at mixed line patterns. The amount of erosion can be measured in a similar manner to the measurement of dishing. At each line, local dishing (recess) may be expected to occur as illustrated in Fig. 30.6. The amount of erosion increases with line density [1, 28]; therefore, erosion is evaluated typically at higher line density regions, such as 9 μm/1 μm line (9-μm line width, 1-μm line space, i.e., 90% line density). The requirement of the IC industry specifies a dishing of less than 20 nm over 100 μm in line width; and an erosion of less than 10 nm in 90% dense line arrays [22]. In this regard, the copper loss caused by dishing is typically several percent of the 500-nm thickness of the original copper line.

For the evaluation of dishing and erosion, nanometer-scale lateral and vertical resolution is required. A mechanical stylus profiler or an AFM technique [5, 29] can be used for this application. However, for a stylus profiler, its limited lateral resolution due to the tip geometry constraints makes the measurement more challenging, especially for small line widths. In addition, the lateral force generated by a stylus during the scan may cause mechanical damage to the wafer surface. The challenge of using an atomic force microscope is its limited scan length. Though the scan length required for the measurement of dishing and erosion is in the millimeter range, the allowable maximum scan length for an atomic force microscope is typically less than 100 μm as regulated by the piezo system used. To overcome this difficulty, an atomic force profiler has been developed with a long scan length capability (up to 100 mm) and AFM-level resolution [30, 31]. A sample stage is driven in the translated direction with a profiling speed up to 200 μm/s, while a scanning head, which is the same as that used in an atomic force microscope, is in operation. The surface height profile is monitored versus scan length. A scanning tip operated in the tapping mode significantly reduces the drag force on a sample surface, leading to the sample damage being minimized. Recent AFP systems have more advanced features. For example, the Vx series systems (Veeco Instruments) are equipped with automated tip replacement, pattern recognition and automated data analysis capabilities [32]. These technical improvements extend AFP/AFM to applications in production lines as well as laboratory usage.

Figure 30.7 presents examples of line profiles measured using an atomic force profiler (Vx310, Veeco Instruments) for 854 Sematech test wafers. The structure contains two isolated lines on its left side and a 50% density array region on its right side (10- and 5-μm line width for panels a and b in Fig. 30.7, respectively). Schematic illustrations of the expected cross-sectional structure are also shown together with the profiles at the bottom (not to scale). The examples shown in Fig. 30.7, panels a and b were obtained for the test wafers polished with two different kinds of barrier slurry. Since the slurry used in the example shown in Fig. 30.7, panel a had a large copper polishing rate, the metal lines are recessed with respect to the silicon dioxide surface, resulting in severe dishing/erosion. In contrast, the slurry used in the example in Fig. 30.7, panel b had a low copper and moderate silicon dioxide removal rates; and as a result, the copper protrudes as indicated by the spikes rising above the reference regions on the plot. The slurry formulation can be optimized on the basis of these topographic data obtained using AFP.

Total metal loss is a sum of dielectric loss, erosion and local dishing at dense metal line regions (Fig. 30.6). The erosion and dishing are measured using AFP as described. The dielectric loss can be measured using an optical tester (e. g., UV1050, KLA-Tencor). The sum of those provides the total metal loss (mechanical method). Here, a typical optical tester uses a monochromatic ray as a probe and detects reflected light. The intensity of the reflected light shows constructive/destructive patterns as the wavelength of the ray is swept. The pattern obtained specifies the film thickness.

As the total metal loss becomes large by the CMP process, the resistance of the metal line increases. The resistance can be generally measured using a four-point

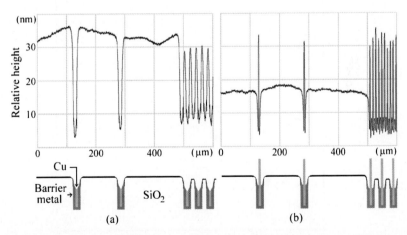

Fig. 30.7. Surface topography profiles acquired by an atomic force profiler (AFP) for 854 Sematech test wafers polished with two different types of barrier slurry. The expected cross-sectional structures are also shown at the *bottom* (not to scale). The structure contains two isolated lines on its left side and a 50% density array region on its right side (10- and 5-μm line width for **a** and **b**, respectively). The wafers were polished with **a** high and **b** low copper removal rate slurries. The copper lines are recessed with respect to the silicon dioxide surface in **a**; those protrude as indicated by the spikes rising above the reference regions in **b**, indicating the different performance of the CMP slurry used

probe electrical measurement technique [33]. Sematech test wafers are designed for resistance measurement at various line patterns. Sheet resistance (R_s) is often used for comparison among various line patterns, and is calculated with line resistance together with geometries of the metal line:

$$R_s = \frac{R}{l/w} = \frac{\rho}{t}, \quad (30.2)$$

where R is the line resistance, l is the line length, w is the line width, t is the line thickness and ρ is the resistivity of the metal. From (30.2), the metal line thickness, t, can be evaluated from R_s. If the thickness of the metal line initially deposited is known, e.g., 500 nm for an 854 Sematech test wafer, the total metal loss can be obtained (electrical method).

The total metal losses evaluated using the mechanical and electrical techniques were compared as follows. Polishing experiments were conducted on 854 Sematech test wafers with four kinds of slurry. Those slurries had various removal rates for copper, barrier metal and silicon dioxide, so the amounts of dielectric loss, erosion, local dishing and resulting metal loss were expected to be different. The total copper losses calculated using the two methods mentioned above were compared at a 10 μm/10 μm copper line. As shown in Fig. 30.8, they exhibit good agreement. This result suggests reliable total metal loss data provided by AFP in combination with an optical tester.

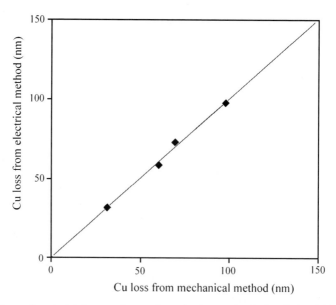

Fig. 30.8. Comparison in total copper loss evaluated using mechanical and electrical characterization techniques. The total copper loss (see Fig. 30.6) is the sum of oxide loss (by an optical tester) and erosion plus local dishing (by an AFP). It is also evaluated using a four-point probe electrical measurement technique. The good agreement suggests that reliable total metal loss data are provided by the AFP used in combination with an optical tester

30.3
Surface Planarization and Roughness Characterization in CMP Using AFM

Figure 30.9 illustrates the shape of a step before and after CMP. Planarization is evaluated using two parameters: step height reduction ratio (SHRR) and planarization efficiency (EFF), defined as [11]

$$\text{SHRR} = 1 - \frac{\text{SH}_{\text{final}}}{\text{SH}_{\text{initial}}} = \frac{\Delta}{\text{SH}_{\text{initial}}}, \tag{30.3}$$

$$\text{EFF} = 1 - \frac{T_{\text{down}}}{T_{\text{up}}} = \frac{\Delta}{T_{\text{up}}}, \tag{30.4}$$

$$\Delta = T_{\text{up}} - T_{\text{down}} = \text{SH}_{\text{initial}} - \text{SH}_{\text{final}}, \tag{30.5}$$

where $\text{SH}_{\text{initial}}$ and SH_{final} are the step height before and after CMP, and T_{up} and T_{down} refer to the thickness of the material removed from topographically elevated and recessed areas, respectively. In the case of a perfectly planarized surface, $\text{SH}_{\text{final}} = 0$ and SHRR = 1. Typically, the value of SHRR ranges from 0.95 to 1.0. EFF describes how much material has been removed when the required step height reduction is achieved. For a simplification, suppose that $\text{SH}_{\text{final}} = 0$, which gives the relationship $\text{SH}_{\text{initial}} = T_{\text{up}} - T_{\text{down}}$. If T_{down} is large, T_{up} should be large enough to achieve $\text{SH}_{\text{initial}}$ so that the surface is planarized. The larger T_{up} leads to more polishing time being required. This is the reason the parameter represents efficiency. The values of EFF range typically from 0.4 to 0.8 depending on the line density [11]. Note that the removal rate at elevated regions decreases with polishing time. In contrast, the removal rate at recessed regions increases with polishing time. Those rates become closer as the surface is being planarized. Both SHRR and EFF depend on multiple factors, such as step geometries, slurry and pad type, polishing machine parameters and materials to be removed. AFM is used to measure the step height, allowing a precise evaluation of SHRR and EFF. As the feature size becomes smaller, AFM-level resolution is required to obtain well-resolved measurements.

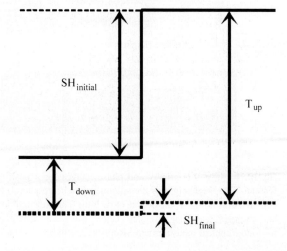

Fig. 30.9. Change in step height profile to define the step height reduction ratio (SHRR) and the planarization efficiency (EFF). Typically, SHRR ranges from 0.95 to 1.0. EFF represents the efficiency of planarization by describing how much material is removed to achieve the required step height reduction

In an STI formation process, the SHRR is critical and is evaluated using an atomic force microscope. Figure 30.10 shows AFM topography and phase images for STI pattern features after CMP with the nitride layer stripped [34]. The STI areas, made of silicon dioxide, correspond to the darker contrast in topography in Fig. 30.10a compared with the remaining silicon area. This feature is defined here as a positive step. On the other hand, the STI regions are topographically elevated and then the boundary forms a negative step in Fig. 30.10b. This positive/negative step formation results from a nitride layer deposited on the silicon area. As mentioned in Sect. 30.1.2, during an STI CMP, silicon dioxide is removed until the silicon nitride layer appears (Fig. 30.5). Further material removal is suppressed at the silicon nitride layer. If polishing is continued, the STI area, made of silicon dioxide, can be overpolished, leading to dishing. The nitride layer is chemically etched away after CMP, and if the dishing exceeds the thickness of the silicon nitride layer, a positive step is formed, and if not, a negative step is likely to occur. The step height of the silicon area with respect to the STI becomes either positive or negative depending on the degree of dishing [35, 36]. These two cases are represented in Fig. 30.10 by opposite contrasts in the topography images. The figure also shows phase images that can be obtained simultaneously with topographic images. The contrast represents a phase lag between sinusoidal driving voltage input to the atomic force microscope tip and a corresponding response of the tip vibration during scanning. It is dependent, typically, on viscoelastic properties of samples [37,38] and generally not on topography. However, phase signals can be affected by localized abrupt topographical variations, such as a step [39]. As shown in Fig. 30.10, clear and identical fringe patterns at both positive and negative steps are seen in the phase images. Lagus and Hand [34] used these phase images to identify the locations of the STI and silicon areas. Then the height data averaged out within the two regions

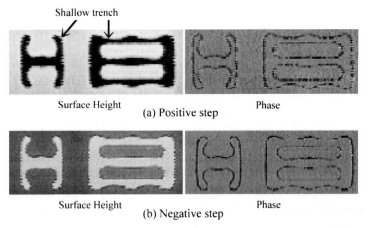

Fig. 30.10. AFM topography and phase images for **a** positive and **b** negative STI test pattern features. The step height of the silicon area with respect to the STI becomes either positive or negative depending on the degree of dishing. The clear and identical fringes of the phase images at both positive and negative steps can be used as a recognized pattern to find specific locations for topographical measurement. (Reprinted with permission from [34])

generated the step height. They performed STI step height measurements in several locations on a wafer for statistical analysis.

The average topographic height variation in an area is usually represented as either the arithmetic roughness (R_a) or the root mean square roughness (R_q), and is classified as roughness, waviness or flatness depending on the spatial frequency ranges [29]. AFM is the most practical technique that provides roughness at nanometer-scale lateral and z-height resolution, in spite of the recent technological improvements made in scatterometry, laser doppler vibrometry and laser reflection spectrometry [40]. Those optical techniques have a limited spatial resolution regulated by the spot size or the wavelength of the beam used.

Improvements in surface roughness, as measured by AFM, are also a concern in the CMP of IC wafers. The effect of polishing time, slurry particle type and size on roughness has been extensively studied [41]. For rigid disks used in hard disk drives, roughness is a key parameter affecting device performance [4, 29]. The high frequency roughness, ranging from 0.1 to 5 μm in spatial wavelength, significantly affects the physical spacing between the read/write head and the disk surface. It, therefore, needs to be low enough to achieve mechanical reliability of the system [4]. The high frequency roughness can be measured as R_a using an atomic force microscope and is often referred as AFM R_a. The CMP process plays a significant role in the control of AFM R_a. Figure 30.11 depicts the effect of abrasive particle size on AFM R_a. With use of smaller particles, AFM R_a can be effectively reduced. It is noted that percentage of solid shows no clear evidence for it contributing to the reduction of roughness, though it affects the removal rate.

Technological improvements have allowed more convenient use of AFM in this application. Characterization of defects such as scratches and pits on a wafer using an

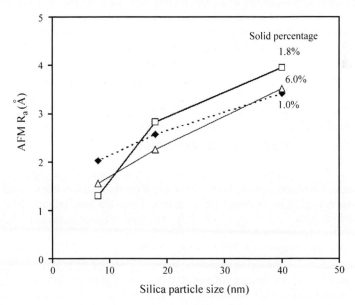

Fig. 30.11. Effect of abrasive particle size on AFM R_a. Samples are rigid disks polished with a CMP slurry. The use of smaller particles reduces AFM R_a effectively

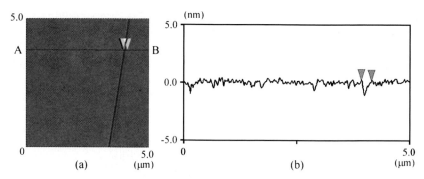

Fig. 30.12. a AFM image of a scratch on a rigid disk after CMP and b section analysis along the line *AB* of the scratch indicated by *arrows*. The width and the depth of micro scratches are found to be typically approximately 1 µm and several nanometers, respectively

atomic force microscope is of primary interest, but it is hard to identify the locations owing to the limited field of view of an atomic force microscope. An advanced stage navigation system for the atomic force microscope combined with a laser-assisted defect-detection system (e. g., KLA Tencor SP-1 and Candela) allows locations of interest to be assessed easily. The example presented in Fig. 30.12 shows an AFM image of a scratch on a rigid disk after CMP. The width and the depth of micro scratches are typically approximately 1 µm and several nanometers, respectively.

AFM has also been used in combination with near-field scanning optical microscopy (NSOM). The NSOM technique uses a subwavelength (20–200-nm) aperture driven by a piezo scanner and maintained in the near-field of a sample. High lateral resolution images are available mainly regulated by the aperture size [42]. It is nondestructive and sensitive to changes in optical properties such as index of refraction. For post-CMP surfaces which show almost flat features with some pitting by an AFM image, an NSOM image can reveal more detailed surface structures, such as micro scratches, because these features have sensible variations in the index of refraction which are emphasized in the NSOM images [43].

30.4
Use of Modified Atomic Force Microscope Tips for Fundamental Studies of CMP Mechanisms

An atomic force microscope can be used to study fundamental aspects of CMP processes at the microscale and the nanoscale. Recent advances allow single-phase abrasive particles to be mounted on an atomic force microscope cantilever. Using this modified atomic force microscope tip, one can examine adhesion, friction and wear via the interaction between the particles and counter materials in a well-controlled environment. An SEM image of ceria nanoparticles (about 50 nm in diameter) attached to the top of an atomic force microscope tip is given in Fig. 30.13, as an example [44]. In the case of simulating a silica particle, an oxidized, single-crystal silicon or silicon nitride atomic force microscope tip can be used, because the contact radius of an atomic force microscope tip is on the order of the size of a silica particle [45]. For a silicon nitride tip, the tip becomes hydrolyzed in an aqueous

Fig. 30.13. SEM image of ceria nanoparticles attached to the top of an atomic force microscope tip. Using this modified atomic force microscope tip, one can examine adhesion, friction and wear via the interaction between the particles and counter materials in a well-controlled environment. (Courtesy of Igor Sokolov, Clarkson University)

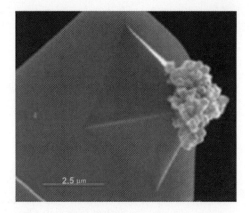

solution, and has a layer of silicon dioxide on the top. Standard AFM methods have been well established for the measurement supported by data interpretation techniques and associated contact models [5, 10, 26, 27, 46].

The physical and chemical properties of two surfaces in a liquid phase, as well as associated structural and chemical changes during interaction of the surfaces, contribute to the interaction forces generated between the surfaces in contact or near contact below a few nanometers. In the case of the modified-tip techniques in a CMP environment, the van der Waals attraction is generally reduced, and long-range electrostatic forces and other short-range forces, mainly originating from chemical bonding, can predominate [5, 27]. Electrostatic attractive/repulsive interaction forces can increase/decrease friction and wear between two surfaces. The charging of particles plays an important role in determining the electrostatic forces generated on the single particle modified atomic force microscope tip. A particle mounted on a cantilever generates attractive or repulsive forces when the atomic force microscope cantilever is driven vertically toward or against the surface. The forces can be estimated by measuring the deflection of the cantilever [5, 47, 48].

Sokolov et al. [49] examined the interfacial forces exerted on silicon nitride and silica covered atomic force microscope tips against polyurethane surfaces in aqueous solutions at different pH. Polyurethane is a material typically used for a polishing pad, so the interaction between the pad material and abrasive particles is of interest. Figure 30.14 shows the estimated surface potential of the polyurethane pad derived from the force measurements and the zeta potential obtained using a streaming potential method over the same pH range. Surface potential originates from the concentration profiles of ions and electrons at the interface between the bulk material and the surrounding environment [50]. It depends on various factors, such as surface condition and additional charges present on the surface [51, 52]. Zeta potential is the potential generated at the boundary between charged particles and surrounding liquid, generally used as a macroscopic indicator of electrostatic forces exerted on charged particles [1]. It can be obtained by measuring the mobility of particles under externally applied electric field. For example, when the zeta potential is negative, the particle surfaces are interpreted as being negatively charged. The pH of a liquid is known to significantly affect the zeta potential, which becomes more positive at lower pH and is zero at the isoelectric point. The similar trend between

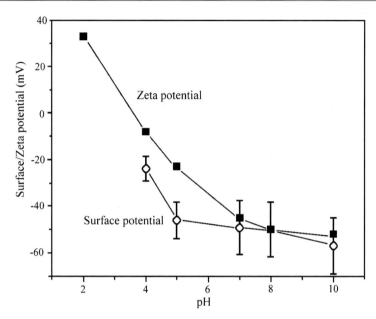

Fig. 30.14. Variation with pH of the estimated surface potential of a polyurethane pad derived from the modified atomic force microscope tip force measurement and the zeta potential. Surface potential is affected by the surface charge and zeta potential is the potential generated at the boundary between charged particles and the surrounding liquid, generally used as a macroscopic indicator of electrostatic forces exerted on charged particles. The similar trend confirms the validation of the microscopic atomic force microscope tip force measurement technique. (Reprinted with permission from [49])

the surface potential and the zeta potential in Fig. 30.14 confirms the validation of the microscopic atomic force microscope tip force measurement.

Hong et al. [53] found a significant change in zeta potential for alumina particles when citric acid was added to the liquid. The zeta potential was originally positive in deionized water at neutral pH, which indicated that the particles were positively charged. The potential became negative in the presence of citric acid, suggesting conversion of the surface charge. The adsorption of citrate ions on the alumina surfaces is the likely cause of the conversion of the surface charge. Microscopic scale tests using an alumina-mounted atomic force microscope tip against a copper surface were then conducted. The results showed a significant reduction in adhesion in the presence of citric acid. Since the surface of copper tends to be negatively charged in a liquid [54], strong attractive force would be generated in the presence of positively charged particles. If the conversion of surface charge occurs from positive to negative for the particles, the associated repulsive force would lead to a reduction in adhesion. It was also reported that a reduction in the macroscale friction and scratch counts was found on the copper surface polished with a slurry containing citric acid and alumina particles. For silica particles, the zeta potential exhibited a steady decrease with an increase in pH. More repulsive forces were found between a silica particle mounted atomic force microscope tip and a silicon dioxide wafer with pH.

Macroscopic friction force measurements showed a decrease in friction, consistent with the decrease in the zeta potential [55].

Similar studies have been reported in the case of ceria-based slurries. These slurries are commonly used for silicon dioxide CMP in STI fabrication. The removal rate is found to be significantly greater than that obtained using silica-based slurry [56, 57]. Abiade et al. [58] found a strong attractive force at pH 4.5 between a ceria thin film and a silicon dioxide particle attached to an atomic force microscope cantilever. They additionally examined a macroscopic friction force generated between a silicon dioxide wafer and a polishing pad in CMP with the presence of either silica-based or ceria-based slurry. The friction force profile showed a sharp peak at pH 4.5 for the ceria-based slurry, but a monotonic decrease for the silica-based slurry. It was concluded that the peak in the friction force occurred possibly owing to the electrostatic attraction between ceria particles and silicon dioxide on a wafer. On the basis of this finding, a mechanical removal process assisted by electrostatic interactions was hypothesized in this pH range as the primary material removal mechanism, rather than the chemical reaction processes previously proposed [57].

The surface charging of particles is also important in the cleaning of wafers after polishing. If the surface and the particles are oppositely charged, the particles may stick to the wafer. Detachment of chemically or physically adsorbed colloidal silica particles from a wafer sample can be accomplished more easily under higher-pH conditions, owing to the particle charging issue [1]. Lee et al. [59] found a smaller adhesive force between a silica-mounted atomic force microscope tip and a low-k dielectric material, compared with other material, such as silicon dioxide, copper and barrier metal. The order of adhesion force measured correlated with the cleaning ability of the surface after CMP.

With addition to the adhesive force measurement, an atomic force microscope allows investigation regarding fundamental aspects of material removal mechanisms in CMP, as described in the literature. An enhanced removal rate of deposited aluminum films was seen using a scanning atomic force microscope tip operated in a corrosive solution environment [60]. DeVecchio et al. [61] extended this technique and found an increased removal rate on an aluminum surface as a function of down force and the concentration of an oxidizing agent. These findings are consistent with general results observed in macroscopic CMP experiments. Sokolev et al. [62] used this atomic force microscope tip technique to examine copper removal processes. The removal rate increased with a decrease in pH, which was consistent with CMP results reported previously. Katsuki et al. [63] found an equivalent molar wear volume for a silicon tip sliding against silicon dioxide film. They claimed that these results provided strong evidence of Si–O–Si bridge formation and its breakage during the material removal in CMP.

In the characterization of CMP polished surfaces, the scanning KPM technique has been recognized as a robust technique. The method measures the surface potential, influenced by associated chemical and/or structural changes on the surface. DeVeccio and Bhushan [64] applied this technique to detect early stages of wear. They abraded a polished single-crystal Al(100) sample using a diamond tip at loads of 1 and 9 µN, and scanned the surface with the KPM and the standard surface height AFM modes. As shown in Fig. 30.15, the surface height map shows a clear wear mark at the region abraded under 9-µN load, but no detectable changes at the area scratched under 1 µN. On the other hand, the surface potential map reveals an apparent potential contrast (approximately 0.17 V) with respect to the nonabraded area in both regions, clearly indicated in the bottom figure, closeup images of the upper (1-µN load) wear region. The KPM technique can be also used for identifying intermetallic particles on an aluminum matrix which is not readily recognized in AFM topography images. Schumtz and Frankel [65] used this technique and found that abrasion of a 2024-T3 aluminum alloy sample by an atomic force microscope tip in the contact mode resulted in an immediate dissolution of the Al–Cu–Mg inter-

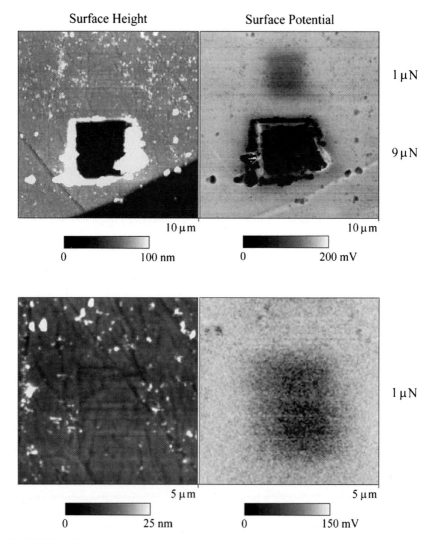

Fig. 30.15. *Top*: Surface height and surface potential maps of wear regions generated at 1 µN and 9 µN on a single-crystal aluminum sample. Noticeable change is found in the surface potential but not in the surface height for the worn region under 1-µN load. *Bottom*: Closeup of the upper (1-µN load) wear region. (Reprinted with permission from [64])

metallic particles. As suggested, one can study material removal initiation process and localized corrosion by applying the KPM technique to metal CMP processes.

To date, computer simulation techniques have been enormously advanced in various fields. Chandra et al. [66] used molecular dynamics simulations that allow the prediction of the atomic-scale asperity interaction with a surface. They examined the abrasion between an atomic force microscope size scale cutting tip and a copper surface. The simulations indicated that a smooth and defect-free surface would result when an atomic force microscope tip was swept across the substrate in the presence of chemical dissolution agents. The result suggests that a certain level of chemical reaction is necessary to minimize surface damage during planarization.

30.5 Conclusions

The AFP technique capable of both AFM-level resolution and large scan size has shown its extensive applicability for characterizing the surface topography of post-CMP IC wafers, such as dishing and erosion. For rigid disk polishing, AFM plays an important role in evaluating surface roughness, in which angstrom-level height resolution is required. AFM is also useful for identifying surface defects such as scratches and pits. The topography information obtained is essential, especially for CMP slurry development.

With the use of a modified atomic force microscope cantilever tip, one can examine microscopic scale attractive/repulsive forces with a substrate material in a controlled environment. The results exhibit good agreement with associated macroscopic CMP properties such as the zeta potential of abrasive particles, friction and material removal rate. On the basis of the wealth of representative data for a variety of systems, SPM provide interesting insights into the CMP process and allow the interpretation of the macroscopic phenomena in terms of microscopic mechanisms.

SPM have derived techniques found to be useful in CMP. Examples include the phase-imaging mode that can be used for pattern recognition of the STI post-CMP surface and the KPM technique that is sensitive to localized corrosion and very early stages of wear. The development of other capabilities/applications using SPM, such as electrical applications briefly given in the chapter, will be a key for the further improvement of CMP technology.

References

1. Steigerwald JM, Murarka SP, Gutmann RJ (1997) Chemical mechanical planarization of microelectronic materials. Wiley, New York
2. Toenshoff HK, Schmieden WV, Inasaki I, Koenig W, Spur G (1990) CIRP Ann 39:621–636
3. Ahearne E, Byrne G (2004) Proc Inst Mech Eng Part B 218:253–267
4. Bhushan B (1996) Tribology and mechanics of magnetic storage systems, 2nd edn. Springer, New York
5. Bhushan B (ed) (1999) Handbook of micro/nanotribology, 2nd edn. CRC, Boca Raton
6. Fang M, Wang S (2002) US Patent 6,468,137B1
7. Hagihara T, Naito K, Fujii S (2002) US Patent 6,454,820B2

8. Kasai T (2007) Tribol Int 41:111–118
9. Eaton WP, Smith JH (1996) Proc SPIE 2882:259–265
10. Bhushan B (ed) (2007) Springer handbook of nanotechnology, 2nd edn. Springer, Heidelberg
11. Oliver MR (ed) (2004) Chemical mechanical planarization of semiconductor materials. Springer, Heidelberg
12. Patrick W, Guthrie WL, Standley CL, Schiable PM (1991) J Electrochem Soc 138:1778–1784
13. Xie Y, Bhushan B (1996) Wear 200:281–295
14. Xie Y, Bhushan B (1996) Wear 202:3–16
15. Luo J, Dornfeld D (2003) IEEE Trans Semicond Manuf 16:469–476
16. Remsen EE, Anjur S, Boldridge D, Kamiti M, Li S, Johns T, Dowell C, Kasthurirangan J, Feeney P (2006) J Electrochem Soc 153:G453–G461
17. Kaufman FB, Thompson DB, Broadie RE, Jaso MA, Guthrie WL, Pearson DJ, Small MB (1991) J Electrochem Soc 138:3460–3465
18. Patri UB, Pandija S, Babu SV (2005) Mater Res Soc Symp Proc W1.11
19. Kim IK, Kang YJ, Hong YK, Park JG (2005) Mater Res Soc Symp Proc W1.3
20. Luo J, Dornfeld D (2004) Integrated modeling of chemical mechanical planarization for sub-micron IC fabrication. Springer, Heidelberg
21. Konishi K (1975) US Patent 3,895,391
22. Singer P (2004) Semicond Int 27(10):38–42
23. Dominget A, Farkas J, Szunerits S (2006) Appl Surf Sci 252:7760–7765
24. Pan JT, Ouma D, Li P, Boning D, Redeker F, Chung J, Whitby J (1998) VMIC Proc, pp 467–472
25. Fiorenza JG (2003) IEEE Electron Device Lett 24:698–700
26. Bhushan B (2005) Wear 259:1507–1531
27. Bhushan B (ed) (2005) Nanotribology and nanomechanics–an introduction. Springer, Heidelberg
28. Chen Z, Sun F, Vacassy R (2006) J Electrochem Soc 153:G582–G586
29. Bhushan B (1996) IEEE Trans Magn 32:1819–1825
30. Ge LM, Cunningham T, Dawson DJ, Serry FM, Heaton M (1999) CMP-MIC Proc, pp 69–72
31. Cunningham T, Dawson D, Ge L, Heaton MG, Serry FM (2001) Atomic force profiler for characterization of chemical mechanical planarization. http://www.veeco.com/library/
32. Braun AE (2004) Semicond Int 27(6):38
33. Li SH, Miller RM, Willardson RK, Weber ER (eds) (1999) Chemical mechanical polishing in silicon processing. Academic, San Diego
34. Lagus ME, Hand SM (2001) VMIC Proc, pp 181–183
35. Lim DS, Ahn JW, Park HS, Shin JH (2005) Surf Coat Technol 200:1751–1754
36. Nam CW, Shen J (2006) CMP-MIC Proc, pp 186–189
37. Scott WW, Bhushan B (2003) Ultramicroscopy 97:151–169
38. Kasai T, Bhushan B, Huang L, Su C (2004) Nanotechnology 15:731–742
39. Schmitz I, Schreiner M, Friedbacher G, Grasserbauer M (1997) Appl Surf Sci 115:190–198
40. Garnaes J, Kofod N, Kuehle A, Nielsen C, Dirscherl K, Blunt L (2003) Precision Eng 27:91–98
41. Chandrasekaran N, Ramarajan S, Lee W, Sabde GM, Meikle S (2004) J Electrochem Soc 151:G882–G889
42. Weston KD, Buratto SK (1997) J Phys Chem B 101:5684–5691
43. Lewis A, Shambrot E, Radko A, Lieberman K, Ezekiel S, Veinger D, Yampolski G (2000) Proc IEEE 88:1471–1479
44. Ong QK, Sokolov I (2007) J Colloid Interface Sci 310:385–390

45. Stevens F, Langford S, Dickinson JT (2003) Mater Res Soc Symp Proc 767:F2.1
46. Kim TW, Bhushan B (2006) ASME J Tribol 128:851–864
47. Lin XY, Creuzet F, Arribart H (1993) J Phys Chem 97:7272–7276
48. Cooper K, Eichenlaub S, Gupta A, Beaudoin S (2002) J Electrochem Soc 149:G239–G244
49. Sokolov I, Ong QK, Shodiev H, Chechik N, James D, Oliver M (2006) J Colloid Interface Sci 300:475–481
50. Badiali JP, Rosinberg ML, Goodisman J (1981) J Electroanal Chem 130:31–45
51. Bockris JO, Khan SUM (1993) Surface electrochemistry: a molecular level approach. Plenum, New York
52. Kasai T (2001) Using an *in-situ* Kelvin probe to study changes in sliding metal surfaces. PhD thesis, The Ohio State University
53. Hong YK, Han JH, Song JH, Park JG (2005) Mater Res Soc Symp Proc 867:W6.2
54. Jones DA (1996) Principles and prevention of corrosion, 2nd edn. Prentice Hall, Upper Saddle River
55. Choi W, Lee SM, Singh RK (2004) Electrochem Solid-State Lett 7:G141–G144
56. Chandrasekaran N (2004) Mater Res Soc Symp Proc 816:K9.2
57. Evans DR (2004) Mater Res Soc Symp Proc 816:K9.1
58. Abiade JT, Yeruva S, Choi W, Moudgil BM, Kumar D, Singh RK (2006) J Electrochem Soc 153:G1001–G1004
59. Lee SY, Lee SH, Park JG (2003) J Electrochem Soc 150:G327–G332
60. Chen L, Guay D (1994) J Electrochem Soc 141:L43–L45
61. DeVecchio D, Schmutz P, Frankel GS (2000) Electrochem Solid-State Lett 3:90–92
62. Berdyyeva TK, Emery SB, Sokolov IY (2003) Electrochem Solid-State Lett 6:G91–G94
63. Katsuki F, Kamei K, Saguchi A, Takahashi W, Watanabe J (2000) J Electrochem Soc 147:2328–2331
64. DeVecchio D, Bhushan B (1998) Rev Sci Instrum 69:3618–3624
65. Schmutz P, Frankel GS (1998) J Electrochem Soc 145:2295–2306
66. Ye YY, Biswas R, Bastawros A, Chandra A (2003) Mater Res Soc Symp Proc 767:K1.8

31 Scanning Probe Microscope Application for Single Molecules in a π-Conjugated Polymer Toward Molecular Devices Based on Polymer Chemistry

Ken-ichi Shinohara

Abstract. π-conjugated polymers are useful substances with excellent performances. To date, we have mainly conducted the following studies. First, we imaged the higher-order structure of a single molecule of a chiral helical π-conjugated polymer. To be more precise, we elucidated the higher-order structure by directly measuring the main-chain chiral helical structure using a scanning tunneling microscope. Second, we discovered a new phenomenon, which is that the color of the fluorescence of a single polymer molecule changes on the order of seconds, by directly observing the dynamic function of photon emission based on the thermal fluctuation of a single molecule of a fluorescent π-conjugated polymer using a total internal reflection fluorescence microscope (TIRFM), which we developed. In addition, we simultaneously imaged the structure and function of a fluorescent π-conjugated polymer. This was achieved using a combination of an atomic force microscope (AFM) and an objective-type TIRFM(AFM-TIRFM). Afterwards, we achieved a high-speed (fast-scan) atomic force microscopy image of the random movement of a single π-conjugated polymer chain in a solution at room temperature. On the basis of an analysis, we have demonstrated that movement, which complies with Einstein's law of Brownian motion, is based on thermal energy. We anticipate that our studies will be the basis for creating innovative molecular devices such as molecular motors or molecular electronics/photonics materials that utilize thermal energy as a driving source.

Key words: Single molecule, π-Conjugated polymer, Polymer synthesis, Scanning probe microscope, Total internal reflection fluorescence microscope, Fast-scan atomic force microscope

31.1 Introduction

What structure does a single molecule of synthetic polymer form at the nanometer level? Does it form a highly ordered structure similar to a biomacromolecule? We, as human beings, do not fully understand the essence of a synthetic polymer, and hence, have yet to take full advantage of synthetic polymers.

Synthetic polymers are very useful materials that display many excellent properties, and they have become indispensable in maintaining and developing our current way of life. Nevertheless, it is difficult to discuss the correlation between their structures and functions at the molecular level, since these are diverse, dynamic, and can be very complex. If the structure and functions of a polymer could be directly observed, with minimal inferences or hypotheses, the relationship between polymer structures and functions could be clarified. Consequently, molecular devices of a polymer might be created based on new design concepts and new working principles.

In the meantime, proteins, biomacromolecules that exist in organisms as the ultimate high-function molecules, play important roles in the expression of life functions. The superb and flexible functions of proteins are considered to be expressed essentially by the flexible higher-order structure and their supramolecular structures. Accordingly, a new study to directly observe single molecules of polymers with various higher-order structures and supramolecular structures has commenced. This study is based on the idea that polymers with superb functions and practical durability may be created if higher-order structures similar to those of proteins can be formed. This chapter chiefly discusses our studies on the single molecule of π-conjugated polymers having peculiar electronic and optical properties that give electron conduction and photon emission functions among other synthetic polymers.

To date, we have mainly conducted the following studies. First, we imaged the higher-order structure of a single molecule of a chiral helical π-conjugated polymer. To be more precise, we elucidated the higher-order structure by directly measuring the main-chain chiral right-handed helical structure on a nanometer scale using a scanning tunneling microscope (STM) [1, 2]. Second, we discovered a new phenomenon, which is that the color of the fluorescence of a single polymer molecule changes on the order of seconds, by directly observing the dynamic function of photon emission based on the thermal fluctuation of a single molecule of fluorescent π-conjugated polymer using a total internal reflection fluorescence microscope (TIRFM), which we developed [3, 4]. In addition, we simultaneously imaged the structure and function of a fluorescent π-conjugated polymer [5]. This was achieved using a combination of an AFM and an objective-type TIRFM. This new microscope can observe long-distance optical communication on the order of micrometers and is based on the interaction of a nanostructure of a π-conjugated polymer and photons in near-field. We also discovered a long periodic structure of a single molecule of a chiral helical π-conjugated polymer using an AFM. We then verified that each polymer chain has distinctive features and simultaneously elucidated its diversity [6]. Afterwards, we achieved a high-speed (fast-scan) AFM image of the random movement of a single π-conjugated polymer chain in a solution at room temperature [7]. On the basis of an analysis, we have demonstrated that movement, which complies with Einstein's law of Brownian motion, is based on thermal energy.

Thus, we have developed a science to elucidate the essence of single molecules of π-conjugated polymer in a flexible and innovative manner without constrictions of the existing academic framework. We anticipate that our studies will be the basis for creating innovative molecular devices such as molecular motors or molecular electronics/photonics materials that utilize thermal energy as a driving source.

31.2
Chiral Helical π-Conjugated Polymer

Polyacetylene is a π-conjugated polymer. As the chemical structures in Fig. 31.1 show, polyacetylene is a polymer where the main chain alternates between double

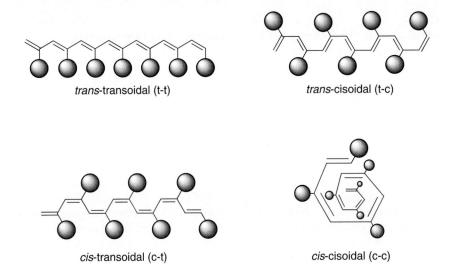

Fig. 31.1. Four conformers of π-conjugated polyacetylene. The *circles* indicate the substitute group

bonds and single bonds. The figure shows the four theoretically predicted types of main-chain structures of polyacetylene, which are the *trans*-transoidal (t-t) structure, the *trans*-cisoidal (t–c) structure, the *cis*-transoidal (c–t) structure, and the *cis*-cisoidal (c–c) structure, respectively. Among these, the t–t and c–t main-chain structures of polyacetylene have been confirmed. On the other hand, substituted polyacetylene with a pendant group introduced has either a t–c or a c–c structure. Because the carbon atoms of polyacetylene main chains all form sp^2 hybridized orbitals, a single p orbital exists on each carbon atom, which stands vertical to the plane formed by the hybridized orbital. Because the p orbitals of adjacent carbon atoms interact, the π electrons that exist on the orbitals are conjugated and can be delocalized, which reduces the energy gap between the highest occupied molecular orbital (HOMO) and the lowest unoccupied molecular orbital (LUMO) of polyacetylene. Hence, an electron state which shows physical properties unique to the π-conjugated polymer is formed. If polyacetylene is combined with a substitute group such as a pendant group, the π-conjugated main chains get twisted owing to steric hindrance. The structure of the substitute group introduced can control the average twist angle, which consequently controls the energy gap between the HOMO and the LUMO. Polyacetylene is easily oxidized and deteriorates in the atmosphere. However, substituted polyacetylene is relatively stable and has better solubility in a solvent. In this section, we discuss substituted polyacetylene. Polyacetylene to which a pinanyl group, a menthyl group, or another bulky optically active group is introduced as a substitute group forms a chiral helical π-conjugated polymer where the main-chain π-conjugation system is twisted in a one-handed helix.

31.2.1
Helical Chirality of a π-Conjugated Main Chain Induced by Polymerization of Phenylacetylene with Chiral Bulky Groups

31.2.1.1
Overview

The helix is one of the most important fundamental structures in polymers [8]. Many stereoregular polymers, including both naturally occurring and synthetic polymers, are known to take on helical conformations in the solid state [9]. A polymer with a right-handed or left-handed helical conformation is optically active without any other chiral components. However, most isotactic vinyl polymers, such as polystyrene and polypropylene, which lack chiral pendant groups, cannot be optically active in solution because the movement of the main chains at room temperature is extremely fast in solution, so the polymers cannot maintain helical conformations [10]. However, the possibility of obtaining optically active polymers has recently been demonstrated as long as their backbones have stereoregularity or the steric repulsion of their side groups is large enough to maintain a stable conformation. Such stable helical conformations have been realized in a few synthetic polymers, such as polyisocyanides [11], polyisocyanates [12], polychlorals [13], poly(triarylmethyl methacrylate)s [14], and polysilylenes [15].

Chiral polyacetylenes have received much attention because of the potential for applications as optoelectronic and optical materials [16]. We have previously reported that a polymer from phenylacetylene, which has a bulky chiral menthoxy group, exhibits large circular dichroism (CD) signals in the main-chain absorption region and that the bulky chiral group is also very effective for induction of asymmetry to the main chain [17]. We also have found that the bulky pinanyl group is a good substituent for the induction of helical chirality in a polyacetylene backbone [18, 19]. In this study, three π-conjugated chiral helical phenylacetylene polymers bearing bulky chiral pinanyl groups were synthesized, and the main-chain chiralities induced by these chiral groups were examined in these polymers.

31.2.1.2
Sythesis and Polymerization of Pinanyl-Containing Phenylacetylenes

(−)-PSPA and (−)-PDSPA monomers were synthesized with good yields according to the reactions in Fig. 31.2. (−)-PSPA and (−)-PDSPA were polymerized by a Rh catalyst to give a higher molecular weight polymer than with W or Ta catalysts. Although both (−)-poly(PSPA) and (+)-poly(PDSPA) prepared by using the Rh catalyst had a high percentage of the *cis* isomer and a high molecular weight, only (−)-poly(PSPA) exhibited a high chirality. In contrast, (+)-poly(PDSPA) having flexible siloxane spacers between the main chain and the chiral pendant group exhibited a lower chirality. It was found that the presence of a flexible spacer between the main chain and the chiral pendant group had a role in the failure of chiral information transfer during polymerization. A chiral copolymer, (−)-poly(PSPA/PDSPA), was also synthesized using the Rh catalyst in triethylamine solvent.

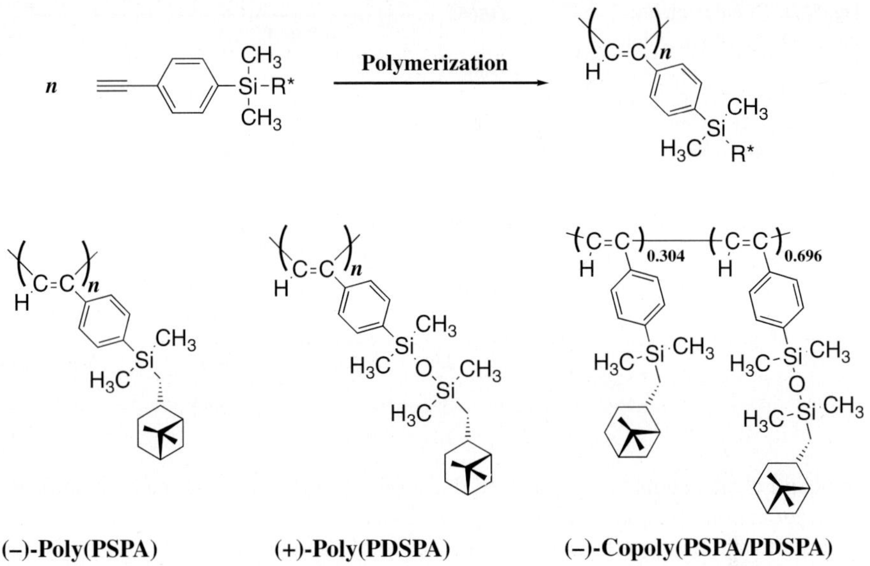

Fig. 31.2. Chemical structures of chiral helical π-conjugated poly(substituted phenylacetylene)s containing a pinanyl group

31.2.1.3
Introduction of Chirality to the Main Chain

In Fig. 31.3, the (−)-poly(PSPA) prepared using the Rh catalyst had large CD signals in the main-chain absorption region similar to (−)-poly(menthoxycarbonylphenyl-acetylene) [17]. Bulky chiral pinanyl groups were found to be effective for induction of asymmetry to the polyphenylacetylene backbone. The two chiral polyphenyl-acetylenes had similar intensities in their CD spectra ($[\theta] \approx 10^4$), and therefore their bulkiness was thought to be important. Since the CD intensity in (−)-poly(PSPA) was stronger than that in (+)-poly{[1-dimethyl(10-pinanyl)silyl]-1-propyne} ($[\theta] \approx 10^3$) [18] and similar to that in (+)-poly[1-{4-[dimethyl(10-pinanyl)silyl]phenyl}-2-phenylacetylene] [19], the phenylene groups in monomers are found to be favorable for the induction of asymmetry in these polymers at least.

31.2.1.4
Effect of a Flexible Spacer between the Chiral Pendant Group and the Polymerizable Group

Since all three polymers showed different CD spectra at the UV–vis band in the main chain, as shown in Fig. 31.3, it was thought that the main chain of these polymers formed a helix of a single-sense excess. (−)-Poly(PSPA) had the highest $[\theta]$ value and (+)-poly(PDSPA) the lowest $[\theta]$ value. Since (+)-poly(PDSPA) has a flexible siloxane spacer between the optically active pinanyl group and the main chain (Fig. 31.2), the transfer of the chiral information from the pinanyl group to

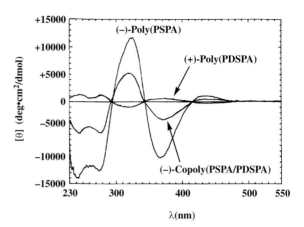

Fig. 31.3. Circular dichroism (CD) spectra in chloroform at room temperature of chiral helical π-conjugated poly(pinanylacetylene)s polymerized using a Rh catalyst

the main chain seems to be disturbed during the formation of the main chain (e.g., polymerization reaction). On the other hand, the (−)-poly(PSPA) main chain is able to accept asymmetric information from the chiral pendant groups. Interestingly, the senses of the chiral helix are reversed between (−)-poly(PSPA) and (+)-poly(PDSPA) in spite of the fact that both have the same optically active pinanyl groups from (−)-pinene. Moreover, a chiral copolymer, (−)-copoly(PSPA/PDSPA), was also obtained from the copolymerization of (−)-PSPA and (−)-PDPSPA, and had a CD spectrum intermediate between that of (−)-poly(PSPA) and that of (+)-poly(PDSPA) (Fig. 31.3).

Since the temperature dependence of the specific rotation for (−)-poly(PSPA) was the highest amongst the three polymers, it was found that the conformation of this chiral main chain was dynamic. In addition, the temperature dependence was observed to be reversible. Moreover, the specific rotation of (+)-poly(PSDPA) was observed to be independent of temperature owing to lower chirality in the main chain. These results indicate that the π-conjugated chiral helix is reversible and thermosensitive. A (−)-poly(PSPA) film cast from chloroform solution also had a CD spectrum similar to that of the solution. This chiral structure obtained in the film state is expected to be useful for application as an optical resolution membrane [18].

31.2.2
Direct Measurement of the Chiral Quaternary Structure in a π-Conjugated Polymer

31.2.2.1
Overview

π-conjugated polymers have been developed as advanced materials for photonic or electronic applications [1, 2]. If the π-conjugated polymer chain can be controlled in the higher-order structure, novel functions at the molecular level will become available owing to the unique π-electron system. Many studies confirming the fact that a π-conjugated polymer has a helical structure have already been completed [17, 20–22]. Most of these studies have provided us with data on molecular aggregates or data on the average of many molecules. Although we now understand that the main chain of the polymer takes the form of a helix, does one chain have both right-handed and left-handed helices? What is the ratio of the right-handed helices to the left-handed ones? What about the regions where the helix is reversed and how does it dynamically change? The answers already provided to all these fundamental questions have been based only on conjecture. Therefore, it is necessary to establish a technique that can determine the structure at the single-molecule level in order to achieve the abovementioned objective. Here we show that a STM [23, 24] allows us to see the π-electron orbital of a chiral quaternary structure, where it was directly observed and its size was even measured with high resolution.

31.2.2.2
Polymer Chemistry of a Chiral Helical π-Conjugated Polymer

We synthesized a π-conjugated polymer, an optically active polyphenylacetylene bearing menthoxycarbonylamino groups [(−)-poly(MtOCAPA), Fig. 31.4a, Scheme 31.5]. It was found that (−)-poly(MtOCAPA) has a high molecular mass, i.e., its molecular mass is on order of 1×10^6 Da, and that the *cis* content of the polymer is over 90 mol%, and the main chain has a c–t high stereoregularity [25]. The primary structure of this polymer was examined using a nuclear magnetic resonance spectrometer. As is apparent from the spectrum of this polymer observed using a CD spectrometer (Fig. 31.4b, top) and a UV–vis absorption spectrometer (Fig. 31.4b, bottom), chirality was detected not only in the side group but also in the main chain, which had expanded π-conjugation. It was found from these observations that the optically active menthyl in the side group induced chirality in the π-conjugated system of the main chain. This CD spectrum shows that the main chain of (−)-poly(MtOCAPA) has a secondary structure having an excess of the one-sense helix structure, which is stabilized by the bulky substituent. Because the intensity of this CD signal increased in inverse proportion to the temperature of the polymer solution, it is presumed that this main chain has a flexible helix structure.

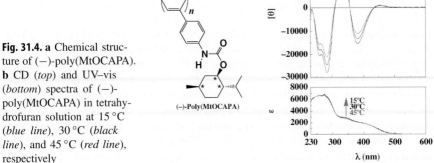

Fig. 31.4. a Chemical structure of (−)-poly(MtOCAPA). **b** CD (*top*) and UV–vis (*bottom*) spectra of (−)-poly(MtOCAPA) in tetrahydrofuran solution at 15 °C (*blue line*), 30 °C (*black line*), and 45 °C (*red line*), respectively

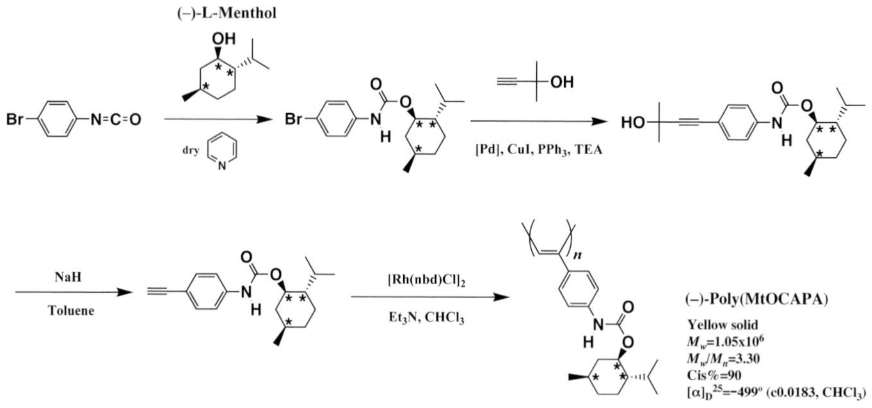

Fig. 31.5. Synthetic route of the chiral helical π-conjugated polymer, (−)-poly(MtOCAPA)

31.2.2.3
Scanning Probe Microscope Imaging of a Chiral Helical π-Conjugated Polymer

Scanning probe microscope imaging was used to directly observe the structure of a single polymer chain. We initially tried to use an atomic force microscope (AFM) image of a polymer placed on a mica substrate under ambient conditions. Each single polymer molecule that formed a chain could be distinguished and intertwined polymer chains could also be observed. An overall image of the polymer molecules could be directly observed using the AFM, but the π-conjugated system in the main chain could not be observed using the AFM since this polymer had a bulky menthyl group and the main chain was hidden behind the side groups at the front. To observe the main chain without being obstructed by the bulky menthyl group, a STM was used, since this allows us to see molecular orbitals having the low energy gap of the HOMO–LUMO on which a tunneling current flows [26–28]. Figure 31.6a shows a typical low-magnification image of low-current scanning tunneling microscopy of (−)-poly(MtOCAPA) placed on a highly oriented pyrolytic graphite substrate under ambient conditions. During our observation, the sample bias voltage and the tunneling current were maintained at +20.0 mV and 30.5 pA, respectively, and a STM probe using a Pt/Ir tip was operated at a scanning rate of 3.05 Hz. In Fig. 31.6a, it can be observed that the two π-conjugated polymer chains are intertwined to form a right-handed double-helix structure, as in the case of the model shown in Fig. 31.6b, and the right-handed helix structure can also be observed. The analytical result is shown in Figs. 31.7 and 31.8. The width of one right-handed helix chain was 0.9 nm (Fig. 31.7b). This width of 0.9 nm matched the width of the main-chain backbone of polyphenylacetylene in the π-electronic system of the c–t main-chain structure 20-mer which was obtained by optimization using molecular mechanics calculation (Fig. 31.7b). This corroborated the observations that the STM image displays the π-electron orbital of the main chain of a polymer and that the helix of the secondary structure is a superhelix tertiary structure (Fig. 31.8a). As a result of the analyses (Fig. 31.8b), the pitch of this superhelix is 2 nm. It is shown to be a superhelix with a close-packed structure, because the pitch agrees with the width of the model

Fig. 31.6. a Low-current scanning tunneling microscope (STM) height image of two intertwined (−)-poly(MtOCAPA) chains on highly oriented pyrolytic graphite at room temperature ($V_s = +20.0$ mV and $I_t = 30.5$ pA). b String model of the intertwining polymer chains

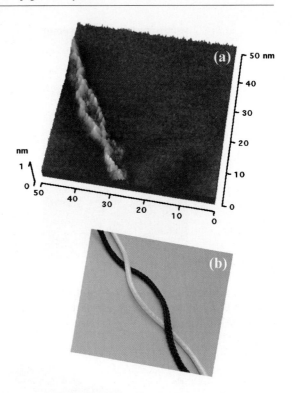

in Fig. 31.7b. In addition, it is shown in Fig. 31.8b that the superhelix width is 2 nm, and that the helix sense is right-handed, and that this precisely controlled superhelix tertiary structure extends over a range of more than 10 nm. When the polymer structure in Fig. 31.6a was turned over, it could be observed that the two π-conjugated polymer chains are intertwined to form a right-handed double-helix structure, as in the case of the model shown in Fig. 31.6b. This substantiated the presence of a quaternary structure, which is much greater in the chiral hierarchical structure. It was also found that the quaternary structure is so soft that its form can be changed during probe scanning. That is, we proved that the main chain of (−)-poly(MtOCAPA) was flexible in spite of the π-conjugated system.

In the present study, it was shown that it was possible to determine the structure of a single polymer chain using a STM. We observed that it has a higher-order structure, and the fine structure of the single π-conjugated polymer chains was confirmed for the first time. However, in our current method, the molecule itself is changed by the STM measurement, and the helical structure becomes loosened. Therefore, in order to determine the electronic structures and the ratio of the two different helical senses from the effect of the chiral pendant groups, we need to use an approach where we stabilize the STM measurements by fixing the molecules on the substrate, which is currently in progress. Finally, this unique higher-order structure of the π-electronic system is expected to provide novel photonic or electronic functions that will lead us toward molecular devices.

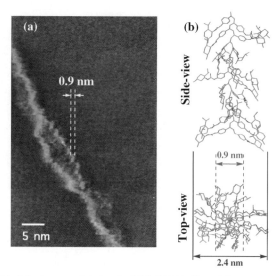

Fig. 31.7. a STM height image of (−)-poly(MtOCAPA) (*bar* 5.0 nm, $V_s = +20.0$ mV and $I_t = 30.5$ pA). **b** Molecular mechanics calculation optimized model of a 20-mer of *cis*-transoidal (−)-poly(MtOCAPA)

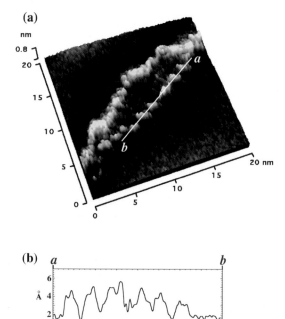

Fig. 31.8. a A closeup view of a STM height image of (−)-poly(MtOCAPA) ($V_s = +20.0$ mV and $I_t = 30.5$ pA). **b** Analytical result of superhelix pitch in the cross section of the STM image

31.2.3
Direct Measurement of Structural Diversity in Single Molecules of a Chiral Helical π-Conjugated Polymer

Figure 31.9 shows a model of the right-handed helix of a single polymer chain. Fig. 31.9a is a molecular mechanics (SYBYL) calculated optimized model of a 20-mer of c–t (−)-poly(MtOCAPA), and Fig. 31.9b is its ribbon model. The height of the helical polymer chain of a single molecule was measured using this model (Fig. 31.9a); the measured value was 2.4 nm. In analyzing the images of the polymer molecules obtained by the AFM imaging, we used this value of 2.4 nm as the unit size of a single molecule.

Figure 31.10a shows a noncontact mode AFM image of the surface of a specimen conditioned by spin-casting. The specimen was also studied in contact mode, and the same structure as that seen in noncontact mode was observed. A cross section of a noncontact AFM image of the polymer was analyzed, as shown in Fig. 31.10b. The height of each of the polymer chains was about 2 nm. This shows that each of the chains is a single molecule. The average degree of polymerization was calculated from the results of gel permeation chromatography (GPC) measurements. This turned out to about 800 monomer units long, though it was calculated in terms of the molecular weight of polystyrene. A 20-mer in the polymer was about 4 nm long, on the basis of the molecular model (Fig. 31.9a). Therefore, a chain consisting of 800 monomer units is about 160 nm in length, which is an average value. This result was in good agreement with the average length of single molecules of the polymer identified from the AFM images. All of these results regarding the length of the polymer chains substantiate the notion that single molecules of polymer were observed in this AFM experiment. Each of these polymer chains takes a different form; no polymer chains of identical form were ever observed. This verified that

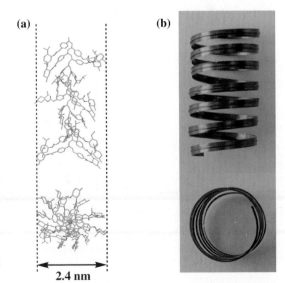

Fig. 31.9. a Molecular mechanics calculated optimized model of a 20-mer of *cis*-transoidal (−)-poly(MtOCAPA). **b** Ribbon model of the helical polymer chain

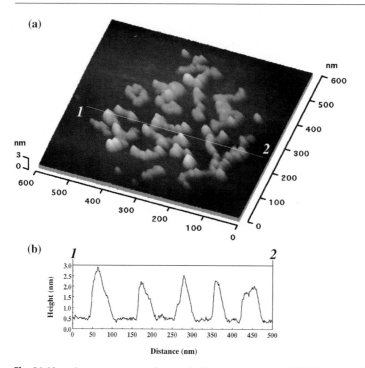

Fig. 31.10. a A noncontact mode atomic force microscope (AFM) image of single molecules of p-conjugated (−)-poly(MtOCAPA) on mica under a high vacuum at room temperature. The frequency shift was −30 Hz. The scan area was 600 nm × 600 nm. b Analytical result of the polymer height in the cross section of the AFM image in **a**

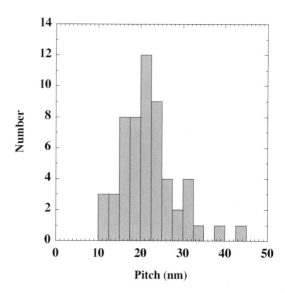

Fig. 31.11. The distribution of helical polymer periods that were measured on the basis of the noncontact mode AFM image in Fig. 31.10a

the single molecules of the polymer have diverse structures. Peaks and troughs were also observed in each polymer chain with a certain periodicity. To examine this periodicity, all of the pitches were measured and a histogram was prepared based on this data (Fig. 31.11). A pitch of 21 nm was a peak in the histogram. On the other hand, it was found that pitches varied greatly in the range from 10 to 45 nm, and the diversity was observed for the periodic structure. This structural characteristic of single molecules was thought to be due to the low affinity between the surface of the polymer chain and the surface of the mica substrate. Specifically, it was elucidated

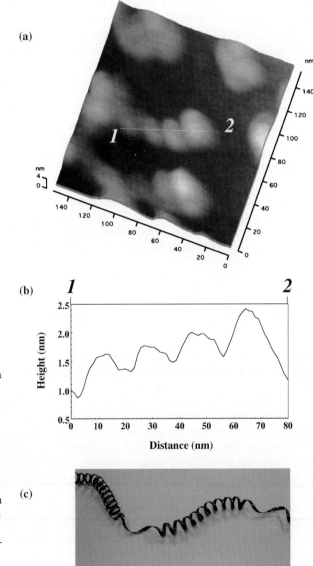

Fig. 31.12. **a** A closeup view of a noncontact mode AFM image of single molecules of π-conjugated (−)-poly(MtOCAPA) on mica under a high vacuum at room temperature. The frequency shift was −30 Hz. the scan area was 153 nm × 153 nm. **b** The cross-sectional profile of a right-handed helix. **c** A ribbon model of a periodic compression structure. This is a coarse and dense structure of a helix, formed by a right-handed helical polymer chain

as follows: When a polymer chain having a surface of low polarity is adsorbed to a mica substrate having a surface of high polarity, an intramolecular aggregation of the polymer occurred in each single molecule to decrease the surface energy. The structure formed in the nonequilibrium process until the solvent evaporated.

To further analyze this periodic structure, a part of a single molecule of the polymer was magnified as shown in the AFM image in Fig. 31.12a. The cross-sectional analysis of this image indicates the presence of a periodic structure of about 20 nm as shown in Fig. 31.12b. A close examination of this structure reveals that each peak is aligned obliquely in the upper-right direction. Although we thought that this could be associated with a right-handed helical structure, it was beyond our understanding how a polymer structure of a helix could have a periodicity as long as 20 nm. To further investigate this structure, we prepared the model shown in Fig. 31.12c. This model is a periodic compression structure, which is a coarse and dense structure of a helix, formed by a right-handed helical polymer chain. This unique structure shows that a local structural disturbance exists in the helical structure. Surprisingly, the long periodicity was even observed with this structural disturbance.

We would not have discovered that a chiral helical π-conjugated polymer of only 2 nm in height could have a nanostructure that is characterized by a periodicity on the surface of a substrate of as much as 20 nm if our study had been weighted toward a structure consisting of repeating monomer units. We hope that our study will provide basic insight into both the polymeric characteristics of single molecules (these are their individual characteristics) and the practical realities of creating a single-molecule device. We expect that the knowledge from our study will contribute to the discovery of novel functions that are made possible by the unique structures formed on the surfaces of substrates, as well as the development of molecular devices.

31.2.4
Dynamic Structure of Single Molecules in a Chiral Helical π-Conjugated Polymer by a High-Speed AFM

31.2.4.1
High-Speed AFM Imaging of a Chiral Helical π-Conjugated Polymer

We observed the movement of a single molecule of a π-conjugated polymer using a high-speed AFM. The high-speed AFM is an apparatus initially developed by our collaborative researcher Toshio Ando of Kanazawa University in Japan and his colleague Noriyuki Kodera. It is a state-of-the-art microscope, which boasts an outstanding performance when observing the movement of protein molecules in water at room temperature [29]. However, we modified the specifications of this high-speed AFM to make it resistant to observations in organic solvents and successfully imaged the movement of a single molecular structure [7] of a chiral helical π-conjugated polymer [1, 2] at room temperature. In addition, we observed the movement and the flexibility of a single polymer chain. Furthermore, we created a high-speed AFM model and quantitatively evaluated the effects of AFM probe scanning on the samples based on the theory of the instrument properties.

We cast a dilute tetrahydrofuran solution of substituted phenylacetylene polymer [(−)-poly(MtOCAPA)] (Fig. 31.4a) on a mica substrate. Then we used tetrahydrofuran to rinse the surface of the mica and n-octylbenzene as an organic solvent for imaging. Afterwards, we conducted high-speed AFM observations in n-octylbenzene at room temperature, and a flexible motion of a string-type substance, which measured approximately 2 nm in height and approximately 200 nm in length, was observed (Fig. 31.13a). The average velocity at the chain-end of the polymer was 67 nm/s at the observation point. Because the size of the string-type substance corresponds to the value of the molecular model calculation, we concluded that the observed substance is a single molecule of the chiral helical π-conjugated polymer. When several molecules are simultaneously observed, their motion is not synchronous, but is random (Fig. 31.13b). We could directly image these motions as folding of a single molecule or a single molecule crooked in the annular form that is extended and moved into a linearlike shape. It is noteworthy that most of the polymer chains were not detached from the substrate surface to diffused in the solvent, but were rolled onto the substrate surface (Fig. 31.13).

We assumed that this is a thermal motion, which reflects the compensation of the energy of solvation and adsorption onto the substrate surface. To validate this assumption, we conducted a detailed motion analysis. We measured all the dimensional displacements of a single polymer chain in the x direction, made a histogram, and fitted it to a Gaussian curve (Fig. 31.14a). The standard deviation values were the time-average displacements at each observation time. The result shows that the mean-square displacement in a single polymer chain is proportional to the time and, hence, complies with Einstein's law of Brownian motion (Fig. 31.14b). This analysis result demonstrates that the motion in a single polymer chain is the micro-Brownian motion excited by thermal energy.

Second, to elucidate the effects of an AFM probe on the motion of a single polymer molecule, we estimated the energy based on the theoretical understanding of the instrument properties. To calculate the maximum energy the probe gives to the

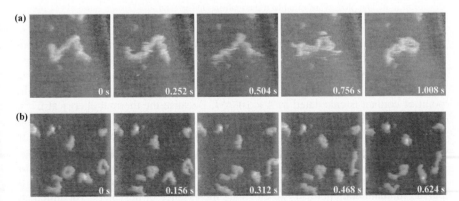

Fig. 31.13. High-speed AFM images in continuous scanning of a chiral helical π-conjugated polymer [(−)-poly(MtOCAPA)] on mica in n-octylbenzene at room temperature. **a** 200 nm × 200 nm. **b** 300 nm × 300 nm

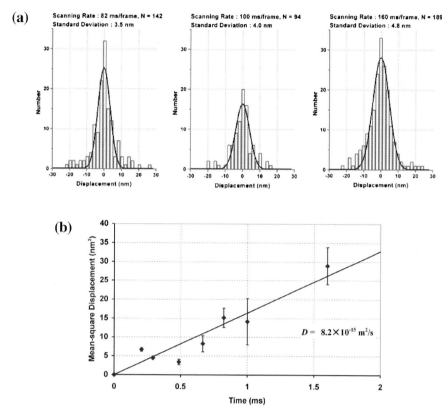

Fig. 31.14. a Analytical results of the high-speed AFM image of (−)-poly(MtOCAPA) of the displacement in single polymer chains at scanning rates of 82, 100, and 160 ms per frame. The *lines* are the best fits of the data to a Gaussian curve; the standard deviations are 3.5, 4.0, and 4.8 nm, respectively. b Mean-square displacement as a function of time of the tip-scanning cycle in the direction of the x-axis. The mean-square displacements were obtained from the standard deviation in **a**. The plots were fitted to the equation, $\langle x^2 \rangle = 2Dt$

sample in the horizontal direction, we assumed that the probe instantaneously adheres to the sample the moment the oscillating cantilever probe touches the sample's surface. In a high-speed AFM, the stage is the x-axis scanned in the direction of the cantilever's length. Consequently, the energy the cantilever probe stored at the point of contact is calculated as 4×10^{-23} J. Because the thermal energy at 27 °C is 4×10^{-21} J, it became clear that the AFM probe contributes merely 1% of the thermal energy. Thus, the motion of single molecules of the polymer observed in our study is based on the thermal energy.

Through this study, we verified the flexibility of a single molecule of a chiral helical π-conjugated polymer based on thermal fluctuations. We are continuing to pursue fundamental chemistry to elucidate the essence of polymers. We anticipate that our studies will be the basis for creating innovative molecular devices such as molecular motors or molecular electronics/photonics materials that utilize thermal energy as a driving source.

31.3 Supramolecular Chiral π-Conjugated Polymer

31.3.1 Simultaneous Imaging of Structure and Fluorescence of a Supramolecular Chiral π-Conjugated Polymer

In this study, the author succeeded in the simultaneous imaging of the structure and the fluorescence function of a π-conjugated polymer. Figure 31.15 shows a chemical structure and a molecular model of a π-conjugated polyrotaxane (+)-poly[(9,10-anthracenediyl-ethynylene-1,4-phenylene-ethynylene)-rotaxa-(α-cyclodextrin)] {(+)-poly[AEPE-rotaxa-(α-CyD)]} that we synthesized. Figure 31.16 show a schematic view and the exterior of the experimental set-up that we built to simultaneously observe the structure and functions of polymer on a molecular level, respectively. The set-up consists of an AFM [30] set above a substrate on which the specimen was mounted and an objective-type TIRFM [31] set below the substrate in air at room

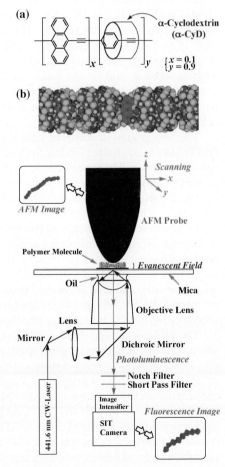

Fig. 31.15. a Chemical structure of a polyrotaxane: (+)-poly[(9,10-anthracenediyl-ethynylene-1,4-phenylene-ethynylene)-rotaxa-(α-cyclodextrin)] {(+)-poly[AEPE-rotaxa-(α-CyD)]}. **b** A part of the molecular mechanics calculation optimized model of a polymer

Fig. 31.16. The experimental set-up of the AFM–total internal reflection fluorescence microscope instrument that we built to simultaneously observe the structure and functions of a polymer (not to scale). The notch filter is to block incident light (441.6 nm). The short pass filter is to cut out the laser light (670 nm) from the AFM detection system

temperature. The AFM can image the structure of the polymer at a molecular level and the TIRFM is able to image the fluorescence of the polymer at video rate, so both the structure and fluorescence function of the polymer can be simultaneously imaged at the substrate surface.

31.3.1.1
Instrumentation

The TIRFM was self-made by modifying an inverted fluorescence microscope (Infinity optical system; IX70, Olympus, Japan) as shown in Fig. 31.16. An objective lens (PlanApo, ×60, oil TIRFM, Olympus, Japan) with a high numerical aperture (NA) of 1.45 was attached to this TIRFM. The laser beam from a He–Cd laser (wavelength 441.6 nm, 70 mW, continuous wave, CW, linearly polarized; blue laser, model Liconix M.4.70, Melles Griot, USA) was attenuated by neutral density filters (transmission of 21%) and expanded by a lens (LBE-3, Sigma Koki, Japan). The laser beam was focused by a lens ($f = 350$ mm) on the back focal plane of the objective lens with a NA of 1.45. θ_a (66.6°) is the angle corresponding to the NA ($1.58 - \sin\theta_a = $ NA; 1.58 is the reflective index of mica). The laser beam was reflected by a dichroic mirror (DM455, Olympus, Japan). θ_c (39.3°) is the critical angle of the mica-air interface ($1.00\sin 90° = 1.58\sin\theta_c$; 1.00 is the reflective index of air). When the incident beam was positioned to propagate along the objective edge between θ_a and θ_c, the beam was totally internally reflected, producing an evanescent field at the mica–air interface (1/e penetration depth is about 40 nm at an angle of 60°. A mica film with a thickness of about 20 μm was used as the substrate.

The background of scattered light was rejected with a holographic notch filter (HNPF-441.6-1.0, Kaiser Optical System, USA), whose optical density at 441.6 nm was larger than 6. The scattered light (wavelength 670 nm) from the AFM detection system was reduced with two short pass filters (cutoff wavelength 630 nm). An intermediate image formed by spherical achromatic lenses ($f = 100$ mm) was enlarged by a projection lens (PE ×5, Olympus, Japan).

Nonfluorescent immersion oil (for ordinary use, $n_d = 1.516$, Olympus, Japan) is suitable for single-molecule fluorescence imaging because it emits low fluorescence and has adequate viscosity.

Images were recorded with an SIT camera (C1000 type 12, Hamamatsu Photonics, Japan) coupled to an image intensifier (VS4-1845, Video Scope, USA), and stored on S-VHS videotape. Videotaped images were processed with a digital image processor (Argus-20, Hamamatsu Photonics, Japan).

The probe-scan-type AFM (PicoSPM, Molecular Imaging Corporation, USA) was used in the contact mode using a soft cantilever (BL-RC150V, BioLever, Olympus, Japan) having a spring constant of 6 pN/nm. The pressure of tip contact was set as low as possible for imaging of a soft contact. This probe-scan-type AFM was combined with the objective-type TIRFM as stated above (Fig. 31.16). In order to image simultaneously with an AFM and a TIRFM, a sample on a stage with an AFM head was used with a TIRFM. First, the TIRFM was focused on a sample's surface. Next, the AFM approached the sample's surface. The best imaging conditions were optimized under observation of both images.

The AFM instrument (JSPM-4210, JEOL, Japan) was also used in a noncontact mode. A cantilever (OMCL-AC160TS, Olympus, Japan) with a resonance frequency of 300 kHz and a spring constant of 42 N/m was used. A high vacuum of 1×10^{-3} Pa was used, and the cantilever was oscillated with constant amplitude at room temperature. Imaging innoncontact mode was carried out by scanning the probe while maintaining a frequency shift (Δf) of -15 Hz.

31.3.1.2
Synthesis of Supramolecular Chiral π-Conjugated Polymer

As shown in Scheme 31.17, to prepare (+)-poly[AEPE-rotaxa-(α-CyD)], p-diiodobenzene, p-diethynylbenzene, and 9,10-diethynylanthracene were added to a saturated α-CyD aqueous solution and irradiated with an ultrasonic wave to form the inclusion complexes (see Fig. 31.18a,b for the α-CyD ring structure). These inclusion complexes were polymerized at 90 °C for 40 h using a palladium(II) acetate and potassium carbonate catalyst system [32] to form a polymer from the constituent complex monomers. After the polymerization reaction, the water-soluble and high molecular weight part was isolated by gel filtration and reprecipitation from an aqueous solution into methanol. The water-soluble (+)-poly[AEPE-rotaxa-(α-CyD)] was a dark-brown solid; $[\alpha]_D^{20} = +24.6°$ (c0.100, H_2O), UV-vis: λ_{max} = 197 nm ($\varepsilon = 6.92 \times 10^3$), $\lambda_{max} = 260$ nm ($\varepsilon = 3.55 \times 10^3$), $\lambda_{cutoff} = 640$ nm ($\varepsilon < 50$), GPC : $M_w = 1.47 \times 10^6$, M_w/M_n(polydisperisty index) = 2.3. A GPC column (TSKgel α-M, Tosoh, Japan) having a high exclusion limit [poly(ethylene oxide), $M_n > 1 \times 10^7$] was used with an eluent of water at 40 °C. In addition, the threaded CyD on an anthracene unit was removed during the process for the purification.

In the ^1H NMR spectrum of the polymer in D_2O, the peaks at 5.1 and 4.0–3.6 ppm were assigned to the protons in the α-CyD (Fig. 31.18c). A broad shoulder peak at 7.8 ppm was assigned to the 1-, 4-, 5-, and 8-position protons in the anthracene ring, and a broad peak at 7.4 ppm was assigned to the 2-, 3-, 6-, and 7-position protons in the anthracene and phenyl rings (Fig. 31.18d).

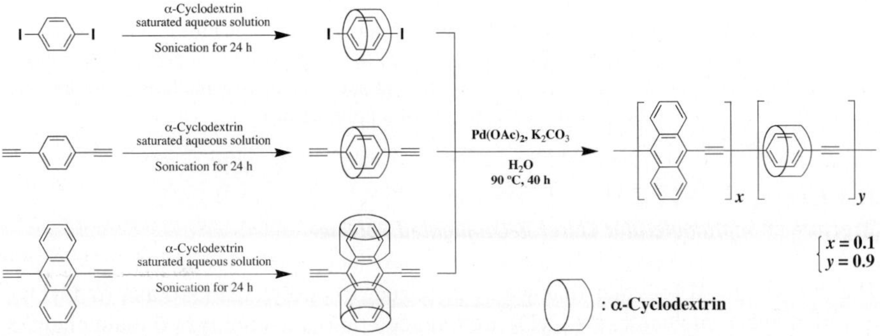

Fig. 31.17. Synthetic route of supramolecular π-conjugated (+)-poly[(9,10-anthracenediyl-ethynylene-1,4-phenylene-ethynylene)-rotaxa-(α-cyclodextrin)]

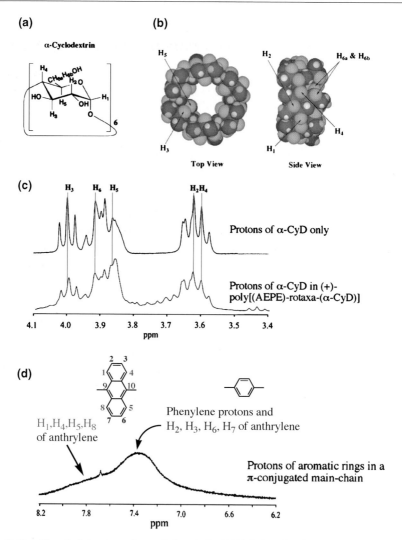

Fig. 31.18. a Chemical structure of α-cyclodextrin (α-CyD). **b** A molecular mechanics calculation optimized model of α-CyD. **c** 400 MHz ^{1}H NMR spectra of a polymer {(+)-poly[AEPE-rotaxa-(α-CyD)]} and α-CyD in D_2O at 27 °C. The external standard is 3-(trimethylsilyl)propionic-2,2,3,3-d_4 acid sodium salt. **d** An aromatic region of a polymer of **c**

31.3.1.3
Structure of Supramolecular Chiral π-Conjugated Polymer

The π-conjugated polymer that we synthesized, (+)-poly[AEPE-rotaxa-(α-CyD)], in which cyclic molecules (α-CyD) are threaded onto a π-conjugated main chain is shown in Fig. 31.15; thus, (+)-poly[AEPE-rotaxa-(α-CyD)] is a kind of polyrotaxane [33].

If substituents such as long-chain alkyl groups are not bonded onto the main chains of a π-conjugated polymer, the polymer chains can interact with each other by π–π interactions and π-stacking occurs, causing aggregates to form and, as a result, the solubility markedly decreases for any solvent. This poses a significant difficulty to overcome in the study of the molecular characterization of polymers at the single-molecule level. The use of a polyrotaxane in this study is based on our conjecture that cyclic molecules would be able to physically block the π-stacking. Consequently, the solubility of such a π-conjugated polymer could be increased (Fig. 31.15b). In ^{1}H NMR measurement (Fig. 31.18), 3-(trimethylsilyl)propionic-2,2,3,3-d_4 acid sodium salt (TSP-d_4, 98 atom% D, Isotec, OH, USA) was used as an external standard. The capillary (a thin glass tube) enclosed the D_2O solution, and it let the sample D_2O solution in. It was used for the measurement in order to avoid the inclusion of a TSP-d_4 molecule in a α-CyD hole.

The results of ^{1}H NMR measurement of (+)-poly[AEPE-rotaxa-(α-CyD)] and α-CyD in D_2O were shown in Fig. 31.18c. Here, H_3 and H_5 protons face toward the inside of the α-CyD holes (Fig. 31.18b). Compared with polyrotaxane and α-CyD,

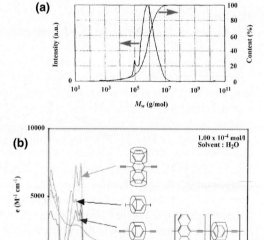

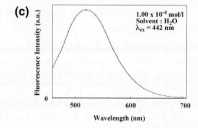

Fig. 31.19. a Gel permeation chromatography chart of a polymer at 40 °C. A UV detector was used; wavelength 190 nm. b UV–vis spectra of a polymer and monomers in H_2O at room temperature. c Photoluminescence spectrum of a polymer in H_2O at room temperature. Excitation wavelength 442 nm

the chemical shift values of the H_3 and H_5 protons changed more noticeably than those of the other protons (H_2, H_4 and H_6). Here, the protons of H_2, H_4 and H_6 face toward the outside of the α-CyD holes. This significant change shows the formation of an inclusion complex [34]. Broad peaks in the lower magnetic field were assigned to the protons of the aromatic rings in the main chains of the π-conjugated polymer. A copolymerization ratio was calculated from the analytical results of aromatic protons and α-CyD protons in the ^{1}H NMR spectrum. In addition, the molecular mass is greater than 10^6 g/mol as measured by GPC using water as the eluent (Fig. 31.19a), and the UV–vis absorption spectrum of this polymer in water shows that the absorption shifted significantly toward the longer wavelengths when compared with the complex monomers, as far as 640 nm at the cutoff wavelength ($\varepsilon < 50$) (Fig. 31.19b). It was concluded that the π-conjugated system had been expanded on the main chain. We determined from the results of the measurements that (+)-poly[AEPE-rotaxa-(α-CyD)] had been successfully synthesized. Furthermore, this polymer fluoresced with a peak wavelength of 520 nm when excited at 442 nm in water (Fig. 31.19c), which is the excitation light wavelength (441.6 nm CW laser) of the TIRFM. This shows that (+)-poly[AEPE-rotaxa-(α-CyD)] can be observed using TIRFM imaging.

31.3.1.4
AFM Imaging of Single Molecules in a Supramolecular Chiral π-Conjugated Polymer

Figure 31.20a shows a noncontact mode AFM image of polymer chains extended outward from a molecular globule. These chains were measured to be a micrometer-long structure. This is a new finding for a macromolecular structure. In Fig. 31.20b a model structure of a polyrotaxane is proposed. This model shows that the size of a single molecule is 1.5 nm. The height of a molecular globule was approximately 4 nm as shown in Fig. 31.20c; this was the height of three molecules. Each chain extending outward from a globule was the size of a single molecule as shown in Fig. 31.20d. Although a micrometer-long chain was observed directly in an AFM measurement, the long polymer chain could not be synthesized by a cross-coupling reaction [32] in this study. Hence, we thought that nanometer-long π-conjugated polyrotaxane chains coupled to form a micrometer-long supramolecular chain owing to intermolecular interaction at the chain ends like a train of nanometer scale as shown in Fig. 31.20b. Actually, a number of short substances of nanometer length were observed in line where imaged by an AFM at the other large area, which had the height of a single molecule of 1.5 nm. Therefore, we thought that these short substances were chains of (+)-poly[(AEPE)-rotaxa-(α-CyD)] without intermolecular interaction. It is a very interesting and a new finding that the polymer chains formed a micrometer-long coupled supramolecular structure by a self-assembly event without forming the disordered aggregate.

31.3.1.5
Simultaneous AFM–TIRFM Imaging of a Supramolecular Chiral π-Conjugated Polymer

Figure 31.20e shows a contact mode AFM image of the intertwined network of the polymer chains. The cross sections of these polymer strands were analyzed and the height of each polymer strand was found to be about 1 nm, as shown in Fig. 31.20f.

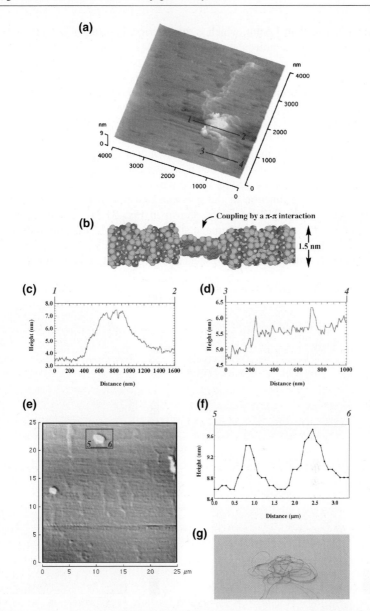

Fig. 31.20. a A noncontact mode AFM image of single chains from a molecular globule of (+)-poly[AEPE-rotaxa-(α-CyD)] on mica at room temperature. A high vacuum of 1×10^{-3} Pa was used, and the cantilever was oscillated with constant amplitude at room temperature. Imaging in the noncontact mode was carried out by scanning the probe while maintaining a frequency shift (Δf) of -15 Hz. b A proposed model of the polymer coupling by π–π interaction at the chain ends. c An analytical result of a cross section of the *line from 1 to 2* indicated in the AFM image (a). d A cross-sectional analysis of the *line from 3 to 4* indicated in (a). e A contact mode AFM image of the intertwined network as a nanostructure. f An analytical result of a cross section of the *line from 5 to 6* indicated in the AFM image (e). g The string model of the polymer chains in the region shown in the *square* in the AFM image (e)

This value is less than 1.5 nm, which is the external diameter of α-CyD in a single polyrotaxane chain—a molecular mechanics (MM2) calculated optimized model is shown in Fig. 31.20b. However, it should not be considered the actual value, because α-CyD molecules threaded onto a polymer chain are squeezed down and deformed by contact pressure from the probe with the cantilever. Therefore, we concluded that the strands of a π-conjugated polymer had been observed on a single-molecule level. It was found from these observations that each spot in the AFM image consists of a globule in which the polymer chains are intertwined like a wound-up ball of string, and that the strands of polymer chains extend outward from these globules, just as string can come loose from a ball when it is unwound, as shown in Fig. 31.20g.

Figure 31.21 shows the results of the simultaneous imaging of the structure and the fluorescence function of the π-conjugated polymer. The positions of the brightest spots in the TIRFM image shown in Fig. 31.21a match the positions of the highest spots in the AFM image shown in Fig. 31.21b. This indicates that simultaneous direct imaging is successful using this TIRFM–AFM set-up, and that some high spots emit a strong fluorescence. In Fig. 31.21c and d, showing control

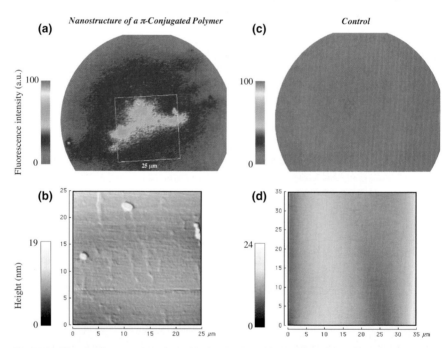

Fig. 31.21. The simultaneous imaging with the TIRFM and the AFM of the intertwined network, i.e., a nanostructure of (+)-poly[AEPE-rotaxa-(α-CyD)] on mica in air at room temperature. **a** Fluorescence image observed by the TIRFM in which the size of the *square* is 25 μm × 25 μm imaged by the AFM. The excitation wavelength was 441.6 nm using a continuous-wave He–Cd laser and the observed fluorescence wavelength was over 475 nm. **b** Topographic image observed by An AFM in contact mode with an area of 25 μm × 25 μm in which the central part was observed by a TIRFM (**a**). **c** Fluorescence image of a mica substrate as a control experiment for TIRFM imaging. **d** Topographic image of the surface of a mica substrate as a control experiment fro AFM imaging; imaging area 35 μm × 35 μm

experiments, no fluorescence and no absorbate are observed on the mica substrate, and it is also shown that the simultaneously imaging of a π-conjugated polymer is successful. In Fig. 31.21a, a relatively weak fluorescence is also observed in the region around the three very bright spots. In the case of the TIRFM, which is an optical microscope, there is an absolute limit to the image resolution. With use of the AFM, however, some stringlike objects that are deemed to be aggregated polymer chains on a molecular level can be observed in the same area where the formation of a nanostructured network by the loose connection of the assembled π-conjugated polymer chains can also be observed, as shown in Fig. 31.21b. Therefore the TIRFM image shown in Fig. 31.21a verifies that the light emitted from the brightest spots and their surrounding area is the fluorescence emitted by the polymer. In addition, the bleaching was only observe over 15 min during the imaging. The polymer strands were extended from the highest spot in the AFM image as shown in Fig. 31.21b.

It is most interesting to note that the fluorescence is brighter in the area formed by connecting the three brightest spots, as shown in the TIRFM image in Fig. 31.21a. This is the fluorescence due to a long-range interaction of a π-conjugated polymer and the photons. The molecular globules have many π-electrons in an excited state (π^* state) in the near-field and excitation energy transfer via a loosely connected π-conjugated polymer intertwined network as a nanostructure. We thought that the fluorescence is a light emission based on not only the combined phenomena of energy migration [35], energy transfer and energy hopping between many chromophores but also on an effect of the total internal reflection. Thus, the photons in the direction of a wave number vector of the Goos–Hänschen shift in the total internal reflection are those scattered at the molecular globules of a π-conjugated polymer. Surprisingly, the photons propagated to a π-conjugated polymer intertwined network over a distance of 5 μm or longer. As the result, the fluorescence was brighter in this area. In other words, this phenomenon can be thought of as optical communication over a long range. This novel phenomenon suggests the possibility of using the polymers to create an intertwined molecular network device in the future.

As a new chemistry, it is worthy that the relationship of a structure and a function could be directly discussed at the molecular level.

31.3.2
Dynamic Structure of a Supramolecular Chiral π-Conjugated Polymer by a High-Speed AFM

31.3.2.1
Single Molecule Bearing Driven by Thermal Energy

Rigid rod-like π-conjugated polymers may have flexibility because they are organic molecules [36]. We used a high-speed AFM to observe their flexibility and motion [37]. The observed polymer is the same as in the previous section where the molecular structure and functions were simultaneously observed (Fig. 31.20b). We have more examples of high-speed AFM imaging. Figure 31.22a–c shows the flexible motion of a single chain of a supramolecular polymer. Various studies are trying to apply rigid-rod like π-conjugated polymers to a molecular wire for molecular

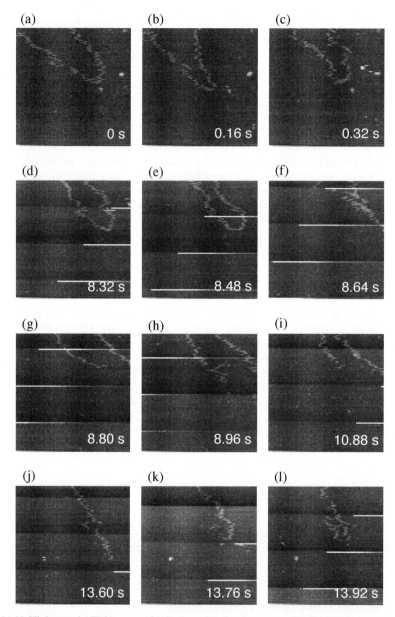

Fig. 31.22. High-speed AFM images of a single molecule of a supramolecular chiral π-conjugated polymer on mica in aqueous solution at room temperature. **a–c** The flexible motion of a single chain of supramolecular polymer. **d–l** Pulsed laser of 355 nm radiating 20 nJ at 20 Hz. **j–l** Dynamic behavior of folding of the end part in a single polymer chain. **e** The imaging of chain cutting in a single molecule. Scan size 400 nm × 400 nm. Scan speed 160 ms per frame

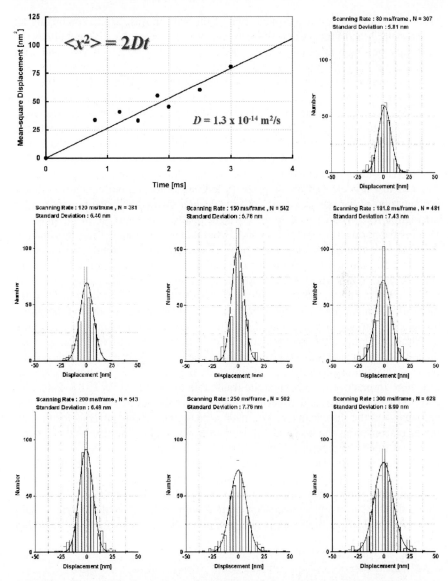

Fig. 31.23. (*Upper left*) A plots of mean-square displacement as a function of time of tip-scanning cycle in the direction of the x-axis. The mean-square displacements were obtained from the standard deviation in the below. The plots were fitted to equation, $\langle x^2 \rangle = 2Dt$. (*Below*) Analytical results of the high-speed AFM image of (+)-poly[AEPE-rotaxa-(α-CyD)] of the displacement in a single polymer chains at scanning rates of 80, 120, 150, 181.8, 200, 250, and 300 ms/frame, respectively. *Lines* are the best fits of the data to a gaussian curve; the standard deviation are 5.81, 6.40, 5.76, 7.43, 6.49, 7.76, and 8.99 nm, respectively

electronics devices, but it is troublesome if the wire moves. While conducting an experiment, I envisioned a new design concept for a molecular device that has the flexibility of single molecules. We collected images for a total of 14 s, and with irradiation from a UV pulsed laser during the last 6 s (imaging time from 8 to 14 s) as shown in Fig. 31.22. During the pulsed laser irradiation (Fig. 31.22d–l), the cantilever was bent by the light pressure of the laser beam, and the probe pressed the observation surface for an instant. The scanning line, which looks like a scratch, shows this state in Fig. 31.22d–l. Hence, we observed the probe pressing onto the single polymer chain with a high probe tip pressure while irradiating, which subsequently cut the single polymer molecule as shown in Fig. 31.22e and f. We also observed a folding phenomenon, i.e., the cut end folded dynamically (Fig. 31.22j–l). The fact that it is cleaved by the probe tip pressure supports our conclusion that it is a supramolecular polymer coupled by molecular interaction because a molecular interaction force is weaker than the covalent bonds. This cleavage of the polymer chain is based on a scission of a secondary bond in a main chain. Thus, information in terms of structural chemistry can be obtained. Because its height, which is approximately 1 nm, corresponds to the height of α-CyD, the polymer chain is clearly a single molecule. Figure 31.22 shows the image in solution. We consider that this is a micro-Brownian motion, which is excited by the Brownian motion of water molecules. Unlike the STM, the AFM uses the AC (tapping) mode and the probe touches the surface. It is speculated that the molecules move mainly owing to the touching probe tip, although there may also be Brownian motion. Hence, we conducted a detailed analysis. We changed the probe scanning speed, and analyzed the one-dimensional displacement of a single polymer molecule in the probe scanning direction. In practice, we changed the frame rate to 80, 120, 150, 181.8, 200, 250, and 300 ms per frame, and collected images under the seven different conditions, and analyzed the displacement. We made a histogram of this analysis result with the displacement as the horizontal axis and the observation number as the vertical axis, and fitted the data to a Gaussian curve. The standard deviation values were the time-average displacements at each observation time. Finally, we plotted the square of the standard deviation of this distribution (i.e., mean-square displacement, $\langle x^2 \rangle$) and the probe scanning cycle time (t) as shown in Fig. 31.23. The result shows that the mean-square displacement in a single supramolecular polymer chain is proportional to the time and, hence, complies with Einstein's law of Brownian motion

$$\langle x^2 \rangle = 2Dt .$$

Here D is a diffusion coefficient. In this result, D is calculated as 1.3×10^{-14} m^2/s in the one-dimensional motion. Hence, this motion is the Brownian motion, and the motion in a single chain of supramolecular polymer is a micro-Brownian motion, i.e., thermal motion.

We anticipate that our studies will be the basis for creating innovative molecular devices such as molecular motors or molecular electronics/photonics materials that utilize thermal energy as a driving source.

I would like to understand the essence of polymers through a deeper study on the science of a single polymer chain. This research on single molecules of π-conjugated polymer should reveal a new frontier of polymer science.

Acknowledgements. I am grateful to my cooperative researchers. I would like to thank Toshio Ando and Noriyuki Kodera of Kanazawa University for the high-speed AFM study, Hideo Higuchi of Tohoku University for the TIRFM study, and Hidemi Shigekawa and Satoshi Yasuda of Tsukuba University for the STM study.

References

1. Shinohara K, Yasuda S, Kato G, Fujita M, Shigekawa H (2001) J Am Chem Soc 123:3619–3620
2. Anonymous (2001) Science 292:15
3. Shinohara K, Yamaguchi S, Higuchi H (2000) Polym J 32:977
4. Shinohara K, Yamaguchi S, Wazawa T (2001) Polymer 42:7915
5. Shinohara K, Suzuki T, Kitami T, Yamaguchi S (2006) J Polym Sci Part A Polym Chem 44:801–809
6. Shinohara K, Kitami T, Nakamae K (2004) J Polym Sci Part A Polym Chem 42:3930–3935
7. Shinohara K (2006) Chem Ind 57:620–625
8. Shinohara K, Aoki T, Kaneko T (2002) J Polym Sci Part A Polym Chem 40:1689–1697
9. Tadokoro H (1979) Structure of crystalline polymers. Wiley, New York
10. Frisch HL, Schuerch C, Szwarc M (1953) J Polym Sci 11:559
11. Takei F, Hayashi H, Onitsuka K, Takahashi S (2001) Polym J 33:310
12. Green MM, Park J-W, Sato T, Teramoto A, Lifson S, Selinger RLB, Selinger JV (1999) Angew Chem Int Ed Engl 38:3138
13. Zhang J, Jaycox GD, Vogl O (1987) Polym J 19:603
14. Okamoto Y, Nakano T (1994) Chem Rev 94:349
15. Fujiki M, Koe JR, Motonaga M, Nakashima H, Terao K, Teramoto A (2001) J Am Chem Soc 123:6253
16. Pu L (1997) Acta Polym 48:116–141
17. Aoki T, Kokai M, Shinohara K, Oikawa E (1993) Chem Lett 2009
18. Aoki T, Shinohara K, Kaneko T, Oikawa E (1996) Macromolecules 29:4192
19. Aoki T, Kobayashi Y, Kaneko T, Oikawa E, Yamamura Y, Fujita Y, Teraguchi M, Nomura R, Masuda T (1999) Macromolecules 32:79–85
20. Green MM, Peterson NC, Sato T, Teramoto A, Cook R, Lifson S (1995) Science 268:1860–1866
21. Akagi K, Piao G, Kaneto S, Sakamaki K, Shirakawa H, Kyotani M (1998) Science 282:1683–1686
22. Yashima E, Maeda K, Okamoto Y (1999) Nature 399:449–451
23. Driscoll R, Youngquist M, Baldeshwieler J (1990) Nature 346:294–296
24. Shigekawa H, Miyake K, Sumaoka J, Harada A, Komiyama M (2000) J Am Chem Soc 122:5411–5412
25. Lee D-H, Lee D-H, Soga K (1990) Makromol Chem Rapid Commun 11:559–563
26. Porath D, Bezryadln A, de Vries S, Dekker C (2000) Nature 403:635–638
27. Magonov SN, Bar G, Cantow H-J, Paradis J, Ren J, Whangbo M-H, Yagubskii EB (1993) J Phys Chem 97:9170–9176
28. Yoshimura M, Shigekawa H, Nejoh H, Saito G, Saito Y, Kawazu A (1991) Phys Rev B 43:13590–13593
29. Ando T, Kodera N, Takai E, Maruyama D, Saito K, Toda A (2001) Proc Natl Acad Sci USA 98:12468–12472
30. Binning G, Quate CF, Gerber C (1986) Phys Rev Lett 56:930
31. Tokunaga M, Kitamura K, Saito K, Hikikoshi-Iwane A, Yanagida T (1997) Biochem Biophys Res Commun 235:47–53

32. Bumagin NA, More PG, Beletskaya IP (1989) J Organomet Chem 371:397
33. Harada A, Li J, Kamachi M (1992) Nature 356:325–327
34. Wood DJ, Hruska FE, Saenger W (1977) J Am Chem Soc 99:1735
35. Grage MM-L, Wood PW, Ruseckas A, Pullerits T, Mitchell W, Burn PL, Samuel IDW, Sundström V (2003) J Chem Phys 118:7644–7650
36. Shinohara K (2005) In: The Society of Polymer Science (ed) Single molecular science of polymers. NTS, Tokyo, pp 1–32
37. Shinohara K, Kodera N, Ando T (2007) Chem Lett (in press)

32 Scanning Probe Microscopy on Polymer Solar Cells

Joachim Loos · Alexander Alexeev

Abstract. Polymer solar cells have the potential to become a major electrical power generating tool in the twenty-first century. Research and development endeavors are focusing on continuous roll-to-roll printing of polymeric or organic compounds from solution—like newspapers—to produce flexible and lightweight devices at low cost. It is recognized, though, that besides the functional properties of the compounds, the organization of structures on the nanometre level—forced and controlled mainly by the processing conditions applied—determines the performance of state-of-the-art polymer solar cells. In such devices the photoactive layer is composed of at least two functional materials that form nanoscale interpenetrating phases with specific functionalities, a so-called bulk heterojunction. In this study, we discuss our current knowledge of the main factors determining the morphology formation and evolution—based on systematic scanning probe microscopy studies—and gaps in our understanding of nanoscale structure–property relations in the field of high-performance polymer solar cells are addressed.

Key words: Polymer solar cells, AFM, Conductive AFM, SNOM, Morphology, Nanoscale functional properties

Abbreviations

AFM	Atomic force microscopy
C-AFM	Conductive atomic force microscopy
EFTEM	Energy-filtered transmission electron microscopy
HOMO	Highest occupied molecular orbital
ITO	Indium tin oxide
LUMO	Lowest unoccupied molecular orbital
MEH-PPV	Poly[2-methoxy-5-(2′-ethylhexyloxy)-1,4-phenylene vinylene]
MDMO-PPV	Poly[2-methoxy-5-(3′,7′-dimethyloctyloxy)-1,4-phenylene vinylene]
P3HT	Poly(3-hexylthiophene)
PCBM	[6,6]-Phenyl C_{61} butyric acid methyl ester
PCNEPV	Poly[oxa-1,4-phenylene-(1-cyano-1,2-vinylene)-(2-methoxy-5-(3,7-dimethyloctyloxy)-1,4-phenylene)-1,2-(2-cyanovinylene)-1,4-phenylene]
PEDOT-PSS	Poly(ethylenedioxythiophene)–poly(styrene sulfonate)
PSC	Polymer solar cell
rms	Root mean square
SNOM	Scanning near-field optical microscopy
SPM	Scanning probe microscopy
STM	Scanning tunnelling microscopy
TEM	Transmission electron microscopy

32.1
Brief Introduction to Polymer Solar Cells

Organic electronics has the potential to become one of the major industries of the twenty-first century. Research and development endeavours are focusing on continuous roll-to-roll printing of polymeric or organic compounds from solution—like newspapers—to produce flexible and lightweight devices at low cost. In particular, polymeric semiconductor-based solar cells are currently in investigation as potential low-cost devices for sustainable solar energy conversion. Because they are large-area electronic devices, readily processed polymeric semiconductors from solution have an enormous cost advantage over inorganic semiconductors. Further benefits are the low weight and flexibility of the resulting thin-film devices (Fig. 32.1). Despite the progress made in the field, it is clear that polymer solar cells (PSCs) are still in their early research and development stage. Several issues must be addressed before PSCs will become practical devices. These includes further understanding of operation and stability of these cells, and the control of the morphology formation—mainly the morphology of the active layer—which is directly linked to the performance of devices.

Figure 32.2 shows the characteristic architecture of a device. It is constructed of an indium tin oxide (ITO) back electrode deposed on glass, a poly(ethylenedioxythiophene)–poly(styrene sulfonate) (PEDOT-PSS) layer, the photoactive layer composed—in the present case—of the C_{60}-derivatized methanofullerene [6,6]-phenyl C_{61} butyric acid methyl ester (PCBM) blended with poly[2-methoxy-5-(3',7'-dimethyloctyloxy)-1,4-phenylene vinylene] (MDMO-PPV), and the top aluminium electrode. During processing of the device, the surface topography of each individual layer was measured. Both the PEDOT-PSS and the MDMO-PPV/PCBM layers are rather smooth, while the ITO and the thermally evaporated Al have a large roughness.

One of the main differences between inorganic and organic semiconductors is the magnitude of the exciton binding energy (an exciton is a bound electron–hole pair). In many inorganic semiconductors, the binding energy is small compared to the thermal energy at room temperature and therefore free charges are created under ambient conditions upon excitation across the band gap [2]. An organic semiconductor, on the other hand, typically possesses an exciton binding energy that exceeds kT

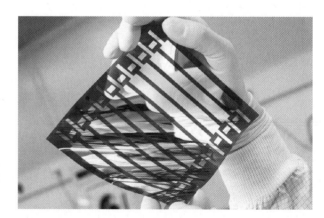

Fig. 32.1. Large-area, flexible and printable polymer solar cell demonstrated by Siemens, Germany

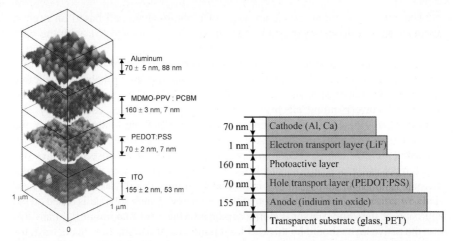

Fig. 32.2. *Left*: Atomic force microscopy (AFM) images of the surface of the different layers of a glass/ITO/PEDOT-PSS/MDMO-PPV/PCBM (1:4 wt%)/Al polymer solar cell. For each layer the thickness as determined by a surface profiler, the root mean square roughness of $1 \times 1\ \mu m^2$ area, and the peak-to-peak variations as determined by AFM are indicated. *Right*: A polymer solar cell. *ITO* indium tin oxide, *PEDOT* poly(ethylenedioxythiophene), *PSS* poly(styrene sulfonate), *MDMO-PPV* poly[2-methoxy-5-(3′,7′-dimethyloctyloxy)-1,4-phenylene vinylene], *PCBM* [6,6]-phenyl C_{61} butyric acid methyl ester, *PET* poly(ethylene terephthalate). (Reprinted with permission from [1]. Copyright 2002 Wiley-VCH)

(roughly by more than an order of magnitude) [3]. As a consequence, excitons are formed upon excitation instead of free charges. This difference between inorganic and organic semiconductors is of critical importance in PSCs. While in conventional, inorganic solar cells, free charges are created upon light absorption, PSCs need an additional mechanism to dissociate the excitons.

A successful method to dissociate bound electron–hole pairs in organic semiconductors is the so-called donor/acceptor interface. This interface is formed between two organic semiconductors with different valence and conduction bands, or equivalently dissimilar highest occupied molecular orbital (HOMO) and lowest unoccupied molecular orbital (LUMO) levels, respectively. The donor material is the material with the lowest ionization potential, and the acceptor material the one with the largest electron affinity. If an exciton is created in the photoactive layer and reaches the donor/acceptor interface, the electron will be transferred to the acceptor material and the hole will recede in the donor material. Afterwards, both charge carriers move to their respective electrode when electrode materials are chosen with the right work functions.

Considerable photovoltaic effects of organic semiconductors applying the heterojunction approach were demonstrated first by Tang [4] in the 1980s. A thin-film, two-layer organic photovoltaic cell has been fabricated that showed a power conversion efficiency of about 1% and large fill factor, which is the ratio of the maximum power ($V_{mp}J_{mp}$) divided by the short-circuit current (I_{sc}) and the open-circuit voltage (V_{oc}) in the light current density–voltage (I–V) characteristics, of 0.65 under simulated AM2 illumination (two suns equivalent).

The external quantum efficiency η_{EQE} of a photovoltaic cell based on exciton dissociation at a donor/acceptor interface is $\eta_{EQE} = \eta_A \eta_{ED} \eta_{CC}$ [5], with the light absorption efficiency η_A, the exciton diffusion efficiency η_{ED}, which is the fraction of photogenerated excitons that reaches a donor/acceptor interface before recombining, and the carrier collection efficiency η_{CC}, which is the probability that a free carrier that is generated at a donor/acceptor interface by dissociation of an exciton reaches its corresponding electrode. Donor/acceptor interfaces can be very efficient in separating excitons: systems are known in which the forward reaction, the charge-generation process, takes place on the femtosecond time scale, while the reverse reaction, the charge-recombination step, occurs in the microsecond range [6]. The typical exciton diffusion length in organic semiconductors, and in particular in conjugated polymers, however, is limited to approximately 10 nm [7–9]. These characteristics result in the limitation of the efficiency of such solar cells based on the approach of thin-film bilayer donor/acceptor interface: for maximum light absorption the active layer should be thick, at least hundreds of nanometres; in contrast, for charge generation the interface between donor and acceptor should be maximized and located near the place where the excitons are formed. Both requirements cannot be fulfilled at the same time by bilayer heterojunctions.

Independently, Yu et al. and Halls et al. have addressed the problem of limited exciton diffusion length by intermixing two conjugated polymers with different electron affinities [11, 12], or a conjugated polymer with C_{60} molecules or their methanofullerene derivatives [13]. Because phase separation occurs between the two components, a large internal interface is created so that most excitons would be formed near the interface, and can be dissociated at the interface. In case of the polymer/polymer intermixed film, evidence for the success of the approach has been found in the observation that the photoluminescence from each of the polymers was quenched. This implies that the excitons generated in one polymer within the intermixed film reach the interface with the other polymer and dissociate before recombining. This device structure, a so-called bulk heterojunction (Fig. 32.3), provides a route by which nearly all photogenerated excitons in the film can be split into free charges. At present bulk heterojunction structures are the main candidates for high-efficiency PSCs. Tailoring their morphology towards optimized performance is a challenge, though.

Several requirements for the photoactive layer of a PSC can be summarized (Table 32.1). Firstly, since the range of the absorption spectrum of the currently used donor materials is insufficient for optimum utilization of the incident light, photoinduced absorption of the cell should better fit the solar spectrum, which requires, e.g., lower band gap polymers to capture more near-IR photons. To enhance absorption capability, the photoactive layer should be thick enough to absorb more photons from incident light, which needs layers with a thickness of hundreds of nanometres. To achieve a bulk-heterojunction structure in the photoactive layer with appropriate thickness via solvent-based thin film deposition technology, a high solubility or well dispersivity of all the constituents in the solvent is a prerequisite. During solidification of the constituents from the solution and thus formation of a thin film, a few methods could be used to tune the morphology of the thin blend film towards the optimum bulk heterojunction, applying both thermodynamic and kinetic aspects. The Flory–Huggins parameter χ between the compounds describes

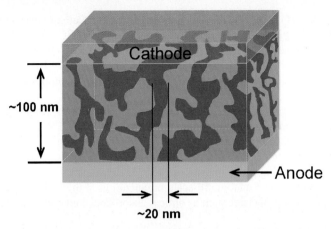

Fig. 32.3. Three-dimensional representation of a bulk heterojunction (electron donor and acceptor compounds in *different colours*) with top and back electrodes. (Reprinted with permission from [10]. Copyright 2007 American Chemical Society)

Table 32.1. Molecular and morphology requirements for the photoactive layer of a high-performance polymer solar cell

Requirement	Influenced by molecular architecture	Morphology
Utilization of incident light	Molecule design to tune band gap(s)	Layer thickness, roughness of interfaces
Exciton dissociation	Match of band properties between donor and acceptor	Maximum interface Small acceptor/donor phases within exciton diffusion range
Charge transport	Molecule design with high charge carrier mobility	Short and continuous pathways to the electrodes Ordered (crystalline) transportation pathways

the main driving force for phase segregation in a blend from the thermodynamics point of view, in which the ratio between the compounds and the conformation of the conjugated polymers in the solution are the key aspects to determine the length scale of phase separation. The kinetic issues also have a significant influence on the morphology of the thin blend film eventually obtained. For instance, by applying spin-coating, one obtains a thin blend film with rather homogeneous morphology; for preparation methods like film-casting, on the other hand, the same blend usually undergoes large-scale phase separation and a more equilibrium organization is achieved during film formation [14]. Therefore, both thermodynamic aspects as well as kinetics determine the organization of the bulk-heterojunction photoactive layer in PSCs, i. e. by controlled tuning of the formation process of the active layer, one can form a nanoscale interpenetrating network with, probably, crystalline order of both constituents [15, 16]. Further, an appropriate length scale of phase separation in the photoactive layer is the key point for high-performing PSCs. To realize this "appropriate" length scale of phase separation, one has to make a compromise between the interface area and efficient pathways for the transportation of free charge carriers towards correct electrodes so as to reduce charge carrier recombination within the photoactive layer.

In the following sections we introduce preparation approaches and routes towards creating the desired morphology of the photoactive layer of a PSC. In particular, we discuss in detail the influence of the constituents and solvent used, composition, and annealing treatments on morphology formation. Moreover, we highlight the importance of scanning probe microscopy (SPM) techniques to provide the required information on the local nanometre-scale organization of the active layer as well as its local functional properties such as optical absorbance and conductivity.

32.2
Sample Preparation and Characterization Techniques

Various blends based on the following compounds have been used as active layers for PSCs: blends of MDMO-PPV [17] as an electron donor and the C_{60} derivative PCBM [18] or poly[oxa-1,4-phenylene-(1-cyano-1,2-vinylene)-(2-methoxy-5-(3,7-dimethyloctyloxy)-1,4-phenylene)-1,2-(2-cyanovinylene)-1,4-phenylene] (PC-NEPV) as an electron acceptor. Further, blends of poly(3-hexylthiophene) (P3HT) with PCBM have been investigated. The average molecular masses of the MDMO-PPV, PCNEPV and P3HT used were 570, 113.5 and 100 kg/mol, respectively, as determined by gel permeation chromatography using polystyrene standards. The chemical structures of the polymers are presented in Fig. 32.4. PEDOT-PSS was purchased from Bayer, Germany.

For preparation of the active layer the compounds were dissolved in suitable solvents (e. g. 1,2-dichlorobenzene) and deposited by spin-coating on glass substrates covered with about 100-nm-thick thick layers of ITO and PEDOT-PSS that form the electrode for hole collection. Spin-coating conditions were adjusted such that a film thickness of the active layer of about 50–200 nm could be established. Such samples represent working photovoltaic devices, except for the missing metal back electrode. For conventional device characterization, an Al top electrode was used.

Thermal annealing was performed on complete devices, i.e. with the photoactive layer between electrodes, or on structures without a top electrode at various temperatures and times; annealing experiments were conducted in situ and ex situ, i.e. at corresponding annealing temperature or after annealing and cooling down to room temperature. More details on the compounds used as well as on layer or device preparation can be found in [19, 20].

SPM measurements were performed with the commercial microscopes Solver P47H, Solver LS and NTEGRA Aura (all NT-MDT, Moscow, Russia) equipped with optical microscopes. The cantilevers used were CSC12 (Micromash) and NSG11 (NT-MDT), and conductive tips of both types were used with an additional Au coating. Height measurements were calibrated using a 25-nm-height standard grating produced by NT-MDT. The typical force constant of the cantilever used for electrical

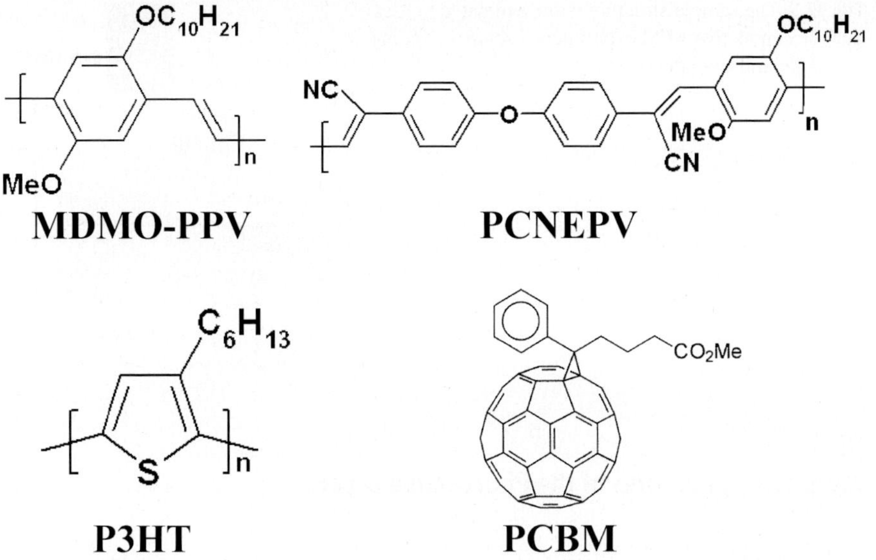

Fig. 32.4. Some suitable conjugated polymers and the fullerene derivative PCBM applied in polymer solar cells. *PCNEPV* poly[oxa-1,4-phenylene-(1-cyano-1,2-vinylene)-(2-methoxy-5-(3,7-dimethyloctyloxy)-1,4-phenylene)-1,2-(2-cyanovinylene)-1,4-phenylene], *P3HT* poly(3-hexylthiophene)

measurements was about 0.65 N/m, and the radius was below 50 nm. For annealing experiments, the Solver P47H instrument was operated in intermittent-contact mode under ambient conditions, and its integrated high-temperature heating stage was employed to acquire the abovementioned in situ and ex situ data. The temperature stability of the hot stage was controlled to within 0.1 °C.

Since the polymers are sensitive to oxygen, some atomic force microscopy (AFM) measurements were performed in a glove box (Unilab, M. Braun) having a nitrogen atmosphere with oxygen and water levels below 1 ppm. For conductive AFM (C-AFM) characterization the ITO layer was grounded during all SPM measurements. Current–voltage (I–V) characteristics were measured with and without illumination. The Fermi levels for ITO, PEDOT-PSS and gold, respectively, are 4.7, 5.2 and 5.1 eV [21, 22]. For reason of better illustration, Fig. 32.5 shows a scheme of the experimental setup for C-AFM measurements.

For scanning near-field optical microscopy (SNOM) measurements, the active layers of PSCs were transferred to glass slides. Measurements were performed using Nanofinder and NTEGRA Spectra instruments (NT-MDT, Moscow, Russia, and Tokyo Instruments, Japan). The scanning near-field optical microscope head with a single-mode optical fibre coated with aluminium was operated in transmission mode. The probes with a tip aperture of 100 nm were prepared by chemical etching and have a transmission efficiency of 10^{-3}–10^{-4} and maximum output power of 5 µW. Two lasers, He–Ne ($\lambda = 632.8$ nm) and Nd–YAG ($\lambda = 532$ nm) were employed as radiation sources. The transmitted light was collected with a ×20 objective and detected with a photomultiplier tube. For each line scan 256 data points were taken with a line scan frequency of 0.8–1.0 Hz.

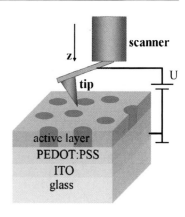

Fig. 32.5. The sample structure with segregated phases and the conductive AFM experimental setup, including a scanner and a conductive tip

32.3 Morphology Features of the Photoactive Layer

32.3.1 Influence of Composition and Solvents on the Morphology of the Active Layer

Intensive morphology studies have been performed on polymer/methanofullerene systems, in which the C_{60} derivative PCBM was applied [18, 23]. So far PCBM is the most widely used electron acceptor and the most successful PSCs are obtained by mixing it with the donor polymers MDMO-PPV and other PPV derivatives [24, 25], or with P3HT [26–30]. PCBM has an electron mobility of 2×10^{-3} cm^2/V/s [31], and compared with C_{60}, the solubility of PCBM in organic solvents is greatly improved, which allows the utilization of film-deposition techniques requiring highly concentrated solutions. After applying chlorobenzene as a solvent Shaheen et al. [24] found a dramatic increase in power conversion efficiency of spin-cast MDMO-PPV/PCBM to 2.5%, whereas for the same preparation conditions but using toluene as the solvent only 0.9% power conversion efficiency was obtained. Recently, a more comprehensive study on the influence of solvents on morphology formation was performed by Rispens et al. [32]. They compared the surface topography of MDMO-PPV/PCBM active layers by varying the solvent from xylene through chlorobenzene to 1,2-dichlorobenzene and found a decrease in phase separation from xylene through chlorobenzene to 1,2-dichlorobenzene (Fig. 32.6).

Besides the solvent used and the evaporation rate applied, the overall compound concentration and the ratio between the two compounds in the solution are important parameters controlling morphology formation; e. g. high compound concentrations induce large-scale phase segregation upon film formation [33]. For the systems MDMO-PPV/PCBM and poly[2-methoxy-5-(2′-ethylhexyloxy)-1,4-phenylene vinylene] (MEH-PPV)/PCBM the optimum ratio of the compounds is about 1:4. Initial studies with C_{60} show that the photoluminescence of PPV could be quenched for much lower C_{60} concentrations [9]. It has been demonstrated that with increasing PCBM concentration the cluster size increases accordingly [14, 34]. Recently van Duren et al. [19] introduced a comprehensive study relating the morphology of MDMO-PPV/PCBM blends to solar cell performance. The amounts of

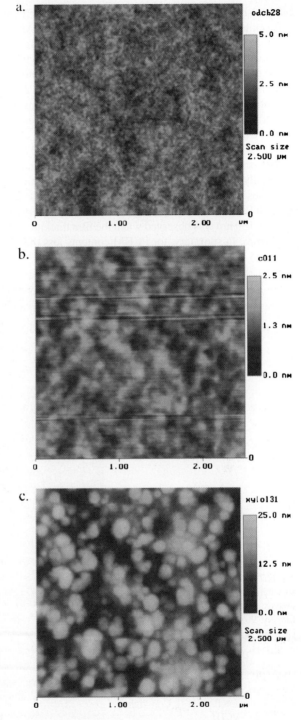

Fig. 32.6. Surface of MDMO-PPV/PCBM (1:4 weight ratio) composites, spin-cast from 0.4 polymer wt% solutions. **a** 1,2-Dichlorobenzene; **b** chlorobenzene; **c** xylene. (Reprinted with permission from [32]. Copyright 2003 Royal Society of Chemistry)

MDMO-PPV and PCBM spanned a wide range. Figure 32.7 shows the height and corresponding phase contrast images obtained by AFM for blend films for four different compositions. Apart from the compositions for the images shown in Fig. 32.7 compositions of 0, 33, 50, 67, 75, 80, 90 and 100 wt% PCBM in MDMO-PPV have been studied. The height images reveal extremely smooth surfaces for the pure compounds and blends with PCBM concentrations of 2–50 wt% with a peak-to-valley roughness of about 3 nm and root-mean-square (rms) values of 0.4 nm for an investigated area of 2 μm × 2 μm. The surface becomes increasingly rough for 67–90 wt% PCBM, with a peak-to-valley roughness of about 3–22 nm and rms values of 0.4–3.3 for the same investigated area size of 2 μm × 2 μm, and a reproducible phase contrast appears. Separate domains of one phase in a matrix of another phase can easily be recognized at these higher concentrations of PCBM. The domain size increases from 45–65 nm for 67 wt% PCBM to 110–200 nm for 90 wt% PCBM. For 80 wt% PCBM blend films, a gradual but small increase in domain size from 60–80 to 100–130 nm was observed when the film thickness was increased in steps from 65 to 270 nm. In summary, for PCBM contents less than 50 wt%, rather homogeneous

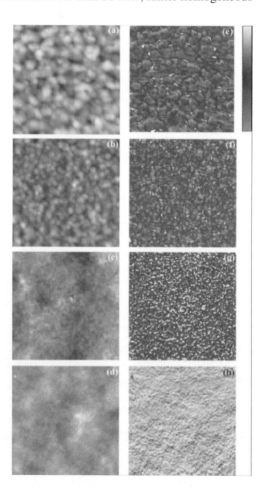

Fig. 32.7. The AFM height (**a–d**) and simultaneously taken phase (**e–h**) images of the MDMO-PPV/PCBM films of 90 wt % (**a, e**), 80 wt % (**b, f**), 67 wt % (**c, g**), and 50 wt % (**d, h**) PCBM. The *height bar* (maximum peak-to-valley) represents 20 nm (**a**), 10 nm (**b**) and 3 nm (**c, d**). The size of the images is 2 × 2 μm². (Reprinted with permission from [19]. Copyright 2004 Wiley-VCH)

film morphology is observed, and for 50 wt% or more photoluminescence is already quenched, indicating complete exciton dissociation. For concentrations around 67 wt% and higher a rather abrupt improvement in the device properties is found with the observed onset of phase separation. Thus, in general it is concluded that charge transportation rather than charge separation is the limiting factor determining the performance of the corresponding device. Only above the critical concentration of about 67 wt% PCBM forms a nanoscale percolating network within the PPV matrix.

32.3.2
Influence of Annealing

Another method to influence the morphology of the active layer of PSCs is application of a controlled thermal posttreatment. The purpose of such treatment is to probe, on the one hand, the long-term stability of the morphology. Improvement of the long-term stability of PSCs in an ambient atmosphere is currently still a challenge. However, an acceptable lifetime is a key point for PSCs to compete with traditional photovoltaic technology and is a prerequisite for commercialization.

In general, the stability of PSCs is limited by two factors. One is the degradation of materials, in particular the conjugated polymers, upon being exposed to oxygen, water or UV radiation. The other limitation comes from the possible morphology instability of the photoactive layer during operation of devices at high temperature (exposed to sunlight, which means at least 60–80 °C is possible!). For the system MDMO-PPV/PCBM annealing always results in large-scale phase separation and dominant formation of large PCBM single crystals, even for short times or low annealing temperatures below the glass-transition temperature of MDMO-PPV of about 80 °C [14, 35], which ultimately corresponds to a significant drop of the efficiency of the corresponding solar cells.

In order to get three-dimensional information of the morphology evolution of the thin blend film on the substrate, and in particular on the PCBM diffusion behaviour, AFM performed in intermittent-contact mode is used to acquire topography images during an annealing process [36]. Figure 32.8 shows a series of AFM topography images of MDMO-PPV/PCBM films recorded in situ upon annealing at 130 °C for different times. For the pristine film before any thermal treatment (as shown in Fig. 32.8a), a homogeneous morphology within a relatively large area is observed at the scanning resolution we used. Actually, the fresh film is composed of PCBM-rich domains with an average size of 80 nm distributed in the relatively PCBM-poor MDMO-PPV/PCBM matrix. For the whole film, PCBM is condensed as nanocrystals adopting various crystallographic orientations as known from corresponding transmission electron microscopy (TEM) studies [14, 37].

Upon annealing, PCBM single crystals grow gradually with annealing time and stick out of the film plane (as shown in Fig. 32.8b–f). Notably, in these AFM topography images, the bright domains are PCBM single crystals (marked as A in Fig. 32.8d); and the dark areas (depletion zones, marked as B in Fig. 32.8d) initially surrounding the PCBM crystals reflect thinner regions of the film, being composed of almost pure MDMO-PPV (i. e. depleted of PCBM). To clearly monitor the evolution of both the PCBM crystals and the depletion zones with time during annealing, a set of cross-sectional profiles across a PCBM single crystal and the surrounding

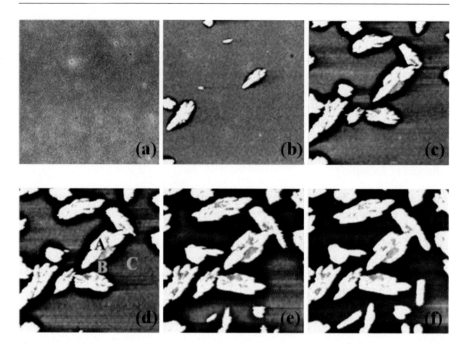

Fig. 32.8. AFM topography images of MDMO-PPV/PCBM blend films (MDMO-PPV:PCBM 1:4 by weight) recorded in situ upon annealing at 130 °C for **a** pristine film, **b** 12 min, **c** 22 min, **d** 27 min;,**e** 38 min and **f** 73 min. Scan size 15×15 μm^2; height range (from peak to valley) 200 nm. *A*, *B* and *C* in **d** represent the region where the PCBM nucleates and the crystal grows (*A*), the depletion zone that is formed owing to moving out of PCBM material towards the growing crystal (*B*), and the initial blend film still consisting of both MDMO-PPV and PCBM (*C*), respectively. (Reprinted with permission from [36]. Copyright 2004 American Chemical Society)

depletion zone are shown in Fig. 32.9. After a PCBM single crystal sticks out from the film, the growth continues in both lateral and perpendicular directions. The depletion zone around it becomes broader and deeper as annealing goes on. However, by simply comparing these topography images, we are not able to judge whether there is a difference in growth kinetics between the PCBM crystals and the depletion zones around them, which reflects how the PCBM molecules diffuse within the composite film and finally contribute to the formation of PCBM single crystals. In order to acquire exact growth kinetics for both the PCBM crystals and the depletion zones, volume quantification calculations were applied to topographic images from the composite film annealed for different times similar to the areas shown in Fig. 32.8. but this time the scan size was 100 μm × 100 μm so that detailed volume evolution of either the PCBM single crystals or the depletion zones could be resolved. Since the calculations were carried out based on quite large areas of the composite films, the results make statistical sense.

In an AFM topography image, the main information that each pixel actually carries is a relative height value. For an image with dimensions of 512×512 pixels, correspondingly, there are 262 144 height values with the image. If a specific resolution of the height value is given, a plot can be created which visualizes the total number of pixels that have a specific height value within the given resolution versus the height value. This plot actually gives a height distribution histogram for the whole topography image. As an example, Fig. 32.10 shows the histogram of height counts from the image of Fig. 32.8d. Three dominant peaks can be resolved from this histogram by using Gaussian distribution fitting with good fitting quality of $R^2 = 0.9847$.

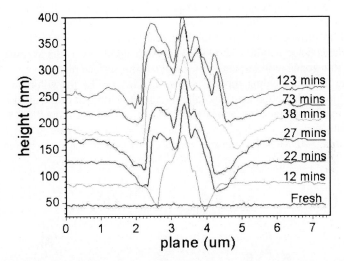

Fig. 32.9. Cross-sectional profiles across a PCBM single-crystal cluster and the depletion zone around it recorded in situ during annealing at 130 °C for the given times. The curves have been shifted along the Y-axis to give a clearer demonstration. (The sketch is similar to Fig. 2 in [36])

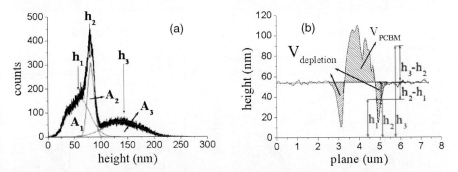

Fig. 32.10. a Height distribution histogram obtained from the AFM topography image shown in Fig. 32.8d, which is resolved into three parts A_1, A_2 and A_3. A_1 denotes the area of the PCBM single crystals, A_2 the area of the depletion zones surrounding the PCBM crystals and A_3 the area of the original film plane. h_1, h_2 and h_3 represent the peak positions of Gaussian-fitted curves corresponding to the PCBM single crystals, the depletion zones and the original film plane, respectively. **b** Cross-sectional profile across a PCBM crystal and the depletion zone around it. The values of h_1, h_2 and h_3 were obtained from **a**. (The sketches are similar to Fig. 3 in [36])

As shown in Fig. 32.10a, in a typical height-distribution histogram obtained from a topography image the three Gaussian distribution fitted curves can be assigned to the PCBM single crystals (A_1), depletion zones surrounding the PCBM crystals (A_2) and the original film surface (A_3), respectively. During the posttreatment of each topography image, the smallest height value measured within the whole film is set to zero and the height values of the other pixels are recalculated with respect to this reference to produce a relative height value for each pixel measured. Therefore, as shown in the cross-sectional profile of Fig. 32.10b, h_1, h_2, and h_3, the height values associated with the three peaks corresponding to regions A, B and C as shown in Fig. 32.8d, represent the relative average height of the depletion zones, of the original film surface plane and of the PCBM crystals, respectively. Hence, $h_1 - h_3$ denotes the average height of the PCBM crystals and $h_3 - h_2$ the average depth of the depletion zones with respect to the original film surface plane.

At the initial annealing time, as shown in Fig. 32.11a, the volume of the film collapsed in the depletion area is smaller compared with that of the diffused PCBM inserted in the crystals. As annealing goes on, more and more PCBM diffuses towards the crystals, which ultimately causes a sudden collapse of large areas of the remaining MDMO-PPV matrix, which contributes to the rapid increase in the

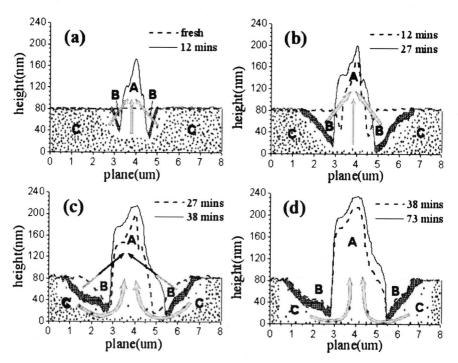

Fig. 32.11. Detailed morphology evolutions of thin MDMO-PPV/PCBM composite film upon thermal treatment. The *dots* represent PCBM molecules/nanocrystals and the density of the dots represents the richness of PCBM; the *diamond outlined regions* represent the depletion zones after PCBM material moved out for crystal growth, in which the density of the diamond outlines represents the richness of MDMO-PPV. (The sketches are similar to Fig. 5 in [36])

R value (R is the volume depletion/volume PCBM) for this period (Fig. 32.11b). With increased annealing time, the R value rapidly decreases from its maximum. This behaviour reflects the fact that almost all the MDMO-PPV matrix film has collapsed; however, PCBM diffusion and crystal growth still continue (Fig. 32.11c). Finally, the diffusion rate of PCBM within the whole film decreases, reaching its equilibrium state, as shown in Fig. 32.11d.

The prominent morphology evolution of thin MDMO-PPV/PCBM films at elevated temperature is a typical phenomenon observed at various annealing temperatures (even as low as 60 °C for freestanding thin films), with different PCBM ratios in the composite and under various spatial confinements. This morphological change of the film is ascribed to the diffusion of PCBM molecules within the MDMO-PPV matrix even at temperatures below the glass-transition temperature of the MDMO-PPV matrix of about 80 °C and subsequent crystallization of PCBM molecules into large-scale crystals. However, for the high-performing PSC, the phase separation between electron donor and electron acceptor components should be controlled within a designed range to ensure a large interface for excitons to dissociate efficiently. Large-scale phase separation enormously reduces the size of this interface area, which causes significantly decreased performance or even leads to failure of the device. Therefore, large phase separation between donor and acceptor compounds should be prevented during both device fabrication and device operation, particularly at elevated temperature.

The purpose of annealing treatments, on the other hand, is to force re-organization of the active layer towards the desired film morphology as used in high-performance PSCs, in particular when one or all compounds of the bulk heterojunction have the ability to crystallize. One example is the system P3HT/PCBM as active layer for PSCs. The high efficiency of these devices can be related to the intrinsic properties of the two components. Regioregular P3HT self-organizes into a microcrystalline structure [38] and because of efficient interchain transport of charge carriers, the (hole) mobility in P3HT is high (up to approximately $0.1 \text{ cm}^2/\text{V/s}$) [39–41]. Moreover, in thin films interchain interactions cause a redshift of the optical absorption of P3HT, which provides an improved overlap with the solar emission. Also PCBM can crystallize, and control of nucleation and crystallization kinetics allows the adjustment of the crystal size [37]. However, continuous crystallization may result in single crystals with micrometre sizes, which is not beneficial for efficient exciton dissociation, as discussed above.

Interestingly, the efficiency of solar cells based on P3HT and fullerenes was shown to depend strongly on the processing conditions and to be improved particularly by a thermal annealing step (Fig. 32.12) [26, 42, 43]. It was demonstrated that crystallization and demixing induced by the thermal annealing controls the nanoscale organization of P3HT—which forms long nanowires—and PCBM in the photoactive layer towards a morphology in which both components have a large interfacial area for efficient charge generation, have attained crystalline order that improves charge transport, and form continuous paths for charge transport to the respective electrodes (Figs. 32.12, 32.13). We infer that the long, thin fibrillar crystals of P3HT in a homogeneous nanocrystalline PCBM layer are the key to the high device performance, because they are beneficial for charge transport and control the degree of demixing [16].

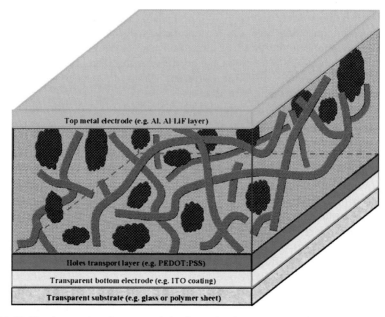

Fig. 32.12. The interpenetrating network in the active layer composed of P3HT and PCBM established after annealing

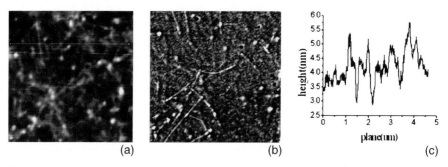

Fig. 32.13. AFM height images (**a**), phase image (**b**) and corresponding cross-section (**c**) of P3HT/C_{60} (PCBM-like) blends; scan size 5×5 μm^2

Another aspect related to annealing experiments is reorganization with or without confinement of the active layer of PSCs, i. e. annealing performed with or without a top electrode. It is well known that especially for thin films the conformation of polymer chains on the surface or at interfaces may differ from that in the "bulk" state. It is recognized that a very thin layer is present on the surface of thin polymer films, in which tie molecules are enriched and therefore the local free volume is high and chain aggregation becomes loose [44, 45]. Moreover, the presence of surfaces/interfaces influences the organization and reorganization of thin-film samples. It has been shown that, depending on the type of confinement of the active layer, the morphology evolution of MDMO-PPV/PCBM composite films is different upon

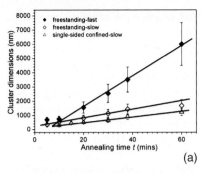
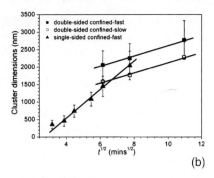

Fig. 32.14. Growth rates of PCBM single crystals from thin blend films with MDMO-PPV annealed at 130 °C for various conditions of confinement. **a** Growth rate versus annealing time t in the fast (*closed diamonds*) and slow (*open diamonds*) growth directions of the PCBM crystals for the freestanding film and in the slow (*triangles*) growth direction for the single-sided confined film. **b** Growth rate versus $t^{1/2}$ in both the fast (*closed squares*) and slow (*open squares*) growth directions for the double-sided confined film and in the fast (*triangles*) growth direction for the single-sided confined film. The *solid lines* are the linear fittings. (Reprinted with permission from [46]. Copyright 2005 American Chemical Society)

thermal annealing (Fig. 32.14) [46]: for no confinement, which corresponds to freestanding films that most of the time are used for TEM investigations, or single-sided confinement, in which the films are deposited on a substrate (common situation for AFM investigations), large elongated PCBM single crystals are formed. In case of double-sided or sandwichlike confinement (situation in the final device), in which the deposited films are additionally covered by a top layer, the top electrode, the low mobility of PCBM results in diffusion rate dominated and slow growth of PCBM crystals within the layer. As a consequence, the kinetics of phase segregation upon thermal annealing for the double-side confined thin film is slower than that in freestanding films or substrate-supported films [46]. This suppressed kinetics of phase separation could benefit the performance and stability of PSCs. However, one bears the risk of interface damage resulting from relaxation of polymer chains in the thin film, or dewetting between the photoactive layer and the metal contact if the thermal annealing is performed with the presence of a top-metal layer. These results demonstrate that annealing at different stages in device processing may result in different morphologies of the active layer, and thus in different performance of the final device.

32.3.3
All-Polymer Solar Cells

One of the main drawbacks of PSCs based on polymer/PCBM photoactive layers is the high amount of PCBM required for charge carrying; a large amount of PCBM (50–80 wt%) is required to form networks for charge transport to the electrode. Unfortunately, because of its symmetrical molecular architecture, PCBM contributes only insignificantly to light absorption. In contrast, polymer/polymer or so-called all-polymer solar cells have the specific advantage that both compounds are able to absorb light. By tuning the molecular architecture of the polymer, like by substitution of ligands, a broad absorption range can be achieved, even reaching the IR region. A few examples of such blends are known from the literature, e.g. a blend of a poly(phenylenevinylene) derivative (MEH-PPV) with the corresponding cyano-substituted variant (CN-PPV) of this polymer, or a blend of p-type and n-type fluorine derivatives [11, 12, 20, 47–50]. In general, all the above-discussed processing parameters influencing morphology formation are also valid for all-polymer systems: solvent used, evaporation rate, compound ratio and concentration in the solution, as well as annealing. Moreover, by choosing the proper synthesis route, one can tailor the molecular weight and branching of the polymers so that processability as well as device performance are improved. After optimization of the polymer architecture as well as the processing conditions for such all-polymer solar cell systems, recently, power conversion efficiencies of about 1.5% have been reported [51].

For the system MDMO-PPV/PCNEPV dewetting of the active layer on PEDOT-PSS has been identified by AFM studies to be an important feature influencing the morphology formation during processing. In AFM height-contrast images of the MDMO-PPV/PCNEPV blend after spin-coating onto PEDOT-PSS-coated glass substrates, the appearance of the film is rather heterogeneous, and large regions with different contrast can be distinguished. As revealed by AFM investigations of the topography of the blend layer and by corresponding thickness maps obtained by energy-filtered TEM (EFTEM), dewetting of PCNEPV with the substrate might cause these thickness variations, and a homogenized active layer only can be created through subsequent annealing [52].

To prove this assumption, we investigated the interface between PEDOT-PSS and the MDMO-PPV/PCNEPV layer; a float-off technique was used to remove the polymer layers from the glass/ITO substrate. By doing so, the PEDOT-PSS layer dissolved and the photoactive layer floated on the water. It is assumed that this procedure does not alter the morphology of the active layer. The backside of the pristine polymer blend film was surprisingly rough; patterns in the form of a network and dropletlike structures were found (Fig. 32.15). This structure is attributed to dewetting of the photoactive blend on top of the PEDOT-PSS layer. To rule out the presence of PEDOT-PSS left after floating, we also probed films that were peeled from the substrate, and these surfaces yielded similar AFM images. Further, we note that the spin-coated PEDOT-PSS layer is flat (rms roughness less than 2 nm) and consequently does not act as a mould for the observed rough layer. The height of the structures is about 12 nm, which is considerable compared with the 40-nm thickness of the measured film.

The backsides of films annealed on PEDOT-PSS were completely flat (rms roughness less than 2 nm). Moreover, temperature-dependent AFM experiments on (initially pristine) films revealed the disappearance of the dewetting structure at about 70 °C. Blends spun on bare glass substrates do not show this dewetting behaviour and form immediately flat films. The same holds for films of the pure materials (MDMO-PPV or PCNEPV) on either PEDOT-PSS or glass, which proves the assumption that dewetting of the active layer onto PEDOT-PSS causes the heterogeneous film morphology, and thus a low efficiency of the corresponding device. After annealing, fortunately, one observes considerable efficiency increase.

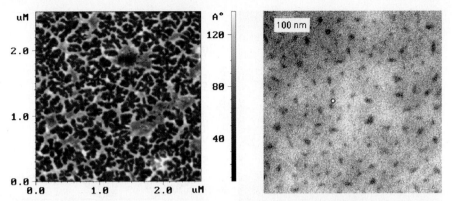

Fig. 32.15. *Left*: AFM height-contrast image of the interface between the MDMO-PPV/PCNEPV active layer and the PEDOT-PSS layer showing high roughness probably caused by dewetting. *Right*: Energy-filtered transmission electron microscopy image showing the homogeneous phase separation after annealing (*black dots* correspond to PCNEPV domains, and *brighter matrix* corresponds to MDMO-PPV; more details on the phase separation of this system can be found in Sect. 32.4.2). (Reprinted with permission from [20], copyright 2004 American Chemical Society, and [52], copyright 2005 Wiley)

32.4
Nanoscale Characterization of Properties of the Active Layer

32.4.1
Local Optical Properties As Measured by Scanning Near-Field Optical Microscopy

As for the system MDMO-PPV/PCBM, the system P3HT/PCBM may show in certain circumstances large-scale phase segregation of PCBM. Better understanding of such diffusion features of PCBM is required to control phase segregation towards the desired morphology, which ultimately offers good device performance. Thus, in the following we demonstrate that the application of SNOM is able to provide insights related to such features.

In a near-field optical measurement a single-mode optical fibre with a hole of around 100 nm in diameter on the sharp end is placed into the near field (distance of 2–4 nm) of the sample investigated. A near-field probe generates an evanescent field, the wavelength of which is shorter than that of the propagating field and thus the diffraction limit of around one half of the wavelength of the source light is circumvented [53]. The spatial resolution is then defined by the diameter of the hole. High-quality optical contrast images were obtained by this technique. It is especially important for analysis of materials on the nanometre scale. As an example, in a recent SNOM study properties of MEH-PPV/PCBM systems were studied [54].

Near-field microscopic measurements provide a range of information about the P3HT/PCBM films studied [55]. In Fig. 32.16 a topography contrast image and the corresponding cross-sectional plot are shown. Different grey levels are consistent with the height of the features as can be seen from the corresponding cross-sectional plot. Three different regions can be distinguished in the sample: (1) a bright crystalline feature sticking out of the film plane and having a size of about 20 μm in

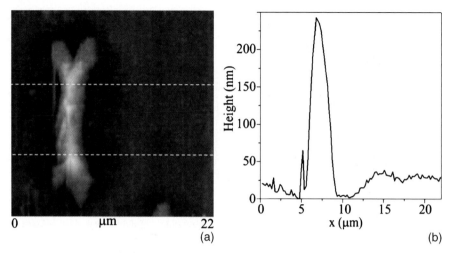

Fig. 32.16. Topography image (**a**) and average profile of selected area (**b**) obtained from scanning near-field optical microscopy (SNOM) measurements. (Reprinted with permission from [55]. Copyright 2004 American Chemical Society)

length and 5 μm in width, and a typical height of about 200 nm; (2) a dark zone surrounding this crystal; and (3) the remaining area representing the initial film plane. From the cross-sectional plot, we see that the zone surrounding the crystal is substantially deeper when compared with the initial film plane having a depth on the order of 30–50 nm, which is a first indication for depletion of PCBM similar to in the MDMO-PPV/PCBM system [14].

Optical contrast SNOM images taken in transmission mode are shown in Fig. 32.17. The images were acquired at two different radiation wavelengths and show different contrast behaviour. For our experimental setup, simultaneously topography and light absorption measurements with a He–Ne laser of wavelength 632.8 nm were performed. Figure 32.17a shows the optical absorption image of the same area as in Fig. 32.16a. For irradiation with such laser light the area with smaller thickness surrounding the crystal demonstrates a higher absorbance value compared with that of the initial film. However, at a wavelength of 532 nm the absorbance behaviour inverts: the absorbance of the initial film exceeds the absorbance of the crystal-close region. The crystals themselves have comparable transmission at both the 632.8- and 532-nm wavelengths. Unfortunately, our experimental setup does not allow simple change of the laser source, so we were not able to perform the optical absorption measurement on the same sample area.

These differences in optical absorption can be explained by monitoring the absorption spectra of pure P3HT and a 1:1 by weight P3HT/PCBM mixture obtained using a conventional UV–vis spectrometer (Fig. 32.18). After mixing the P3HT with PCBM, the absorption maximum in the visible spectrum shifted from 570 nm to shorter wavelengths. Assuming that the areas surrounding the PCBM crystals are depleted of PCBM, we can explain the different absorption characteristics depending on the irradiation wavelength used. For both wavelengths, maximum absorption is found for the pure PCBM crystals, which mainly is related to their large thickness

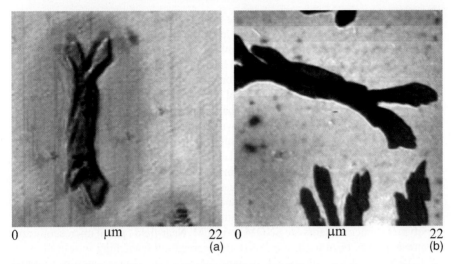

Fig. 32.17. Optical contrast images obtained in SNOM transmission mode using monochromatic laser radiation with wavelengths of 632.8 nm (**a**) and 532 nm (**b**). (Reprinted with permission from [55]. Copyright 2004 American Chemical Society)

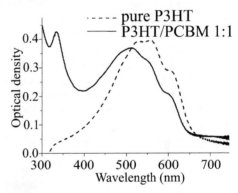

Fig. 32.18. UV–vis absorption spectra in the visible region of pure P3HT and its 1:1 by weight mixture with PCBM. (Reprinted with permission from [55]. Copyright 2004 American Chemical Society)

of several hundreds of nanometres. Comparing the initial film area, still composed of 1:1 P3HT/PCBM, with the area probably depleted by PCBM, the relatively higher absorption of 632.8-nm irradiation of the latter is related to the dominantly larger absorption of pure P3HT for such a wavelength. In contrast, the slight higher absorption of the initial film for 532-nm irradiation is caused by the somewhat similar absorption characteristics of pure P3HT and the P3HT/PCBM mixture for such a wavelength; however, the depletion zone is thinner and thus results in less total absorption.

32.4.2 Characterization of Nanoscale Electrical Properties

In general, performance measurements of PSCs are carried out on operational devices having at least the size of square millimetres to centimetres. On the other hand, the characteristic length scale determining the functional behaviour of the active layer is on the order of 10 nm (exciton diffusion length) to about 100 nm (layer thickness). Moreover, it is believed that the local organization of nanostructures dominantly controls the electrical behaviour of devices. Thus, it is necessary to obtain property data of nanostructures with nanometre resolution to be able to establish structure–property relations that link length scales from local nanostructures to large-scale devices.

In this respect, a very useful analytical technique is SPM. In previous studies, scanning tunnelling microscopy (STM) was used for investigation of semiconducting polymers [56–58]. In particular, the current–voltage (I–V) characteristics at the surface of PPV samples have been studied and modelled. However, in STM measurements, variations of the topography and information on the electrical behaviour are superimposed, and especially for electrically heterogeneous samples like bulk heterojunctions, separation of electrical data from topography information is difficult. Other SPM techniques probably better suited for analysis of the local functionality of polymer semiconductors are SNOM and AFM. Near-field optical microscopy and spectroscopy has been used, e. g., to study aggregation quenching in thin films of MEH-PPV [57]. The results obtained suggest that the size of aggregates in thin films must be smaller than the resolution limit of SNOM, roughly 50–100 nm. Further, SNOM has been applied to map topography and photocurrent of the active layer of some organic photovoltaic devices [59]; however, the spatial resolution reported in that study was only about 200 nm. Our results, as reported earlier, show a similar lateral resolution of about 100 nm.

One of the electrical methods of AFM is scanning Kelvin probe microscopy (SKPM), which is used to obtain the distribution of potentials at the sample surface. In fact, SKPM determines the difference between the work function of the tip and the sample surface. Measuring inside a glove box allows us to reduce the influence of adsorbed water on the surface potential measurements. We have used gold-coated tips to measure surface potential variations within MDMO-PPV/PCNEPV blends. The surface potential distribution image obtained (Fig. 32.19) shows that the work

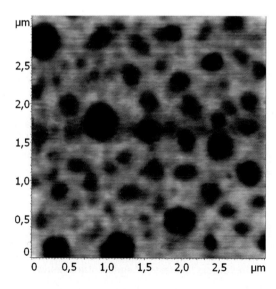

Fig. 32.19. Surface potential distribution measured with an Au-coated tip on a 3.1 × 3.1 μm^2 area of a MDMO-PPV/PCNEPV blend sample. The measured colour scale corresponds to a difference of 80 mV

function of the domains is different from the work function of the matrix, which proves the existence of phase separation inside the active layer. The difference in surface potential between domains and matrix of the sample is up to 0.12 V. Since morphology influences the measured potential, absolute determination of the work function of the components in the blend is difficult [60]. The lateral resolution of SKPM is about 50 nm [61].

AFM using a conductive probe, so-called C-AFM [62, 63], is able to overcome the abovementioned problem of STM and provides a higher spatial resolution than SNOM. Because AFM uses the interaction force between the probe and the sample surface as a feedback signal, both topography and conductivity of the sample can be mapped independently. Theoretically, the resolution of C-AFM is as small as the tip–sample contact area, which can be less than 20 nm. C-AFM is widely used for the characterization of electrical properties of organic semiconductors. For example, single crystals of sexithiophene have been studied [64], where the $I-V$ characteristics of the samples were measured. Several electrical parameters, such as grain resistivity and tip–sample barrier height, were determined from these data. In another study, the hole transport in thin films of MEH-PPV was investigated and the spatial current distribution and $I-V$ characteristics of the samples were discussed [22].

As mentioned above, for analysing our active layer samples by means of C-AFM, we operate the atomic force microscope in a glove box to protect the samples from degeneration. It is well known that the performance of most PSCs as well as of organic electronic systems, in general, drops dramatically after short-time exposure to oxygen, especially when they are illuminated by light [65]. In contrast to characterization of whole devices in an inert atmosphere, C-AFM measurements are commonly performed, however, in air at ambient conditions. It is the purpose this section to clearly demonstrate that the local electrical properties of nanostructures in the active layer of PSCs are changed for several reasons when C-AFM measurements are performed at ambient conditions; and as a consequence the results obtained are not comparable with data gained from device characterization. At the same time, we would like to illustrate that C-AFM measurements performed in the inert atmosphere of a glove box provide reliable information on electrical properties of organic nanostructures and allow the creation structure–processing–property relations of functional polymer systems.

MDMO-PPV/PCBM is one of the systems best studied for applications as an active layer in high-efficiency PSCs; efficiencies of about 2.5% have been reported for optimized preparation conditions [66]. Further, it has been demonstrated that the performance of devices having these compounds as a blend in their active layers decreases immediately when they are exposed to air [67]. For this reason, we chose the blend MDMO-PPV/PCBM as a model system for our C-AFM experiments.

Figure 32.20 shows a series of C-AFM images obtained at ambient conditions in air of a thin PCBM/MDMO-PPV film spin-coated from toluene solution. All images were acquired with a tip coated with a gold layer. For such preparation conditions, PCBM and MDMO-PPV segregate, and PCBM forms large nanocrystalline domains imbedded in the MDMO-PPV matrix [14, 68]. The topography image (Fig. 32.20a) shows that the PCBM domains (bright areas) have maximum diameters of about 500 nm. Phase segregation is responsible for the high roughness of the film: the PCBM domains stick a few tenths of a nanometre out of the film plane.

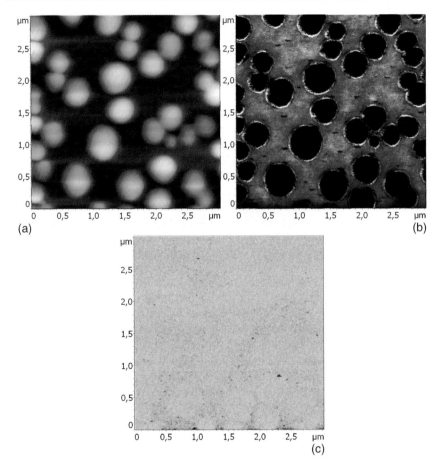

Fig. 32.20. Conductive AFM (C-AFM) image series acquired at ambient conditions in air of a thin PCBM/MDMO-PPV film spin-coated from toluene solution: **a** topography image (colour scale represents 70-nm height variations), and current distribution images of the same area for **b** a negative bias at the tip of −2.3 V (colour values represent 1.2-nA current variation) and for **c** a positive bias at the tip of +10 V (colour values represent 250-pA current variation)

Figure 32.20b and c represents the current distribution image for bias voltages at the tip of −2.3 V (Fig. 32.20b) and +10 V (Fig. 32.20c), respectively, measured at the same sample area as the topography image. For negative bias, good contrast is obtained between the electron donor (p-type semiconductor) and the electron acceptor (n-type semiconductor) materials in the sample. From the corresponding energy level diagram (Fig. 32.21) it follows that the difference between the HOMO level of MDMO-PPV and the Fermi levels of both electrodes (ITO/PEDOT-PSS and the Au tip) is rather small, so we expect ohmic contacts for hole injection and strong energy barriers for electrons [21, 64]. Therefore, a hole-only current through the MDMO-PPV is expected for both polarities of voltage in an ITO/PEDOT-PSS/MDMO-PPV/Au-tip structure. On the other hand, we can conclude that areas

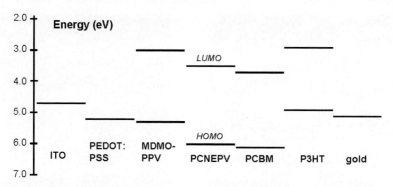

Fig. 32.21. Energy level diagram for a few materials commonly used in polymer solar cells. *HUMO* highest occupied molecular orbital, *LUMO* lowest unoccupied molecular orbital

of low current level correspond to the electron-acceptor materials, i.e. PCBM, which is in accordance with the abovementioned topographical observations. For positive bias, however, no differences between the two phases can be obtained, and the measured overall current level is below the noise level of our experimental setup.

Since the compounds under investigation are sensitive to oxygen and at ambient conditions the always-present water layer on top of the sample surface interferes with nanometre-scale electrical measurements, which has been reported recently for the case of surface potential measurements [69], we performed additional C-AFM experiments on the same MDMO-PPV/PCBM system but in an inert atmosphere. We assembled the C-AFM setup in a glove box filled with a nitrogen atmosphere and an oxygen and water level below 1 ppm.

Figure 32.22 presents C-AFM measurements of such a MDMO-PPV/PCBM thin-film sample. The main difference is that the sample was sealed in a vessel after preparation in another glove box, and was subsequently transported to and mounted in the C-AFM setup assembled in a glove box. This procedure guarantees that the sample is not exposed to oxygen or water, and imitates the common procedure of handling devices in an inert atmosphere. For the negative bias of -10 V similar features as discussed for Fig. 32.20b can be seen (Fig. 32.22a). The dark areas represent PCBM domains with low current imbedded in the bright MDMO-PPV matrix showing high current. The current level scaling of the image of about 20 nA is larger than for Fig. 32.20b because of the higher bias applied, and probably the different contact load of the tip on the sample surface. In contrast to measurements performed in air, also for the positive bias of $+10$ V good contrast between the two phase-segregated compounds can be obtained (Fig. 32.22a). Besides some little drift of the area probed during the two successive C-AFM measurements with positive and negative biases, PCBM domains can be recognized as dark areas embedded in the bright MDMO-PPV matrix. The current level scaling of the image is lower as for the case of positive bias; however, the higher noise level results in a larger total current level scaling of about 20 nA. These results demonstrate the importance of performing C-AFM analysis in an inert and oxygen-free and water-free atmosphere to protect samples from degeneration and to save their full functionality for reliable data sets.

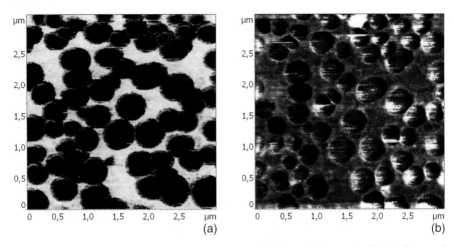

Fig. 32.22. C-AFM images of a PCBM/MDMO-PPV thin film sample acquired in the inert and water-free atmosphere of a glove box showing the current distribution of the same sample area for **a** a positive bias at the tip of +10.0 V (colour values represent 18-nA current variation) and for **b** a negative bias at the tip of −10 V (colour values represent 20-nA current variation)

On the basis of these results, recently the first study on the spatial distribution of electrical properties of realistic bulk heterojunctions was performed by applying C-AFM with lateral resolution better than 20 nm [70]. For this study the MDMO-PPV/PCNEPV system was chosen. Measurements of the electrical current distribution over the sample surface were performed with a Au-coated tip. Again, in such an experiment the tip plays the role of the back electrode but having a much more localized contact area. A voltage was applied to the tip and the ITO front electrode was grounded (Fig. 32.5). For C-AFM measurements the tip was kept in contact with the sample surface while the current through the tip was measured. In contrast to operation in intermittent-contact mode, contact mode is characterized by a strong tip–sample interaction that can lead to destruction of the surface, especially in the case of soft polymer samples. Therefore, the load applied to the tip during C-AFM has to be small enough to reduce sample destruction and, at the same time, it must provide a reliable electric contact. We usually operated with a load of about 10–20 nN. The contact cantilevers used for C-AFM are suitable for operation in intermittent-contact mode as well as in contact mode, so nondestructive testing of the sample surface could be performed before and after the C-AFM measurements. C-AFM measurements of the same sample area were done several times and resulted in completely reproducible data. Subsequent analysis of the surface performed in intermittent-contact mode showed almost no destruction of the sample surface; only minor changes were detected from time to time.

A topography image and the corresponding current distribution measured at +8 and −8 V on the tip are shown in Fig. 32.23. All images were acquired subsequently, so some drift occurred. All pronounced domains in the topography image (Fig. 32.23a) correlate with regions of minimal current in the C-AFM image (dark areas in Fig. 32.23b).

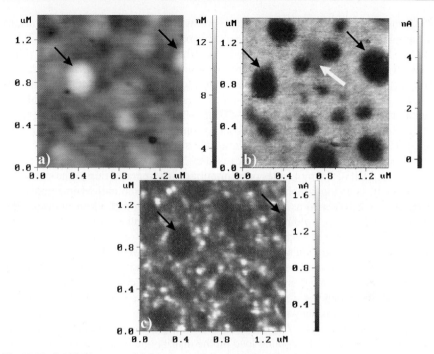

Fig. 32.23. C-AFM images of the same area of a MDMO-PPV/PCNEPV active layer: **a** topography, **b** current distribution image with a positive bias at $U_{tip} = +8$ V(the *white arrow* in **b** indicates a domain with reduced current) and **c** current distribution image with a negative bias at $U_{tip} = -8$ V. *Black arrows* indicate the same domains for reason of easy identification. (Reprinted with permission from [70]. Copyright 2006 Elsevier)

From the corresponding energy level diagram (Fig. 32.21), it follows that the difference between the HOMO level of MDMO-PPV and the Fermi level of both electrodes is rather small, so we expect ohmic contacts for a hole injection and strong energy barriers for electrons. Therefore, a hole-only current through the MDMO-PPV is expected for both polarities of voltage in an ITO/PEDOT-PSS/MDMO-PPV/Au-tip system. The energy difference between the HOMO and the LUMO of PCNEPV and the Fermi levels of both electrodes is about 1 eV, which means that a large barrier for electron injection exists in the structure ITO/PEDOT-PSS/PCNEPV/Au tip (some changes of barrier heights are possible when contact between metal electrodes and organic material occurs [21, 64]). Because the hole mobility of an n-type polymer is typically smaller than that of a p-type polymer, a hole-only current through the MDMO-PPV is larger than for PCNEPV in both bulk and contact-limited regimes. Therefore, we assume that the observed contrast in Fig. 32.23b is due to a hole current, flowing through the MDMO-PPV-rich phase (see also Fig. 32.15, EFTEM image of a comparable system).

However, C-AFM measurement also shows regions with a current value lying between that of the MDMO-PPV matrix and the PCNEPV domains. An arrow in Fig. 32.23b marks one of these regions. These areas might be assigned as PCNEPV domains inside the active layer that are possibly covered by MDMO-PPV.

It has been reported that the electrical contrast measured by C-AFM at the surface of samples depends on the sign of the voltage applied [64]. As shown in Fig. 32.23c, the C-AFM measurements at negative bias on the tip showed drastic changes of the contrast in the current images compared with positive bias (Fig. 32.22b). PCNEPV domains again showed only little current at low load; however, MDMO-PPV showed a heterogeneous spatial current distribution. These electrical heterogeneities indicate small grains with a typical size of 20–50 nm, which differs depending on the value of current. A similar structure was observed on MEH-PPV films [22]. In the case of MEH-PPV, the authors attribute these substructures to a special and very local organization of the film that could be caused by local crystallization of stereoregular parts of the MEH-PPV molecules or by impurities incorporated during the synthesis.

In addition to topography and current-sensing analysis, current-imaging spectroscopy was performed as well. The procedure for such measurements is similar to the so-called force volume technique [71], which implies measurements of the force–distance curve at each point of a scan in order to get complete information about lateral distribution of mechanical properties at the surface. Here, we extend this method to measurements of electrical properties of the sample [72]. Current–distance (I–z, at constant voltage) and current–voltage (I–V, for constant distance; always in contact) dependencies were collected at each point of a scan. The procedure for such measurements is shown schematically in Fig. 32.24.

An array of 128×128 I–z curves at $+8$ V on the tip was obtained in order to study the influence of the applied load on the distribution of current. An I–z curve at one point was obtained by movement of the tip by the scanner in the z-direction. In our experiment first we bring the tip in contact with the sample surface, then the scanner with the attached tip is moved up 60 nm, and finally moves down for 160 nm while simultaneous measurements of the local current are performed (Fig. 32.25). The change of the z-position of the scanner, when tip–sample contact occurs, leads to both the bending of the lever and penetration of the tip in the sample. Additional experiments showed that the penetration depth in the range of the load used is much smaller than lever bending, i.e. the load can be roughly estimated from the z movement of the scanner: $F = k\Delta z$, where k is the cantilever force constant and Δz is the scanner displacement calculated from the first point where tip–sample contact occurs. The cross-section of the array of I–z curves obtained shows the distribution of the current at the sample surface for a certain z-position of the scanner (or load). Movement from point to point was executed in contact mode under small load (less than 20 nN in our experiments). As a result the destructive action of the lateral force is reduced significantly. It is also possible to perform such measurements in intermittent-contact mode, thus making the measurements of polymer topography

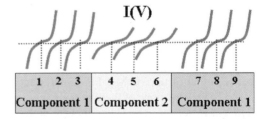

Fig. 32.24. I–V measurements at nine consecutive points. The *light blue area* represents a PCNEPV-rich domain, the *beige areas* the MDMO-PPV-rich matrix

more precise. For the sample studied here, we obtained similar results with current imaging spectroscopy performed in both contact and intermittent-contact mode.

Figure 32.25 demonstrates I–z curves measured in the PCNEPV domains and the MDMO-PPV matrix. For loads in the range 10–30 nN the current is approximately constant, but for an increased load of more than 30 nN the current starts to rise rapidly. In the case of PCNEPV a similar trend could be measured; however, because PCNEPV is a less good hole conductor, current values are on a lower level compared with MDMO-PPV. This probably means that current measurements on PCNEPV need a greater tip–sample interaction compared with those on MDMO-PPV. Even a strong penetration of the sample surface by the tip may be required.

In contrast, Fig. 32.26 shows the typical I–V behaviour of this system, and the corresponding C-AFM current distribution images for various biases when they are measured in an inert atmosphere. The data were obtained by application of so-called I–V spectroscopy, which means that for a matrix of 128×128 points on the surface of the sample full I–V curves were measured for biases from -10 to $+10$ V. Three different types of I–V characteristics can be found in the sample depending on the location of the measurement. For domains of the electron-acceptor compound PCNEPV (see corresponding EFTEM data, Fig. 32.15b) the current is always low, and the general contrast of the current distribution images depends only on the bias applied. In the case of the electron-donor matrix compound MDMO-PPV two different I–V characteristics can be obtained showing almost the same behaviour for positive bias but varying significantly for negative bias. The current distribution images of Fig. 32.26 demonstrate this behaviour and provide additional information about lateral sizes of the MDMO-PPV heterogeneities. The point resolution of the images is about 15 nm. Some heterogeneities can be recognized with sizes as small as about 20 nm as follows from C-AFM measurements (Fig. 32.23). The origin of the heterogeneities in MDMO-PPV still is under investigation; however, it is assumed that either small crystalline domains are formed or impurities originating from the synthesis process are detected. The C-AFM results obtained on pure MDMO-PPV film are identical to that obtained on the matrix in the heterogeneous film MDMO-

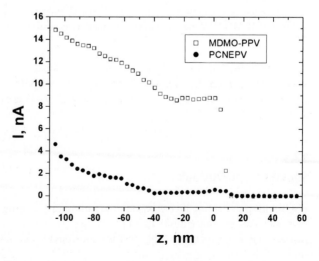

Fig. 32.25. Current–distance behaviour of the MDMO-PPV matrix and PCNEPV domains. The direction from right to the left corresponds to the approaching of the tip to the substrate surface. (Reprinted with permission from [70]. Copyright 2006 Elsevier)

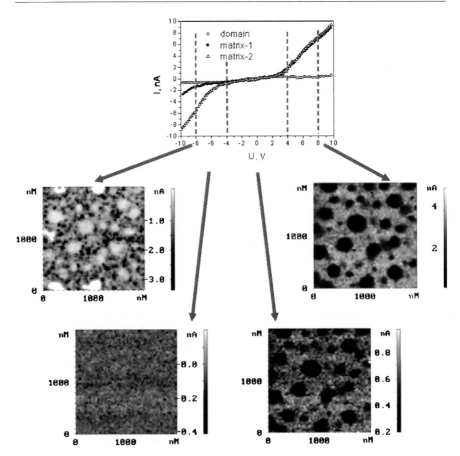

Fig. 32.26. *Top*: Typical $I–V$ curves as measured for each point of the $I–V$ spectroscopy scan (128 × 128 pixels) demonstrating the heterogeneous $I–V$ characteristics of the MDMO-PPV matrix. *Bottom*: For four biases the corresponding current distribution images are shown demonstrating the obvious contrast between PCNEPV and MDMO-PPV for high bias as well as contrast within the MDMO-PPV matrix with heterogeneities as small as a few tens of nanometres

PPV/PCNEPV. The log I–log V plot of data obtained shows quadratic dependence of the current I on the voltage V measured on MDMO-PPV. This implies space-charge-limited current that is in agreement with $I–V$ measurements on complete devices [73, 74].

32.5
Summary and Outlook

PSCs with photoactive layers based on compounds forming intermixed bulk heterojunctions are becoming more and more attractive for commercial application. In this respect, one key aspect for high-performance devices is fundamental knowledge of structure–processing–property relations, on the nanometre length scale. Intensively, studies have been performed on the influence of solvent used, the composition of the solutions, thin-film preparation conditions, crystallization behaviour and post-treatments, to name but a few, on the functional properties of the photoactive layer. Nevertheless, our insights into features influenced by the local organization of functional structures, such as the interface organization at the junctions between phases, are sparse and need improvement. Only with the availability of such data, novel approaches for further improving and designing the morphology of the bulk heterojunction photoactive layer can be developed.

Currently, by choosing the optimum solvent, compound composition and preparation conditions nanoscale phase separation between electron-donor and electron-acceptor constituents can be achieved, and continuous pathways for charge carriers to the electrodes are provided. However, from theoretical considerations the efficiency of PSCs should reach at least 10%, which is twice the highest efficiency reported recently. Such performance would make PSCs commercially very attractive and comparable with amorphous silicon solar cells, but having the strong advantage of easy processing. So, we constantly have to continue closing our knowledge gaps on structure–processing–property relations on the nanoscale to ultimately reach the highest performance of PSCs.

As demonstrated by the present study, SPM techniques as tools to explore the nanoworld have played and continuously play an important role in the research field of PSCs. In particular, operation modes analysing simultaneously the local morphology of the photoactive layer and measuring functional (optical and electrical) properties with nanometre resolution at the same spot contribute in establishing new insights required to commercially produce the high-performance PSCs of the future.

Acknowledgements. We would like to thank Xiaoniu Yang, Evgueny Klimov, Denis Ovchinnikov, Wei Zhen Li and Hanfang Zhong for their important assistance in major parts of the work presented. We thank René Janssen, Martijn Wienk, Jeroen van Duren, Sjoerd Veenstra and Marc Koetse for fruitful discussions. Part of the work is embedded in the research program of the Dutch Polymer Institute (DPI projects 326 and 524), and additional financial support is appreciated from the Ministry of Economic Affairs of the Netherlands via the Technologische Samenwerkings project QUANAP (SenterNovem TSGE3108).

References

1. van Duren JKJ, Loos J, Morrissey F, Leewis CM, Kivits KPH, van IJzendoorn LJ, Rispens MT, Hummelen JC, Janssen RAJ (2002) Adv Funct Mater 12:665
2. Bube RH (1992) Photoelectronic properties of semiconductors. Cambridge University Press Cambridge
3. Pope M, Swenberg CE (1999) Electronic processes in organic crystals and polymers. Oxford University Press, Oxford
4. Tang CW (1998) Appl Phys Lett 48:183
5. Peumans P, Yakimov A, Forrest SR (2003) J Appl Phys 93:3693
6. Smilowitz L, Sariciftci NS, Wu R, Gettinger C, Heeger AJ, Wudl F (1993) Phys Rev B 47:13835
7. Yoshino K, Hong YX, Muro K, Kiyomatsu S, Morita S, Zakhidov AA, Noguchi T, Ohnishi T (1993) Jpn J Appl Phys Part 2 32:L357

8. Halls JJM, Pichler K, Friend RH, Moratti SC, Holmes AB (1996) Appl Phys Lett 68:3120
9. Haugeneder A, Neges M, Kallinger C, Spirkl W, Lemmer U, Feldmann (1999) J Phys Rev B 59:15346
10. Yang X, Loos J (2007) Macromolecules 40:1353
11. Yu G, Heeger AJ (1995) J Appl Phys 78:4510
12. Halls JJM, Walsh CA, Greenham NC, Marseglia EA, Friend RH, Moratti SC, Holmes AB (1995) Nature 376:498
13. Yu G, Gao J, Hummelen JC, Wudl F, Heeger AJ (1995) Science 270:1789
14. Yang X, van Duren JKJ, Janssen RAJ, Michels MAJ, Loos J (2004) Macromolecules 37:2151
15. Schmidt-Mende L, Fechtenkötter A, Müllen K, Moons E, Friend RH, MacKenzie JD (2001) Science 293:1119
16. Yang X, Loos J, Veenstra SC, Verhees WJH, Wienk MM, Kroon JM, Michels MAJ, Janssen RAJ (2005) Nano Lett 5:579
17. Becker H, Spreitzer H, Kreuder W, Kluge E, Schenk H, Parker I, Cao Y (2000) Adv Mater 12:42
18. Hummelen JC, Knight BW, LePeq F, Wudl F, Yao J, Wilkins CL (1995) J Org Chem 60:532
19. van Duren JKJ, Yang X, Loos J, Bulle-Lieuwma CWT, Sieval AB, Hummelen JC, Janssen RAJ (2004) Adv Funct Mater 14:425
20. Veenstra SC, Verhees WJH, Kroon JM, Koetse MM, Sweelssen J, Bastiaansen JJAM, Schoo HFM, Yang X, Alexeev A, Loos J, Schubert US, Wienk MM (2004) Chem Mater 16:2503
21. Koch N, Elschner A, Schwartz J, Kahn A (2003) Appl Phys Lett 82:2281
22. Lin H-N, Lin H-L, Wang S-S, Yu L-S, Perng G-Y, Chen S-A, Chen S-H (2002) Appl Phys Lett 81:2572
23. Wudl F (1992) Acc Chem Res 25:157
24. Shaheen SE, Brabec CJ, Sariciftci NS, Padinger F, Fromherz T, Hummelen JC (2001) Appl Phys Lett 78:841
25. Mozer A, Denk P, Scharber M, Neugebauer H, Sariciftci NS, Wagner P, Lutsen L, Venderzande D (2004) J Phys Chem 108:5235
26. Padinger F, Rittberger RS, Sariciftci NS (2003) Adv Funct Mater 13:85
27. Waldauf C, Schilinsky P, Hauch J, Brabec CJ (2004) Thin Solid Films 503:451
28. Al-Ibrahim M, Ambacher O, Sensfuss S, Gobsch G (2005) Appl Phys Lett 86:201120
29. Reyes-Reyes M, Kim K, Carrolla DL (2005) Appl Phys Lett 87:083506
30. Ma W, Yang C, Gong X, Lee K, Heeger AJ (2005) Adv Funct Mater 15:1617
31. Mihailetchi VD, van Duren JKJ, Blom PWM, Hummelen JC, Janssen RAJ, Kroon JM, Rispens MT, Verhees WJH, Wienk MM (2003) Adv Funct Mater 13:43
32. Rispens MT, Meetsma A, Rittberger R, Brabec CJ, Sariciftci NS, Hummelen JC (2003) Chem Commun 2116
33. Hoppe H, Niggemann M, Winder C, Kraut J, Hiesgen R, Hinsch A, Meissner D, Sariciftci NS (2004) Adv Funct Mater 14:1005
34. Martens T, D'Haen J, Munters T, Beelen Z, Goris L, Manca J, D'Olieslaeger M, Vanderzande D, Schepper LD, Andriessen R (2003) Synth Met 138:243
35. Hoppe H, Drees M, Schwinger W, Schäffler F, Sariciftci NS (2005) Synth Met 152:117
36. Zhong H, Yang X, deWith B, Loos J (2006) Macromolecules 39:218
37. Yang X, van Duren JKJ, Rispens MT, Hummelen JC, Michels MAJ, Loos J (2004) Adv Mater 16:802
38. Prosa TJ, Winokur MJ, Moulton J, Smith P, Heeger AJ (1992) Macromolecules 25:4364
39. Bao Z, Dodabalapur A, Lovinger A (1996) Appl Phys Lett 69:4108
40. Sirringhaus H, Brown PJ, Friend RH, Nielsen MM, Bechgaard K, Langeveld-Voss BMW, Spiering AJH, Janssen RAJ, Meijer EW, Herwig P, de Leeuw DM (1999) Nature 401:685

41. Sirringhaus H, Tessler N, Friend RH (1998) Science 280:1741
42. Camaioni N, Ridolfi G, Casalbore-Miceli G, Possamai G, Maggini M (2002) Adv Mater 14:1735
43. Chirvase D, Parisi J, Hummelen JC, Dyakonov V (2004) Nanotechnology 15:1317
44. Toney MF, Russell TP, Logan JA, Kikuchi H, Sands JM, Kumar SK (1995) Nature 374:709
45. Hayakawa T, Wang J, Xiang M, Li X, Ueda M, Ober CK, Genzer J, Sivaniah E, Kramer EJ, Fisher D A (2000) Macromolecules 33:8012
46. Yang X, Alexeev A, Michels MAJ, Loos J (2005) Macromolecules 38:4289
47. Halls JJM, Arias AC, MacKenzie JD, Wu W, Inbasekaran M, Woo WP, Friend RH (2000) Adv Mater 12:498
48. Breeze AJ, Schlesinger Z, Carter SA, Tillmann H, Hörhold H-H (2004) Sol Energy Mater Sol Cells 83:263
49. Stalmach U, de Boer B, Videlot C, van Hutten PF, Hadziioannou G (2000) J Am Chem Soc 122:5464
50. Zhang F, Jonforsen M, Johansson DM, Andersson MR, Inganäs O (2003) Synth Met 138:555
51. Koetse MM, Sweelssen J, Hoekerd KT, Schoo HFM, Veenstra SC, Kroon JM, Yang X, Loos J (2006) Appl Phys Lett 88:083504
52. Loos J, Yang X, Koetse MM, Sweelssen J, Schoo HFM, Veenstra SC, Grogger W, Kothleitner G, Hofer F (2005) J Appl Polym Sci 97:1001
53. Kawata S, Inouye Y (2002) In: Chalmers J, Griffiths P (eds) Handbook of vibrational spectroscopy, vol 1. Wiley-VCH, Chichester, p 1460
54. McNeill CR, Frohne H, Holdsworth JL, Frust JE, King BV, Dastoor P C (2004) Nano Lett 4:219
55. Klimov E, Li W, Yang X, Hoffmann GG, Loos J (2006) Macromolecules 39:4493
56. Alvarado SF, Rieß W, Seidler PF, Strohriegl P (1997) Phys Rev B 56:1269
57. Rinaldi R, Cingolani R, Jones KM, Baski AA, Morkoc H, Di Carlo A, Widany J, Della Sala F, Lugli P (2001) Phys Rev B 63:075311
58. Kemerink M, Alvarado SF, Müller P, Koenraad PM, Salemink HWM, Wolter JH, Janssen RAJ (2004) Phys Rev B 70:045202
59. Huser T, Yan M (2001) Synth Met 116:333
60. Hoppe H, Glatzel T, Niggemann M, Hinsch A, Lux-Steiner MC, Sariciftci NS (2005) Nano Lett 5:269
61. Nonnenmacher M, O'Boyle MP, Wickramasinghe HK (1991) Appl Phys Lett 58:2921
62. Shafai C, Thomson DJ, Simard-Normandin M, Mattiussi G, Scanlon P (1994) J Appl Phys Lett 64:342
63. De Wolf P, Snauwaert J, Clarysse T, Vandervorst W, Hellemans L (1995) Appl Phys Lett 66:1530
64. Kelley TW, Frisbie CD (2000) J Vac Sci Technol B 18:632
65. Morgado J, Friend RH, Cacialli F (2000) Synt Met 114:189
66. Brabec CJ, Sariciftci NS, Hummelen JC (2001) Adv Funct Mater 11:15
67. Padinger F, Fromherz T, Denk P, Brabec C, Zettner J, Hierl T, Sariciftci N (2001) Synth Met 121:1605
68. Hoppe H, Sariciftci NS (2006) J Mater Chem 16:45
69. Sugimura H, Ishida Y, Hayashi K, Takai O, Nakagiri N (2002) Appl Phys Lett 80:1459
70. Alexeev A, Loos J, Koetse MM (2006) Ultramicroscopy 106:191
71. Heinz WF, Hoh JH (1999) Trends Biotechnol 17:143
72. Eyben P, Xu M, Duhayon N, Clarysse T, Callewaert S, Vandervorst W (2002) J Vac Sci Technol B 20:471
73. Blom PWM, de Jong MJM, Vleggaar JJM (1996) Appl Phys Lett 68:3308
74. Tanase C, Blom PWM, de Leeuw DM (2004) Phys Rev B 70:193202

33 Scanning Probe Anodization for Nanopatterning

Hiroyuki Sugimura

Abstract. Local oxidation of material surfaces has attracted great attention in the research field of scanning probe microscope (SPM) based nanofabrication since the first demonstrations on nanopatterning of semiconductor surfaces in 1990. This method has become the crucial technology for nanofabrication and nanolithography applicable to patterning of a wide variety of materials, including metals, semiconductors, inorganic compounds and organic materials. The mechanism of the SPM-induced local oxidation is ascribed to electrochemical oxidation, i.e., anodization, in the presence of adsorbed water at the SPM tip–sample junction. Besides oxidation, such scanning probe electrochemistry has been extended to various chemical surface modifications other than oxidation, e. g., nitrization, carbonization and electrochemical reduction. This chapter reviews its historical background, demonstrated applications and technical developments in these nearly two decades.

Key words: Oxidation, Anodization, Electrochemistry, Patterning, Lithography

33.1 Introduction

Microscope technologies, which enable one to obtain enlarged images of a sample's surface, are close relations to microfabrication processes, by which micrometer-scale to nanometer-scale patterns are generated and replicated. Typical examples are the relation of optical microscopy with photolithography and that of electron microscopy with electron-beam lithography. Hence, the development of a new microscope technique has improved microfabrication/nanofabrication processes. Scanning probe microscopy (SPM) has been a powerful means to investigate material surfaces with high spatial resolutions at nanometer scales and, in favorable cases, at atomic to molecular scales. As well as other microscopes, SPM technologies have inevitably attracted much attention as nanoprocessing tools since their birth [1–14].

Many reports been published describing nanofabrication using SPM on the basis of various surface-modification mechanisms. Among such mechanisms, the local "oxidation" of a material surface induced at the scanning probe microscope tip–sample junction [3, 10, 13, 14] is a promising technique for the following reasons. This technique is applicable to drawing nanopatterns on a wide variety of materials, including metals, semiconductors, inorganic compounds and organic materials. In particular, nano-oxidation of silicon surfaces, which is the most important material for microelectronics, is of central interest. Furthermore, many pattern-transfer processes, e. g., etching, lift-off and templating, are available using SPM-drawn oxide nanopatterns as masks for those processes. This feature is important for the development of nanolithography. In addition, the local "oxidation" can be conducted in an ambient condition by simply applying a bias voltage to a tip–sample junction.

In this chapter, we review this SPM probe-induced local "oxidation" of material surfaces.

33.2
Electrochemical Origin of SPM-Based Local Oxidation

In 1990, the local oxidation of semiconductor surfaces induced by the tip scanning of a scanning tunneling microscope (STM) was reported for the first time [15, 16]. Dagata et al. [15] modified hydrogen-terminated silicon (Si–H) surfaces using an STM operated in a normal atmosphere. The oxide growth on Si–H due to the tip-scanning was confirmed by secondary ion mass spectroscopy. Nagahara et al. [16] conducted STM oxidation of Si and GaAs in a dilute HF solution. Immediately following the STM oxidation, the same HF solution etched the oxidized areas, resulting in the fabrication of narrow grooves. The demonstration of the STM-stimulated oxidation of semiconductor surfaces by Dagata et al. [15] and Nagahara et al. [16] had the particular feature that an STM tip was biased positively with respect to a sample substrate; thus, the substrate worked as a *cathode,* as illustrated schematically in Fig. 33.1.

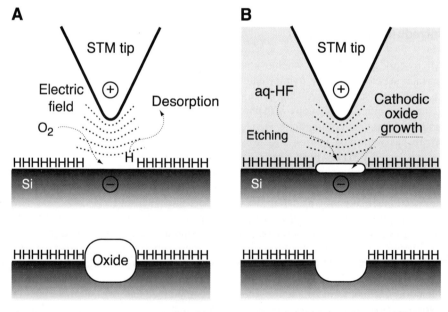

Fig. 33.1. Scanning tunneling microscope (*STM*) oxidation on Si. **A** Field-enhanced desorption of a surface with terminating hydrogen and subsequent oxidation with ambient oxygen. **B** Cathodic oxidation of Si followed by chemical etching with HF

Therefore, as pointed out by Nagahara et al. [16], the oxidation chemistry is not identified as anodization, which occurs on an *anode* and is the general electrochemistry for the mechanism of oxide growth on metals and semiconductors [17]. Dagata et al. [15] suggested a field-enhanced oxidation chemistry in the presence of oxygen as the mechanism of STM-induced cathodic oxidation. As shown in Fig. 33.1a, assisted by the intense electric filed at the STM tip–sample junction, surface H atoms terminating the Si substrate desorb, resulting in the oxidation of this H-desorbed area with ambient oxygen molecules. They noticed that "water" could take roles in the oxidation as well as "oxygen" and demonstrated that the oxidation rate was enhanced in a humid atmosphere, that is, ambient air. On the other hand, Nagahara et al. considered the mechanism to be cathodic electrochemical oxidation [18]. Although, the true oxidation mechanism of these STM-stimulated cathodic oxidations has not been clearly explained yet, this type of scanning probe oxidation of Si conducted on a cathodically biased sample surface was recognized to be useful as a nanofabrication tool and was followed by several research groups [19, 20].

Besides semiconductor surfaces, the STM-based surface oxidation of a metal substrate has been reported by Thundat et al. [21]. They observed morphology changes in a Ta substrate surface during STM-tip scanning in air or alkaline solution. They confirmed oxide growth on the Ta surface by X-ray photoelectron spectroscopy. In contrast to [15, 16], the metal substrate was biased anodically against the STM tip. Although this is an important result as the first demonstration of SPM-controlled oxidation on the anodically polarized substrate, its possible application to nanopatterning was not shown.

On the other hand, the formation of etch pits on a graphite surface was demonstrated by Alberecht et al. [22]. They applied a pulsed bias voltage to a graphite substrate placed in air or another humid atmosphere by using an STM tip as a point counter electrode. In this case, the graphite substrate was polarized positively with respect to the STM tip so that the substrate worked as an anode. They reported that water vapor was crucial for the etch pit formation. McCarley et al. [23] studied this phenomenon in detail under constant STM current conditions. They proposed electrochemical oxidation for the etching mechanism, that is, anodic oxidation. When a sample substrate and an SPM tip are located in air or another humid atmosphere, a thin layer of adsorbed water is inevitably formed on both the sample and the tip surfaces. Owing to the capillarity of water, these adsorbed water layers are connected and form a thin water column at the SPM tip–sample junction [24, 25]. They suggested that such a water column worked as a minute electrochemical cell. Faradaic current in addition to tunneling current flowed through the water column, resulting in the promotion of electrochemical etching of graphite.

Considering the basic knowledge of electrochemistry, SPM-based oxidation on anodically polarized substrates is thought to be more fruitful than that of the cathodic condition, since it proceeds more naturally and may be applicable to a wide variety of materials other than the Si–H surface, including metals, semiconductors, other inorganic substances and organic ones.

Indeed, we have reported for the first time that nanoscale oxide patterns were successfully fabricated on a Ti surface using an STM with a positive sample bias, and proposed the electrochemical oxidation reaction locally induced beneath the tip, i.e., *scanning probe anodization*, as the oxidation mechanism [26], as illustrated schematically in Fig. 33.2.

The net electrochemical reaction of the interfaces of water–oxide and oxide–substrate is shown in (33.1).

$$\text{Sample (anode) reaction}: \quad M + nH_2O \rightarrow MO_n + 2nH^+ + 2ne^- \quad (33.1)$$

Hydrogen generation most certainly occurs at the tip electrode as the counter electrochemical reaction shown in (33.2).

$$\text{Tip (cathode) reaction}: \quad 2nH_2O + 2ne^- \rightarrow nH_2 + 2nOH^- \quad (33.2)$$

A typical example of scanning probe anodization of Si–H is shown in Fig. 33.2.

Nanoscale oxide patterns were fabricated on a Si–H surface using atomic force microscopy (AFM). The oxide lines are observed as the protruded features owing to the volume expansion accompanied with the chemical conversion from Si to SiO_x. When Si oxidizes to SiO_2, the molar volume increases from $12.0\,cm^3/mol$ (Si) to $19.8\,cm^3/mol$ (SiO_2).

It is well known that oxide films are grown on the surfaces of various substrate materials, when the substrate is polarized anodically in an electrochemical cell [17]. This electrochemical oxidation reaction is called *anodization*. Anodic oxide films are put to practical use for hardening, coloring, corrosion protection, preparation of dielectric or semiconducting materials, etc. During anodization, either cations or anions, or both, must migrate across the thickening oxide film, although the proportion of migrating ion species is dependent on various factors, such as substrate material and current density [27]. Such ion transport across the anodic oxide is much

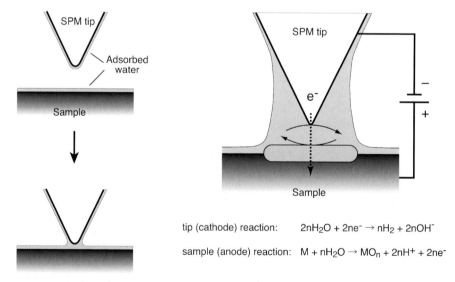

Fig. 33.2. Scanning probe surface modification: electrochemistry in the adsorbed water nanocolumn formed at the scanning probe microscope (*SPM*) probe–sample junction

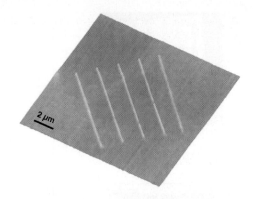

Fig. 33.3. Oxide lines on a Si–H surface fabricated by atomic force microscopy (AFM) based scanning probe anodization conducted in air. A sample bias of 10 V was applied. The conductive AFM probe was scanned at a rate of 0.2 μm/s. The oxide lines protrude about 2 nm in height from the surrounding Si–H surface

accelerated by the intense electric field produced inside the oxide by the applied electrode potential. This is the reason why relatively thick oxide films, compared with native oxide films, can be grown by anodization at a low temperature, even room temperature. The growth rate of the anodic oxide film is governed by the transport rate of the ions, which is determined by the electric field strength on the order of 1 V/nm and the nature of the oxide grown. When a sample is anodized at a constant potential, the strength of the electric field generated in the oxide film decreases with the progress of anodization, since the electric field is in inverse proportion to the thickness of the anodic oxide film. The growth of the film is therefore fast in the early stage of anodization because of the relatively strong electric field generated in the extremely thin oxide film. As the oxide layer grows on the substrate, the effective electric field becomes weaker, and accordingly the growth rate decreases. Finally, the oxide thickness saturates at a certain value determined by the potential. This thickness has a linear relationship with the applied potential, as long as the oxide layer remains within several tens of nanometers. The thickness of the anodic oxide can therefore be controlled by the applied potential.

In principle, the oxide growth behavior in scanning probe anodization is considered to follow that in the macroscale anodization as described above. In research on scanning probe anodization of Ti and Si–H surfaces following the first result for STM oxidation of Ti [28–31], we proved that the oxidation mechanism was truly anodization in the adsorbed water cell formed at the tip–sample junction. We confirmed the saturation of oxide thickness depending on the substrate bias and the linear relation between the oxide thickness and the sample bias [29, 30]. Furthermore, the process strongly depends on the amount of adsorbed water, therefore atmospheric humidity [26, 27], although atmospheric oxygen had contributed to some extent in the anodization [29]. The spatial resolution of the scanning probe anodization worsened with increasing humidity, since the thickness of the water column formed around the tip–sample junction increases with an increase in humidity. Evidence of oxide growth on the sample due to the tip-scanning was confirmed by Auger electron spectroscopy [31]. The difference between scanning probe anodization and the cathodic oxidation of a Si–H surface is shown in Fig. 33.4.

Figure 33.4a shows an STM image of a sample surface with an oxide pattern. The oxide area is expected to protrude from the surrounding Si–H surface similarly to that shown in Fig. 33.3; nevertheless, the oxidized Si pattern is observed as a depressed

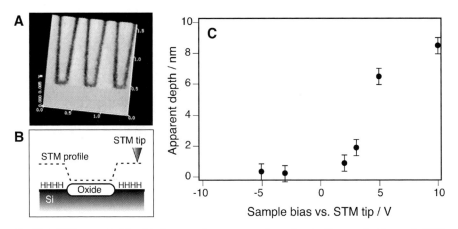

Fig. 33.4. Difference in Si oxidation rates between anodic and cathodic sample biases. **A** STM image of a Si–H surface acquired with a constant-current mode at a sample bias of − 2.0 V and a reference current of 0.1 nA. Prior to this STM imaging, the sample surface was scanned with a sample bias of + 5.0 V and a reference current of 0.1 nA at a tip-scanning rate of 9 nm/s. **B** An STM tip trace on the patterned Si–H sample surface. **C** STM image depth versus sample bias for oxidation

structure in the STM image. As illustrated schematically in Fig. 33.3b, this is simply because the position of the STM tip is lowered until the tunneling current reaches a reference value when the tip is scanned over the oxide surface, which is less conductive than the Si–H surface. Such an apparent STM image depth is not the same with the thickness of the grown oxide, but can be used as an indicator of the oxide thickness. Results with various sample bias voltages are summarized in Fig. 33.4c. Since all these experiments were conducted in the absence of ambient oxygen, it is proved that both anodic and cathodic oxidation of Si–H could be promoted only with adsorbed water. In addition, at the same bias voltage, the thickness of anodic oxide is 3–5 times greater than that of cathodic oxide [30]. The advantage of local anodic oxidation using scanning tunneling microscopy and AFM was recognized by several research groups as well [32–36]. Similar results, that is, the oxide-pattern formation on Si–H and oxide-covered Si surfaces by applying a positive sample bias, were obtained although there were no detailed discussions on the electrochemical origin of the oxidation and the role of adsorbed water.

Local oxidation based on SPM is sometimes called *"field-enhanced oxidation"*, which means that an intense electric field promotes some chemical reactions on the sample's surface. This term was frequently used by defining the unusually intense electric field as a magic tool for chemistry without any electrochemical considerations. However, the field strength at the scanning probe microscope tip–sample junction during local oxidation is not surprisingly intense but is a similar level to that usually generated at an electric double layer on a sample's surface immersed in an electrolyte solution. Such a double-layer field is well known to promote a variety of electrochemical reactions on the sample's surface. As described in this section, anodization depends on the electric field generated in the grown oxide layer. Thus, the term "field enhance" may be a correct one for explaining the mechanism of the

SPM-based local oxidation. However, anodization and its mechanism are common scientific knowledge and, consequently, "anodization" is now widely accepted as the word for the electrochemical oxide growth. Therefore, the terms "anodization" or "anodic oxidation" should be used for the SPM-based oxidation with positive sample biasing in the presence of surface-adsorbed water.

33.3
Variation in Scanning Probe Anodization

33.3.1
Patternable Materials in Scanning Probe Anodization

Anodic electrochemical reactions including the surface oxide formation on material surfaces and degradation/etching of substances have been widely applied to surface modification and microfabrication of a wide variety of materials so far. Therefore, scanning probe anodization is expected to be applicable to nanopatterning of various material surfaces besides Si and Ti. Indeed, there have been many reports on this issue [37–77] as summarized in Table 33.1.

Scanning probe anodization of metals and semiconductors has attracted central interest. On an anodically polarized substrate of metals and semiconductors, an oxide film locally grows beneath a scanning probe microscope tip, resulting in the formation of an oxide nanopattern replicating the scanning probe microscope tip scanning trace. Compound semiconductors, e.g., GaAs, InP and SiGe, have been employed as substrates for nanopatterning experiments [16, 40, 49, 57, 67, 68] as well as Si. Although Ti is most frequently treated as a sample material for studies on scanning probe anodization of metals, a variety of metals other have been given attention. Similar oxide nanopatterns were successfully fabricated on such metals [37, 44, 55, 61, 70, 71, 73, 74].

Inorganic compounds have also been used as sample substrates for experiments of scanning probe anodization. When a metal nitride is used as a substrate for scanning probe anodization, nitrogen atoms desorb from the surface and the remaining metal atoms are oxidized [46, 58, 71, 73]. For example, TiN and ZrN are converted to TiO_2 and ZrO_2, respectively. In the case of Si_3N_4 [46, 58], which is an insulator, the substance is formed as a thin film on a conductive substrate, mainly Si, where a sample bias for scanning probe anodization is applied. Si_3N_4 is converted to SiO_2 by anodization. A nanopatterning of another Si compound, that is, SiC [71], was demonstrated as well by converting SiC to SiO_2 by scanning probe anodization. In this case, a SiC single crystal was employed as a sample substrate. A chalcogenide film of $GeSb_2Te_4$ was found to be oxidized and nanopatterned by scanning probe anodization [61]. In the case of inorganic oxides, the surface-modification behavior is different from that of other inorganic materials, since those materials are not further oxidized in general. Hung et al. [51] have reported that a surface of a conducting thallium(III) oxide was locally etched with STM-induced anodic oxidation in the presence of adsorbed water. They attributed the etching mechanism to the generation of H^+ ions through electrochemical oxidation of water molecules at the sample's surface. Consequently, the adsorbed water became acidic at the sample's surface beneath the STM tip, so thallium(III) oxide was dissolved as Tl^{3+},

Table 33.1. Materials patternable in scanning probe anodization

Materials	References
Semiconductors	
Si, Si-H, SiO$_2$/Si	[15, 16, 19, 20, 30, 32–36]
SiGe	[67]
GaAs, GaAlAs	[16, 40, 57]
InP, InAs	[49, 68]
Metals	
Al	[44]
Co	[74]
Cr	[37]
Hf	[73]
Mo	[63]
Nb	[55]
Ta	[70]
Ti	[26, 28, 29, 31, 41]
Ni	[74]
Zr	[70, 73]
Inorganic compounds	
Si$_3$N$_4$	[46, 58]
HfN, TiN	[73]
ZrN	[71, 73]
SiC	[72]
GeSb$_2$Te$_4$	[61]
SrTiO$_3$	[76, 77]
Tl$_2$O$_3$	[51]
Oxide superconductor	[38, 52, 59]
Carbon	
Graphite	[23]
Amorphous carbon	[53, 54]
Diamond	[60, 66]
Organic materials	
Self-assembled monolayers	[42, 44, 47, 48, 50, 56, 64, 75]
Langmuir–Blodgett films	[65]
Polyorganic silane films	[43, 69]
Organic super conductor	[62]

resulting in the formation of an etched hole. Li et al. [76] have applied local anodic oxidation to a conductive perovskite oxide, that is, Nb-doped SrTiO$_3$. They found that protruded features were formed along the SPM tip scanning trace. Pellegrino et al. [77] have observed a similar mound formation on SrTiO$_{3-x}$. They assumed that such a protrusion was caused by a local decomposition of the oxide surface due to anodization.

From the viewpoint of the fabrication of superconductor nanodevices, nanopatterning of high-temperature superconducting oxide has attracted attention as well. Thomson et al. [38] have fabricated narrow grooves on an yttrium barium copper oxide (YBCO) substrate by the use of an STM. Boneberg et al. [52] have reported nanostucturing of YBCO by AFM with a conductive probe. They showed that the YBCO surface was locally corroded and became nonconducting by an electric current with a positive sample bias. Furthermore, repeated line scans, while current was being injected, resulted in the formation of grooves. Namely, a mechanoelectrochemical nanofabrication, that is, a combination of electrochemical reaction and mechanical abrasion, was demonstrated. Besides the difference in the instruments used for the experiments, the particular difference between these two results was that Thomson et al. used an atmosphere containing CO_2. This CO_2 effect is discussed in the following section. In addition, Song et al. [59] have reported that protruded lines were fabricated on a YBCO thin film by a conductive AFM probe in air.

As described in Sect. 33.2, the SPM-based etch pit formation on graphite in the presence of adsorbed water was ascribed to anodic oxidation of carbon with water. The graphite surface etching was attained with a depth resolution of one graphite sheet [23]. Similarly to graphite, carbon-based materials and organic molecular materials are expected to be patternable by scanning probe anodization. Indeed, nanopatterning of amorphous carbon (a-C) thin films has been achieved [53, 54]. Besides nanopatterning of a-C, scanning probe anodization of diamond, an attractive electronic material, was also conducted [60, 66]. When a diamond thin film, which is generally synthesized by chemical vapor deposition, is treated in hydrogen plasma, its surface becomes terminated with hydrogen and shows a distinct surface electrical conductivity [78]. The thickness of such a surface conductive layer has been estimated to be less than 10 nm. Tachiki et al. [60] have reported that a hydrogen-terminated diamond surface was electrochemically converted to an oxygen-terminated surface by an atomic force microscope tip scanning on an anodically polarized diamond substrate, while no modification was proceeded when the diamond substrate was polarized cathodically. Namely, the diamond surface with the surface C–H layer before anodization was converted to a surface terminated with C–OH, C–O–C and >C=O groups by anodization. The modified region covered with the oxygen-terminated surface was confirmed to lose its electrical conductivity. A similar phenomenon was observed in scanning probe anodization of a boron-doped diamond surface [66].

Patterning of organic molecular films using SPM has attracted much attention owing to the role of organic thin films as resist films for lithography. In particular, a self-assembled monolayer (SAM) formed on various substrates has been demonstrated to be very useful for lithographic applications of SPM patterning, since a wide variety of pattern-transfer processes are available. Although, it was shown that SAMs could be patterned in a vacuum for the first time through irradiation with electrons emitted from an STM tip [79], nanopatterning of SAMs in air or another humid atmosphere has been reported as well [42,44,47,48,50,56,62,75]. In the case of SAMs on Si substrates covalently fixed through Si–O–Si or Si–C bonds, C–C and C–H bonds in the alkyl chains of the organic molecules were decomposed through anodic oxidation, resulting in the etching of the SAMs [42,48,80] similarly to the anodic SPM etching of graphite [23]. On the other hand, when alkylthiol SAMs on Au substrates were anodized, S–Au bonds connecting the SAMs to the substrates were broken through the oxidation of the S atoms to $-SO_2^-$ or $-SO_2H$, resulting in the detachment of the thiol molecules [44]. Besides desorption and etching of the molecules consisting of SAMs, electrochemical conversion of surface-terminating functional groups on SAMs has been demonstrated. For example, amino (NH_2) groups on the top surface of an aminosilane SAM were deactivated, namely, degraded in chemical activity, through their anodic oxidation [50]. The chemical reaction behind this result is probably oxidation of $-NH_2$ to $-NO$ or $-NO_2$ [75]. Furthermore, anodic oxidation of $-CH_2=CH_2$ and $-CH_3$ to oxygen-containing functional groups, such as $-COOH$ and $-CHO$, has been achieved as well [56, 75]. Langmuir–Blodgett films, which are one type of organized organic molecular films, were employed as patterning resist films in scanning probe lithography based on anodic oxidation of organic substances [44, 65]. Besides monolayers, spin-cast polyorganosilane films were successfully demonstrated to be patternable by scanning probe anodization as well [43, 69]. Another interesting example of anodic oxidation nanopatterning of organic materials is local modification of an organic super conductor [62]. A tetramethyltetraselenafulvalene salt single crystal was anodically biased and, then, the crystal surface was scanned with a conductive AFM probe. A distinctly insulating pattern was formed along the tip-trace with a slight depression. The conductivity of the anodized region was less than 10^{-4} times that of the original crystal surface.

33.3.2
Environment Control in Scanning Probe Anodization

One important aspect of scanning probe anodization is that the process can be conducted in air. However, in order to ensure its reproducibility and controllability, the control of the environment for scanning probe anodization is crucial. Furthermore, through the environmental control, we can expect to promote electrochemical reactions other than the oxide layer growth and to add some chemical reactions to the surface-modification mechanisms. Among various environmental factors affecting scanning probe anodization, humidity, that is, the partial pressure of water molecules in the atmosphere, is of primary importance. The increase in humidity further promotes supplying water molecules to accelerate the growth rate of anodic oxide [29] and the degradation rate of an organic monolayer [80].

Furthermore, as illustrated schematically in Fig. 33.5c, when the environmental humidity is high, the diameter of the water column formed at the scanning probe microscope tip–sample junction becomes large, resulting in the degradation of patterning resolution [28, 29, 81] (Fig. 33.5d). However, such a humidity effect in patterning resolution depends strongly on the wettability of sample's surface as exaggeratedly illustrated in Fig. 33.6.

In the case of a sample with a hydrophilic surface such as Ti, the surface of which is covered with a hydrophilic thin oxide layer, a relatively thick water layer is present on the sample's surface and the adsorbed water column is spread at its bottom owing to the low water contact angle of the hydrophilic surface (Fig. 33.6a). Consequently, a marked humidity effect appeared for the hydrophilic surface. Fabricated oxide patterns were no longer confined beneath the scanning probe microscope tip. In some cases, such a pattern-enlarging effect was on the order of up to a few micrometers [29]. In contrast, in the case of a sample with a hydrophobic surface, there is little adsorbed water on the surface. The adsorbed water column is relatively confined owing to the high water contact angle of the surface. Indeed, when an organosilane SAM surface, the surface of which is terminated with $-CH_3$ groups and

33 Scanning Probe Anodization for Nanopatterning

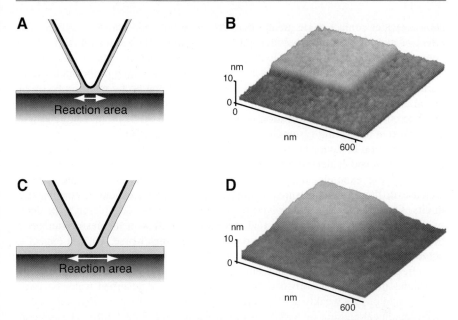

Fig. 33.5. Humidity effect of scanning probe anodization on patterning resolution. **A** A small water nanocolumn confined at an SPM tip–sample junction formed under a low-humidity condition. **B** An STM image of the edges of an anodized square fabricated on Ti under a low humidity of less than 25%. The clear edges were successfully fabricated. **C** An enlarged water nanocolumn formed under high humidity. **D** An STM image of the edges of an anodized square fabricated on Ti under a high humidity of 90%. The pattern edges became blurred, showing the patterning resolution was degraded under the high-humidity condition

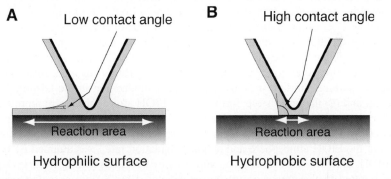

Fig. 33.6. Reaction area size dependent on the wettability of the sample's surface. **A** On a hydrophilic sample surface: the contact area of the adsorbed water column is spread widely owing to the low water contact angle of the hydrophilic surface. **B** On a hydrophobic sample surface: the contact area of the adsorbed water column is confined beneath the SPM tip owing to the high water contact angle of the hydrophobic surface

shows a water contact angle greater than 90°, was pattered, there was only a small effect of humidity on the resolution [80]. For a similar surface a wettability effect between hydrophobic Si–H and hydrophilic oxide-covered Si surfaces has been reported [82]. These humidity effects are key factors for scanning probe anodization. Further detailed studies in which the kinetics of scanning probe microscope tip induced anodization of Si–H and its humidity effects were elucidated by measuring faradaic current during anodization using a conductive AFM probe equipped with a low-current measuring system have been conducted [83, 84].

Anodization behavior becomes different, when some gas species are dissolved into the adsorbed water column, from the anodization behavior with water vapor alone. For example, the anodic oxide growth rate on Ti increased with an increase in the partial pressure of atmospheric oxygen even under constant humidity. Oxygen molecules dissolved in adsorbed water are considered to promote oxidation of Ti. The anodization rate can be increased with increasing humidity; however, in this case, the patterning resolution becomes worse. We can expect to obtain a higher patterning rate without degrading the resolution by use of this oxygen effect. When CO_2 molecules are intentionally added to the atmosphere, the molecules dissolve in adsorbed water; consequently, the adsorbed water becomes weakly acidic. This weakly acidic water was found to be effective in promoting the electrochemical etching of YBCO as described in Sect. 33.3.1 [38]. The pH control of adsorbed water might provide a new aspect to scanning probe anodization. For example, if NH_3 molecules are added instead of CO_2, adsorbed water becomes basic.

If an adsorbed liquid nanocolumn consisting of a substance other than water is formed at the scanning probe microscope tip–sample junction and is provided with an electrochemical cell, it is expected to be able to perform electrochemical nanofabrication other than oxide growth and etching. Interesting results have been reported by Tello et al. [85, 86] as explained as follows. In the presence of ethyl alcohol (EtOH) vapor in the atmosphere for SPM nanofabrication, a liquid column composed of EtOH was thought to be formed at the scanning probe microscope tip–sample junction. They performed scanning probe anodization experiments of Si–H under this condition. They reported that (1) protrusions were formed on the sample's surface similarly to what occurred in anodization in the adsorbed water column and (2) the protrusions formed in the EtOH atmosphere were more 5 times higher than those formed with adsorbed water when the same bias voltage was applied for the fabrication [85]. They first reported that silicon oxide was grown even in the EtOH atmosphere and a "giant oxidation rate" was attained by the use of EtOH. However they finally concluded that carbon originating from EtOH was incorporated in the grown protrusion as estimated by X-ray photoelectron spectroscopy. This conclusion sounds natural considering the basic knowledge of anodic oxidation. Since the thickness of an anodic oxide film is definitely determined by the applied bias voltage and the anodized material, the thicknesses of the anodic oxide films are identical regardless of the environment if the applied voltage and the material are same. The marked height difference between the features fabricated at identical potentials but under different environments means that the chemical compositions of the fabricated features were quite different. Aside from the vagueness of electrochemical understanding of the results, the concept of electrochemical nanofabrication using an "adsorbed liquid nanocolumn" other than water is very curious and may open a new approach of scanning probe anodization. From the same research group, the use of a hydrocarbon liquid has been reported [87]. They fabricated carboneous nanopatterns on Si though anodic reactions of hydrocarbon molecules. Kim et al. [88] have used a liquid nanocolumn of EtOH containing a small amount of HF for AFM-based anodization. In the HF–EtOH nanocolumn, anodic reactions of Si and chemical etching of the Si compounds generated, probably consisting of Si, C and O, proceeded simultaneously, resulting in the formation of an etched groove along a scanning trace of the atomic force microscope tip.

Here, other two interesting reposts related to scanning probe anodization are introduced. Kim et al. [89] have reported a positive charge effect of AFM-based anodization of Si. Prior to conducting AFM-based anodization of Si, they treated an oxide-covered Si substrate with phosphoric acid molecules and Zr^{4+} ions, in that order. Consequently, the substrate was terminated with a monolayer of a phosphoric acid and Zr^{4+} ion complex which was positively charged. This positively charged monolayer promoted the electron injection from the atomic force microscope tip to the substrate, namely, anodization of the substrate. The threshold voltage was lowered and a higher patterning speed was attained with the positively charged monolayer. On the other hand, Pyle et al. [90] have reported the formation of silicon nitride on a Si–H substrate locally promoted by AFM. The substrate was located in a vacuum chamber which was evacuated and, then, filled with a reduced pressure of NH_3. By applying a bias voltage between the substrate and the conductive AFM probe so that the sample became an anode, they found that nitride lines could be formed. In this case, no liquid column was probably formed at the tip–sample junction. However, a negligible number of NH_3 molecules were considered to be sandwiched between the sample and tip surfaces to participate in the electrochemical reactions proceeding at the junction. Note that the reactions are also anodic oxidation since the oxidation number of Si increases from 0 to +4 when Si is converted to Si_3N_4.

33.3.3
Electrochemical Scanning Surface Modification Using Cathodic Reactions

When a bias voltage is applied to a scanning probe microscope tip–sample junction so that the sample is negatively polarized, cathodic reactions proceed on the sample's surface. If we can use cathodic reactions in addition to anodic reactions for surface modification, more sophisticated SPM nanofabrications are expected to be achieved. However, except for the first two examples of the SPM-promoted cathodic oxidation of Si–H [15, 16], the SPM-based electrochemical surface modification of material surfaces has been mainly performed on anodically polarized samples, namely, by the use of anodic reactions. There have been a number of reports comparing anodic and cathodic surface modifications by SPM. Hung et al. [51] have reported that Tl(III) cations in thallium(III) oxide could be electrochemically reduced to Tl(I) ions at the surface under an STM tip so that the reduced ions were dissolved in adsorbed water, resulting in the formation of an etched hole. This cathodic etching was found to be faster than the anodic etching of thallium(III) oxide as described in Sect. 33.3.1. In many other reports, it was demonstrated that SPM nanopatterning was more effective or more reliable in the anodic cases than in the cathodic ones [29, 30, 48, 62, 86].

Distinctive examples in which cathodic reactions were intentionally used have not been reported so much. Here, several interesting results on cathodic nanofabrication on the bias of SPM electrochemistry are explained. As introduced in Sect. 33.3.1, Sugawara et al. [61] reported oxide growth on a calcogenide substrate via scanning probe anodization. In the same paper, they also reported that etched grooves formed on the same calcogenide surface by STM tip scanning with a sample bias less than -5 V in a humid atmosphere with a relative humidity of more than 50%. The etching mechanism was attributed to some cathodic electrochemical reactions in the presence of adsorbed water. Li et al. [91] have found that in SPM nanofabrication of a perovskite manganite thin film, that is, $La_{0.8}Ba_{0.2}MnO_3$, the sample's surface was more reproducibly nanopatterned at negative sample biases than at positive sample biases. Regions reduced by the AFM probe scanning with a negative sample bias were protruded and became more corrosive in an etching solution. Maoz et al. [92] have reported the fabrication of Ag nanostructures.

As illustrated schematically in Fig. 33.7, they developed a skilled process. First, a monolayer of Ag^+ ions was assembled on a SH-terminated SAM surface through the affinity of S to Ag. This Ag^+ monolayer was locally reduced to a Ag^0 layer by an AFM probe tip. Next, this Ag^0 pattern was developed by a sliver enhancer kit, resulting in the growth of Ag thin films on the AFM-reduced areas. This technique can be understood as an analogy of the photograph system using silver salt.

A further enriched SPM nanofabrication might be attainable, if both anodic and cathodic electrochemical reactions are manipulated.

As shown in Fig. 33.8, reversible nanochemical conversion, in which electrochemical oxidation and reduction reactions of organic molecules were reversibly

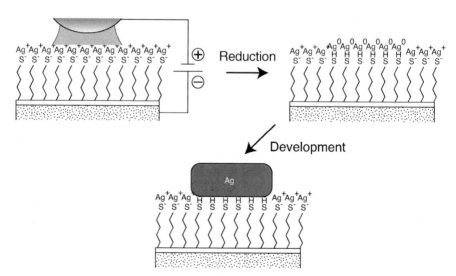

Fig. 33.7. Scanning probe reduction: electrochemical reduction of Ag^+ ions to Ag^0 atoms followed by the development in a silver enhancer kit. Desorption of metallic silver proceeds area-selectively on the probe-reduced region

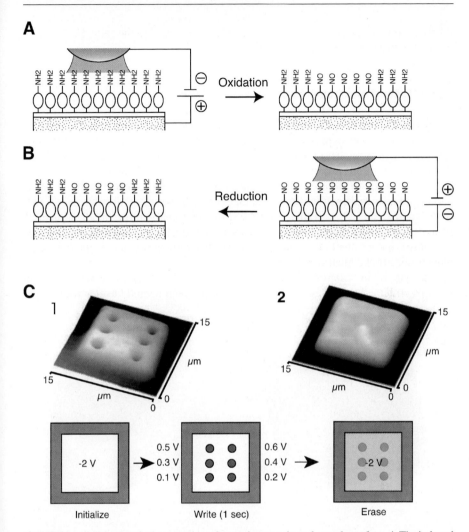

Fig. 33.8. Reversible chemical conversion of an amino-terminated sample surface. **A** Tip-induced anodization of an amino-terminated self-assembled monolayer (SAM) surface. **B** Tip-induced reduction of the oxidized surface. **C** Reversible writing and erasing. *1* Kelvin-probe force microscopy (KFM) image of an initialized 10 μm × 10 μm square at a DC bias of −2 V and six dots written in the initialized square by oxidation with 1-s pulses. *2* KFM image of the same area acquired after erasing with a reductive overscanning at a DC bias of −2 V

conducted at an AFM probe–sample junction, has been demonstrated using an amino-terminated SAM as a sample [93]. In the presence of adsorbed water on the monolayer, surface functional groups were oxidized from –NH_2 to –NO and were reduced from –NO back to –NH_2. Owing to the change in the molecular dipole moment accompanied with the electrochemical reactions, the surface potential of the SAM increased with oxidation and reduced with reduction as confirmed by Kelvin-probe force microscopy (KFM).

33.4
Progress in Scanning Probe Anodization

33.4.1
From STM-Based Anodization to AFM-Based Anodization

In the early stage of the research on scanning probe anodization and related techniques, an STM was mainly used as a tool to locally induce surface electrochemical reactions. However, the use of AFM with a conductive probe was proposed and demonstrated [32–35] right after the research on STM-based anodization had been conducted. The AFM-based method has become a main tool for scanning probe anodization, since there are many advantages of AFM-based anodization over STM-based anodization. The advantages of AFM-based scanning probe anodization arise from one particular feature of AFM. Namely, the feedback system for AFM probe scanning is independent of the current flowing through the tip–sample junction which induces electrochemical reactions.

In contrast, in scanning tunneling microscopy, the tip–sample junction current governs both the feedback control of the STM tip scanning and the electrochemical

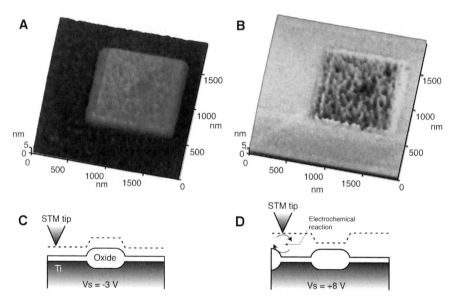

Fig. 33.9. Effect of faradaic current on the STM feedback. **A** STM image of a square pattern (1000 nm × 1000 nm) fabricated on a Ti surface at a sample bias $V_s = +8.0$ V, and acquired at $V_s = -3.0$ V and a reference current $I = 0.1$ nA. The average height of the pattern is estimated to be 6 nm from the STM image. **B** STM image of the square region (2000 nm × 2000 nm) with constant $V_s = +8.0$ V and $I = 0.1$ nA. At these conditions, the area was anodized once again. **C** STM tip trace under the imaging condition ($V_s = -3.0$ V and $I = 0.1$ nA). **D** STM tip trace under the anodization condition ($V_s = +8.0$ V and $I = 0.1$ nA). The difference in the tip-to-sample distance between the tip scans on the native and the anodized areas is estimated to be approximately 7 nm from these STM images

surface modification. A part of the junction current consists of faradaic current when electrochemical reactions proceed on the sample's surface [23, 29]. The effect of faradic current on the STM feedback system during anodization of Ti is demonstrated in Fig. 33.9 [29].

The anodized square pattern, which is observed as the swelled structure in the STM image (Fig. 33.9a) acquired under the condition in which anodization of Ti does not proceed is recognized as a depressed pattern with an average depth of approximately 1 nm (Fig. 33.9b), indicating that the tip-to-sample distance varies during anodization as illustrated schematically in Fig. 33.9d. When the tip is scanned over the Ti surface under the anodization condition, the faradic current accompanied by the anodization flows between the tip and the sample. When the tip is scanned on the surface that has already been anodized, on the other hand, electrochemical reactions scarcely proceed. To attain enough junction current, therefore, the tip approaches the sample's surface until electron tunneling or a field-emission process occurs between the tip and the sample.

33.4.2
Versatility of AFM-Based Scanning Probe Anodization

Since in AFM-based anodization, the surface modification and the feedback control are independent, its primary advantage is the applicability to materials with low electrical conductivity. Of course, in order to promote anodization, a certain amount of the probe junction current is required, even though it is in the range of picoamperes [48, 84]. Thus, scanning probe anodization is not applicable to the surface modification of insulating bulk materials. However, if an insulator is formed as a thin film on a conductive substrate, it can be nanopatterned by AFM-based anodization. A variety of insulator thin film/conductive substrate samples, e.g., oxide-covered Si substrates [32, 33], Si_3N_4 films on Si [46, 58], organic films on Si or Au substrates [44, 48, 50, 56, 63, 69], have been used for nanopatterning of the insulating layer and/or the substrate through the insulating layer.

Besides the versatility in materials applicable in AFM-based anodization, its patterning speed is another noticeable advantage. Different from STM-based anodization, AFM-based anodization does not rely on the feedback system for the tunneling current regulation, the response speed of which is relatively slow. Furthermore, the AFM probe tip is located on a cantilever spring. These two features of AFM-based anodization provide a low risk of tip crashing. Thus, in AFM-based anodization, higher patterning rates are achievable than in STM-based anodization. Indeed, several results of high-speed patterning in the range from millimeters per second to centimeters per second by AFM-based anodization have been reported [59, 94–96].

33.4.3
In Situ Characterization of Anodized Structures by AFM-Based Methods

In research on SPM-based nanofabrication, most of the nanostructures fabricated are very minute; thus, techniques to observe and evaluate such nanostructures are of fundamental importance in order to develop SPM-based nanofabrication processes.

Topographic imaging by SPM was naturally applied for such a purpose and has played a central role. However, in the case of STM-based anodization, the STM image obtained of a fabricated nanostructure was sometimes distorted by electric properties of the nanostructure and was not a reflection of the true topography as demonstrated in Fig. 33.4. Moreover, during such STM imaging, the junction current inevitably flows; thus, imaging conditions must be carefully selected in order to avoid the sample's surface being modified by the junction current. In contrast, a true topographic image of fabricated structures is available using AFM as shown in Fig. 33.3. In this case, undesirable surface modification is avoided completely by switching off the electric circuit for biasing the AFM tip–sample junction. This imaging ability of AFM without undesirable surface modification is quite useful for lithographic purposes as described in Sect. 33.5, since an AFM probe can be located on a defined point on a sample surface.

Besides the ability of topographic imaging, techniques to measure a wide variety of physical and chemical properties on the nanoscale have been provided based on AFM. Such functions of AFM are certainly useful for imaging nanopatterns fabricated by scanning probe anodization and for probing their particular properties. For example, lateral force microscopy (LFM) images are well known to provide some chemical information of material surfaces, so LFM was applied to study chemical changes of the surfaces of samples induced with SPM-promoted anodic reactions. Teuschler et al. [97] fabricated oxide patterns on Si–H fabricated by AFM-based anodization and observed the patterns by LFM. They showed that the anodized area, which was imaged as a protrusion, indicating that oxide was formed in the area, had a higher frictional force compared with the Si–H surface that was not anodized. This LFM contrast is ascribed to being caused by a difference in hydrophilicity between the anodic oxide and Si–H surfaces. Si–H surfaces are relatively hydrophobic, while anodic oxide surfaces are very hydrophilic and are wetted well with water [42]. On a hydrophilic surface, a lateral force becomes large because of the stronger adhesion of a scanning probe microscope tip to the surface. Moreover, capillary force increases when an atomic force microscope tip is scanned over a hydrophilic surface, since the amount of adsorbed water on the surface is greater than that on a hydrophobic surface [98]. A capillary force contrast certainly exists between the anodized and the Si–H surfaces and is considered to increase the lateral force contrast to some extent. LFM observation has been proved to be powerful for studies on AFM-based anodization of Si substrates covered with an organic monolayer [99].

As demonstrated in Fig. 33.10, LFM successfully provides chemical information on AFM-based anodization of an alkylsilane SAM formed on a Si substrate [99]. As clearly seen in LFM and topographic images (Fig. 33.10b,c, respectively), the dimensions of these dot features depend on the bias duration time. The dot features grow both in the lateral and in the vertical directions with the increases in duration time and bias voltage. Lateral force contrast increases with an increase in bias duration time. However, it seems to be constant around 100–140 mV at certain duration times depending on V_s. In addition, it is noteworthy that the topographic height is still found to increase even after the lateral force contrast becomes constant. The surface-modification processes that occurred at the probe–sample junction is explained as follows. In the initial stage, the organic molecules are chemically modified, decomposed and probably removed in part of the probe-scanned region. Consequently, the

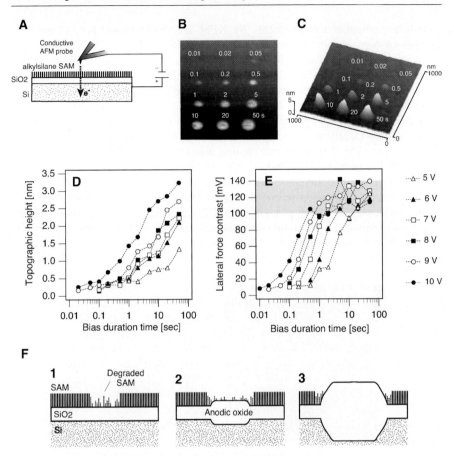

Fig. 33.10. AFM-based anodization of an alkylsilane SAM on Si. **A** Dot features were fabricated on another octadecylsilyl–Si sample by placing an AFM probe at each of the positions and applying a bias voltage from 0.01 to 50 s. **B** Lateral force microscopy (LFM) and **C** topographic images of the dot features fabricated on the sample. The bias (+10 V) duration times for each of the dot features are indicated. Summarized dot height (**D**) and lateral force contrast (**E**) versus bias duration time. **F** The initial (*1*), middle (*2*) and final (*3*) stages for the modification of the sample's surface beneath the AFM probe

underlaying oxide surface partly appears as illustrated in Fig. 33.10d, stage 1. Since the surface of silicon oxide has a frictional coefficient larger than that of the SAM and the capillary force effect on such a hydrophilic oxide surface is larger than that on the hydrophobic organic monolayer surface, a lateral force contrast between the modified and unmodified regions is generated. When the surface modification progresses further, anodization of the substrate Si begins. However, some parts of the organic monolayer still remain on the surface as shown in Fig. 33.10d, stage 2. In the final stage (Fig. 33.10d, stage 3), the organic molecules have been decomposed completely and only the Si anodization proceeds further in this stage. Therefore, lateral force contrast does not change since the modified surface consists of silicon oxide throughout the final stage, although it protrudes owing to the progress of the Si anodization.

Surface potential measurements by KFM are applicable to detect chemical changes in organic molecular surfaces as demonstrated in Fig. 33.8. In general, when an organic molecular film is oxidized by scanning probe anodization, polar functional groups containing oxygen are formed on the surface at the very beginning stage of anodization before starting the degradation of the film as illustrated Fig. 33.10d, stage 1. In the case of a CH_3-terminated molecular surface, –COOH, –CHO or –OH groups are thought to be formed. Consequently, the anodized pattern clearly appeared in the KFM image as a region with a lower surface potential [75]. This chemical change was so slight that it could not be imaged at all by other SPM techniques, e.g., topographic imaging and LFM. When a semiconductor substrate is provided as a sample for scanning probe anodization, scanning capacitance microscopy (SCM) becomes a powerful tool to characterize the surface-modification process, since SCM can sense electrical properties of an interface between a substrate semiconductor and its surface oxide layer [100]. One example is shown in Fig. 33.11 (T. Yamamoto and N. Nakagiri, unpublished results).

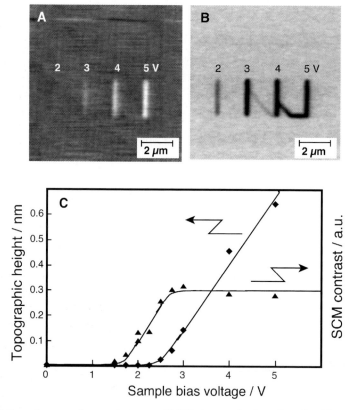

Fig. 33.11. Scanning capacitance microscopy (*SCM*) characterization of anodized features on Si. **A** Topographic and **B** SCM images of anodic oxide lines on a Si sample fabricated with each of the bias voltages indicated. **C** Topographic height and SCM contrast versus anodization bias voltage

Figure 33.11a shows a topographic AFM image of a Si substrate covered with native oxide. Anodic oxide lines fabricated on the substrate appear as protruded lines similarly to those in Fig. 33.3. These oxide lines can be recognized as dark lines in an SCM image (Fig. 33.11b). The origin of these SCM contrasts is most likely changes in oxide thickness. Note that, as shown in Fig. 33.11c, there are no topographic contrasts at anodization voltages less than 2 V, indicating that the oxide thickness has not apparently increased. Nevertheless, SCM contrasts are certainly present in this bias range. This preanodization behavior is considered to be interdiffusion of Si and O atoms and/or charge trapping at the oxide–Si interface.

33.4.4
Technical Development of Scanning Probe Anodization

In scanning probe anodization, the width of the pattern lines depends on the junction current. The line width sometimes changes if the current is not stable. In addition, line drawing with faster probe-scan rates is achieved by increasing the junction current through applying a higher bias voltage. However, the controllability of the current is not satisfactory when using the constant-bias mode. The relationship between the junction current and the bias voltage depends on several factors, such as the age and the idiosyncrasies of the particular probe used. Although, in STM-based anodization, the junction current is controlled so as to be constant by its feedback system, the current cannot be precisely controlled by simply applying a defined bias voltage in AFM-based anodization. In order to overcome the shortcomings, constant-current AFM anodization has been proposed [95], and a current-regulation circuit was developed in order to improve the controllability of the current [101]. A special conductive AFM cantilever, in which a metal oxide semiconductor (MOS) field-effect transistor (FET) as a constant-current source is integrated, designed and fabricated by Wilder et al. [102], will be useful in order to construct a patterning system based on scanning probe anodization.

When AFM was first developed it was operated in the contact mode. In contact-mode AFM (C-AFM), the cantilever bending of an AFM probe is kept constant by its feedback system. The AFM probe tip always touches sample's surface at a certain contact force kept constant in the range of nanonewtons. Since C-AFM is operated with a relatively high contact force and lateral dragging of a sample's surface with an atomic force microscope tip is inevitable, the sample and the tip are frequently damaged. In order to reduce damage both to the sample and to the atomic force microscope tip, dynamic-mode AFM, that is, dynamic force microscopy (DFM) was developed. In DFM, an AFM cantilever is vibrated at a frequency around its resonance frequency and the tip–sample force is measured by detecting the amplitude or the frequency of the cantilever vibration. As well as C-AFM, DFM has been successfully applied in AFM-based scanning probe anodization [35, 103–105]. Although, there are many types of DFMs ranging from intermittent-contact AFM (IC-AFM) to noncontact AFM, DFM-based anodization is considered to work in IC-AFM mode, since actual contact between the AFM probe and the sample's surface is achieved in IC-AFM and is crucial in order to form the adsorbed water column as illustrated in Fig. 33.2.

When DFM is used for scanning probe anodization, a conductive AFM probe tip periodically touches the sample's surface. This means that a bias voltage is applied periodically as well. Such a periodic current injection may have some effects on anodization. Effects of voltage modulation on AFM-based anodization have been reported as well [106–108]. In these experiments, instead of a DC voltage, an alternating voltage pulse was applied to the contact-mode atomic force microscope tip–sample junction. Pérez-Murano et al. [106] fabricated oxide nanostructures on Si–H by applying an alternating voltage pulse in which positive and negative sample bias voltages were repeated at a frequency up to 500 Hz. They reported that fabricated oxide patterns had a higher aspect ratio than those fabricated using the same DC voltage with the positive voltage of the alternating pulse. The method was successfully applied for AFM-based anodization of Ti [107] and diamond [108], resulting in the fabrication of finer lines.

The combination of DFM with the AC biasing method is also attractive [109–113]. Legrand and Stievenard [111] used a pulsed bias with a duration time around 1 μs or less with a vibrating DFM cantilever with a vibration cycle of 3.6 μs. They found that the anodized patten width varied from 25 to 16 nm depending on the phase of the pulsed bias with a top voltage of 12 V with respect to the cantilever oscillation, while the patten width was 21 nm when fabricated with a DC bias with the same voltage. This result indicates that the pulsed bias method based on DFM improves patterning resolution and the importance of phase matching between the pulsed bias wave and the cantilever oscillation. García et al. [109, 110] have studied in detail the formation of a water nanocolumn at the tip–sample junction of the dynamic force

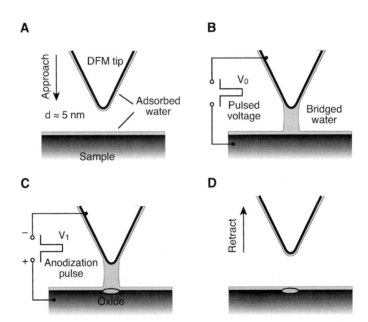

Fig. 33.12. Electrostatically controlled water bridge formation between a dynamic force microscope (*DFM*) tip and a sample's surface. **A** Approach of a DFM tip to a sample's surface. **B** Application of a pulsed bias between the tip and the sample. A water nanocolumn is electrostatically formed by the pulsed bias voltage. **C** Application of the next bias pulse for anodization. **D** Retraction of the DFM tip to break the bridged water column

microscope and developed a highly controllable method to form the water column as illustrated schematically in Fig. 33.12.

When a DFM tip approaches the sample's surface with a distance of about 5 nm, a pulsed bias V_0 is applied to the tip–sample junction. The adsorbed water layers are polarized by this applied electric field and, then, jump into a state where they are connected owing to the electrostatic force. The tip and sample's surface are bridged with a water column. They have confirmed this water bridging by measuring the cantilever vibration of DFM. The amplitude of the cantilever vibration at a frequency of about 330 kHz was reduced to approximately 2 nm from approximately 7 nm, which was the vibration amplitude without the water bridge. Then, a next pulsed bias of V_1 is applied to the junction in order to promote anodization. Typically, V_1 is smaller than V_0. Finally, the dynamic force microscope tip is retracted and the water bridge breaks. This electrostatic water bridge formation has been extended to bridging with a liquid other than water [85–87].

33.5
Lithographic Applications of Scanning Probe Anodization

33.5.1
Device Prototyping

Lithographic applications of scanning probe anodization have attracted central attention mainly because of expectations regarding the fabrication of electronic nanodevices based on the patternability of semiconductor materials such as Si and GaAs. However, in scanning probe lithography, the probe tip must write lines one by one to construct a whole pattern. Consequently, its throughput is not satisfactorily high. Therefore, at present, scanning probe lithography is not suitable for mass production. Nevertheless, scanning probe lithography based on anodization has been accepted as a powerful tool for prototyping of electronic nanodevices, since the method has a high spatial resolution down to around 10 nm and requires no complicated processes, e. g., resist coating, curing, development, etching, resist removal, which are indispensable for usual lithographic techniques.

Fayfield and Higman [114] have applied STM-based anodization in order to fabricate Si MOSFET devices. They reported that the FET with STM-grown anodic oxide successfully worked; however, the anodic oxide caused a slight mobility decrease as compared with the mobility for an FET without the STM process. Minne et al. [115] have fabricated a Si MOSFET with a gate width of 100 nm. They brought a process step based on AFM-based anodization lithography into a series of photolithographic processes in order to fabricate MOSFET devices. They used AFM lithography in order to fabricate a 100-nm-wide gate electrode made of amorphous Si (a-Si). The fabricated MOSFET was confirmed to work correctly. On the other hand, Campbell et al. [116] have reported the fabrication of a side-gated Si FET by means of AFM-based anodization lithography. They used silicon-on-insulator (SOI) as a sample substrate. The single-crystal Si layer on a SOI substrate was anodized by AFM and, then, etched using the patterned anodic oxide as an etching mask. The fabrication of FET devices was attained with a critical feature size of 30 nm.

Although the examples introduced in the previous paragraph were successful demonstrations of scanning probe anodization lithography, they required other process steps, for example, etching. Matsumoto et al. [41, 107, 117] have reported an ingenious method to construct electronic nanodevices on the basis of scanning probe anodization. They fabricated a nanodevice in a Ti film with a thickness of 3 nm. This very small thickness was crucial, since the Ti film was locally anodized from its top surface to the bottom interface with its substrate. Accordingly, the Ti film was divided into two parts separated by one anodic oxide line, which was employed as an electric insulating part of the nanodevice. Matsumoto et al. [41] have fabricated a planar-type metal–insulator–metal diode and even a single-electron transistor (SET) working at room temperature using the anodic oxide/Ti system [117]. Moreover, by improving the surface roughness of a Ti film by depositing it onto an atomically flat sapphire substrate and employing AFM-based anodization with voltage modulation, they succeeded in fabricating a room-temperature single -electron memory [107].

The strategy that employs anodic oxide patterns as insulating parts in electronic devices has been extended to fabricate many nanoelectronics devices on a variety of material surfaces other than Ti. For example, compound semiconductors, e. g., GaAs, have attracted attention as base substrates for the fabrication of SETs and other nanodevices [40, 57, 118–120]. Besides electronic applications, a photonics application of scanning probe anodization has been proposed. Hennessy et al. [121] have achieved high-precision tuning of photonic crystal nanocavities by AFM-based anodization. A photonic crystal which was an array of minute holes made of GaAs was partially anodized in order to modify the hole-array pattern. Metal films other than Ti were also used for the fabrication of single-electron devices via scanning probe anodization. Snow et al. [45] have fabricated a single atom contact in an Al film. Shirakashi et al. [122] have fabricated a side-gate SET in a Nb film working at room temperature, similar to that reported in [117]. Magnetic nanostructures have also been fabricated by scanning probe anodization of Ni [74] and NiFe [123].

Diamond has also been applied as a device material to be fabricated by scanning probe anodization. Tachiki et al. [124] have reported the fabrication of single hole transistors working at 77 K on a hydrogen-terminated diamond surface by creating oxygen-terminated surface regions through AFM-based anodization. Such a device was constructed using hydrogen- and oxygen-terminated diamond surfaces as conductive and insulating parts, respectively. Fabrication of functional elements on the surface of oxide materials has been reported as well. Examples are the fabrication of a Josephson junction on a YBCO substrate [59], a $SrTiO_3$-based FET [125] and a magnetoresistive structure on a perovskite manganite [91].

33.5.2
Pattern Transfer from Anodic Oxide to Other Materials

Scanning probe anodization both based on scanning tunneling microscopy and based on AFM has been proved to be a powerful means for the fabrication of oxide nanopatterns on substrates made of a variety of materials. In order to accomplish "nanolithography" based on this technology, pattern-transfer processes compatible to oxide nanostructures are crucial. The transferability of SPM-drawn patterns, that is, the capability of those patterns to serve as either masks or templates in subsequent processes, is a key factor for nanolithography. The SPM-based oxidation techniques both under anodic and under cathodic conditions offer great promise, since various pattern-transfer processes have been successfully demonstrated so far.

33.5.2.1
Pattern-Transfer Process for Oxide Nanopatterns Fabricated on Si–H

First, we discuss pattern-transfer techniques from experimental results on Si–H, which is the most general material for scanning probe oxidation. Both by scanning tunneling microscopy and by AFM and both under anodic and under cathodic polarization conditions, a Si–H surface can be locally oxidized as illustrated in Figure 33.13, step A. Since silicon oxide rapidly dissolves in an aqueous solution of HF, while Si is hardly etched in aqueous HF, the SPM-grown oxide is selec-

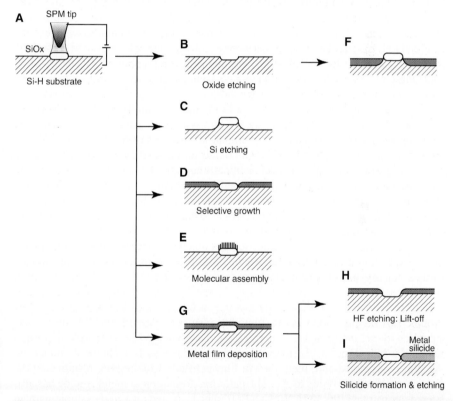

Fig. 33.13. Patten-transfer process for oxide nanopatterns fabricated on Si–H. **A** Local oxide growth by scanning probe oxidation. **B** Oxide etching with HF. **C** Si etching using the oxide layer as an etch mask. **D** Selective material deposition on the Si–H surface. **E** Molecular self-assembly proceeding area-selectively on the oxide surface. **F** Selective material deposition on the etched Si surface. **G** Deposition of a metal thin film. **H** Lift-off process: the metal film on the oxide is lifted off when the oxide pattern is etched in HF. **I** Area-selective silicide formation

tively etched as shown in Fig. 33.13, step B [32, 126, 127]. This process is crucial for studies on kinetics and mechanisms of scanning probe oxidation of Si–H and has been used to determine that fabricated features certainly consisted of silicon oxide by confirming the formation of recessed features with HF etching and to measure volumes or thicknesses of oxide features by comparing topographic profiles acquired before and after HF etching (Fig. 33.13, steps A and B). In contrast to HF, alkaline solutions have chemical activities to Si, while silicon oxide is more stable in solution. Thus, as illustrated in Fig. 33.13, step C, a Si nanostructure is formed through selective etching using the oxide pattern as an etch mask. Typically, KOH [34, 81, 94] and tetramethylammonium hydroxide (TMAH) [128] are used for the etching. Plasma etching using a mixture of SF_6/C_2F_5Cl or Cl_2/O_2 as the etching gas was also successful for selective Si etching using anodic silicon oxide as the masking material [115, 129]. Besides etching, area-selective nucleation and deposition, as illustrated in Fig. 33.13, steps D and E, are promising pattern-transfer processes as well. Dagata et al. [130] demonstrated this technique for the first time. They found by means of molecular-beam epitaxy that a GaAs layer was area-selectively deposited on the Si–H surface, while the oxide surface remained undeposited. Electroless plating was successfully applied to area-selective metallization [131, 132]. Gold nanocrystals were selectively nucleated on the Si–H surface (Fig. 33.13, step D) or the etched Si surface (Fig. 33.13, step F). Recently, Si–H was found by Yoshinobu et al. [133] to have chemical affinity toward some types of protein molecules, while the molecules did not adsorb on oxide. They accordingly fabricated adsorbed ferritin patterns by a lithographic method based on AFM-based anodization of Si. In other pattern-transfer demonstrations using a chemical contrast between Si–H and oxide surfaces, chemical affinity of organosilane molecules is employed. In this case, Si–H and anodic oxide surfaces act as inactive and reactive sites, respectively. As illustrated schematically in Fig. 33.13, step E, since there are a considerable number of –OH groups on the hydrophilic oxide surface, organosilane molecules preferentially react with the –OH groups, resulting in the formation of an organosilane monolayer on the oxide surface [134]. Such chemical affinity of SPM-produced anodic oxide was reported to depend on the atmosphere used for anodization. An anodic oxide surface formed in a wet environment with a humidity of 88% was more reactive to organosilane molecules, while an anodic oxide surface formed in a dry atmosphere with a humidity of less than 1% showed little reactivity to the molecules [135]. In the latter case, the primary species of oxidation is considered to be oxygen rather than water and the surface is rarely terminated with –OH groups. Lift-off patterning of Au films has also been demonstrated using a Si–H sample with an anodic oxide pattern [136]. The patterned Si–H sample deposited with a Au film (Fig. 33.13, step G) is etched in aqueous HF. The oxide under the Au film is dissolved and, then, the Au film is lifted of the region, resulting in the pattern transfer from oxide to Au. The patterned formation of metal silicide has also been demonstrated [96, 137]. When a Pt film is deposited on a Si–H sample with an anodic oxide pattern,chemical reaction of Si with Pt to form PtSi is promoted faster on the Si–H surface than on the oxide surface. After platinum silicide has formed on the Si–H surface, unreacted Pt remaining on the oxide surface is etched off and, then, the sample is annealed to complete the silicide formation. The oxide pattern is transferred to the PtSi pattern.

33.5.2.2
Patten-Transfer Process Using Conductive Thin Films

There are reports of scanning probe lithography using metallic resist films [31, 63]. In these demonstrations, thin films of anodically oxidizable metals were used as resist films for nanolithography based on scanning probe anodization. One example is illustrated schematically in Fig. 33.14, steps A–C.

In this process, a Mo film of 4-nm thickness formed on Si and polymer substrates was used as a resist film. The resist was patterned by AFM-based anodization in which the junction bias was applied between the atomic force microscope tip and the Mo film. The film was locally anodized to MoO_3 (Fi. 33.14, step A). This anodic Mo oxide was etched in a KOH solution, while the unanodized Mo film remained unetched. This KOH etching step corresponds to the resist development process. When the KOH etching was further prolonged, the substrate Si was etched in the anodized region. In the case of the polymer substrate, a sample with the KOH-developed Mo resist film was treated in oxygen plasma in order to etch the polymer substrate. In both cases, the anodized patterns were successfully transferred as etched grooves (Fig. 33.14, step C). The combination of Mo and KOH etching worked as a positive-tone resist process. With use of nanopatterned polymer films fabricated by the Mo resist process and the lift-off technique, various nanostructures have been fabricated. A negative-tone metal resist process has been demonstrated using Ti thin films [31]. A Ti film of 50-nm thickness was anodized so that an unanodized part remained under the anodic oxide as illustrated in Fig. 33.14, step E. The patterned Ti films was etched in a dilute HF solution of 0.1%. The Ti film was not etched in the anodized regions, showing that this was a negative-tone resist process (Fig. 33.14, step F). Similar metallic resist processes might be achieved using a Cr resist, since pattern-transfer processes for anodic chromium oxide via chemical etching have been successfully developed [37]. In this report on chemical etching techniques both positive and negative tone pattern transfers were demonstrated.

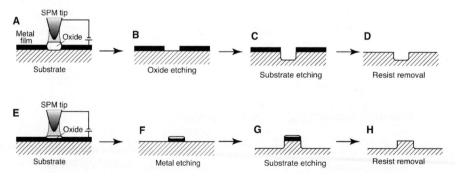

Fig. 33.14. Pattern-transfer processes for metal resist. **A** Local oxidation of a metal film by scanning probe anodization. **B** Oxide etching. **C** Substrate etching though the hole fabricated in the metal resist film. **D** Resist removal. **E** Local oxidation of the top surface of a metal film by scanning probe anodization. **F** Metal etching masked by the anodic oxide. **G** Substrate etching masked by the metal film. **H** Resist removal

Besides metallic thin films, Si films are applicable to conductive resist films as well. Snow et al. [20] reported nanofabrication of GaAs substrates using a thin Si film of 5-nm thickness epitaxially grown on the substrates as a resist film for STM-based oxidation lithography. Kramer et al. [138] have reported an application of hydrogenated a-Si (a-Si:H) films formed by plasma chemical vapor deposition to resist films for scanning probe anodization lithography. They formed an a-Si:H film on a PtIr substrate. The film was highly doped in order to attain a satisfactory conductivity for STM-based anodization. First, anodization patterning of the a-Si:H resist is conducted (Fig. 33.14, step E). Next, the patterned resist is developed by TMAH etching (Fig. 33.14, step F). Then, the resist pattern is transferred onto the substrate PtIr by Ar^+ ion etching (Fig. 33.14, step G). Finally, the remaining resist film is removed by dipping in HF for a short time and subsequent TMAH etching (Fig. 33.14, step H). The particular advantage of a-Si:H resist films is that the films can be deposited on almost any surface at low temperature.

Note that these conductive-resist techniques are applicable to the fabrication of metal and Si nanostructures if the process is stopped at the resist development step (Fig. 33.14. steps B and F). Indeed, Si electrodes for nanoelectronic devices as discussed in Sect. 33.5.1 were fabricated by scanning probe anodization lithography and chemical etching techniques [115, 116]. Fabication of Ti nanowires has been reported as well using chemical etching of anodic titatium oxide with NaOH [139]. The process corresponds to Fig. 33.14, steps A and B.

33.5.2.3
Patten-Transfer Process Using Insulating Thin Films on Conductive Substrates

Si_3N_4 has been proved as a resist material for scanning probe anodization nanolithography [140]. Since Si_3N_4 is an electrical insulator, it must be formed as a thin film on a conductive substrate, mainly on Si.

As illustrated in Fig. 33.15, step A, when a Si_3N_4/Si sample prepared by means of a low-pressure chemical vapor deposition method is anodized by applying an anodization bias voltage between a SPM tip and the Si substrate, Si_3N_4 is chemically converted to silicon oxide. The anodized depth can be controlled so as to remain within the nitride film or to be expanded to the Si substrate. When the SPM-anodized

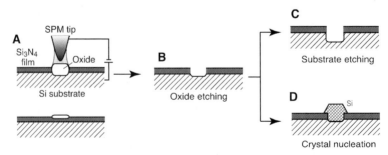

Fig. 33.15. Pattern-transfer processes for Si_3N_4 resist on Si. **A** Local anodization of a Si_3N_4 layer formed on a Si substrate. **B** HF etching of the anodic oxide. **C** Substrate etching. **D** Epitaxial growth of a Si crystal masked with the Si_3N_4 resist

Si_3N_4/Si sample is dipped into aqueous HF, the oxide dissolves faster than Si_3N_4, so a hole is made in the resist film. The Si_3N_4 resist film is regarded as being developed (Fig. 33.15, step B). Since the etch selectivity of Si to Si_3N_4 in a KOH solution is as high as 1000, nanoholes or grooves can be fabricated on the Si substrate using the Si_3N_4 resist film as an etch mask (Fig. 33.15, step C). A promising application of the nanopatterned Si_3N_4 mask is area-selective epitaxial growth of Si nanocrystals demonstrated by Yasuda et al. [141]. They coated top and sidewall surfaces of the Si_3N_4 mask, but exposed a clean Si surface at the bottom of the hole. The sample was treated by chemical vapor deposition using Si_2H_6 as the source gas. The nucleation of Si proceeded selectively at the bottom of the etched hole where Si was exposed. Accordingly, one Si crystal was successfully grown in each of the holes on the Si_3N_4 resist film, resulting in the formation of a Si nanocrystal array.

Among a variety of resist materials, organic thin films are practically important and have been frequently applied for lithography. Thus, the patterning of organic thin films by SPM is of special importance. However, in order to attain high spatial resolution on nanometer scale, resist films must be prepared in a thin and uniform layer. Furthermore, the films must be compatible with pattern-transfer processes, particularly with chemical etching. Organic SAMs fulfill the requirements for resist films for scanning probe anodization lithography including thickness, uniformity, patternability and compatibility to various pattern-transfer processes and, therefore, are one of the most promising candidates [42, 47, 48, 50, 64]. As described in Sect. 33.3.1, when a SAM is locally anodized, organic molecules can be decomposed in or detached from the anodized area; consequently, an etched hole is formed by scanning probe anodization (Fig. 33.16, steps A and B) [47, 48, 99].

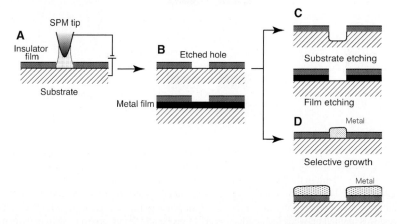

Fig. 33.16. Pattern-transfer processes for a SAM resist on a conductive substrate or a metal thin film. **A** SAM patterning by scanning probe anodization. **B** Patterned SAM resist. (If an anodically oxidizable material, e.g., Si or Ti, is used as the substrate, the growth of an anodic oxide layer, which is not indicated, is inevitable in the anodized region as illustrated in Fig. 33.10c.) **C** Substrate and metal film etching. **D** Selective metal deposition with positive and negative tones

The resist-developing process is not necessarily different from that for the metal and Si_3N_4 resist films. Namely, the anodization patterning of SAMs can be considered as "self-developing." For example, Si nanostructures have been successfully fabricated using alkylsilane SAMs as resist films by scanning probe anodization lithography and the subsequent chemical etching processes (Fig. 33.16, step C) [42, 99]. For these etching processes, HF and a mixture of NH_4F and H_2O_2 were employed as etching chemicals. When an anodically oxidizable metallic film is used for a sample substrate, a metal nanostructure can be fabricated as illustrated in Fig. 33.16. The nanoprocessing of Ti thin films has been reported [142]. Besides the use of SAMs, a-C films have been employed as semiconducting resist films [54]. An a-C layer prepared on a PdAu film surface was patterned by AFM-base anodization. The film was decomposed in the anodized region, similarly to scanning probe anodization of graphite [23], so an etch hole was created beneath the atomic force microscope tip. The substrate PdAu film was etched by Ar^+ ion etching using the patterned a-C layer

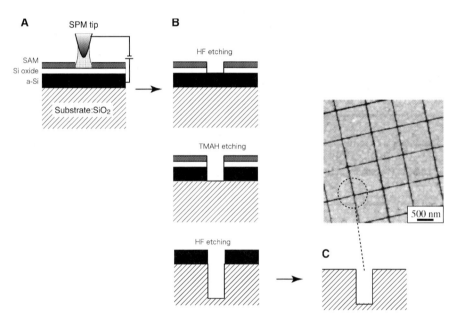

Fig. 33.17A–C. Nanoprocessing of an insulator based on scanning probe anodization lithography. The sample substrate was a single-crystal Si plate with a thermally grown oxide layer approximately 20 nm in thickness. A triple-layered resist film, a thin Si film of 20-nm thickness, a photochemically grown Si oxide layer of 2-nm thickness and an octadecylsilyl SAM of 2-nm in thickness was prepared on this Si substrate. In order to inject current into the SAM, a bias voltage was applied between the conductive AFM probe and the amorphous Si layer, which was positively polarized. The patterning was conducted using the constant-current mode. The resist development was a two-step chemical etching conducted at room temperature, that is, etching in 0.5% aqueous HF and 25% aqueous tetramethylammonium hydroxide (*TMAH*), in that order. The sample was etched in the HF solution again in order to transfer the resist pattern to the SiO_2 substrate. After this pattern transfer step had been completed, the resist film was etched again in the TMAH solution. The AFM image shows a grid pattern at horizontal and vertical intervals of approximately 500 nm fabricated on the SiO_2 substrate

as a mask. Finally, the remaining a-C film was removed by oxygen plasma etching. Consequently, PdAu nanostructures were fabricated. Considering the semiconducting property of the a-C films, the a-C resist films are not restricted to nanofabrication of conducting substrates, but might be applicable to nanofabrication of insulating substrates.

Selective metallization is possible using scanning probe anodization lithography and SAM resist films as illustrated in Fig. 33.16, step D. Electroless plating has been employed in order to fabricate positive-tone and negative-tone metal patterns on patterned SAM samples. The former was achieved using an alkylsilane SAM as a resist film [143]. The SAM was coated on a Si substrate and patterned by AFM-based anodization. In the anodized area, the substrate Si was exposed by HF etching. The exposed Si surface worked as a nucleation site for Au electroless plating through the galvanic displacement of Si by Au ions, resulting in the AFM-drawn pattern being transferred to a Au nanopattern. In contrast, in the latter case, an amino-terminated organosilane SAM was used as a resist film [50]. Since amino-functional groups have an affinity to Pd-based catalysts, the nonanodized area on the patterned aminosilane SAM was area-selectively catalyzed and, then, metallized with electroless Ni deposits. On the other hand, Schoer et al. [144] have reported the fabrication of Cu nanopatterns down to 50 nm in width. They used an alkylthiol SAM on Au and patterned the SAM by scanning tunneling microscopy. The patterned sample was further treated with a metal–organic chemical vapor deposition method in order to deposit Cu on the probe-scanned area, resulting in the fabrication of positive-tone metallic nanostructures.

Figure 33.17 demonstrates the capability of AFM-based anodization lithography using mutilayer resist films for nanoprocessing of SiO_2, which is a key insulating material for Si-based microdevices [145]. The resist consists of an alkylsilane SAM and an a-Si film which serve as an imaging layer and a current-passing layer, respectively. The developing process for the exposed resist film was a two-step chemical etching conducted at room temperature. Nanostructures were fabricated on the substrate SiO_2 through chemical etching using the developed resist film as an etching mask. The AFM image shown in Fig. 33.17 demonstrates that fine grooves less than 50 nm in width were successfully fabricated on the thermal SiO_2 substrate.

33.5.3 Integration of Scanning Probe Lithography with Other High-Throughput Lithographies

Although, scanning probe lithography is a powerful technique for nanofabrication, its patterning speed is relatively slow compared with that of electron-beam lithography even though scanning probe lithography has attained a pattern drawing speed in the range of centimeters per second. This is a serious disadvantage in practical applications. One approach to improving the total throughput of scanning probe lithography is to complement this slow but high-resolution lithography with a high-speed, low-resolution pattern-transfer technique. The integration of photolithography and scanning probe lithography is the most promising way. In order to perform such an integration in practice, two key technologies are needed. The first one is a resist material which is commonly usable in both photolithography and scanning probe lithography. The second one is an accurate alignment of an SPM-generated pattern with a photolithographically transferred pattern on the resist. The latter problem can be solved in AFM-based lithography, since AFM has excellent abilities in high-resolution surface imaging. Prior to AFM lithography, photoprinted patterns on a resist film could be imaged without rarely causing damage to the patterns. The AFM probe is readily located at a required position using the acquired image as guidance.

As a resist material patternable both in photolithogaphy and in AFM-based anodization lithography, a-Si films were first proposed [146,147]. With use of the light beam from an Ar^+ ion laser of 488-nm in wavelength, a hydrogen-terminated a-Si surface was locally oxidized with a minimum spot size of 500 nm. Next, an additional pattern was drawn on the sample surface by AFM-based anodization by aligning both the laser-drawn and the AFM-drawn patterns. Finally, the oxide patterns were transferred to the a-Si film via KOH etching. This integrated lithography was further extended to use Al thin layers as resist films [148]. On the other hand, another approach for the integration of photolithography and AFM-based anodization lithography has been reported [149,150]. In this integrated lithography, organosilane SAMs were used as resist films for photolithography and AFM-based anodization lithography and LFM imaging was used for the pattern alignment. Since, in general, most of the organosilane SAMs did not have the photosensitivities required for photolithography, a novel photolithographic technology using vacuum-ultraviolet light of 172-nm wavelength was developed for the integration. In addition, the integration of AFM-based anodization with electron-beam lithography is also promising, since special SAM resists patternable with both lithographies have been developed [151].

33.5.4
Chemical Manipulation of Nano-objects by the Use of a Nanotemplate Prepared by Scanning Probe Anodization

Most lithographic technologies depend on etching processes in which materials are removed selectively from certain locations with defined patterns. Therefore, the compatibility of such etching processes is of primary importance for the lithographic applications of scanning probe anodization as described in Sects. 33.5.2 and 33.5.3. Another important approach for constructing nanostructures and devices is to build up minute objects onto particular positions on a substrate. Scanning probe anodization is, thus, expected to be applied to this approach, that is, "chemical manipulation of nano-objects" as illustrated schematically in Fig. 33.18.

The strategy of chemical nanomanipulation consists of (1) preparation of a patterned substrate, (2) provision of chemical contrasts to the patterned substrate and (3) programmed arrangement of nano-objects promoted by the chemical contrasts. The chemical contrast means the presence of chemical affinity and reactivity between adjacent regions. Free spaces themselves are also regarded as nano-objects and are very useful parts for constructing nano-systems and devices. As described in Sect. 33.5.2, there are a few chemical contrasts between Si–H and SiO_2 surfaces and such contrasts are useful for the chemical manipulation [132, 134]. However, there is a serious disadvantage of that system. The Si–H surface is not stable for long

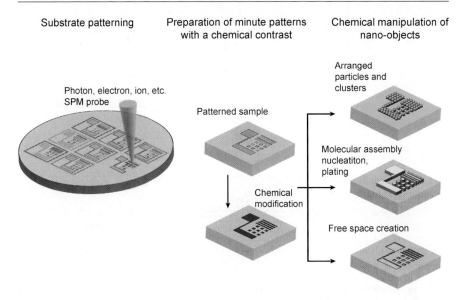

Fig. 33.18. Chemical manipulation of nano-objects

exposure to air and in some chemical environments. This lack of chemical durability of Si–H creates some restrictions for the use of Si–H in chemical manipulation processes. As an alternative, a SAM is a fruitful candidate as the base material for chemical manipulation, since some types of SAMs are much more durable than Si–H substrates and a wide variety of chemical functions can be provided by the use of SAMs consisting of various types of molecules. Furthermore, scanning probe anodization lithography is the most promising method for the first step of chemical manipulation, since it can be conducted under ambient conditions and, thus, does not require any hard environments in which some types of SAMs are deactivated or even decomposed.

Figure 33.19 illustrates a process chart of chemical manipulation based on organosilane SAMs and scanning probe anodization lithography [152]. When an alkylsilane SAM is patterned by scanning probe anodization (Fig. 33.19, step A), silicon oxide is exposed in the anodized region [99]. A distinct chemical contrast exists between the anodized and unanodized regions, the surfaces of which are hydrophobic and hydrophilic owing to the surface termination with $-CH_3$ and $-OH$ groups, respectively (Fig. 33.19, step B). This chemical contrast might be applicable to the manipulation of minute objects. For example, the condensation of water on the hydrophilic region has been observed (Fig. 33.19, step C) [42]. Other polar liquids and molecules or water-soluble substances might be delivered on the hydrophilic region. Some types of protein molecules are expected to be immobilized on the hydrophobic region though hydrophobic interaction similarly to the demonstration conducted on an oxide-pattern/Si–H sample [134], since the wettability contrast between $-CH_3$ and $-OH$ terminated surfaces is larger than that between Si–H and –OH.

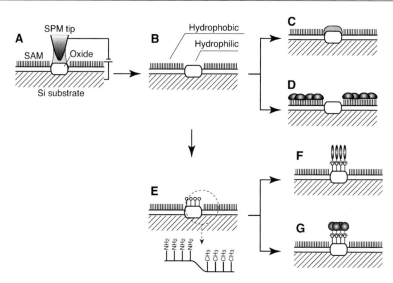

Fig. 33.19. Template fabrication and chemical manipulation on organosilane SAM covered Si substrates structured by scanning probe anodization lithography. **A** Anodization lithography of a SAM/Si sample. **B** Template with a wettability contrast. Selective assembly of substances on the **C** hydrophilic and **D** hydrophobic surfaces. **E** Chemical modification of the template. **F** Molecular assembling and **G** nanoparticle immobilization on the modified oxide surface

It is well known that an oxide surface terminated with –OH groups can serves as an adsorption site of organosilane molecules; hence, another organosilane SAM with chemical properties different from those of the alkylsilane SAM surrounding the anodic oxide patten can be formed on the oxide (Fig. 33.19, step E) [152, 153]. More sophisticated chemical manipulations will be attainable though this approach. Typically, an aminosilane is employed for preparing the second SAM on the anodic oxide pattern. Indeed, protein molecules were immobilized on such an aminosilane SAM pattern using a proper cross-linker connecting $-NH_2$ groups and the protein molecules (Fig. 33.16, step F) [152, 154]. Moreover, arrays of latex and Au nanoparticle have been constructed [152, 155]. As illustrated in Fig. 33.20, step A, the first SAM was locally removed by scanning probe microscope tip induced anodization in order to prepare chemically reactive sites for the second SAM formation in the demonstrations described above. The removal of the first SAM is not absolutely necessary in order to fabricate chemical nanotemplates. Maoz et al. [56] have fabricated templates having a specific chemical contrast by chemically converting SAM surfaces instead of removing SAMs. They demonstrated that an organosilane SAM surface terminated with vinyl groups could be oxidized to –COOH groups by AFM-base anodization and such a COOH-terminated surface served as a molecular self-assembly site. Fresco et al. [156] have developed a precursor molecule for a SAM compatible with the chemical conversion of surface functional groups by AFM-based anodization. They fabricated SH-terminated sites in order to fix Au nanoparticles. One advantage of this surface chemical conversion method is that both cathodic and anodic reactions are available for attaining chemical contrasts [92, 157].

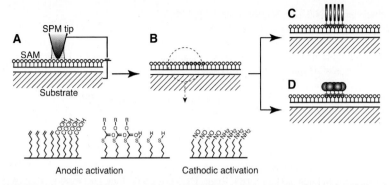

Fig. 33.20. Surface chemical conversion of a SAM by scanning probe electrochemistry. **A** Formation of chemical contrast by SPM tip induced electrochemical reactions. **B** Template with chemical contrast created owing to a chemical change of the surface functional groups on the SAM. **C** Molecular assembly and **D** nanoparticle immobilization on the chemically converted region

Although available chemical routes for generating chemical contrasts are limited in the surface chemical conversion approach, the method certainly provides additional versatility to the chemical manipulation strategy.

33.6
Conclusion

Since 1990, local oxidation of material surfaces has attracted primary attention in the research field of nanofabrication based on SPM. In this chapter, this SPM-induced local oxidation in the presence of adsorbed water at the scanning probe microscope tip–sample junction was discussed mainly from the viewpoint of electrochemistry, in addition to a historical review and technical developments. Although it started from the first demonstrations on semiconductor surfaces, at present the method has become crucial technology for nanofabrication and nanolithography applicable to patterning of a wide variety of materials ranging from metals to organic thin films. In these nearly two decades, significant technological developments have been achieved in probe-motion control, junction-current regulation, characterization of fabricated features, pattern-transfer processes and so forth. In particular, the method has been extended to promote chemical reactions other than oxidation, e. g., nitrization, carbonization and electrochemical reduction. Namely, from the local oxidation with adsorbed water, it has become the local electrochemical modification with an adsorbed substance. These technical developments have enriched the versatility and applicability of the method.

References

1. Rohrer H (1993) Jpn J Appl Phys 32:1335
2. Avoris P (1995) Acc Chem Res 28:95

3. Sugimura H, Nakagiri N (1995) Jpn J Appl Phys 34:3406
4. Nyffebegger RM, Penner RM (1997) Chem Rev 4:1195
5. Quate CF (1997) Surf Sci 386:259
6. Soh HT, Guarini KW, Quate CF (2001) Scanning probe lithography. Kluwer, Boston
7. Sattler K (2003) Jpn J Appl Phys 42:4825
8. Krämer S, Fuierer RR, Gorma CB (2003) Chem Rev 103:4367
9. Wouters D, Schubert US (2004) Angew Chem Int Ed Engl 43:2480
10. Sugimura H (2005) Int J Nanotechnol 2:314
11. Kolb DM, Simeone FC (2005) Electrochim Acta 50:2989
12. Tseng AA, Notargiacomo A, Chen TP (2005) J Vac Sci Technol B 23:877
13. Garcia R, Martinez RV, Martinez J (2006) Chem Soc Rev 35:29
14. Stiévenard D, Legrand B, (2006) Prog Surf Sci 81:112
15. Dagata JA, Schneir J, Harary HH, Evans CJ, Postek MT, Bennett J (1990) Appl Phys Lett 56:2001
16. Nagahara LA, Thundat T, Lindsay SM (1990) Appl Phys Lett 57:270
17. Young L (1961) Anodic oxide films. Academic, New York
18. Gerischer H, Mindt W (1968) Electrochim Acta 13:1329
19. Barniol N, Prez-Murano F (1992) Appl Phys Lett 61:462
20. Snow ES, Campbell PM, Shanabrook BV (1993) Appl Phys Lett 63:3488
21. Thundat T, Nagahara LA, Oden PI, Lindsay SM, George MA, Glaunsinger WS (1990) J Vac Sci Technol A 8:3537
22. Albrecht TR, Dovek MM, Kirk MD, Lang CA, Quate CF, Smith DPE (1989) Appl Phys Lett 55:1727
23. McCarley RL, Hendricks SA, Bard AJ (1992) J Phys Chem 96:10089
24. Grigg DA, Russell PE, Griffith (1992) J Vac Sci Technol A 10:680
25. Weeks BL, Vaughn MW (2005) Langmuir 21:8096
26. Sugimura H, Uchida T, Kitamura N, Masuhara H (1993) Jpn J Appl Phys 32:L553
27. Hurlen T, Gulbandsen E (1994) Electrochim Acta 39:2169
28. Sugimura H, Uchida T, Kitamura N, Masuhara H (1993) Appl Phys Lett 63:1288
29. Sugimura H, Uchida T, Kitamura N, Masuhara H (1994) J Phys Chem 98:4352
30. Sugimura H, Kitamura N, Masuhara H (1994) Jpn J Appl Phys 33:L143
31. Sugimura H, Uchida T, Kitamura N, Masuhara H (1994) J Vac Sci Technol B 12:2884
32. Day HC, Allee DR (1993) Appl Phys Lett 62:2691
33. Yasutake M, Ejiri Y, Hattori Y (1993) Jpn J Appl Phys 32:L1021
34. Snow ES, Campbell PM, MacMarr PJ (1993) Appl Phys Lett 63:749
35. Wang D, Tsau L, Wang KL (1994) Appl Phys Lett 65:1415
36. Fay P, Brockenbrough RT, Abeln G, Scott P, Agarwala S, Adesida I, Lyding JW (1994) J Appl Phys 75:7545
37. Song HJ, Rack MJ, Abugharbieh K, Lee SY, Khan V, Ferry DK, Allee DR (1994) J Vac Sci Technol 12:3720
38. Thomson RE, Moreland J, Roshko A (1994) Nanotechnology 2:57
39. Wang D, Tsau L, Wang KL, Chow P (1995) Appl Phys Lett 9:1295
40. Ishii M, Matsumoto K (1995) Jpn J Appl Phys 34:1329
41. Matsumoto K, Takahashi S, Ishii M, Hoshi M, Kurokawa A, Ichimura S, Ando (1995) Jpn J Appl Phys 34:1387
42. Sugimura H, Nakagiri N (1995) Langmuir 10:3623
43. Park SW, Soh HT, Quate CF, Park SI (1995) Appl Phys Lett 67:2415
44. Xu LS, Allee DR (1995) J Vac Sci Technol B 13:2837
45. Snow ES, Park D, Campbell PM (1996) Appl Phys Lett 69:269
46. Day HC, Allee DR (1996) Nanotechnology 7:106
47. Schoer JK, Zamborini FP, Crooks RM (1996) J Phys Chem 100:11086

48. Sugimura H, Okiguchi K, Nakagiri N Miyashita M (1996) J Vac Sci Technol B 14:4140
49. Sasa S, Ikeda T, Dohno C, Inoue M (1997) Jpn J Appl Phys 36:4065
50. Brandow SL, Calvert JM, Snow E, Campbell PM (1997) J Vac Sci Technol A 15:1445
51. Hung CJ, Gui J, Switzer JA (1997) Appl Phys Lett 71:1637
52. Boneberg J, Böhmisch M, Ochmann M, Leiderer P (1997) Appl Phys Lett 71:3805
53. Avramescu A, Ueta A, Uesugi K, Suemune I (1998) Appl Phys Lett 72:716
54. Mühl T, Brückl H, Kraut D, Kretz J, Mönch I, Reiss G (1998) J Vac Sci Technol B 16:3879
55. Shirakashi J, Ishii M, Matsumoto K, Miura N, Konagai M (1996) Jpn J Appl Phys 35:L1524
56. Maoz R, Cohen SR, Sagiv J (1999) Adv Mater 11:55
57. Lüscher S, Fuhrer A, Held R, Heinzel T, Ensslin K, Wegscheider W (1999) Appl Phys Lett 75:2452
58. Chien FSS, Chang JW, Lin SW, Chou YC, Chen TT, Gwo S, Chao TS, Hsieh WF (2000) Appl Phys Lett 76:360
59. Song I, Kim BM, Park G (2000) Appl Phys Lett 76:601
60. Tachiki M, Fukuda T, Sugata K, Seo H, Umezawa H, Kawarada H (2000) Jpn J Appl Phys 39:4631
61. Sugawara K, Gotoh T, Tanaka K (2001) Appl Phys Lett 79:1549
62. Schneegans O, Moradpour A, Houzé F, Angelova A, Henry de Villeneuve C, Allongue P, Chrétien P (2001) J Am Chem Soc 123:11486
63. Rolandi M, Quate CF, Da H (2002) Adv Mater 14:191
64. Ara M, Graaf H, Tada H (2002) Appl Phys Lett 80:2565
65. Ahn SJ, Jang YK, Lee H, Lee H (2002) Appl Phys Lett 80:2592
66. Kondo T, Yanagisawa M, Jiang L, Tryk DA, Fujishima A (2002) Diamond Relat Mater 11:1788
67. Bo XZ, Rokhinson LP, Yin H, Tsui DC, Sturm JC (2002) Appl Phys Lett 81:3263
68. Xu MW, Eyben P, Hantschel T, Vandervorst W (2002) Jpn J Appl Phys 41:1048
69. Okazaki A, Akita S, Nakayama Y (2002) Jpn J Appl Phys 41:4973
70. Kim YH, Zhao J, Uosaki K (2003) J Appl Phys 94:7733
71. Farkas N, Zhang G, Evans EA, Ramsier RD, Dagata JA (2003) J Vac Sci Technol A 21:1188
72. Xie XN, Chung HJ, Xu H, Xu X, Sow CH, Wee ATS (2004) J Am Chem Soc 126:7665
73. Farkas N, Tokash JC, Zhang G, Evans EA, Ramsier RD, Dagata JA (2004) J Vac Sci Technol A 22:1879
74. Watanabe K, Takemura Y, Shimazu Y, Shirakashi J (2004) Nanotechnology 15:S566
75. Sugimura H (2004) Jpn J Appl Phys 43:4477
76. Li RW, Kanki T, Hirooka M, Takagi A, Matsumoto T, Tanaka H, Kawai T (2004) Appl Phys Lett 84:2670
77. Pellegrino L, Bellingeri E, Siri AS, Marré D (2005) Appl Phys Lett 87:064102
78. Maki T, Shikama S, Komori M, Sakaguchi Y, Sakuta K, Kobayashi T (1992) Jpn J Appl Phys 31:L1446
79. Marrian CRK, Perkins FK, Brandow SL, Koloski TS, Dobisz EA, Calvert JM (1994) Appl Phys Lett 64:390
80. Sugimura H, Okiguchi K, Nakagiri N (1996) Jpn J Appl Phys 35:3749
81. Sugimura H, Yamamoto T, Nakagiri N, Miyashita T, Onuki T (1994) Appl Phys Lett 65:1569
82. Moon WC, Yoshinobu T, Iwasaki H (2002) Jpn J Appl Phys 41:4754
83. Avouris P, Hertel T, Martel R (1997) Appl Phys Lett 71:285
84. Kuramochi H, Pérez-Murano F, Dagata JA, Yokoyama H (2004) Nanotechnology 15:297
85. Tello M, García R (2003) Appl Phys Lett 83:2529
86. Tello M, Garcia R, Martín-Gago JA, Martínez NF, Martín-González MS, Aballe L, Baranov A, Gregoratti L (2005) Adv Mater 17:1480
87. Martinez RV, García R, (2005) Nano Lett 5:1161

88. Kim Y, Kang SK, Choi I, Lee J, Yi J (2005) 127:9380
89. Kim SM, Ahn SJ, Lee H, Kim ER, Lee H (2002) Ultramicroscopy 91:165
90. Pyle JL, Ruskell TG, Workman RK, Yao X, Sarid D (1997) J Vac Sci Technol B 15:38
91. Li RW, Kanki T, Tohyama H, Hirooka M, Tanaka H, Kawai T (2005) Nanotechnology 16:28
92. Maoz R, Frydman E, Cohen SR, Sagiv J (2000) Adv Mater 12:424
93. Sugimura H, Sito N, Lee SH, Takai O (2004) J Vac Sci Technol B 22:L44
94. Snow ES, Campbell PM (1993) Appl Phys Lett 64:1932
95. Sugimura H, Nakagiri N (1997) Nanotechnology 8:A15
96. Snow ES, Campbell PM, Perkins FK (1999) Appl Phys Lett 75:1476
97. Teuschler T, Mahr K, Miyazaki S, Hundhausen M, Ley L (1995) Appl Phys Lett 66:2499
98. Fujihira M, Aoki D, Okabe Y, akano H, Hokari H, Frommer J, Nagatani Y, Sakai F (1996) Chem Lett 499
99. Sugimura H, Hanji T, Hayashi K, Takai O (2002) Ultramicroscopy 91:221
100. Nakagiri N, Yamamoto T, Sugimura S, Suzuki Y, Miyashita M, Watanabe S (1997) Nanotechnology 8:A32
101. Miyazaki T, Kobayashi K, Yamada H, Horiuchi T, Matsushige K (2002) Jpn J Appl Phys 41:4948
102. Wilder K, Quate CF (1999) J Vac Sci Technol B 17:3256
103. Servat J, Gorostiza, Sanz F, Pérez-Murano F, Bamiol N, Abadal G, Aymerich X (1996) J Vac Sci Technol A 14:1208
104. Irmer B, Kehrle M, Lorenz H, Kotthaus P (1997) Appl Phys Lett 71:1733
105. Kuramochi H, Ando K, Tokizaki T, Yokoyama H (2006) 88:093109
106. Pérez-Murano F, Birkelund K, Morimoto K, Dagata JA (1999) Appl Phys Lett 75:199
107. Matsumoto K, Gotoh Y, Maeda T, Dagata JA, Harris JS (2000) Appl Phys Lett 76:239
108. Sugata K, Tachiki M, Fukuda T, Seo H, Kawarada H (2002) Jpn J Appl Phys 41:4983
109. García R, Calleja M, Pérez-Murano F (1998) Appl Phys Lett 72:2295
110. García R, Calleja M, Rohrer H (1999) J Appl Phys 86:1898
111. Legrand B, Stievenard D (1999) Appl Phys Lett 74:4049
112. Calleja M, García R (2000) Appl Phys Lett 76:3427
113. Graf G, Frommenwiler M, Studerus P, Ihn T, Ensslin K, Driscoll DC, Gossard AC (2006) J Appl Phys 99:053707
114. Fayfield T, Higman TK (1994) J Vac Sci Technol B 12:3731
115. Minne SC, Soh HT, Flueckiger P, Quate CF (1995) Appl Phys Lett 66:703
116. Campbell PM, Snow ES, McMarr PJ (1995) Appl Phys Lett 66:1388
117. Matsumoto K, Ishii M, Segawa K, Oka Y (1996) Appl Phys Lett 68:34
118. Held R, Heinzel T, Studrus P, Ensslin K, Holland M (1997) Appl Phys Lett 71:2689
119. Held R, Lüscher S, Heinzel T, Ensslin K, Wegscheider W (1999) Appl Phys Lett 75:1134
120. Nemutudi R, Kataoka M, Ford CJB, Appleyard NJ, Pepper M, Ritchie DA, Jones GAC (2004) 95:2557
121. Hennessy K, Högerle C, Hu E, Badolato A, Imaoglu A (2006) Appl Phys Lett 89:041118
122. Shirakashi J, Matsumoto K, Miura N, Konagai M (1998) Appl Phys Lett 72:1893
123. Watanabe K, Koizumi S, Yamada T, Takemura Y, Shirakashi J (2005) J Vac Sci Technol B 23:2390
124. Tachiki M, Seo H, Banno T, Sumikawa Y, Umezawa H, Kawarada H (2002) Appl Phys Lett 81:2854
125. Pellegrino L, Pallecchi I, Marré D, Bellingeri, Siri AS (2002) 81:3849
126. Tsau L, Wang D, Wang KL (1994) Appl Phys Lett 64:2133
127. Hattori T, Ejiri Y, Saito K, Yasutake M (1994) J Vac Sci Technol A 12:2586
128. Sugimura H, Nakagiri N (1995) Nanotechnology 6:29
129. Snow ES, Juan WH, Pang SW, Campbell PM (1995) Appl Phys Lett 66:1729

130. Dagata JA, Schneir J, Harary HH, Bennett J, Tseng W (1991) J Vac Sci Technol B 9:1384
131. Sugimura H, Nakagiri N (1995) Appl Phys Lett 66:1430
132. Sugimura H, Nakagiri N (1995) J Vac Sci Technol B 13:1933
133. Yoshinobu Y, Suzuki J, Kurooka H, Moon WC, Iwasaki H (2003) Electrochim Acta 48:3131
134. Sugimura H, Nakagiri N, Ichinose N (1995) Appl Phys Lett 66:3686
135. Inoue A, Ishida T, Choi N, Mizutani W, Tokumoto H (1988) 73:1976
136. Moon WC, Yoshinobu T, Iwasaki H (2000) Ultramicroscopy 82:119
137. Snow ES, Campbell PM, Twigg M, Perkins FK (2001) Appl Phys Lett 79:1109
138. Kramer N, Birk H, Jorritsma J Schönenberger C (1995) Appl Phys Lett 66:1325
139. Yukiya T, Aizawa K, Fujihashi C (2004) Jpn J Appl Phys 43:1660
140. Gwo S (2001) J Phys Chem Solid 62:1673
141. Yasura T, Yamasaki S, Gwo S (2000) Appl Phys Lett 77:3917
142. Sugimura H, Nakagiri N (1996) J Vac Sci Technol A 14:1223
143. Sugimura H, Takai O, Nakagiri N (1999) J Electroanal Chem 473:230
144. Schoer JK, Ross CB, Crooks RM, Corbit TS, Hampden-Smith MJ (1994) Langmuir 10:615
145. Sugimura H, Takai O, Nakagiri N (1999) J Vac Sci Technol B 17:1605
146. Müllenborn M, Birkelund K, Grey F, Madsen S (1996) Appl Phys Lett 69:3043
147. Birkelund K, Thomsen EV, Rasmussen JP, Hansen O, Tang PT, Möller P, Grey F (1997) J Vac Sci Technol B 15:2912
148. Boisen A, Birkelund K, Hansen O, Grey F (1998) J Vac Sci Technol 16:2997
149. Sugimura H, Nakaigir N (1997) Jpn J Appl Phys 36:L968
150. Sugimura H, Nakagiri N (1998) Appl Phys A 66:S427
151. Wang X, Hu W, Ramasubramaniam R, Bernstein GH, Snider G, Lieberman M (2003) Langmuir 19:9748
152. Sugimura H, Nakagiri N (1997) J Am Chem Soc 119:9226
153. Sugimura H, Hanji T, Hayashi K, Takai O (2002) Adv Mater 14:524
154. Suzuki J, Yoshinobu T, Moon W, Shanmugam K, Iwasaki H (2006) Electrochemistry 74:131
155. Zheng J, Zhu Z, Chen H, Liu Z (2000) Langmuir 16:4409
156. Fresco ZM, Fréchet MJ (2005) J Am Chem Soc 127:8302
157. Saito N, Maeda N, Sugimura H, Takai O (2004) Langmuir 20:5182

34 Tissue Engineering: Nanoscale Contacts in Cell Adhesion to Substrates

Mario D'Acunto · Paolo Giusti · Franco Maria Montevecchi · Gianluca Ciardelli

Abstract. Tissue engineering is the exploitation of a combination of cells, engineered materials, and suitable biochemical and mechanical factors and processes to improve or replace biological functions. A number of questions in tissue engineering involve cell dynamics and proliferation. Motility is the hallmark of life, and mechanical forces—foremost adhesion and friction forces—play a fundamental role in cell migration, cell positioning, and cell-to-cell binding and tissue stabilization when contractile forces are generated within the cell and pull the cell body forward. In living systems such as cells, force interactions are more complex than those of the inorganic world because living systems continuously adapt the forces required for their movements by processing a number of internal and external signals and by converting different forms of energy into mechanical energy. Animal cells have an average size of 10–40 µm, but the adhesion with substrates is limited to sites whose dimensions fall in the nanometer range. As a consequence, a basic understanding on the nanoscale level is needed to have satisfactory knowledge of cell frictional and adhesion fundamental properties. Different techniques has been used to measure or to investigate how living cells adhere to other cells, to the extracellular matrix, or biocompatible scaffolds in their native environment. With the advent of the scanning probe microscopy family, it has been made possible both to study adhesive fundamental properties and features simulating the interaction between living systems, and to shed light on some processes occurring in complex living matter on the nanoscale making use of the atomic force microscopy based technique. This chapter describes summarily the role of the adhesion mechanisms occurring in cell dynamics and discusses some experimental results for adhesion forces between cells and scaffolding substrates on which they spread and proliferate.

Key words: Tissue engineering, Cell spreading, Nanoscale contacts, Focal adhesion

34.1
Tissue Engineering: A Brief Introduction

The main objective of regenerative medicine is the replacement, repair, or functional enhancement of tissues and organs. Tissue engineering provides the 3D assembly of cell species and extracellular matrix (ECM) able to restore, maintain, or enhance the function of tissues and has recently emerged as the most advanced therapeutic option presently available in regenerative medicine [1–5].

A damaged tissue is regenerated by the migration of cells to the injured area, induced by the release of chemotactic signals from the damaged tissue (Fig. 34.1). Tissue regeneration and healing is eventually accomplished by a combination of cell proliferation and cell matrix synthesis to form the regenerated tissue [5, 6]. The field of tissue engineering shows enormous potential for reparative medicine,

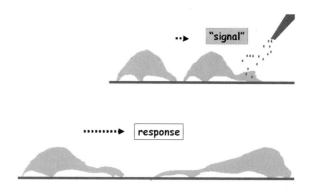

Fig. 34.1. Release of chemotactic signals and subsequent cell migration

but full achievement of this target may not be possible for many decades, although a number of related goals could be reached earlier. Challenging objectives are to initiate and control the regeneration of pathological tissue, and to treat, modify, and prevent disabling chronic disorders. It is a further challenge for regenerative medicine to deliver the disease-modifying benefits of tissue-engineered products to a wide patient population, in a cost-effective way. Again, since force generation and processing is a key component of cell behavior, it will be crucial to know how cells dynamically interact with substrates and their subsequent proliferations and differentiations. Besides, interactions occurring between biopolymers involve mechanics on the molecular and supramolecular scales. As a consequence, a basic understanding on the nanoscale level is needed to have satisfactory knowledge of cell frictional and adhesion fundamental properties.

The ultimate goal of regenerative medicine will require ECM engineering, a task that necessitates more learning about self-assembly, supramolecular chemistry and physics, and biomimetic material properties and, as far as contacts between living systems are concerned, particular features of tribology on the typical molecular or supramolecular scale of protein and molecular motors should be addressed.

One of the major goals will be to effectively exploit the enormous self-repair potential that has been observed in adult stem cells. Today's tissue engineering therapies are based on the autologous reimplantation of culture-expanded differentiated cells; next-generation therapies will need to build on the progress made with tissue engineering in understanding the huge potential for cell-based therapies which involve undifferentiated cells.

As the basic unit of life, cells are complex biological systems (Fig. 34.2). Cells must express genetic information to perform their specialized functions: synthesize, modify, sort, store, and transport biomolecules, convert different forms of energy, transduce signals, maintain internal structures, and respond to external environments. All of these processes involve mechanical, chemical, and physical processes. A major goal of tissue engineering is to generate living cellular constructs with 3D, tissuelike organization of cells and matrices characterized by cell migration, activation of signaling pathways, induction of growth factors, differentiation, tissue remodeling, and morphogenesis. Motility and mechanical forces, foremost adhesion and friction forces, play a fundamental role in cell migration, cell positioning, and cell-to-cell binding and tissue stabilization when contractile forces are generated within the

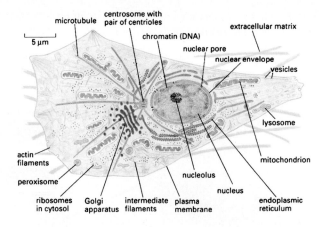

Fig. 34.2. An animal eukaryotic cell, which has an average size of 10–40 μm

cell. In protein secretion, for example, protein molecules are packaged in vesicles and are transported to the cell membrane by means of a molecular motor running along filaments and cells. On the other hand, mechanical forces and deformations induce biological response in cells, and many normal and diseased conditions of cells are dependent upon or regulated by their mechanical environment [7]. The effects of applied forces depend on the type of cells and how the forces are applied to, transmitted into, and distributed within cells.

A cell is formed by a cytoskeleton wrapped by the plasma membrane and trapping inside a nucleus surrounded by the souplike cytoplasm. The cytoskeleton is a system of protein filaments—microtubules, actin filaments, and intermediate filaments—that give to the cell shape and capability for directed movements (Fig. 34.3). The cytoskeleton contains three distinct filamentous biopolymers, the microtubules, microfilaments, and intermediate filaments. The basic structural elements of these three filaments are linear polymers of the proteins tubulin, actin, and vimentin or another related intermediate filament protein, respectively. The viscoelastic properties of cytoskeleton filaments are likely to be relevant to their biologic function, because their extreme length and rodlike structure dominate the rheologic behavior of cytoplasm.

Cells are attached to the ECM, a complex network of polysaccharides and proteins secreted by the cells that serves as a structural element in tissues. Proteins in the ECM, including collagen, elastin, fibronectin, vitronectin, and laminin, also play a regulatory role in cellular function through binding to various receptor proteins, found on the cell surface. Some of these receptors are members of the integrin family of transmembrane proteins. They are composed of two units, α and β, and are expressed on the membrane of a wide variety of cells. When attached to the cell cytoskeleton, integrins are critical to the mechanical stability of cell adhesion to the ECM and to other cells. Integrins also serve as biochemical signaling molecules in normal and diseased states of cells, and are involved in regulating cytoskeletal organization.

Adhesion plays a fundamental role in cell motion, proliferation, and homeostasis, both in the interaction with the external environment and in cell–cell interaction. For example, epithelial cells that form tissues that cover the internal and external

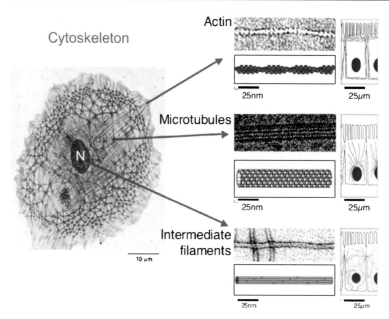

Fig. 34.3. Cell cytoskeleton consists of microtubules (25 nm in diameter), actin filaments (5–7 nm in diameter), intermediate filaments (8–12 nm in diameter), and other binding proteins

surfaces of organs, such as skin cells, the lining of the lungs, and the lining of the intestines, must adhere to substrates under a wide variety of conditions. Their adhesion properties can be regulated by the cell—or the system of cells—which simultaneously senses the chemical and mechanical properties of the environment. While these mechanical processes of biological cells must ultimately be described by the known laws regulating tribological systems, there are important differences between conventional inorganic ("dead") materials and wet living matter. A complex combination of shear forces and adhesive features is the key for understanding the ability of cells to proliferate, the phenotype characterization, and the ultimate ability to regenerate a tissue. Recent experiments show that, contrarily to artificial vesicles that exert only normal forces when adhering to a substrate, adhering cells show both normal and lateral forces [8, 9]. The normal forces arise from the action of specific adhesion molecules, van der Waals interactions, or macromolecular adsorption, while the lateral force arise from elastic deformations of the adhesion region by cytoskeletal forces. These lateral forces regulate the size and shape of the finite-sized, discrete adhesion regions (focal adhesion) and allow a cell to probe and to adjust the strength of the adhesion to its physical environment (Fig. 34.4). For mechanically active cells like fibroblasts, there could be hundreds of focal adhesions. The forces associated with the sites distributed along the cell rim keep the cell under tension. The forces that arise from the tension in the actin cytoskeleton tend to polarize the tense actin filaments. It is possible to sum over all the focal adhesions and model such an adhering cell as a pair of nearly equal and oppositely directed contraction forces (elastic force dipoles) with typical forces of 100 nN over a scale of tens of microns.

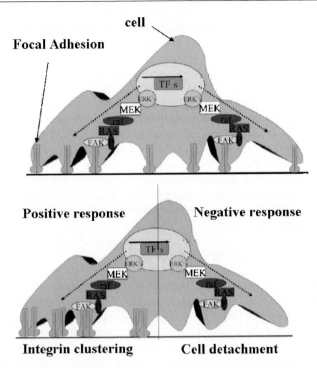

Fig. 34.4. How integrin-mediated activation of the signal transduction pathway may regulate the cell–substrate interaction. Once the cell comes into contact with the substrate it forms focal adhesions. The integrins are thought to relay signals to the nucleus though the mitogen-activated protein kinase pathway that is a cascade of focal adhesion kinase (*FAK*), phosphorylation (*P*), transcription factors (*TFs*), RAS and raf monomeric GTPases, extracellular signal related kinases (*ERKs*), and mitogen-activated protein kinase (*MEK*). In the *bottom image*, it is shown that if the signals relayed by the focal adhesions to the nucleus are positive, the integrin clustering occurs, increasing the area of cell adhesion to the substrate; if the signals are negative, then matrix metalloproteinases are released, causing integrin substrate detachment, decreasing the area of cell attachment. (Adapted from [25], with permission)

This chapter is organized as follows. After this introduction, in Sects. 34.2 and 34.2.1, basic mechanisms for cell–substrate adhesion, including the importance of the mechanical forces, in cell dynamics and proliferation are discussed. In Sect. 34.3, the nanoscale-level contacts between biomembranes and biopolymers through scanning probe microscopy techniques are reviewed, with attention focusing on experimental strategies for the measurements of cell–substrate (ECM or biomimetic scaffolds) adhesion force. At the end of the chapter, there is brief glossary to help the reader with biochemical terms.

34.2
Fundamental Features of Cell Motility and Cell–Substrates Adhesion

Phenomena involved in the movement of cells and of their internal components are the subject of a great number of investigations, owing to the intrinsic interest in the processes and because of the medical importance not limited only to tissue engineering [10]. Most cancers, for example, are not life-threatening until they metastasize and spread throughout the body. Metastasis occurs when previously sessile cells in a tumor acquire the ability to move and invade nearby tissues and circulate in the bloodstream or lymphatic system. A treatment that could impede the ability of tumor cells to acquire motility would largely prevent metastasis. The investigation of motility and its connection with adhesion mechanism processes offers a growing interdisciplary area in which a detailed knowledge of the mechanisms of cell motility might prove useful. Cells have spent several billion years developing highly efficient machinery to generate forces in the piconewton-to-nanonewton range that operate over distances of nanometers to micrometers and function in an aqueous environment. The better we understand the mechanics of cell motility, the more we will be able to adapt the cell movement machinery for medical treatments and, in general, for bioartificial systems.

A wide variety of cell movements have been characterized by biologists and biophysicists in relation to different degrees of molecular and mechanical features [12]. Movements of whole cells can be roughly divided into two functional categories: swimming, when the movement is through liquid water, and crawling, when the movement is across a rigid surface. Since viscous forces are many orders of magnitude stronger than inertial forces at the speeds, viscosities, and length scales experienced by swimming cells, for bacterial cells, the rotation of a helical or corkscrew-shaped flagellum in bacterial cells has been particularly well characterized from the biophysics perspective [13]. The flagellum is a long filament constructed by the noncovalent polymerization of hundreds of identical protein subunits, called flagellin. The speeds of flagellar swimming range from about 10 to 100 μm/s. Some unicellular eukaryotes swim by gradually changing the contour of their surface. This movement, called *metaboly* [14], shows that the reshifting of cellular content to generate a local increase in drag that propagates from the front of the cell to the back is sufficient to pull the cell center of mass forward through the viscous aqueous environment.

The best-characterized movement of a cell across a rigid surface is the crawling motility, or amoeboid motility. It is a general process where a cell attached to a rigid substrate extends forward a projection at its leading edge that then attaches to the

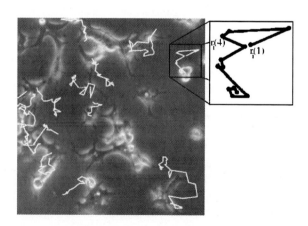

Fig. 34.5. Cell trajectories superposed on the initial frame. The *inset* shows 13 subsequent positions along the trajectory of the cell examined. (Adapted from [11])

substrate. Long, thin projections are called filopodia; flat, veil-shaped projections are called lamellipodia; and thick, knobby projections are called pseudopodia. All three types of projections are filled with assemblies of cytoskeletal actin filaments. Some differences can be observed in the cytoskeleton strategy for movement: lamellipodia are often associated with continuous advancing owing to a rolling mechanism, while filopodia and pseudopodia are characterized by a protrusion, sticking, and pulling strategy. After protrusion and attachment, the crawling cell then contracts to move the cell body forward, and movement continues as a threadmilling cycle of front protusion and rear retraction. The speed of such amoeboid movement can range from less than $1\,\mu m/h$ to more than $1\,\mu m/s$, depending on the cell type and its degree of stimulation. Several kinds of movements in which cells slide across a rigid substrate are known as gliding movements. Gliding appears in some bacteria to be driven by a low Reynolds number analogous to jet propulsion, in which a sticky and cohesive slime is extruded backward to push the cell forward [15].

Molecular motors denote a biological mechanism that converts chemical energy into mechanical energy, used by a cell to generate directed motion [10, 16–18]. Cytoskeletal motors bind to the filaments of the cytoskeleton and then walk along these filaments in a directed fashion. This class of motors is essential for intracellular transport, cell division, and cell locomotion. Cells generally store chemical energy in two forms: high-energy chemical bonds, such as the phosphoanhydride bonds in adenosine triphosphate (ATP), and asymmetric ion gradients across membranes, such as the electrical potential seen in nerve cells, see Sect. 34.5 for details. These sources of chemical energy drive all cell processes, from metabolism through to DNA replication. The subset of cell proteins and macromolecular complexes that convert chemical energy into mechanical forces are generally called molecular motors. Their wide variety reflects the diversity of cell movements necessary for life. Known biological molecular motors may be divided into three principal groups: (1) rotatory motors; (2) linear stepper motors; (3) assembly and disassembly motors [13, 18].

All the various cell movements are performed by ensembles of molecular motors that fall into these categories. One of the best-characterized motors in the bacterial species is the tiny rotary motor that enables bacteria to swim [20]. This motor uses ion flux along an electrochemical gradient to drive the rotation of the long, thin helical flagellum at a frequency of about 100 Hz. All known biological rotary motors use energy stored in an ion gradient to produce torque [21]. Most use the gradients of hydrogen ions that are found across the membranes of living cells.

Linear stepper motors are much more common in eukaryotic forms of motility. These motors move along preassembled linear tracks by coupling binding to the track, ATP hydrolysis, and a large-scale protein conformational change (see Sect. 34.5 for the connection between ATP hydrolysis and molecular motors). The first linear stepper motor to be characterized was myosin, the motor that drives filament sliding in skeletal muscle contraction [22]. The track for myosin is the actin filament, a helical polymer formed by noncovalent self-association of identical globular subunits. Currently, there are at least 18 different classes of myosins known, and each class may comprise dozens of different members even in a single organism. Figure 34.6 shows a schematic representation of the myosin V double arm acting between an actin filament and a cargo and the corresponding simulated steplike motion. Various forms of myosin in humans are responsible for biological movements

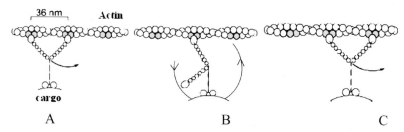

Fig. 34.6. Sequence for a swinging lever arm model proposed for the walking of a myosin V dimer along actin filaments [19]. Changes in the actin-binding domain during the ATPase cycle make possible the rotation of the lever arm, resulting in movement. The binding sites on the actin filament (8-nm diameter) are marked in *gray*

as diverse as muscle contraction, cell division, pigment granule transport in the skin, and sound adaptation in the hair cells of the inner ear. The exact mechanical features of each type of myosin motor appear to be carefully tuned to their biological functions. Similar functional tuning of motor mechanical properties is found in another abundant and diverse family of linear stepper motors, the kinesins, which use microtubules rather than actin filaments as a track [21]. Kinesins are involved in many types of intracellular transport, including transport of organelles along nerve axons and chromosome segregation. Another family of linear stepper motors that walk on microtubules, the dyneins, is less well-characterized.

Microtubules and actin filaments can assemble and disassemble rapidly to change the shape of the cell and to produce force on their own [22]. In these forms of biological force generation, the chemical energy comes from nonequilibrium protein polymerization, although ultimately the cellular pools of polymerizing actin and tubulin subunits are maintained in a steady state far from chemical equilibrium due to a coupling between protein polymerization and ATP hydrolysis. Force generated by actin polymerization is responsible for the movement of certain kinds of bacteria pathogens and the major driving force for cell protrusion at the leading edge in amoeboid motility. Whereas rotary motors and linear stepper motors have been characterized in great physical detail and force measurements have been performed on single molecules of each class, assembly and disassembly motors are poorly understood. Only a few measurements of the amount of force generated by single microtubules have yet been made, and there is to date no direct measurement of the force generated by polymerization of a single actin filament.

Cells adhere to a surface initially by attaching to a preadsorbed protein network called the extracellular matrix (ECM) or to neighboring cells. The cells spread out and their shapes are influenced by the surface topography and contribute to their phenotypic behavior. The nature of the ECM influences major cellular perspectives of growth, differentiation, and apoptosis and its composition will ultimately determine which cellular functions will be selected. In Fig. 34.4, it is schematically shown how cells feel and react to the ECM by means of integrin-dependent focal adhesion sites. Focal adhesions were first observed to form between cells and solids by Ambrose [23], in 1961, and later, in 1964, Curtis [24] found that their distance of closest approach was approximately 10 nm. The focal adhesion complex is composed of a high density of proteins that attach the extracellular portion of the cell to the intracellular cytoskeleton portion. Transmembrane proteins, such as integrins, attach to the ECM and connect indirectly to the actin filaments through protein assemblies of talin–paxillin–vinculin (Figs. 34.2, 34.4). Recently, it was shown experimentally and with a theoretical approach that mechanical forces can influence the association of integrin with the cytoskeleton to form the focal adhesion complexes [25–29]. In such complexes, integrins are likely a major force transmitter, since they provide the mechanical linkage between the cell cytoskeleton and the ECM. Further, a cell may be sensitive to mechanical forces or deformations through both integrin-mediated cell–ECM interactions and the subsequent force balance within the cytoskeleton. This balance may play a crucial role in regulating the shape, spreading, crawling, and polarity of cells. A still-open question is how mechanical force balance is recognized by cells and transduced into biological responses, and the exact molecular mechanisms responsible for mechanochemical transduction in living cells remain unknown. Different mechanisms have been proposed. One proposed mechanism considers the ion transport in the cell membranes. Ion transport can be changed by mechanical forces, especially tension in the cell membrane, thus changing the biochemical processes in cells [30]. Another proposed mechanism is based on the experimental evidence that cell cytoskeleton components, such as actin filaments and microtubulues, deform under mechanical forces, inducing conformational changes of other proteins attached to them and altering their functions. Moreover, ECM biopolymers such as fibronectin may deform under force, changing their interactions with cell-surface receptors, including integrins [31]. Fibronectin is able to contract to a fraction of its original length, and this property serves as a mechanosensitive control of ligand recognition. Existing experimental results suggest that the binding specificity and affinity between a receptor and a ligand can be changed by mechanical forces.

Living cells exert directional, lateral forces on adhesive sites. Since the adhering cells are rather flat, the forces exerted on the substrate can be considered to be tangential to the plane of the substrate surface. These forces originate from the interaction of the contractile cytoskeleton (actin filaments and their associated myosin motors) and the focal adhesion sites. Such adhesion sites respond dynamically to the local stresses: increased contractility leads locally to larger adhesions; in contrast, focal adhesions are disrupted when myosin in inhibited [32]. Typical forces at mature focal adhesions of human fibloblasts were found to vary between 10 and 30 nN [9]. Since each cell can have several hundred focal adhesions, the overall force exerted by a single cell goes up to the micronewton range. A linear relationship between the magnitude of the force and the focal adhesion area was found [33]. The direction of force usually agreed with focal adhesion elongation: for larger areas, force increases in proportion to area, with a stress constant of 5.5 nN/μm^2. The relation between force and size of the focal adhesion indicates that they act as mechanosensors: forces are used to actively probe the mechanical properties of the environment.

Several models have been introduced to describe within physical condensed-matter schemes the anchorage-dependent cells constantly assembling and disassembling focal adhesion sites, thereby probing the mechanical properties of their environment. For example, the concepts of force dipoles has been used to model cells in an elastic environment [34, 35] (Fig. 34.7).

The force dipoles model applied to the cellular case can be summarized as follows: anchorage-dependent cells probe the mechanical properties of the soft en-

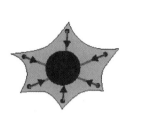

Fig. 34.7. Cellular force dipoles. *Left*: Cells feel the mechanical properties of the soft environment through their contractile sensors. Actin stress fibers (*lines*) are contracted by myosin II molecular motors and are connected to the environment though focal adhesions (*dots*). Different stress fibers probe different directions of space and compete with each other for stabilization of the corresponding focal adhesion. Such a probing process can be modeled as an anisotropic force contraction dipole. *Right*: The cell morphology becomes elongated in response to anisotropic external stimuli, during locomotion or spontaneously during times of strong mechanical activity. As a consequence, most stress fibers run in parallel and the whole cell acts as an anisotropic force contraction dipole. (Reprinted from [9], copyright 2005, with permission from Elsevier Science)

vironment though their contractile machinery. Actin stress fibers (lines in Fig. 34.7, left) are contracted by myosin II molecular motors and are connected to the environment through focal adhesions (dots in Fig. 34.7, left). Independent of the cell shape, different stress fibers probe different directions of space and compete with each other for stabilization of the corresponding focal adhesion. As a consequence, the probing process can be modeled as an anisotropic force contraction dipole. Cell morphology becomes elongated in response to anisotropic external stimuli during locomotion or spontaneously during times of strong mechanical activity (Fig. 34.7, right). Then most stress fibers run in parallel and the whole cell acts as an anisotropic force contraction dipole. One main advantage of such a model is that since cells are modeled as anisotropic force dipoles, the calculations are, in general, similar to calculations for isotropic force dipoles. Minimizing the interaction energy, one can predict in some cases that cells orient parallel and perpendicular to soft or stiff domains of the substrates, respectively, as observed in many experiments. This is fundamental for a rationale design of biomimetic scaffolds in tissue engineering. Recently, Schwarz et al. [36] proposed a simple two-spring model to make predictions regarding the way cells perceive extracellular rigidity (Fig. 34.8). A focal adhesion complex is a structure based on an adhesion cluster that can be schematically represented as certain number of bonds. The rupture and rebinding mechanisms of such an adhesion cluster play a fundamental role in mature focal adhesion formation.

In the two-spring model, the ECM and the force-bearing intracellular structures are represented by harmonic springs with spring constants K_e and K_i, respectively. Because the springs are in series, the overall stiffness is mainly determined by the softer spring, which in a physiological situation should be the ECM. Tension in the actin stress fibers is generated by myosin II molecular motors. As the motors pull, the springs get strained; for the static situation, the stored energy is $W = F^2/2K$, where K is the overall system stiffness. For the dynamic situation, the power dW/dt

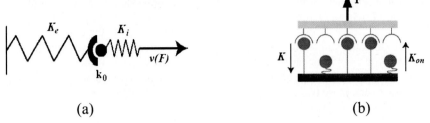

Fig. 34.8. a Two-spring model for a mature focal adhesion site. K_e represents extracellular elasticity and the spring constant K_i represents the mechanical properties of the intracellular structure. Force generation by the actin cytoskeleton is represented by the linearized force–velocity relation $v(F)$ for a single myosin II molecular motor. The internal state of the focal adhesions is represented by a biomolecular bond which opens in a stochastic manner with dissociation rate k_0. b An adhesion cluster under force before mature focal adhesion formation. Closed bonds rupture with a force-dependent rate and open bonds close with a force-independent rebinding rate k_{on}. (Reprinted from [36], copyright 2006, with permission from Elsevier Science)

is given by $dW/dt = (F/K)dF/dt = Fv(F)$, where $v(F)$ is the force-velocity for the molecular motors; this implies that the major contribution to the power is generated by the molecular motors. For a typical force–velocity relation such as $v(F) = v_0(1 - F/F_s)$, the expression for the dissipation power can be readily integrated, giving $F = F_s(1 - e^{-t/t_K})$, with $t_K = F_s/v_0 K$. This relation implies that the pulling force saturates at $F = F_s$, but the stiffer is the environment (the larger K), the faster a given threshold for force can be reached. If the cell pulls on a material with a bulk modulus on the order of kilopascals, then the corresponding spring constant K on the molecular level can be expected to be on the order of piconewtons per micron and the typical time scale t_K is in seconds. If the bulk modulus is on the order of megapascals, then K on the order of piconewtons per nanometer falls in the typical range for protein stiffness and the time scale t_K is in milliseconds. In the two-spring model the internal structure of the focal adhesion is represented by one biomolecular bond with unstressed dissociation rate k_0 (Fig. 34.8a). As in the general picture of receptor–ligand bonds, the rupture under mechanical force of such bonds determines many properties of focal adhesion sites. Rupture dynamics of an adhesion cluster under a pulling force is schematically outlined in Fig. 34.8b. Such a model is a stochastic version of an earlier model by Bell [37]. Briefly, the model assumes that N receptor–ligand bonds have been clustered on opposing surfaces, of which the upper one acts as a rigid transducer which transmits the constant force F homogeneously onto the array of bonds. At each time, if i bonds are closed, $N - i$ bonds are open (Fig. 34.8b). Closed bonds are assumed to rupture with a force-dependent rupture rate $k = k_a e^{F/iF_b}$, where k_a is the unstressed rupture rate (typically around 1/s) and F_b is the internal force scale (typically a few piconewtons) of the adhesion bonds. The exponential dependence between force and rupture rate results from a Kramers-type description of bond rupture as the escape over a transition state barrier. The factor i results because force is assumed to be shared equally between closed bonds, which holds true when the transducer is connected to a soft spring, whereas in the opposite limit of a stiff spring, all bonds feel the same force. Open bonds are assumed to rebind with a force-independent rebinding rate K_{on}.

The probability that i bonds are closed at time t is described by a master equation with appropriate rates for the reverse and forward rates between possible states i. The force F destabilizes the cluster; rebinding stabilizes it. In the case of the two-spring model, the single bond under a stall constant force F_s has an average lifetime $t = (1/k_0)e^{-F_s/F_b}$. For large K, the bond experiences constant loading with stall force F_s. In the case of small K, loading is approximately linear with loading rate F_s/t_K. Between these two limiting cases, the average force that is built up until bond rupture is given by the relation [36]

$$\langle F \rangle = \int_0^\infty p(t) F(t)\, dt = \frac{F_s}{1 + k_0 F_s / v_0 K}, \tag{34.1}$$

where $p(t) = e^{-k_0 t} k_0\, dt$ is the assumed probability that the bond beaks at time t in a interval $t + dt$. It is interesting to note that in (34.1) the level of force reached is essentially determined by the quantity $k_0 F_s / v_0 K$. Since unstressed dissociation k_0, stall force F_s, and maximal motor velocity v_0 are molecular constants, the only relevant quantity in this context is the external stiffness K. By using the preview data, the average force $\langle F \rangle$ is larger by a factor 2 in a stiff environment. If a cell is pulling at several focal adhesions with a similar investment of resources, then those contacts will reach the level of force putatively required for activation of the relevant signaling pathways which experience the largest local stiffness in their environment. As a consequence, growth of contacts in an elastically anisotropic environment might then lead to cell polarization and locomotion in the direction of maximal effective stiffness in the environment, which has been observed in many experimental situations.

34.2.1
Biomimetic Scaffolds, Roughness, and Contact Guidance for Cell Adhesion and Motility

All tissue engineering constructs are composed of two major components: a scaffolding material that provides the mechanical and structural properties required, and site-specific cells. One fundamental target of tissue engineering is the production of biomimetic scaffolds [38–40]. Polymeric fibers and cellular solid scaffolds should be engineered to include the possibility of biomolecular signals [41]. Since information that is introduced on the scaffold material surface is processed as biomechanical and biochemical signals through receptors, which are nanometer-sized entities on cell surfaces, it is important that this information be presented on the same length scale as it occurs in nature. Generally, an assembly of functionalized particles could serve as a versatile tool for imparting texture and chemical functionality on a variety of surfaces. Such modified surfaces can be tuned to posses tethered or covalently adsorbed biomolecules such as peptides, proteins, and biopolymers and can serve as a platform for engineering biomimetic interfaces to modulate cellular behavior toward implants and scaffolds in tissue engineering [42]. In the early twentieth century Weiss [43] observed that cells preferentially orient along ECM fibers, an organization principle he termed *contact guidance*. Further, Weiss observed that two tissue explants reorganize the collagen gel between them into aligned parallel fiber bundles and that cells leaving the explants migrate and orient along the aligned fibers. Contact guidance therefore could serve both as a cue for organization on cellular scales and as a large-scale organization tool in tissue development by guiding motile cells along ECM bundles. More recently, the community of scientists and bioengineers involved in tissue reconstruction and remodeling has agreed that cells use spatial variation of adhesiveness and frictional force to favor orientation of cells along thick fiber bundles [44]. Adherent cells can respond to mechanical properties of their environment, and the use of a sophisticated elastic scaffold has provided strong evidence that cells respond to purely elastic features in their environment [28,45].

One emerging aspect in the realization of biomimetic scaffolds is the role of scaffold roughness for contact guidance on nanopatterned surfaces [46,47]. Surface roughness has an enormous influence, such as in contact mechanisms, friction, sealing, and adhesion, in a large variety of situations [48]; also in cellular dynamics surface roughness plays a fundamental role For single cells, the cellular response to alterations in surface topography, down to the nanometer scale, has been documented especially for grooved topography [49]. Most cells follow the discontinuities of grooves and ridges, and attain an elongated shape owing to surface-induced rearrangements of the cytoskeleton. The development of techniques to produce surface structures in the nanometer range has revealed that cells also respond to such nanostructures (Fig. 34.9). Macrophage-like cells can react to steps in the nanometer range. Endothelial and fibroblast cells are sensitive to patterns with features down to 10 nm, and the importance of symmetry and discontinuity of the nanometer patterns was pointed out.

A range of techniques can be used to create well-defined topographical and chemical cues for cell patterning. Many of these approaches rely on photolithography and reactive ion etching of the substrate and can be also followed by UV and glow

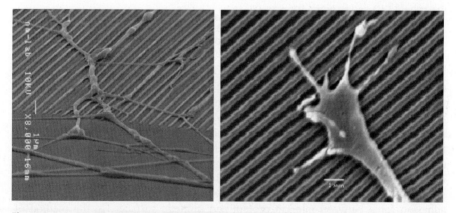

Fig. 34.9. *Left*: Scanning electron microscopy image showing axons that after a few microns align to the ridges. The ridges are 100 nm high and 100 nm wide and they are of the same dimensions as the axons. It is interesting to note the contact guidance role of the ridges: all axons are growing on top of the ridges and not in the grooves between the ridges, unit 1 μm. *Right*: Scanning electron microscopy image showing filopodia extending both perpendicular and parallel to the underlying pattern of the ridges. *Scale bars* 1 μm. (Reprinted from [49], copyright 2006, with permission from Elsevier)

discharge treatments [50]. Microcontact printing is also a popular technique, and other methods, including inkjet printing and diamond cutting, have been successfully exploited [51]. These techniques are generally suitable only for micropatterning. To go down in size, photolithography is limited by light-diffraction effects. In fact, its resolution is on the order of the wavelength of the light used for exposure (typically more than 200 nm). Electron-beam lithography can be used to produce nanoscale patterns, but it is expensive and time-consuming. A simple method has been reported to fabricate nanoislands 13–95 nm in height on a large scale based on phase separation of polystyrene and poly(4-bromostyrene) spin-coated on silicon wafers [52]. However, the reliability to produce nanofeatures of controlled size and geometry based on such a phase-separation phenomenon is relatively poor.

Molecularly tailored surfaces utilizing peptide segments provide a route to mimic cell or ECM environments without the use of an entire protein. Molecular biologists have identified many minimal sequences of amino acids of proteins that are responsible for adhesion to a particular receptor. These minimal peptide sequences are much more stable than entire proteins and the surface concentration of the specific ligand may be much higher. Here, we limit our description to the typical Johnson–Kendall–Roberts (JKR) test performed on such bioactive surfaces [53]. The Langmuir–Blodgett technique is used to create amphilic bilayers of controlled surface and peptide density; a sketch of the contacting surfaces is shown in Fig. 34.10. The contact mechanical method used is that referred to as the JKR method, developed by Johnson et al. [54]. The JKR theory represents a balance of elastic energy associated with compressing the elastic sphere, potential energy associated with displacing a normal load, and surface energy associated with forming a contact area. The contact area will increase under an applied load (P) such that at mechanical equilibrium and energy balance the contact radius (a) is a function of the adhesion energy (G), the elastic modulus of the materials (K), and the radius of curvature of the sphere (R) and is given by

$$a^3 = \frac{R}{K}\left(P + 3\pi GR + \sqrt{6\pi GR + (3\pi GR)^2}\right), \qquad (34.2)$$

where $1/K = 3/4\left[(1-v_1^2)/E_1 + (1-v_2^2)/E_2\right]$ and the radius of curvature of the two spheres is calculated as $1/R = 1/R_1 + 1/R_2$ (generally, for a sphere-contacting flat surface, $R_2 \to \infty$, $R_1 = R_{\text{sphere}}$). The adhesion energy (G) is a function, among other things, of the surface energy and the interfacial energy between the two surfaces. At equilibrium G is equal to the thermodynamic work of adhesion, which is equal to $\gamma_1 + \gamma_2 - \gamma_{12}$. With use of the JKR method, the contact area between an elastic sphere and a rigid flat surface can be measured optically and plotted as a function of the measured normal load. By modifying the surfaces of materials and biomolecules that comply, one can use the JKR method to measure specific adhesion force between integrin protein and mimetic surfaces. In the method of Dillow et al. [53], one can measure the contact area during loading and unloading; any hysteretic differences can be used to distinguish between specific and nonspecific adhesion. The adhesion in the loading process is due mainly to nonspecific van der Waals interactions. Once the solids are in contact, short-range specific interactions may occur (including hydrogen bonds, specific lock-and-key-type bonds associated with ligand–receptor interactions, and molecular rearrangements). Once these bonds

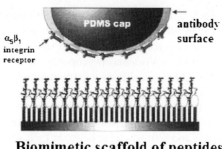

Fig. 34.10. Johnson–Kendall–Roberts apparatus and biomimetic surface in contact with an integrin-immobilized lens. The antibody layer serves to block the integrin receptor. The polydimethylsiloxane (*PDMS*) lens has a radius of curvature of 1–1.5 mm, and applied loads are less than 400 mg, the maximum time of contact is 10 min, the contact area falls in the range 0–2×10^{-3} mm^3. (Reprinted from [53], copyright 2001, with permission from Elsevier)

have formed, additional energy is required to separate the surfaces, which results in a hysteresis in the unloading curve. The degree of hysteresis provides information about the strength of adhesion due to interactions such as molecular rearrangements, hydrogen bonds, and other specific interactions formed at the interface. The method used provides a useful tool to select peptides on the mimetic surface increasing adhesion to integrin and revealed that peptide orientation, spatial arrangement, and binding site accessibility are crucial for effective ligation to integrin receptors.

Different time phases can be distinguished in culture of cells on surface scaffolds. The first one is the initial cell-adhesion phase involving nonspecific electrostatic forces and passive formation of ligand–receptor bonds, followed in the hours following the cell-spreading phase involving receptor recruitment by clustering to anchoring sites and interactions with cytoskeletal elements. The second phase concerns the proliferation and differentiation phases involving the ECM formation. During the first phase, when the ECM is not already formed, the surface properties of the scaffold play a major role for the subsequent proliferation phase. In particular, surface polymer presenting a consistent texture of gradient elasticity can induce contact guidance for cells [55]. Triblock polyurethane family surfaces have been extensively used in tissue engineering recently. They present a wide variety of surface interconnections between harder and softer domains and it was revealed recently that a higher interconnection of phase separation in the material improves cell proliferation [56, 57]. The connection between the increased contact guidance due to phase separation and the increased cell proliferation in the first phase of seeding still has many open questions and there are still many unknown experimental and theoretical mechanisms.

34.3
Experimental Strategies for Cell–ECM Adhesion Force Measurements

Cell adhesion is one of the initial events essential to subsequent proliferation and differentiation of cells before tissue formation. Consequently, many in vitro evaluations of cell adhesion on substrates have been performed in order to discern the main surface properties influencing the cell response to implant surfaces. To determine cell adhesion, many techniques have evolved, such as functionalized latex beads moved with optical tweezers [58], interferometric techniques [59], and centrifugation experiments [60,61]. Viscoelastic properties of cells were measured with atomic force microscopy (AFM) in either force modulation mode or more recently by force volume technique [62,63]. Adhesion between single cells was measured in the past using mechanical methods, such as micropipette manipulations [64,65] or hydrodynamic stress [66,67]. Many new techniques for measurements of adhesion force between a cell and a substrate or between cells or between a ligand and a receptor are based on the combination of traditional optical microscopy with the AFM technique. AFM offers particular advantages in biology: measurements can be carried out in both aqueous and physiological environments, and the dynamics of biological processes in vivo can be studied. Many cellular functions require accurate knowledge of ligand binding to receptors; single ligand–receptor pair measurements include biotin–avidin [68], antibody–antigen [69], and cellular proteins, either isolated or in cell membranes [70]. The general strategy is to bind ligands to atomic force microscope tips and receptors to probe surfaces (or vice versa), respectively. In a force–distance cycle, for example [70], the tip first approaches the surface, upon which receptor–ligand complexes are formed owing to the specific ligand–receptor recognition. During subsequent tip–surface retraction a temporarily increasing force is applied to the ligand–receptor connection until the interaction bond breaks at a critical force (unbinding force). AFM-based investigations on cellular mechanics were performed on elastic and mechanical properties of cells, including platelets, osteoblasts, glial cells, macrophages, endothelial cells, epithelial cells, fibroblasts, and bladder cells [62,63,71–77]. Generally, Young moduli of living cells vary in the range 1–100 kPa depending by the cell type and the softest parts, 1 kPa, are located around the nucleus.

Sato et al. [78] performed an indentation test on endothelial cells exposed to shear stresses of 2 Pa. They obtained a relationship between the external applied force F and the indentation depth δ, as $F = a\delta^2 + b\delta$, where a and b are two constants representative of the mechanical response. Viscoelasticity of leading-edge fibroblast lamellopodia was measured by Mahaffy et al. [79] using an AFM-based microrheology device. They quantified viscoelastic constants such as elastic storage modulus, viscous loss modulus, and the Poisson ratio. Lamellopodia are thin regions (less than 1000 nm) of cells strongly adherent to a substrate with an elastic strength of approximately 1.6 ± 0.2 kPa and with an experimentally determined Poisson ratio of approximately 0.4–0.5.

Many experiments have been carried out to reveal the effect of lateral forces on adhesion sites; one important technique is the so-called elastic substrate method [80]. This technique produces a thin elastic film over a fluid: under cell traction the film shows a wrinkled pattern, which is characteristic of the pattern of forces exerted. From the first semiquantitative tests, quantitative analysis of elastic substrate data was pioneered by Dembo and Wang [81]. Using linear elasticity theory for thin elastic films and numerical algorithms for solving inverse problems, they reconstructed the surface forces exerted by keratocytes. Recently, a novel elastic substrate technique to measure cellular forces at the level of a single focal adhesion was proposed [82]. Wang et al. [83] carried out experimental testing on various adherent cell types cultured on deformable substrates and revealed specific patterns of cell reorientation in response to cyclic stretching of the substrate. They showed that under uniaxial deformations, cells were found to elongate perpendicular to the stretch direction (Fig. 34.11), whereas in cases where the substrate was laterally unrestrained (biaxial deformations), cells were found to elongate at an angle to the stretch direction. The alignment directions in both cases correspond to directions of minimum substrate strain. McGerry et al. [84] performed a finite-element study on a similar system in order to investigate the role of cell viscoelasticity in cell debonding and realignment under the conditions of cyclic substrate stretching. The characteristic length scale used in the simulation was based on the length of the receptor–ligand bonds at the cell–substrate interface, a receptor–ligand bond strength varying in the range 5–40 pN, and a bond density of approximately 50 sites per square micron. Such a 2D computational model revealed that discrete cell–substrate contacts at focal adhesion sites result in the completion of debonding in fewer cycles and that permanent debonding at the cell–substrate interface occurs owing to the accumulation of strain concentrations in the cell.

The role of shear strength on cell-material adhesion has also been investigated by Yamamoto et al. [85] and Athanasioiu et al. [86, 87]. Such a method was based on the design of a cytodetachment instrument able to quantify the force required to displace and to detach cells attached to a substratum (Fig. 34.12). The cytodetacher allows one to directly measure the force required to detach cells from a substratum and to indirectly determine the ability of different substrata to support cell adhesion.

A cell adheres to a material in a medium inside a dish on an XY stage. The stage can be moved at a speed of 20 µm/s in the direction of a tip attached to a cantilever. The distance between the pointed head of the tip and the material's

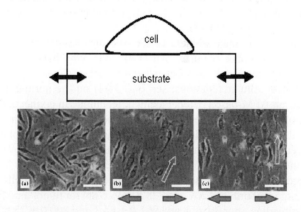

Fig. 34.11. *Top*: Typical setup for cell orientation under cyclic substrate deformation tests, the *arrows* indicate the direction of cyclic substrate deformation. *Bottom*: Representative phase-contrast microphotographs of endothelial cells: unstretched (*a*), after 3 h of simple 10% elongation (*b*), and after 3 h of pure uniaxial stretching along the *x*-axis with a maximum strain 10% and a constant frequency of 0.5 Hz (*c*). Unstretched cells did not appear to orient in any specific direction, but with simple elongation and pure uniaxial stretching, the cells oriented about 70° and 90°, respectively (*arrows* in the photographs). (Reprinted from [84], copyright 2001, with permission from Elsevier Science

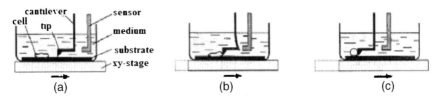

Fig. 34.12. A probe-cell detachment instrument as proposed in [85]. **a** A cell adhered to a material on the XY stage is moved to the tip attached to a cantilever. **b** The cell touches the tip and a lateral load is applied to the cell. The cantilever is deflected corresponding to the shear strength of the cell-material adhesion. **c** The cell is detached from the material. The deflection of the cantilever is measured and recorded. The magnitude of the shear force applied to the cell is given as the product of the force constant of the cantilever and the deflection of the cantilever. (Reprinted from [85], copyright 1998, with permission from Elsevier)

surface is controlled to be nearly 0.2 μm prior to each measurement. Then, the tip touches the cell and a lateral load is applied to the cell. The cantilever is deflected corresponding to the deformation of the cell from the material. Finally, the cell is completely detached from the material. The deflection of the cantilever is measured and the shear force applied to the cell is calculated as the product of the force constant of the cantilever and the deflection of the same. The shear force applied to the cell is recorded as a function of the displacement of the XY stage, which is named as a force–displacement curve. The averages of the cell detachment shear strengths measured on murine fibroblasts after 24 h of incubation of glass by Yamamoto et al. fell in the range 500–750 kPa, and shear force fell in the range 350–580 nN. Athanasiou et al. [86, 87] used the cytodetachment to measure single-cell adhesiveness and the effect of protein on cell adhesiveness. They found that chondrocytes have an adhesion force around 20 nN and that fibronectin increases strongly the adhesion of a cell to the substrate (Fig. 34.13).

An example of the measurement of the role of a single protein in a complex cell-adhesion process has given by Wu et al. [88], who have recently measured the effect of a protein such as focal adhesion kinase (FAK, it is a regulator of integrin-mediated cellular functions; see Fig. 34.4) on the normal adhesion force making use of an AFM-based cytodetachment technique in connection with optical trapping and optical tweezers. They investigated the effects of FAK on adhesion force during several stages, such as initial binding phase (5 s), beginning of cell spreading (30 min), spreading out (12 h), and migration phase (more than 12 h). They found that a high concentration of FAK in a cell culture increased the adhesion forces at all the different stages. In fact, the detachment force required for single cells was 343.2 ± 43.4 nN for large amounts of FAK and 228.8 ± 36.6 nN for minor amounts of FAK at the stage of spreading after 30 min, and 961.0 ± 64.1 and 800.0 ± 75.5, respectively, for major or minor amounts of FAK at the stage of spreading out after 12 h.

Another diffuse technique is the combination of the JKR apparatus with an AFM design. Generally, the standard atomic force microscope tip when used on soft materials such as cells and/or cellular constituents can produce damaging stress; to prevent such unwanted effects, a microsphere bead should be attached to the atomic force microscope cantilever. The advantage is also to study adhesion force with the microbead coated with proteins or components of ECM, like collagen, fibronectin,

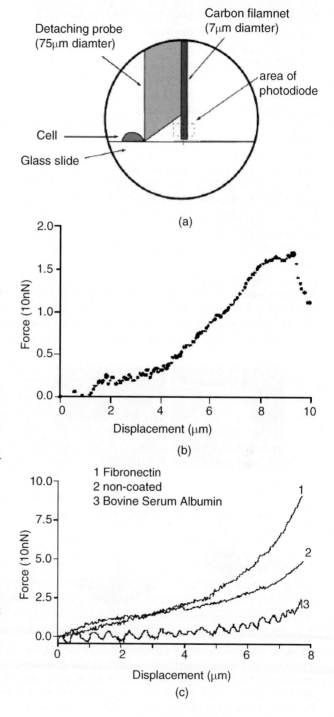

Fig. 34.13. a The cytodetachment probe of Athanasiou et al. [86]. Deflection of the detaching probe is used to calculate the resistance force offered by the cell. Contact of the cell by the detaching probe causes the probe to bend in response to the resistance offered by the cell. The difference between the displacement of the driving arm and the carbon filament marker beam is used to calculate the resistance force given by the cell to the detaching probe. **b** Typical force versus displacement graph of a single chondrocyte under shear force applied by the cytodetacher. Mechanical adhesiveness is defined as the maximum force encountered by the cytodetaching probe. **c** The adhesiveness increases when cells are seeded onto substrata that enhance cellular attachment, such as fibronectin. (Reprinted from [86], copyright 1999, with permission from Elsevier Science)

laminin, or entire cells (Fig. 34.14). This technique is equivalent to that using the JKR apparatus described in the preview section, with the advantage that such AFM-based instruments work on reduced scales because the microsphere radius generally lies in the 1–15-μm range. Making use of such a JKR method, Canetta et al. [89] determined both the local elastic modulus of endothelial cells (0.2–0.8 kPa) and the adhesion energy ($0.3-3.0 \times 10^{-7}$ J) with a bead attached to the cantilever with a radius of 15 μm on endothelial cells.

An AFM-based adhesion force spectrometer combining an AFM design and a light microscope to investigate cell-to-cell interactions in vivo at the molecular level has been proposed by Benoit et al. [90, 91]. Single cells or a monolayer of epithelial cells were immobilized on a cantilever. The cells can be monitored using a light microscope during the entire experiment. Adhesion force measurements were performed via *de-adhesion force versus piezo position traces*, that are analogous to force–distance curves with a cell-functionalized cantilever. The force resolution was less than 20 pN [92].

The adhesion force measured with the AFM-based spectrometer is characterized via de-adhesion force versus piezo position traces [90] (Figs. 34.14, 34.15. The z piezo velocity was typically set within 1 and 7 μm/s. Low velocities can interfere with drift effects mostly caused by cell movement, while at higher velocities hydrodynamics influence the measurement. When a sphere lands on a soft cell surface, the area of interaction increases with the indentation, which leads to an enhancement of the adhesion signal. The adhesion strength is dependent both on the indentation

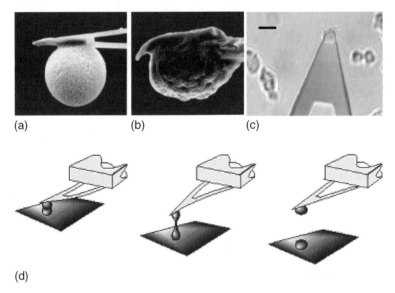

Fig. 34.14. Microbead-mounted cantilevers imaged by scanning electron microscopy. The microbeads glued to the cantilevers were coated both with serum bovine albumin (**a**) or with human trophoblast-type JAR cells (**b**); *scale bars* 10 μm. **c** Light-microscope image of a cantilever-mounted cell before being brought into contact with another cell; *scale bar* 20 μm. **d** The typical approach–retraction cycle for the de-adhesion force spectrometer. (Reprinted from [91,92], with permission from HUMREP and Nature, respectively)

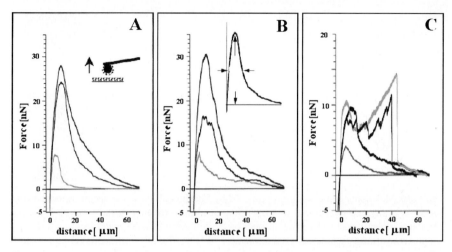

Fig. 34.15. Typical adhesion force curves for human epithelial cells resulting when a bovine serum albumin coated microbead (**a**, **b**) or a JAR-coated microbead (**c**) was retraced after periods of 1–40 min of contact. The x-axis shows the cantilever displacement, the y-axis shows the force acting on the microbead. In **a** and **b**, the rupture events correspond to 1, 10, and 40 min of interaction time, from the *bottom* to the *top*, respectively. In **c** the different rupture events corresponding to 20 and 40 min of interaction time are observed. Such curves were characterized by increased adhesion force at a distance of 20–30 μm from the zero force. (Reprinted from [91], copyright 1998, with permission from HUMREP)

force and on the contact duration. This effect seems to be due to the fact that the cell shape adapts to the sphere's surface geometry, resulting in more and more molecules interacting with the underlying surface during the contact. Benoit et al. observed that a change of the velocity of retraction leads to a fairly linear relation between separation speed and adhesion in the range 2–27 μm/s. The de-adhesion force curves show a variety of more or less pronounced single steps on the order of 100 pN in the region of descending adhesion, indicating ruptures on the molecular level.

Recently, Puech et al. [93] proposed a suitable approach to quantify cell–cell adhesion forces by AFM (Fig. 34.16). They described an approach for cell–cell adhesion experiments, where the sample stage can be moved 100 μm in the z-direction by closed loop linearized piezo elements. Such an approach enables an increase in pulling distance sufficient for the observation of long-distance cell-unbinding events without reducing the imaging capabilities of the atomic force microscope. The atomic force microscope head and the piezo-driven sample stage were installed on an inverted optical microscope fitted with a piezo-driven objective, to allow the monitoring of cell morphology by conventional light microscopy, concomitant with force spectroscopy measurements. In such experiments, they used the WM115 melanoma cell line binding to human umbilical vein endothelial cells to demonstrate the capabilities of this system and the necessity for such an extended pulling range when quantifying cell–cell adhesion events.

Madl et al. [94] made use of a combination of an atomic force microscope and an epifluorescence microscope for positioning of an antibody-modified atomic force microscope tip on Chinese hamster ovary (CHO) cell line. Generally, these

Fig. 34.16. Melanoma cell capture, as proposed by Puech et al. [93]. The lectin-decorated cantilever is positioned over a cell in suspension, at close proximity to the surface. Then, the cantilever is gently pushed for a few seconds on the cell. After this, the cantilever-bound cell and support are vertically separated by approximately 100 μm. This allows the cell to establish a firm adhesion to the cantilever. *Top right*: An optical image of a cell captured on a tipless cantilever. *Left*: Sketches of an adhesion assay of a single captured melanoma cell on an endothelial cell layer. The probe cell is positioned over a zone of interest, which was selected by phase-contrast microscopy. A given force (usually of several hundred piconewtons) is applied for a given time (usually in the range of seconds) on a cell layer. The probe cell is subsequently separated from the surface, and the de-adhesion events are recorded. *Bottom right*: A phase-contrast image of a cantilever-bound melanoma cell (*arrow*) in contact with a layer of human umbilical vein endothelial cells. *Scale bars* 20 μm. (Reprinted from [93], copyright 2006, with permission from Elsevier)

cells are advantageous for AFM imaging as they adhere well to a solid support. In addition, such cell lines tend to be more resistant to the onset of apoptosis, facilitating AFM imaging of live cells over a longer period. In this study, Madl et al. used ldlA/-SRBI cells, i.e., CHO cells lacking the low density lipoprotein receptor (ldlA7 cells) and expressing high levels of recombinant scavenger receptor class B type I (SRBI). In turn, the receptor was fused with green fluorescent protein (eGFP), having an absorption maximum at 488 nm and an emission maximum at 508 nm. Combined optical and AFM measurements are not restricted to fixed cells; simultaneous fluorescence and AFM imaging is also feasible to scan and monitor live cells in contact mode. In turn, we mention a possible relevant use of AFM force spectroscopy for diagnosis of pathogen agents. Beckmann et al. [95] measured cell-surface elasticity on enteroaggregative *Escherichia coli* (EAEC) via force–distance spectroscopy. EAEC is a diarrheal pathogen that has been associated with endemic and epidemic diarrheal illness. Generally, the specific EAEC wild-type strain, called *042pet*, produces a protein, dispersin, that is implicated in pathogenesis. Dispersin is a 10-kDa low molecular mass protein, and it is secreted to the exterior environment of the bacteria and remains noncovalently associated with the cell surface. Beckmann et al. considered a mutant of *042pet* that does not produce dispersin, *042aap*, that was compared with *042pet* to determine if differences in cell elasticity were detectable after strain application. They found some differences in cell rigidity between the *042pet* and the mutant *042aap*, which could possibly be used to monitor the presence or absence of dispersin as a pathogen agent.

34.4
Conclusions

Recent advancements in many biomedical applications require a multidisciplinary approach. The connection between mechanics and biochemistry and structural aspects on molecular and supramolecular scales could provide new significant knowledge in medicine. Strongly related to these biomedical application ambits, tissue engineering plays a fundamental role in improving or replacing biological functions of human tissues. A damaged tissue can be regenerated by a migration of cells to the injured area; as a consequence, many questions in tissue engineering involve cell dynamics and the proliferation of cells on substrates composed of ECM. Animal cells show overall dimensions on the order of 10–30 μm, but the adhesion with substrates is limited to sites (focal adhesion sites) whose dimensions fall in the nanometer range. As a consequence, a basic understanding of the phenomena occurring at the nanoscale level has to be further developed in order to achieve satisfactory knowledge of such processes. All tissue engineering constructs are composed of two major interacting components: a scaffolding material offering suitable mechanical and structural properties and a specific cell layer adhering to and spreading on it. Mechanical forces play a fundamental role in cell migration, where contractile forces are generated within the cell and pull the cell body forward. For living systems such as complex system cells, the mechanical features are more complicated with respect to the inorganic world since living systems tend to continuously adapt the forces required for their positioning and movement by converting different forms of energy into mechanical energy.

34.5
Glossary

ATP: Following Fig. 34.17, in system A, held to constant temperature and pressure, ATP molecules hydrolyze to form the products adenosine diphosphate (ADP) and inorganic phosphate. ATP hydrolysis can occur in solution (straight arrow in system A) or through an enzyme-catalyzed pathway (curved arrow though the motor enzyme system, system B). Independent of the pathway, when 1 mol of ATP molecules is hydrolyzed, the free energy of system A changes by a quantity ΔG_{ATP}. In system B, molecular motors (double ovals) move the cargo (helix) against a force, F, as they catalyze the ATP hydrolysis reaction, generating external work, w_{ext} and heat, q.

Chemotaxis: The directional translocation of cells in a concentration gradient of some chemoattractant or chemorepellent substances.

Contact guidance: The directional translocation of cells in response to some anisotropic elastic property of the substratum.

Extracellular matrix: The class of components (collagen, fibronectin, etc.) present in the extracellular space of tissues that serve to mediate cell adhesion and organization.

Haptotaxis: The tendency of cells to translocate unidirectionally up a steep gradient of increasing adhesiveness of the substratum.

Fig. 34.17. Adenosine triphosphate (*ATP*) hydrolysis and energy transfer. *ADP* adenosine diphosphate, P_i inorganic phosphate

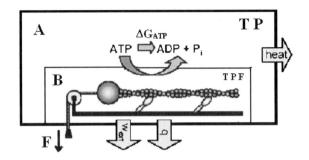

Integrin: Cellular transmembrane protein that acts as a receptor for adhesive ECM proteins such as fibronectin. The tripeptide RDG is the sequence recognized by many integrins.

Motility: Directed motion driven by energy-consuming motors (opposite to mobility, which is passive motion driven by thermal or other external mechanisms).

Molecular motor: Biological mechanism that converts chemical energy into mechanical energy, used by a cell to generate directed motion. A single motor can be as simple as a single polypeptide chain or as complicated as a giant macromolecular complex with hundreds of protein and proteoglycan constituents. Chemical energy used by molecular motors is generally in the form of either a high-energy chemical bond (as in ATP) or an ion gradient across a membrane.

Peptide: Organic compound composed of amino acids linked together chemically by peptide bonds. (Amino acids are a class of 20 simple organic compounds containing carbon, hydrogen, oxygen, nitrogen, and in certain cases sulfur. These compounds are the building blocks of proteins.) The peptide bond always involves a single covalent link between the α-carboxyl (oxygen-bearing carbon) of one amino acid and the amino nitrogen of a second amino acid. In the formation of a peptide bond from two amino acids, a molecule of water is eliminated. Compounds with molecular masses of more than 50–100 amino acids are usually termed proteins.

Taxis: A directed response to some vectorlike property of the environment. Translocation is usually biased unidirectionally, either along the field vector or opposite to it.

Acknowledgements. The authors gratefully acknowledge C. Ascoli and P. Baschieri, IPCF-CNR Pisa, and U. Schwartz, University of Heidelberg, for useful discussions and suggestions.

References

1. Langer R, Vacanti JP (1993) Science 260:920–926
2. Peter SL et al. (1998) J Biomed Mater Res 43:422
3. Griffith M et al. (1999) Science 286:2169
4. Burg KJL, Porter S, Kellam J (2000) Biomaterials 21:234
5. Palsson B, Hubbell JA, Plonsey R, Bronzino JD (2003) Tissue engineering. CRC, London
6. Anderson JM (1993) Cardiovasc Pathol 2:S33
7. Bao G (2002) J Mech Phys Solids 50:2237

8. Geiger B, Bershadsky AD (2002) Cell 110:139
9. Safran SA, Gov N, Nicolas A, Schwarz US, Tulstry T (2005) Physica A 352:171
10. Bray D (2001) Cell movements: from molecules to motility, 2nd edn. Garland, New York
11. Diambra L, Cintra LC, Schubert D, da Cinta LF (2005) http://xxx.lanl.gov/q-bio.CB/0503013
12. Fletcher DA, Theriot JA (2004) Phys Biol 1:T1
13. Jones CJ, Aizawa S (1991) Microb Physiol 32:10
14. Purcell EM (1977) Am J Phys 45:3
15. Wolgemuth C, Hoiczyk E, Kaiser D, Oster G (2002) Curr Biol 12:369
16. Howard J (2001) Mechanics of motor proteins and the cytoskeleton, 1st edn. Sinauer, Sunderland
17. Schliwa M, Woehlke G (2003) Nature 422:759
18. Lipowsky R, Klumpp S (2005) Physica A 352:53
19. Schott DH, Collins RN, Bretscher A (2002) J Cell Biol 156:35
20. Berg HC (2003) Annu Rev Biochem 72:19
21. Kojima S, Yamamoto K, Kamagishi I, Homma M (1999) J Bacteriol 181:1927
22. Theriot JA (2000) Traffic 1:19
23. Ambrose EJ (1961) Exp Cell Res 8:93
24. Curtis ASG (1964) J Cell Biol 20:199
25. Owen GR, Meredith DO, Gwynn I, Richards RG (2005) Eur Cells Mater 9:85
26. Chicurel ME, Chen CS, Ingber DE (1998) Curr Opin Cell Biol 10:232
27. Chicurel ME, Singer RH, Meyer CJ, Ingber DE (1998) Nature 392:73
28. Lo CM, Wang HB, Dembo M, Wang YL (2000) Biophys J 79:144
29. Schwarz US, Balaban NQ, Riveline D, Addadi L, Bershadsky A, Safran SA, Geiger B (2003) Mater Sci Eng C 23:387
30. Hu H, Sachs F (1997) J Mol Cell Cardiol 29:1511
31. Koonce MP, Kölher J, Neujahr P, Schwartz JM, Tikhonenko I, Gerish G (1999) EMBO J 18:6786
32. Riveline D, Zamir E, Balaban NQ, Schwarz US, Ishizaki T, Narumiya S, Kam Z, Geiger B, Bershadsky AD (2001) J Cell Biol 153:1175
33. Merkel R, Nassoy P, Leung A, Ritchie K, Evans E (1999) Nature 397:50
34. Bischofs IB, Schwarz US (2003) Proc Natl Acad Sci USA 100:9274
35. Bischofs IB, Safran SA, Schwarz US (2004) Phys Rev E 69:012911
36. Schwarz US, Erdmann T, Bischofs IB (2006) Biosystems 83:225
37. Bell GI (1978) Science 200:618
38. Ellingsen JE, Lyngstadaas SP (eds) (2003) Bio-implant interface: improving biomaterials and tissue reactions. CRC, Boca Raton
39. Peter SL, Miller MJ, Yasko AW, Yaszemski MJ, Mikos AG (1998) J Biomed Mater Res 43:422
40. Hutmacher DW (2000) Biomaterials 21:2529
41. Curtis A, Riehle M (2001) Phys Med Biol 46:R47
42. Dillow AK, Lowmann AM (2002) Biomimetic materials and design. Dekker, New York
43. Weiss P (1945) J Exp Zool 100:353
44. Schwarz US, Bishofs IB (2005) Med Eng Phys 27:763
45. Wong JY, Velasco A, Rajagopalan P, Pham Q (2003) Langmuir 19:1908
46. Meyer U, Büchter A, Wiesmann HP, Joos U, Jones DB (2005) Eur Cells Mater 9:39
47. Anselme K, Bigerelle M (2006) Biomaterials 27:1187
48. Persson BNJ, Albohr O, Tartaglino U, Volokitin AI, Tosatti E (2005) J Phys 17:R1
49. Johansson F, Carlberg P, Danielsen N, Montelius L, Kanje M (2006) Biomaterials 27:1290
50. Winkelmann M, Gold J, Hauert R, Kasemo B, Spencer ND, Brunette DM, Textor M (2003) Biomaterials 24:1133

51. Curtis A, Wilkinson C (2001) Trends Biotechnol 19:97
52. Dalby MJ, Yarwood SJ, Riehle MO, Johnstone HJH, Affrossman S, Curtis ASG (2002) Exp Cell Res 276:1
53. Dillow A, Ochsenhirt SE, McCarthy JB, Fields GB, Tirrell M (2001) Biomaterials 22:1493
54. Johnson KL, Kendall K, Roberts AD (1971) Proc R Soc Lond Ser A 324:301
55. Brandl F, Sommer F, Goepferich A (2007) Biomaterials 28:134
56. Ciardelli G, Rechichi A, Cerrai P, Tricoli M, Barbani N, Giusti P (2004) Mol Symp 218:261
57. D'Acunto M, Ciardelli G, Narducci P, Rechichi A, Giusti P (2005) Mater Lett 59:1627
58. Cloquet D, Felsenfeld DP, Sheetz MP (1997) Cell 88:39
59. Briunsma G, Behrisch A, Sackmann E (2000) Phys Rev E 61:4253
60. John N, Linke M, Denker HW (1993) In Vitro Cell Dev Biol 29A:461
61. Suter CM, Errante LE, Belotserkovsky V, Foscher PJ (1998) Cell Biol 141:227
62. Radmacher M, Fritz M, Kacher CM, Cleveland JP, Hansma PK (1996) Biophys J 70:556
63. Domke J, Dannöhl S, Parak WJ, Müller O, Aicher WK, Radmacher M (2000) Colloids Surf B 19:367
64. Evans EA (1985) Biophys J 48:185
65. Evans E (1995) In: Lipowsky R (ed) Handbook of biological physics. Elsevier, Amsterdam, p 723
66. Curtis ASG (1970) Symp Zool Soc Lond 25:335
67. Chen S, Springer T (1999) J Cell Biol 144:185
68. Florin EL, Moy VT, Gaub HE (1994) Science 264:415
69. Willemsen OH, Snel MM, van der Werf KO, de Grooth BG, Greve J, Hinterdorfer P, Gruber HJ, Schindler H, van Kooyk Y, Figdor G (1998) Biophys J 75:2220
70. Jena BP, HorberJKH (eds) (2002) Atomic force microscopy in cell biology. Academic, Amsterdam
71. Butt HJ, Cappella B, Kappl M (2005) Surf Sci Rep 59:1
72. Henderson E, Haydon PG, Sakaguchi DS (1992) Science 257:1944
73. Rotsch C, Braet F, Wisse E, Radmacher M (1997) Cell Biol Int 21:685
74. Braet F, Rotsch C, Wisse E, Radmacher M (1997) Appl Phys A 66:S575
75. A-Hassan E, Heinz WF, Antonik MD, D'Costa NP, Nagaswaran S, Schoenenberger CA, Hoh JH (1998) Biophys J 74:1564
76. Rotsch C, Radmacher M (2000) Biophys J 78:520
77. Lekka M, Laidler P, Gil D, Lekki J, Stachura Z, Hrynmiewicz AZ (1999) Eur Biophys J 28:312
78. Sato NK, Kataoka N, Sasaki M, Hake K (2000) J Biomech 33:127
79. Mahaffy RE, Park S, Gerde E, Käs J, Shih CK (2004) Biophys J 86:1777
80. Beningo KA, Wang YL (2002) Trends Cell Biol 12:79
81. Dembo M, Wang YL (1999) Biophys J 76:2307
82. Balaban NQ, Schwarx US, Riveline D, Goichberg P, Tzur G, Sabanay I, Mahalu D, Safran SA, Bershadsky AD, Addadi L, Geiger B (2001) Nature Cell Biol 3:466
83. Wang JHC, Goldschmit-Clermont P, Wille J, Yin FCP (2001) J Biomech 34:1563
84. McGerry JP, Murphy BP, McHugh PE (2005) J Mech Phys Solids 53:2597
85. Yamamoto A, Mishima S, Maruyama N, Sumita M (1998) Biomaterials 19:871
86. Athanasiou KA, Thoma BS, Lanctot DR, Shin D, Agrawal CM, LeBaron RG (1999) Biomaterials 20:2405
87. Hoben H, Huang W, Thoma BS, LeBaron RG, Athanasiou KA (2002) Ann Biomed Eng 30:703
88. Wu CC, Su HW, Lee CC, Tang MJ, Su FC (2005) Biochem Biophys Rese Commun 329:256
89. Canetta E, Leyrat A, Verdier C (2003) Math Comput Model 37:1121
90. Benoit M (2002) In: Jena BP, Horber JKH (eds) Atomic force microscopy in cell biology. Academic, Amsterdam

91. Thie M, Röspel R, Dettmann W, Benoit M, Ludwig M, Gaub HE, Denker HW (1998) Hum Reprod 13:3211
92. Benoit M, Gabriel D, Gerisch G, Gaub HE (2000) Nat Cell Biol 2:313
93. Puech PH, Poole K, Knebel D, Muller DJ (2006) Ultramicroscopy 106:637
94. Madl J, Rhode S, Stangl H, Stockinger H, Hintrerdorfer P, Schütz GJ, Kada G (2006) Ultramicroscopy 106:645
95. Beckmann MA, Venkataraman S, Doktycz MJ, Nataro JO, Sullivan CJ, Morrell-Falvey JL, Allison DP (2006) Ultramicroscopy 106:695

35 Scanning Probe Microscopy in Biological Research

Tatsuo Ushiki · Kazushige Kawabata

Abstract. Scanning probe microscopy (SPM) is especially attractive to biologists, because it has the advantage of obtaining three-dimensional images of sample surfaces at high resolution not only in a vacuum but also in a nonvacuum (i.e., air or liquid) environment. In addition to the visualization of biological structures, SPM has also been applied for measuring the physical properties (i.e., viscoelasticity) of living cells, and as a manipulation tool in biomaterials. The present review describes the application of SPM to biological fields for investigating the structure and function of biomaterials from biomolecules to living cells, and shows that SPM has great potential for broadening and opening up new fields in biomedical research.

Key words: Scanning probe microscopy, Atomic force microscopy, Near-field optic microscopy, Viscoelasticity, Nanomanipulation, Biomolecules, DNA collagen, Chromosomes, Living cells

Abbreviations

AFM	Atomic force microscopy
DNA	Deoxyribonucleic acid
GFP	Green fluorescent protein
HEPES	N-(2-Hydroxyethyl)piperazine-N'-ethanesulfonic acid
SEM	Scanning electron microscopy
SNOM	Scanning near-field optic microscopy
SPM	Scanning probe microscopy
STM	Scanning tunneling microscopy
TEM	Transmission electron microscopy

35.1
Introduction

Since the invention of the scanning tunneling microscope, various types of microscopes using a probing tip have been introduced, and are now referred to as a family of scanning probe microscopes. Among them, the scanning tunneling microscope is the first to be applied to biological fields, but its applications have been limited to observations of some biomolecules such as deoxyribonucleic acid (DNA) [1]. This is because the samples for scanning tunneling microscopy (STM) must be electrically conductive, indicating that STM is not well suited for analyzing biological samples, which are generally not conductive. In contrast with the scanning tunneling microscope, the atomic force microscope invented in 1986 has been widely applied in biological fields [2]. This microscope can create topographic images of samples three-dimensionally, which are similar to those obtained by scanning electron microscopy (SEM). In addition, atomic force microscopy (AFM) can be used for imaging both conductive and nonconductive samples at high resolution from the micrometer scale to the atomic scale. Thus, a number of researchers have been interested in studying the three-dimensional surface structure of biological samples from cells to biological molecules by AFM [3–7].

On the other hand, some other investigators have also been increasing their attention to the application of the atomic force microscope to the study of the physical properties of biological samples, because this microscope can measure the interaction forces between the tip and the sample [6–8]. Thus, AFM has been used for measuring the elasticity of single surface molecules in relation with function. There are also studies measuring the physical properties of the sample's surface, including surface changes, adhesion and elasticity of cells [8, 9].

In addition to collecting surface information of biological structures, the atomic force microscope is expected to be used as a dissecting tool, because it can directly contact a probing tip with the sample [10]. Thus, dissections of biological samples such as chromosomes and DNA have been performed with an atomic force microscope.

In this chapter, we will firstly introduce the applications of scanning probe microscopy (SPM) for visualization of biological samples, secondly show the SPM applications for measuring the physical properties of biomaterials, and finally describe the possibility of SPM as a manipulation tool in biology.

35.2
SPM for Visualization of the Surface of Biomaterials

35.2.1
Advantages of AFM in Biological Studies

Among various types of SPM, AFM has been expected to be an essential tool for visualizing three-dimensionally the surface structure of biological samples. Although the AFM images are similar to those obtained by SEM, there are several advantages of AFM imaging, which are summarized as follows:

1. AFM provides three-dimensional information. Because the atomic force microscope traces the sample's surface by monitoring the interaction force between the tip and the sample, it can create the three-dimensional information on the sample topography. Thus, the AFM images are similar to SEM images.
2. The resolution of AFM is high. By AFM, individual atoms have been observed with a few hard crystalline samples. This means that AFM has the potential to provide three-dimensional information on the surface topography of samples at resolutions from the micrometer scale to the atomic scale, although AFM images of biological samples have not yet shown this level of resolution.
3. Samples do not need to have electron conductivity: Because SEM creates images by scanning the sample's surface with an electron beam, samples need to be conductively stained and/or metal-coated before SEM observation. This

implies that precise information on the surface topography is hidden owing to the presence of the coating and/or staining substances. In contrast, AFM enables the direct observation of nonconductive samples without a metal coating or any conductive treatment.
4. Samples can be observed in nonvacuum environments. In contrast with the scanning electron microscope operated in vacuum conditions, the atomic force microscope has the advantage of creating images not only in a vacuum but also in air or a liquid environment, indicating that functional information can be expected to be obtained directly from living organisms.
5. Quantitative height information can be provided. Usually, SEM images contain no quantitative information on the sample's height. However, AFM provides quantitative information in three dimensions, and the sample's profile containing height information can be easily obtained from AFM images.

35.2.2
AFM of Biomolecules

The structure of biomolecules has been studied by X-ray crystalline analysis. Recent investigators have reported that AFM has the potential for observing a two-dimensional crystalline sheet of molecules [7]. Some of the best examples to date have been obtained by Müller et al. [11,12]; they have succeeded in observing a crystalline array of bacteriorhodopsin on the cytoplasmic surface of the purple membrane. These studies imply that AFM has the potential to analyze the submolecular structure of certain samples. However, in general, AFM images of biomolecules have not shown this level of resolution. This is probably due to the unevenness of randomly distributed samples. On the other hand, the shape of macromolecules such as DNA has been visualized by AFM [13]. Similar images have been reported by transmission electron microscopy (TEM) of negatively stained or rotary shadowed samples. In contrast with the complicated and skillful preparation techniques for TEM, the techniques for AFM are rather simple and consistent, indicating that AFM is expected to become a powerful tool for visualizing macromolecules. Two examples are given in the following sections.

35.2.2.1
DNA Molecules

DNA is a polymer of nucleotides which forms a double helix. Because of the characteristic shape and dimensions of DNA, its visualization by AFM has attracted many investigators. The commonest method for DNA preparation for AFM is deposition of DNA on the substrate by dropping a DNA solution followed by evaporation of the solvent. For the preparation, the following two points are the most important: (1) the choice of substrate and (2) the method of DNA attachment to substrates. Although various materials such as sapphire and graphite have been reported, freshly cleaved mica is widely used as the substrate. Previous studies have been reported that DNA molecules do not bind strongly to mica and are usually moved or swept away by the scanning tip. This is probably because freshly cleaved mica and DNA are both negatively charged. Therefore, the mica surface was often treated with Mg^{2+} and a variety of other divalent and trivalent cations for AFM observation of DNA [13]. According to our experiments, DNA molecules dissolved in N-(2-hydroxyethyl)piperazine-N'-ethanesulfonic acid (HEPES)-Mg^{2+} buffer consistently binds to nontreated freshly cleaved mica, which is enough for AFM observation (Fig. 35.1). The operating mode suitable for DNA observation is a dynamic mode (or intermittent contact mode). With use of these techniques, clear images are consistently obtained in ambient conditions and even in liquid conditions. Our experiments have shown that DNA molecules dissolved in distilled water also consistently bind to freshly cleaved mica enough to image by dynamic mode AFM.

Although AFM images of DNA are comparable to those taken by a rotary shadowing technique for TEM, AFM has the advantage of observing directly the molecule without special staining or shadowing in ambient conditions and even in liquid conditions. AFM images also have the advantage of containing quantitative information on the sample's height: the DNA height in AFM images is usually about 1 nm or less. However, the image width typically varies from 8 to 20 nm. The reason why the width is much larger than the height of DNA is explained by the convolution effect of the shape of the probing tip: the radius of the probing tip is generally about 10 nm when the cantilevers are commercially purchased. To minimize this

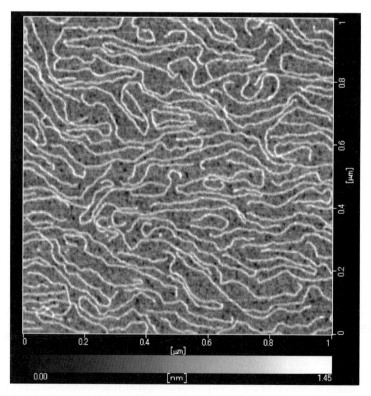

Fig. 35.1. Atomic force microscopy (AFM) image of DNA molecules absorbed on mica obtained in a dynamic mode in ambient conditions

artifact, the application of a carbon nanotube probe has been introduced by some researchers [14].

35.2.2.2
Collagen Type I Molecules

Collagen type I molecules are composed of three separate chains (α chains), each of which contains a characteristic sequence of amino acids twisted in the form of a left-handed helix. Biochemically, the length and the diameter of the molecules are estimated at about 300 and 1.5 nm, respectively. By AFM, the collagen molecules are observed as twisted threads about 280–310 nm long [15] (Fig. 35.2). The height and the width of the imaged molecules were 0.5–1 and 6–10 nm. At high magnification, somewhat globular bulges are usually present at both ends of the molecules. These terminal bulges are probably connected with the structure of the C- and N-terminals, which do not form the triple helix. A prominent depression, on the other hand, is often found at about 70 nm from one end of the molecules, suggesting that it might be a site of a relaxed triple helix which has been chemically studied. A characteristic bending of the molecules is sometimes found at one end of the molecules opposite the depression. Thus, because the surface morphology of collagen molecules is observed by dynamic mode AFM, it is expected to be used for future studies on the structure and function of various types of collagen molecules.

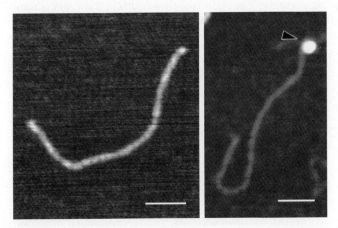

Fig. 35.2. AFM images of a collagen type I molecule (*left*) and a procollagen I molecule (*right*), obtained in a dynamic mode in ambient conditions. The collagen type I molecule is observed as a flexible thread 300 nm long. The procollagen I molecule is characterized by the presence of a globular C-terminal propeptide (*arrowhead*). *Bar* 50 nm. (Reproduced from [20])

35.2.3
AFM of Isolated Intracellular and Extracellular Structures

Various intracellular and extracellular structures have also been visualized by AFM. Among them, AFM imaging of chromosomes and collagen fibrils is introduced in the following sections.

35.2.3.1
Chromosomes

Chromosomes are produced by condensation of chromatin fibers, which consist of DNA and proteins. They appear in eukaryotic cells during cell division. For AFM imaging, the chromosomes are usually prepared in accordance with the standard method for light microscopy. Briefly, chromosome spreads are made by dropping a suspension of peripheral blood cultures (which are treated with a hypotonic solution of potassium chloride and fixed with a mixture of methanol and acetic acid). When these chromosomes are simply dried in air and observed by dynamic mode AFM, the samples are rather flattened apparently owing to the drying artifacts caused by the surface tension of water. On the other hand, the morphology of the critical-point-dried chromosomes was well preserved when observed by dynamic mode AFM in ambient conditions [16]; they were composed of the highly condensed chromatids of a mitotic pair, each of which was characterized by the presence of alternating ridges and grooves (Fig. 35.3). At high magnification, they are composed of tightly packed chromatin fibers about 50–60 nm thick.

AFM images of wet chromosomes can also be obtained using a dynamic mode in a liquid environment [17, 18]. Because of their extreme softness in liquid environments, the interaction force between the tip and the sample was especially carefully adjusted to the minimum degree, otherwise chromosomes were easily deformed during scanning. The surface of chromatids in these wet chromosomes was characterized by the presence of alternating ridges and grooves, as observed in the critical-point-dried chromosomes (Fig. 35.4).

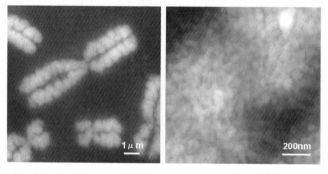

Fig. 35.3. AFM images of human metaphase chromosomes. At high magnification (*right*), the chromosomes were observed as an aggregation of strongly wound chromatin fibers

35.2.3.2
Collagen Fibrils

Collagen fibrils consist of collagen molecules which are assembled into a cylindrical fiber about 30–100 nm in diameter. When collagen fibrils with a diameter of about 40 nm are dried and observed by dynamic mode AFM, they appear to be semicylin-

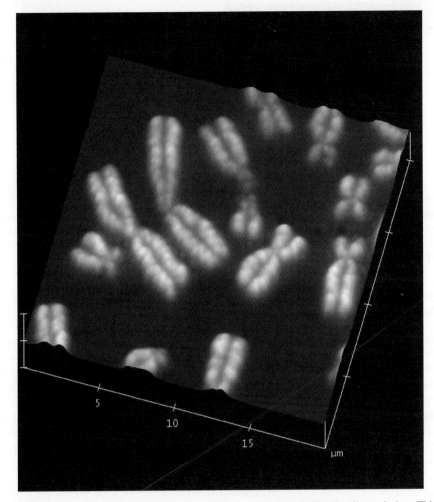

Fig. 35.4. AFM image of a chromosome spread in a phosphate-buffered saline solution. This image was obtained using dynamic mode AFM (Courtesy of Dr. Osamu Hoshi)

drical in profile, about 100 nm in width and 35–40 nm in height [19, 20] (Fig. 35.5); the reason why the width of the fibrils is imaged about 3 times larger than their height is explained by the pyramidal shape of the AFM probe. Most fibrils examined by dynamic mode AFM showed a repeating pattern of grooves and ridges with a period of 60–70 nm, which is characteristic of collagen fibrils.

Very recently, high-speed AFM was introduced by Picco et al. [21]. With use of this technique, collagen fibrils were imaged both at high-speed video rate (30 fps) and at kiloheltz frame rates up to 1300 fps in ambient conditions. Thus, this high-speed AFM method is expected to become a powerful tool for investigating sequential processes of biological events occurring on a time scale of a few milliseconds.

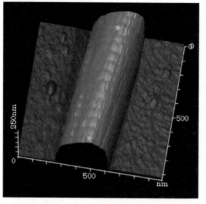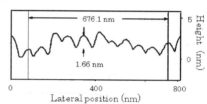

Fig. 35.5. AFM image of a bovine scleral collagen fibril. The fibril has periodic transverse grooves and ridges. Shallow longitudinal grooves are found on the surface of collagen fibrils. The longitudinal section profile of this fibril between the *two asterisks* is shown on the *left*. (Reproduced from [19])

35.2.4
AFM of Tissue Sections

In order to analyze internal structures of cells and tissues *in situ* with an atomic force microscope, relatively flat sectioned or fractured samples should be prepared because of the narrow z-range of the atomic force microscope scanner. Several investigators have attempted to observe the surface of ultrathin sections of LR white or epoxy resin with an atomic force microscope [22,23]. It is true that the contours of cellular structures such as unit membranes and chromatin blocks are traceable even in the AFM images of the resin sections. However, these cellular images are produced merely by depression or elevations of resin, probably owing to the uneven cleavage at the site of cellular structures during ultrathin sectioning. Thus, the possibility to obtain new information exceeding that from TEM observation is not expected with these sections. In this sense, embedment-free tissue sections are useful for AFM studies of cells and tissues because the internal structures can be observed directly at high resolution [24] (Fig. 35.6).

35.2.5
AFM of Living Cells and Their Movement

Because the atomic force microscope can be operated in liquid conditions, many researchers have been interested in visualizing the surface topography of cultured cells by AFM in liquid environments. For AFM imaging, flat cells should be attached firmly to the substrate (i.e., glass cover slip); spherical or domed cells are easily detached from the glass surface during scanning, probably because it is rather difficult to operate the tip accurately at the steep slope of the high sample. Adjustment of the imaging force to the weakest level is very important, otherwise cells are shrunk or damaged owing to the artificial forces overloaded by the scanning tip. When some kinds of cells of an epithelial nature are examined appropriately using contact mode

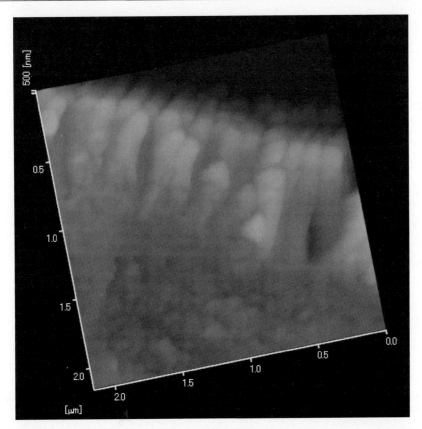

Fig. 35.6. AFM image of the surface of the embedment-free section of the mouse small intestine. Microvilli of the absorption cells are clearly three-dimensionally observed in this micrograph

AFM, cell processes such as lamellipodia are clearly observed at the cell margin of living cells as well as fixed ones [25, 26]. AFM also provides information on the contour of cytoskeletal elements just beneath the cell membrane, especially in living cells. In these studies, the height images are useful for measuring the thickness of these processes, but it may sometimes be difficult to visualize clearly the undulation of the cell surface. This is because the surface details may be hidden in the gradation of the wide range scale of the height images. Thus, variable deflection mode images (or error images) are often used in order to emphasize the contour of the cell surface as well as the shape of the cell processes (Fig. 35.7).

Because AFM can obtain images of living cells in liquid environments, there is the possibility that dynamic events of living cells can be analyzed in real time by AFM. The scan speed for AFM imaging of living cells depends on various factors, including the scan range, scanning mode and instrumentation. For example, in our previous studies of living epithelial cells, the practical scan speed was under 20 μm/s by contact mode AFM using our instrument [26]. This means that it takes 2 min or more to obtain a single image with a 40 μm × 40 μm scan area even when we use a 256 × 128 pixel size, and a series of AFM images with a 40 μm × 40 μm scan area

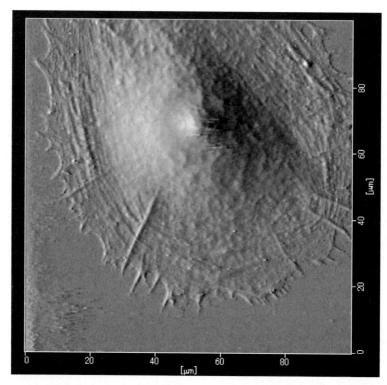

Fig. 35.7. AFM image of a living cultured cell (A549) in a liquid. This image was obtained using contact mode AFM

are acquired at time intervals of 2–8 min. Thus, at least, AFM is useful for examining dynamic events of living cells on a time scale of minutes. In order to observe living mammalian cultured cells by AFM under suitable culture conditions, it is necessary to maintain the pH of the culture medium either by exposing the chamber to 5% CO_2/95% air or by perfusing fresh culture medium into the chamber. Our studies showed that a fluid chamber system is useful for obtaining AFM images of living cells over the period of 1 h in a liquid environment. Several investigators have produced time-lapse photographic records from AFM images. These time-lapse movies are also very useful for realistically visualizing cellular dynamics. High-speed AFM is also expected to investigate cellular dynamics, including cell movements and secretory and excretory events, though these attempts are still awaited for with further studies [21].

35.2.6
Combination of AFM with Scanning Near-Field Optical Microscopy for Imaging Biomaterials

Recent advances in SPM have enabled the simultaneous collection of topographic and other physiological information of samples at the same portion of biological samples. AFM combined with scanning near-field optical microscopy (SNOM), or

SNOM/AFM, is one of the candidates for an attractive tool for the study of biological samples. This is because SNOM/AFM can collect both topographic and fluorescence images of the same portion of the samples simultaneously [27, 28].

There are various types of SNOM/AFM instrumentation, but the SNOM/AFM instrumentation usually has an optical fiber probe or a special cantilever, either of which has a subwavelength (i.e., 50–100-nm) aperture at the tip. During the operation, excitation light is introduced to the aperture either through the optical fiber or from the rear side of the cantilever, resulting in the formation of the evanescent field between the tip and the sample (Fig. 35.8). When the fluorescent substance lies in the evanescent field, far-field light is emitted by the fluorescent substance, and is detected by light microscopy.

Using this instrument, one can observe a single DNA fiber at 100-nm resolution [29]. Several researchers including us have applied SNOM/AFM to chromosome research and have succeeded in simultaneously obtaining topographic and fluorescence images of such samples as fluorescence in situ hybridization treated chromosomes and immunostained chromosomes [28, 30]. For example, in chromosomes immunostained with an anti 5-bromo-2'-deoxyuridine (BrdU) antibody and Alexa Fluor 488 after incorporation of BrdU into DNA, the fluorescence images obtained by SNOM/AFM clearly showed the portion of sister-chromatid exchanges, and the corresponding topographic images provided precise information on the fine structure of those portions [30] (Fig. 35.9). Since the spatial resolution of the fluorescence images obtained is obviously higher than that of those obtained by conventional

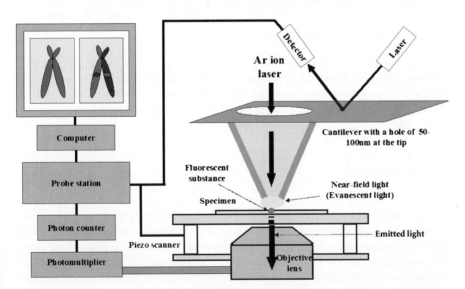

Fig. 35.8. Scanning near-field optical microscopy (SNOM)/AFM. In this system, the cantilever with an aperture at the tip is operated by dynamic mode AFM. During the scan, a focused laser beam is introduced to the aperture of the cantilever from the rear side, which results in the formation of an evanescent field between the tip and the sample. When a fluorescent substance is present in this field, light is emitted by the substance and is detected by a photomultiplier tube through the objective lens below the sample

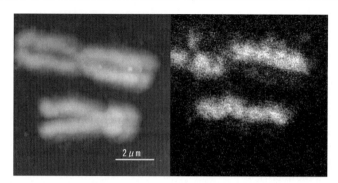

Fig. 35.9. SNOM/AFM images of 5-bromo-2′-deoxyuridine (BrdU) incorporated human chromosomes after differential staining with the anti-BrdU antibody. Topographic (*left*) and fluorescence (*right*) images of the same portion were simultaneously obtained by this technique. One of the paired sister chromatids was stained with fluorescence-labeled antibody. (Reproduced from [30])

fluorescence microscopy, SNOM/AFM is expected to serve as an important tool for future studies of the structure of biological samples in relation to the localization of their components.

35.3
SPM for Measuring Physical Properties of Biomaterials

35.3.1
Evaluation Methods of Viscoelasticity

Another important applications of SPM to biological fields is viscoelasticity mapping on living systems in liquid environments. There are two modes for mapping of mechanical properties over samples with SPM; force mapping mode and force modulation mode. These methods are especially useful for soft materials, such as cells or gels. Here, we will describe these modes in detail.

35.3.1.1
Force Mapping Mode

The force mapping mode is generally used to measure the local Young's modulus of soft samples such as cells and gels [31, 32] and can evaluate its absolute value. In this mode, a force versus distance curve (force curve) is recorded while a cantilever is indenting each point on the surface of the sample. Figure 35.10 shows schematic drawings for measuring force curves. The force (F) detected by the cantilever increases steeply, as the cantilever tip indents the sample by the piezoactuator translation (Z). The indentation depth of the sample δ and F are related by

$$Z - Z_0 = \frac{F}{k} + \delta ,$$

where k is the spring constant of the cantilever and Z_0 is the sample height under zero loading force.

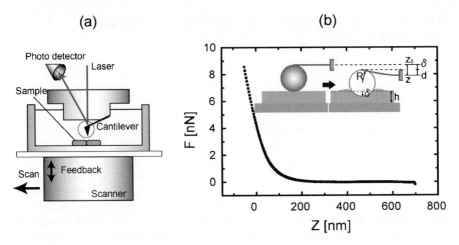

Fig. 35.10. Measuring the force curves in the force mapping mode

With use of nonlinear least squares fitting, the force curve obtained at one point of the sample's surface was fitted to Sneddon's model, a theoretical model in which the elastic contact theory of Hertz is applied to contact between an elastic plane and a rigid cone (the details are described in [33]). The applied loading force F is related to the indentation depth δ by

$$F = \frac{4\sqrt{R}}{3} \frac{E}{1-v^2} \delta^{2/3} \,,$$

where E is the Young's modulus, v is the Poisson ratio of the sample, and R is the radius of the cantilever tip [34, 35]. This fitting allowed the evaluation not only of the Young's modulus but also of the z-position, where the cantilever initially contacts the sample (i.e., the height of the sample at the point). For example, the force curve obtained from the cellular stiffness measurements was fitted in the loading force range 0.1–0.3 nN. The cellular Poisson ratio was assumed to be 0.5. Under the condition that a single force curve was taken for 100 ms, each image consisting of 64 × 64 pixels could be obtained in approximately 40 min. It should be noted that the spring constant of the cantilevers is chosen to be comparable to those of the samples, because this reduces sample damage and retains maximum sensitivity for elasticity measurements. This model is valid only when the indentation is much smaller than the thickness of the sample. For the case when the thickness of the sample is comparable to the indentation of the tip, Dimitriadis et al. [36] proposed an evaluation model that takes into account the finite thickness effect of samples on the estimation of the local elasticity. In this model, the relation between F and δ is modified as follows, where R is the radius of curvature of the cantilever tip and h is the thickness of the sample:

$$F = \frac{16E}{9} R^{\frac{1}{2}} \delta^{\frac{3}{2}} \left(1 + 1.133 \frac{\sqrt{R\delta}}{h} + 1.283 \frac{R\delta}{h^2} + \ldots \right) \,. \tag{35.1}$$

35.3.1.2
Force Modulation Mode

This mode is able to estimate both local elasticity and viscosity of samples by detecting the sinusoidal sample indentation caused by sinusoidal external stress, as shown in Fig. 35.11 [37, 38]. To load stress on the sample, a sinusoidal vibration whose amplitude was about 10 nm was applied to the cantilever (or the piezo scanner) by using a function generator during the contact mode scanning. The periodic strain of the sample was detected as a cantilever deflection. The oscillating component of the cantilever deflection that reflects the viscoelasticity of the sample was extracted through a two-phase lock-in amplifier. The frequency dependence of the viscoelasticity can be measured by varying the oscillating frequency. In the liquid environment, however, the frequency should be set as low as possible, because the viscosity of the medium surrounding the sample cannot be neglected at higher frequencies. On the other hand, at lower frequencies, the temporal resolution becomes poorer. Then, the oscillating frequency was set at 100–500 Hz for viscoelastic imaging in a liquid. For example, a temporal resolution of 6 min for a 128 pixels × 64 lines image at 500 Hz. Since the feedback system for measuring topography cancels the periodic stress applied by the cantilever vibration, a band-elimination filter was used to remove the oscillating component from the cantilever deflection signal sent to the feedback system.

The viscoelasticity of the sample was calculated from amplitude and phase data by the method based on the linear viscoelasticity theory [38]. The output signals from the two-phase lock-in amplifier represent the amplitude ratio β and the phase lag θ_s of the cantilever deflection relative to the reference signals from the function generator. Using the experimental data β and θ_s, one can express Young's modulus E and the viscosity coefficient η of the sample as

$$E = C_1 \frac{\beta(\cos\theta_s - \beta)}{1 - \beta^2 - 2\beta\cos\theta_s}$$

and

$$\eta = C_2 \frac{\beta\sin\theta_s}{1 + \beta^2 - 2\beta\cos\theta_s},$$

where C_1 and C_2 are instrument constants including the spring constant of the cantilever, etc. For purpose of simplicity, we assumed that C_1 and C_2 were equal to

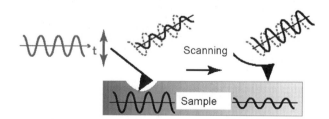

Fig. 35.11. The force modulation mode and a measurement system

1, so the values of E and η were relative quantities. In the case of the mechanical response originating only from the elastic property of samples, E simply represents Young's modulus of the materials.

35.3.2
Examples for Viscoelasticity Mapping Measurements

As examples for measuring physical properties of living systems by SPM, our results obtained from chromosomes, single cells and cell colonies are given in the following sections.

35.3.2.1
Chromosomes

The elasticity mapping of mitotic human chromosomes was carried out in phosphate-buffered saline (PBS) solution using the force mapping mode. The chromosomes were fixed with Carnor's fixative and kept in PBS solution to avoid drying. We used commercially available silicon nitride cantilevers with a length of 85 μm, a width of 20 μm, and a pyramidal tip with a typical radius of curvature of 50 nm. The spring constant of the cantilever was chosen to be 0.5 N/m to obtain the best sensitivity. The force curves obtained were evaluated by the Dimitriadis model taking account of the effect of the thickness of the sample [37]. In this study, we used only the first and second terms in (35.1) to compare the force curve with the model.

Figure 35.12 shows both topographic and elasticity images of the mitotic human chromosome in the PBS solution [39]. The height of the chromosome is about 150–360 nm. An imperceptible uneven surface structure of about 50–100 nm is visible in the morphology. This result agrees qualitatively with the spiral structure of the chromosome morphology previously reported in [16]. The elasticity distribution is not homogeneous over an entire chromosome, but a certain structure is visible. The values vary largely from 5 to 50 kPa.

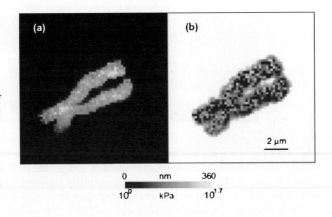

Fig. 35.12. a Topography and **b** elasticity images of a chromosome visualized by force mapping using a model that takes into account the effect of the thickness of the sample on the estimation of local elasticity

35.3.2.2
Living Single Cells

The viscoelastic measurements of living fibroblasts (NIH-3T3) were carried using the force modulation mode [40]. For the SPM imaging, the cell suspension was plated on a glass petri dish precoated with fibronectin, and then incubated for at least one night. To keep the pH constant during the SPM measurements, preheated HEPES buffer (pH 7.2–7.3) was substituted for the culture medium, and the samples were incubated for 1 h before the measurement.

Figure 35.13 shows typical topographic, local stiffness, and viscosity images of a living fibroblast. The cell extends thinly on the substrate and elongates toward the direction of crawling. Its height is about 3 μm. Many stress fibers appear on the cellular surface and align along the direction of movement. In the stiffness image, brighter areas represent greater hardness. We have confirmed that the distribution of the local stiffness measured using the force modulation mode is in good agreement with that obtained using the force mapping mode for an identical cell [38]. Therefore, the stiffness distributions range from several kilopascals to a few hundred kilopascals. It was found that local stiffness is not homogeneous on the cellular surface but varies largely from point to point. The stress fibers seen in the topographic image are harder than the surroundings, and this finding is consistent with the previous results [38,41,42]. In the viscosity image, brighter areas represent higher viscosity. The local viscosity of the area of the nucleus appeared to be lower than that of the periphery. However, further detailed, quantitative analysis will still be needed, because the values of the viscosity on the nanoscale have never been compared with those obtained by other methods. Sequential sets of topographic and local stiffness images of a typical moving cell are shown in Fig. 35.14. The topographic and stiffness images were taken simultaneously. The images were captured every 10 min. The SPM measurements were continued for about 60 min, which corresponds to the average time period of the intermittent cell crawling. The cell is changing its shape and moving downward in the image. A part of the cell is contracting, and the lamellipodium is extending gradually during the measurement. Accompanying the

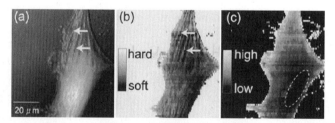

Fig. 35.13. Topographic (**a**), local stiffness (**b**), and viscosity (**c**) images of a fibroblast (NIH3T3) obtained using the force modulation mode. All images are 80 μm square. The topographic image (**a**) is graphically shaded. In the stiffness image (**b**), increased brightness represents greater hardness. The elasticity of the cell is spatially inhomogeneous. *Arrows* indicate stress fibers covering the cellular surface and aligning along the direction of the cell crawling. In the viscosity image (**c**), the brighter regions represent higher viscosity. The nuclear area within the *oval* appears to be less viscous than the periphery

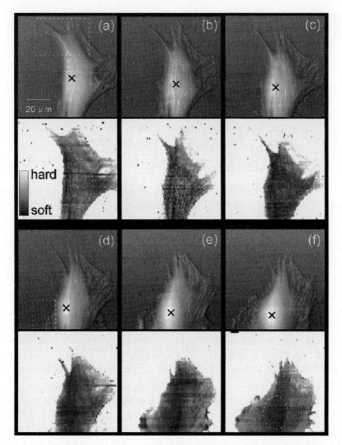

Fig. 35.14. Time-lapse images of topography and local stiffness of a moving cell. All images are 80 μm square. Topographic and stiffness images measured simultaneously are shown as coupled images. The coupled images were taken every 10 min. The posterior portion of the cell body (*dashed rectangle* in **a**) contracted, and a lamellipodiim (*dashed rectangle* in **d**) extended. The position of the nucleus (*cross*) moved downward. **a, b** The nucleus region (marked within an *oval*) is stiffer than the peripheral regions. Once the cell begins to move, the nucleus region becomes drastically softer

change in shape, the position of the nucleus is moving downward in the time course of the images. It should be noted that the local stiffness distribution also changes drastically with the cell migration.

35.3.2.3
Living Colony of Epithelial Cells

Epithelial cells (MDCK) were cultured in low-glucose Dulbecco's modified Eagle's medium containing 10% fetal bovine serum. The cells were grown in an incubator. The culture medium was substituted with HEPES buffer to avoid extreme pH de-

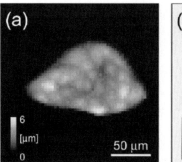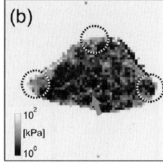

Fig. 35.15. a Topography and **b** stiffness images of an epithelial colony visualized using the force mapping mode. The data acquisition time was 30 min for a 64 × 64 pixel image

viation during the SPM measurements. A wide-range scanning probe microscope based on a conventional scanning probe microscope was developed [43]. In order to expand the measurement area, two conventional cylindrical piezotranslators were joined in tandem. We achieved the maximum scan ranges of 400 μm square in the xy-plane and 16 μm along the z-axis.

We measured the stiffness distribution and time-lapse images of a large epithelial colony using the wide-range scanning probe microscope. The height data imaged using the force mapping mode are shown in Fig. 35.15. The colony consisted of approximately 40 cells spread over 150 μm. The higher cells tend to locate on the boundary of the colony. The stiffness is not uniform in a colony; the average stiffness of each cell varies in the colony. The cell–cell boundaries (arrow) and the leading edges (circles) are stiffer than the surroundings, which is consistent with a previous report [44]. Figure 35.16 shows the time-lapse deflection images, which were measured in the constant force mode and captured every 1 h. The cells deformed cooperatively and induced migration of the whole colony. One portion of the colony (denoted by black circles in the figure) retracted inward, and then another portion (denoted by gray circles) extended outward. These results indicate that an epithelial colony moves collectively like single-cell migration. Recently we found the collective movements of thousands of epithelial cells on soft gels. [45] The present mechanical measurements are expected to provide the mechanism of the collective movements.

35.3.3
Combination of Viscoelasticity Measurement with Other Techniques

Combination of SPM with other techniques has the potential for investigating the physical properties of living systems in relation to their function. For example, in order to examine the response of living cells caused by external mechanical stimuli, stretching devices are very useful [46]. In our experimental systems, elastic chambers of silicone rubber are used to apply rapid, uniaxial deformation; a transparent bottom (20 mm × 20 mm) of 200-μm thickness and a wall of 5-mm height. These are deformable up to 20% along a single direction by the use of a handmade clamping de-

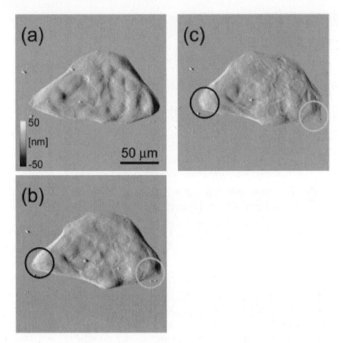

Fig. 35.16. Time-lapse deflection images of an epithelial colony taken every 1 h. The scanning frequency was 0.25 Hz. The posterior portion of the colony (*black circles*) contracted, and then the anterior portion (*gray circles*) elongated with a thin lammellipodium

vice. In order to fix living cells firmly, the silicone rubber substrate is usually coated with 20 μg/ml fibronectin in PBS for 30 min. During the experiment, trypsinized fibroblasts are plated onto either the prestretched or the unstretched chambers and incubated for 10 h. In order to observe a change in the cytoskeleton network, fibroblasts transfected with green fluorescent protein (GFP)–actin are stretched or compressed uniaxially on the silicon substrate, while the GFP fluorescence images are obtained by confocal microscopy.

A time-lapse series of topographic and stiffness images of a fibroblast, before and after the 8% elongation, is shown in Fig. 35.17. The images were captured every 30 min using the force mapping mode. The cell shape and stiffness were stable for 1 h before the elongation, but the height of the cell reduced and the stiffness of the cell increased immediately after the elongation. Many structures corresponding to the stress fibers became much stiffer after the elongation, and then softened gradually during the following 2 h. On the other hand, when the cells are contracted, the stiffness decreases instantly after the contraction and then increases gradually and stabilizes during the following 2 h. To examine whether or not the stress fibers are reconstructed after the deformation, fluorescence observation for GFP–actin was performed. Changes in the distribution of actin filaments beneath the cell membrane were observed in real time, while cells were deformed to a degree of 8%. Figure 35.18 shows a typical time-lapse series of fluorescence micrographs of GFP–actin, when cells were elongated or compressed. New thin stress fibers that did not significantly

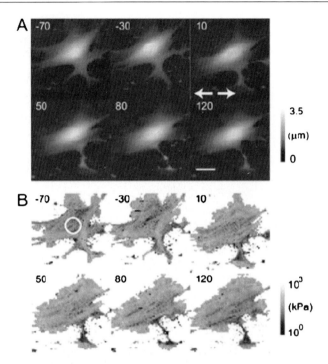

Fig. 35.17. Time-lapse SPM images of **a** topography and **b** stiffness distribution of a fibroblast before and after the application of force. Topographic and stiffness images measured simultaneously are shown as coupled images. *Numbers* represent the relative time (minutes) from the moment of substrate stretch. The *arrows* in the topography image represent the direction of the substrate stretch. The *circle* in the stiffness image represents the typical area used for the averaging of stiffness. *Bar* 20 μm

alter the cellular stiffness observed by SPM appeared slightly. These results indicate that the stiffness responses to the deformation originate from the contractile tension generated by the stress fibers, and the existence of tensional homeostasis at the cellular level.

35.4
SPM as a Manipulation Tool in Biology

In addition to imaging and/or measuring the physical properties of biological samples, the atomic force microscope has the potential to be used as a dissection tool for tiny substances at atomic-scale to submicron-scale levels. For example, with use of the contact mode, DNA molecules immobilized on mica can be dissected both in air and in a liquid environment by increasing the force applied to the AFM probe [47]. The size of these biomolecules is suitable for dissecting and manipulating by contact mode AFM using a standard probe. Dissection of the large biological structures such as a chromosome has also been performed by several investigators.

Fig. 35.18. Temporal behavior of green fluorescent protein–actin expressing fibroblasts, when subjected to elongation or 8% compression. The direction is shown by *arrows. Bar* 20 μm

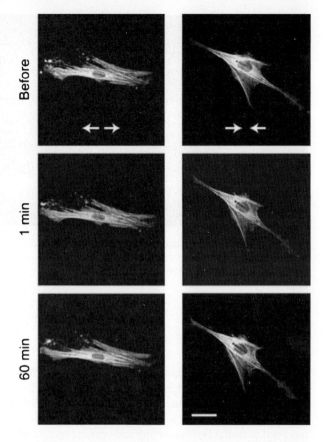

However, it is often difficult to sharply dissect these samples using the contact mode, because samples are crushed or scratched by the scanning probe. Thus, dissection techniques using a dynamic force mode have been introduced for these thick and wide samples [48] (Fig. 35.19).

The development of a suitable and novel AFM probe is also an interesting subject for AFM manipulation of biological samples. For dissection of biomolecules at the nanoscale, a carbon nanotube probe has been applied. More recently, a knife-edged AFM probe was introduced for dissection of DNA and chromosomes [49]. With use of this probe, chromosomes can be cut in a dynamic mode more sharply and consistently than when using the conventional AFM probe. In addition, a tweezer-type AFM probe was also introduced for manipulation of nanomaterials [50]. This device has two thin probes, which are formed in the front of the silicon cantilever, with a vertical triangular shape. One functioned as an AFM imaging probe tip (sensing probe) for a dynamic mode, while the other can be made to work as a tweezer, by closing the vertical face of the probe (the movable probe) owing to the effect of the thermal expansion actuator. This tweezer-type AFM probe can be used for picking up small fragments after dissection of the samples.

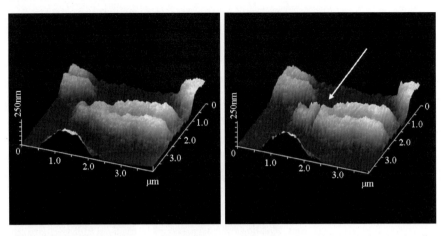

Fig. 35.19. AFM image of a human metaphase chromosome before (*left*) and after (*right*) dissection with an AFM probe using a dynamic mode. The *arrow* indicates the dissection place. A dissected fragment is observed on the left side of the cut part of the chromosome. (Reproduced from [48])

35.5
Conclusion

The present study introduced the application of SPM to biological fields for investigating various structures from biomolecules to living cells. As shown in this chapter, SPM provides much information which cannot obtained with other type of microscopes such as light microscopes and electron microscopes. In this sense, SPM has great potential for broadening and opening up new fields in biomedical research. SPM instrumentation and the technique of specimen preparation for biological studies are advancing, and we are just opening the door to the new microscopic world which will be made available by SPM for solving enigmas in biomedical fields.

References

1. Guckenberger R, Hartmann T, Wiegräbe W, Baumeister W (1992) In: Wiesendanger R, Güntherodt HJ (eds) Scanning tunneling microscopy II. Springer, Berlin, p 51
2. Binnig G, Quate CF, Gerber C (1986) Phys Rev Lett 56:930
3. Lal R, John SA (1994) Am J Physiol 266:C1
4. Ikai A (1996) Surf Sci Rep 26:261
5. Ushiki T, Hitomi J, Ogura S, Umemoto T, Shigeno M (1996) Arch Histol Cytol 59:421
6. Jena BP, Hörber JKH (2002) (eds) Atomic force microscopy in cell biology. Academic, San Diego
7. Hörber JKH, Miles MJ (2003) Science 302:1002
8. Jena BP, Hörber JKH (2006) (eds) Force microscopy. Applications in biology and medicine. Willy-Liss, Hoboken
9. Braga PC, Ricci D (2004) (eds) Atomic force microscopy. Biomedical methods and applications. Humana, Totowa
10. Liu Z, Li Z, Wei G, Song Y, Wang L, Sun, L (2006) Microsc Res Tech 69:998

11. Müller DJ, Heymann JB, Oesterhelt F, Moller C, Gaub H, Buldt G, Engel A (2000) Biochim Biophys Acta 1460:27
12. Müller DJ, Engel A (2002) In: Jena BP, Hörber JKH (eds) Atomic force microscopy in cell biology. Academic, San Diego
13. Zuccheri G, Samorì B (2002) In: Jena BP, Hörber JKH (eds) Atomic force microscopy in cell biology. Academic, San Diego
14. Hohmura KI, Itokazu Y, Yoshimura SH, Mizuguchi G, Masamura YS, Takeyasu K, Shiomi Y, Tsurimoto T, Nishijima H, Akita S, Nakayama Y (2000) J Electron Microsc 49:415
15. Yamamoto S, Nakamura F, Hitomi J, Shigeno M, Sawaguchi S, Abe H, Ushiki T (2000) J Electron Microsc 49:473
16. Ushiki T, Hoshi O, Iwai K, Kimura E, Shigeno M (2002) Arch Histol Cytol 65:377
17. Hoshi O, Owen R, Miles M, Ushiki T (2004) Cytogenet Genome Res 107:28
18. Hoshi O, Shigeno M, Ushiki T (2006) Arch Histol Cytol 69:73
19. Yamamoto S, Hitomi J, Shigeno M, Sawaguchi S, Abe H, Ushiki T (1997) Arch Histol Cytol 60:371
20. Ushiki T (2002) Arch Histol Cytol 65:109
21. Picco LM, Bozec L, Ulcinas A, Engledew DJ, Antognozzi M, Horton MA, Miles MJ (2007) Nanotechnology 18:1
22. Yamamoto A, Tashiro Y (1994) J Histochem Cytochem 42:1463
23. Yamashina S, Shigeno M (1995) J Electron Microsc 44:462
24. Ushiki T, Shigeno M, Abe K (1994) Arch Histol Cytol 57:427
25. Ushiki T, Yamamoto S, Hitomi J, Ogura S, Umemoto T, Shigeno M (2000) Jpn J Appl Phys 39:3761
26. Ushiki T, Hitomi J, Umemoto T, Yamamoto S, Kanazawa H, Shigeno M (1999) Arch Histol Cytol 62:47
27. Muramatsu H, Chiba N, Nakajima K, Ataka T, Fujihira M, Hitomi J, Ushiki T (1996) Scanning Microsc 10:975
28. Ohtani T, Shichiri M, Fukushi D, Sugiyama S, Yoshino T, Kobori T, Hagiwara S, Ushiki T (2002) Arch Histol Cytol 65:425
29. Yoshino T, Sugiyama S, Hagiwara S, Fukushi D, Shichiri M, Nakao H, Kim JM, Hirose T, Muramatsu H, Ohtani T (2003) Ultramicroscopy 97:81
30. Kimura E, Hitomi J, Ushiki T (2002) Arch Histol Cytol 65:435
31. Haga H, Sasaki S, Kawabata K, Ito E, Ushiki T, Sambongi T (2000) Ultramicroscopy 82:253
32. Nitta T, Haga H, Kawabata K, Abe K, Sambongi T (2000) Ultramicroscopy 82:223
33. Vinckier A, Semenza G (1998) FEBS Lett 430:12
34. Munevar S, Wang YL, Dembo M (2001) Mol Biol Cell 12:3947
35. Lauffenburger DA, Horwitz AF (1996) Cell 84:359
36. Dimitriadis EK, Horkay F, Maresca J, Kachar B, Chadwick RS (2002) Biophys J 82:2798
37. Radmacher M, Tillmann RW, Gaub HE (1993) Biophys J 64:735
38. Haga H, Nagayama M, Kawabata K, Ito E, Ushiki T, Sambongi T (2000) J Electron Microsc 49:473
39. Nomura K, Hoshi O, Fukushi D, Ushiki T, Haga H, Kawabata K (2005) Jpn J Appl Phys 44:5421
40. Nagayama M, Haga H, Kawabata K (2001) Cell Motil Cytoskeleton 50:173
41. Kawabata K, Nagayama M, Haga H, Sambongi T (2001) Curr Appl Phys 1:66
42. Rotsch C, Radmacher M (2000) Biophys J 78:520
43. Mizutani T, Haga H, Nemoto K, Kawabata K (2004) Jpn J Appl Phys 43(7B):4525
44. A-Hassan E, Heinz WF, Antonik MD, D'Costa NP, Nageswaran S, Schoenenberger CA, Hoh JH (1998) Biophys J 74:1564
45. Haga H, Irahara C, Kobayashi R, Nakagaki T, Kawabata K (2005) Biophys J 88:2250
46. Mizutani T, Haga H, Kawabata K (2004) Cell Motil Cytoskeleton 59:242

47. Hansma HG, Vesenka J, Siegerist C, Kelderman G, Morrett H, Sinsheimer RL, Elings V, Bustamante C, Hansma PK (1992) Science 256:1180
48. Iwabuchi S, Mori T, Ogawa K, Sato K, Saito M, Morita Y, Ushiki T, Tamiya E (2002) Arch Histol Cytol 65:473
49. Saito M, Nakagawa K, Yamanaka K, Takamura Y, Hashiguchi G, Tamiya E (2006) Sens Actuators A 130–131:616
50. Takekawa T, Nakagawa K, Hashiguchi G, Saito M, Yamanaka K, Tamiya E (2005) IEEJ Trans SM 125:448

36 Novel Nanoindentation Techniques and Their Applications

Jiping Ye

Abstract. Nanoindentation measurement as one tool in the scanning probe microscope family is the most successful means for evaluating the mechanical properties of small-volume materials, such as thin films, microparticles and multiphase materials. This chapter demonstrates that elastic, elastoplastic and viscoelastic contact solutions permit nanoindentation load–displacement curves to be used to evaluate many kinds of mechanical properties on a nanometer scale. More than four different kinds of convenient and novel nanoindentation techniques for practical purposes are described. The primary emphasis is on how to determine the most frequently used mechanical properties such as hardness and modulus, yield stress, stress–strain curve, and viscoelasticity. Focus is also put on how to employ these methods to various kinds of materials in different application fields. This chapter proposes that all kinds of bulk-scale mechanical properties or characteristics will be easily determined on a nanometer scale by using suitable nanoindentation methods in the near future.

Key words: Nanoindentation technique, Indentation load-displacement curve, Nanomechanical properties, Hardness/modulus, Yield stress, Stress-strain curve, Viscoelasticity

Abbreviations

CSM	Continuous stiffness measurement
DMA	Dynamic mechanical analysis
FEM	finite-element method
MoDTC	Molybdenum dithiocarbamate
PAE	Polyarylene ether
SPM	Scanning probe microscope
ZDDP	Zinc dialkylsithiophosphate

36.1 Introduction

Miniaturizing the feature size of components has become a driving force behind the search for new characteristics in material sciences and the acceleration of improvements in the performance of advanced materials in industry. The miniaturization process introduces various small volumes of material, and it results in lowering the mechanical and thermomechanical strength of materials. In the past two decades, a lot of effort has been expended to characterize fundamental mechanical properties on a nanometer scale, not only for optimizing miniaturization processes, but also for microstructure design and life prediction using computer modeling and simulation. The most successful means for evaluating the mechanical properties of small-volume materials, such as thin films, microparticles and multiphase materials, is to use a load-sensing or depth-sensing indentation technique with loads on a micronewton scale and depths on a nanometer scale, referred to as nanoindentation. This nanoindentation measurement can be regarded as one tool in the scanning probe microscope (SPM) family. As illustrated in Fig. 36.1, an indenter acts in the same way as a SPM probe. Interaction between the indenter tip (or SPM probe) and the sample's surface increases greatly as the tip approaches the surface and comes in contact with it, and decreases as the tip moves away from the sample's surface. The mechanics of nanoindentation measurement constitute the same principle as those of a SPM: the interaction changes rapidly depending on the distance of the probe from the surface and the measurement operation can be controlled by detecting the distance when varying the interaction or by detecting the interaction when varying the distance.

Nanoindentation is used to measure mechanical properties from the indentation load–displacement curve based on elastic and elastic/plastic contact theory. Meyer's law gives important information about the relationship between the indentation load and the indentation depth [1–4]. Hertzian contact analysis shows a basic geometric view of elastic contact for spherical indenters and all real pyramidal indenters at small displacements [5, 6]. Sneddon's stiffness equation [3, 4] provides the elastic modulus relative to stiffness and contact area. The von Mises and Tresca criterion shows a simple linear relationship between the contact pressure for yield and yield stress in the elastic-plastic contact regime [7, 8].

On the basis of these fundamental contact theories, several novel nanoindentation techniques have been developed for estimating various properties such as hardness and modulus, yield stress, plasticity, viscoelasticity, creep and relaxation relative to time and temperature, and fracture toughness. Pharr et al. [9, 10] developed a convenient technique for determining hardness and the elastic modulus by extrapolating the contact depth from the indentation unloading curve. This method has been widely adopted in the instrumented indentation field and has also been promoted and refined by many researchers. Ye et al. [11, 12] developed a convenient spherical nanoindentation technique for determining the yield stress of ultrathin films by extrapolating the indentation pressure from the load–displacement hysteretic loop

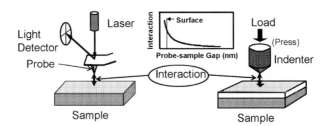

Fig. 36.1. Illustration showing that both scanning probe microscopy and nanoindentation techniques are performed with the same measurement mechanics by controlling the interaction between the probe and the sample or the probe–sample distance, where the interaction changes rapidly with the distance

energy. Bucaille et al. [13], Chollacoop et al. [14] and Ogasawara et al. [15] developed a dual indenter method for determining the stress-strain curve based on a finite element analysis of the influence of the included angle of conical indenters [16, 17]. Huang et al. [18, 19] developed a method for measuring viscoelastic functions in the frequency.

We are concerned with elastic, elastoplastic and viscoelastic contact solutions for the measurement of mechanical properties based on the indentation load–displacement curve. We describe several convenient and novel nanoindentation techniques for practical purposes. The primary emphasis is on how to determine the most frequently used mechanical properties such as hardness and modulus, yield stress, stress–strain curve, and viscoelasticity. Focus is also placed on the latest applications of these methods.

36.2
Basic Principles of Contact

All the recent approaches for determining mechanical properties from the nanoindentation load-displacement curve are based on two principles. One is Meyer's law giving the relationships among load, displacement and contact area. The other is an elastic contact solution, including Hertzian contact analysis and Sneddon's stiffness equation. Although there are many other important contact solutions for elastoplastic and viscoelastic indentation models with different contact geometries, Meyer's law and an elastic contact solution are the starting points for solving a complex contact problem.

36.2.1
Meyer's Law

Meyer's law provides the simplest and most basic relationship between the indentation load and depth [1–4]. This law states that for a ball of fixed diameter, the indentation load P and the diameter d of the remaining indentation are related in such a way that

$$P = kd^n , \qquad (36.1)$$

where k and n are constants for the material under examination. The value of n is experimentally confirmed to be 2–2.5. Mayer experimentally found that when balls of different diameters D are used, the values of k decrease with increasing D, but the constant n is almost independent of D in a way that

$$A = kD^{n-2} , \qquad (36.2)$$

where A is a constant. Thus, the relationship between P and d can be expressed as

$$\frac{P}{d^2} = A \left(\frac{d}{D}\right)^{n-2} . \qquad (36.3)$$

For geometrically similar impressions, the ratio d/D must be constant, and that makes P directly proportional to d^2. Because the principle of geometric similarity is a fundamental physical concept, the general relationship between P and d can be rewritten as

$$\frac{P}{d^2} = f\left(\frac{d}{D}\right), \tag{36.4}$$

where $f(d/D)$ is some function depending on the specimen. On the basis of this viewpoint, Sneddon derived general relationships among the load, displacement and contact area for any indenter that can be described as a smooth, axisymmetric body of revolution [3,4]. The load–displacement relationships for various simple indenter geometries can be summarized in a simple exponent law as

$$P = Ch^m, \tag{36.5}$$

where h is the elastic displacement of the indenter and C and m are constants. Referred to as Meyer's exponent, m can have different values for different types of indenter geometries, i.e., $m = 1$ for flat cylinders, $m = 1.5$ for spheres and $m = 2$ for cones. Recent studies also show that Mayer's law can be extended to elastic-plastic and viscoelastic contact ranges [10, 11, 16].

36.2.2
Elastic Contact Solution

36.2.2.1
Hertzian Solution

For elastic contact between smooth axisymmetric bodies of revolution, two assumptions are made in Hertzian contact analysis [5,6]. One is that each body is considered as an elastic half-space, which means that the contact region is much smaller than the overall dimensions of the body. This assumption simplifies the boundary conditions. The contact stress is highly concentrated close to the contact region and decreases quickly with increasing distance from the contact point. The stress in the contact region is not critically affected by the shape of the bodies away from the contact region. The other assumption is that of frictionless contact, which means that the indentation load acts only in the direction normal to the contact surface and there is no shear stress in the contact region. On the basis of these two assumptions, the relationship between the applied indentation load and the displacement for a spherical indenter in elastic contact can be expressed as

$$P = \frac{4}{3}\sqrt{R}E_r h^{3/2}, \tag{36.6}$$

where R is the indenter radius, h is the elastic displacement of the indenter and E_r is the reduced elastic modulus, defined by

$$\frac{1}{E_r} = \frac{1-v^2}{E} + \frac{1-v_i^2}{E_i}, \tag{36.7}$$

where v and E are Poisson's ratio and the modulus of the indented material, and v_i and E_i are those of the indenter.

36.2.2.2
Sneddon's Solution

For elastic contact between a conical indenter and an elastic half-space, subject to the boundary conditions of frictionless contact, prescribed vertical displacements and no normal stress outside the contact area, the relationship between the applied indentation load and the displacement can be expressed by Sneddon's stiffness equation:

$$S = \frac{dP}{dh} = \frac{2}{\sqrt{\pi}} E_r \sqrt{A} \,. \tag{36.8}$$

Here S is the elastic stiffness of the contact and A is the projected area of the elastic contact [3].

Buylchev et al. [21] and Pharr et al. [9] showed that Sneddon's stiffness equation generally applies to any smooth function of revolution. The equation can be slightly corrected to

$$S = \frac{dP}{dh} = \beta \frac{2}{\sqrt{\pi}} E_r \sqrt{A} \,, \tag{36.9}$$

where β is a constant; $\beta = 1$ for a conical indenter, $\beta = 1.011$ for a pyramidal indenter, $\beta = 1.012$ for a Vickers indenter and $\beta = 1.034$ for a Berkovich indenter.

36.3
Tip Rigidity and Geometry

Equation (36.7) shows that the reduced modulus takes into consideration bidirectional displacements in both the indented material and the indenter. If the indenter is perfectly rigid, that is, if its hardness and elastic modulus are high enough and Poisson's ratio is low enough to avoid any deformation during the indentation process, the contribution of the indenter to the measured displacement is zero. Diamond indenters are generally used in practical applications. Their Young's modulus $E_i = 1141$ GPa is vastly larger and their Poisson's ratio $v_i = 0.07$ is much smaller than the corresponding values of the materials examined [10, 22]. Thus, Young's modulus E_r can be given by

$$E_r = \frac{E}{1 - v^2} \,, \tag{36.10}$$

where v and E are Poisson's ratio and modulus of the material examined.

The geometries of indenters generally take one of two forms [23]. One is a pointed form, including pyramids and cones (such as the Berkovich type, cube-corner type, conical type, etc.). For ideal indenters without any roundness on the top, the ratio of the contact depth to the contact area is constant, being independent of the contact depth. Theoretically, plasticity forms when the indenter first touches the surface; indentation strain is constant and the volume of the affected material grows proportionally to the indent volume during the indentation process [2, 23–25]. The other one is a smooth axisymmetric form, including spheres and paraboloids. The indentation contact is characterized very well by a Hertzian solution [2, 5, 6, 24]. During the indentation process, yield begins below the surface and the strain varies with the contact depth [26, 27].

In fact, the Berkovich indenter is the most frequently used type for measuring hardness and modulus. The centerline-to-face angle is 65.3° and its equivalent cone angle is 70.3°, the same as that of a Vickers indenter. Three sides of the Berkovich indenter easily form a point on the top [23, 28, 29]. The best tip is ground to a radius of 20 nm. These geometric characteristics make it possible to produce full plasticity at a very small load, suitable for hardness and modulus measurements. The cube-corner indenter is another type of three-sided pyramid tip, with a centerline-to-face angle of 35.3° and an equivalent cone angle of 42.3° [23]. The best tip is ground to a radius of 40 nm. This type of indenter can displace more than 3 times the volume of a Berkovich indenter at the same load [30–32]. The cube-corner indenter produces a more severe indent than a Berkovich indenter; namely, it produces much higher stresses and strains in the vicinity of the contact and reduces the cracking threshold. Generally, this type of indenter is more suitable for estimating fracture toughness than for hardness and modulus measurements [25, 33–35].

The third type of pointed-form indenter for practical use is the conical indenter. However, it is logistically very difficult to grind and the best tip is ground to a radius of 500 nm. This indenter is advantageous for testing highly anisotropic or oriented materials and for reducing undesired cracking during the test, as compared with a Berkovich indenter and a cube-corner indenter [23]. Thus, the conical indenter is preferred for doing scratch tests for evaluating the friction coefficient and adhesion or cohesion of thin films [36, 37]. Moreover, the conical indenter is also useful for comparing experimental results to a 2D axisymmetric model, thereby making it possible to evaluate the stress–strain curve based on a finite-element analysis of the influence of the included angle of conical indenters [13–15, 23].

On the other hand, the most frequently used indenter with a smooth axisymmetric form is the spherical one. The tip radius can be varied from 20 nm to 100 μm and the depth of the concentrated stress area can be adjusted with the tip radius. The spherical indenter requires a high load to produce plastic contact, making it easy to evaluate Young's modulus in the elastic deformation range and yield stress near the surface [11, 12]. The spherical indenter also exhibits less abrasion, consistent lateral force and spatial insensitivity during scratch tests [23, 36, 37].

36.4
Hardness and Modulus Measurements

36.4.1
Analysis Method

The indentation hardness of materials is simply defined as the ratio between the peak load P normal to the specimen surface and the projected area A under load:

$$H = \frac{P_{max}}{A}. \tag{36.11}$$

Note that here A is not the residual area of indentation. It is crucial to obtain A at the peak load from the indentation load–displacement curve for determining hardness and modulus.

Pharr et al. [9, 10] presented a widely used method for determining the contact area from the depth of contact and the geometry of the indenter based on two assumptions. First, the elastic modulus is not affected by the indentation process. This means that the elastic modulus can be estimated from the unloading of an indentation, in which no plastic deformation is involved. Second, the hardness impressions have the same shape as the indenter within the plane, but are shallower in depth. On the basis of these assumptions, as shown in Fig. 36.2, the contact depth h_c of the indenter can be expressed as

$$h_c = h - h_s = h - \varepsilon \frac{P}{S}, \tag{36.12}$$

where h is the total displacement under load, h_s is the displacement of the surface at the perimeter of the contact and ε is a constant that differs depending on the indenter geometry; $\varepsilon = 0.72$ for conical indenters, $\varepsilon = 0.75$ for the Berkovich indenter and indenters with paraboloid of revolution, and $\varepsilon = 1.0$ for flat indenters.

In general, the indentation system always has its characteristic compliance under load and the indenter always possesses some finite roundness on its top. As the next step, therefore, the area function $f(h_c)$ for obtaining the contact area from the contact depth has to be decided such that there is less influence from the compliance C_f of the load frame and the roundness at the tip. In the indentation loading system, the compliance C_s of the specimen and C_f constitute the resultant compliance C, and C_s during elastic contact is the inverse of the contact stiffness S_s of the specimen as shown in (36.8). If the reduced modulus is constant as a function of depth, the relationship between C and $A^{-1/2}$ is known to be linear for a given material, being expressed as

$$C = C_f + C_s = C_f + \frac{1}{S_s} = C_f + \frac{\sqrt{\pi}}{2E_r} \frac{1}{\sqrt{A}}. \tag{36.13}$$

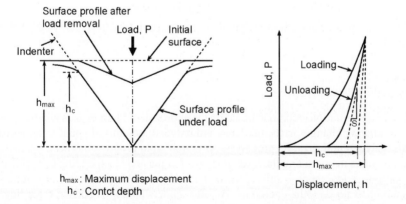

Fig. 36.2. *Left*: Indentation, showing the quantities used in the analysis. *Right*: Load–displacement curves showing the quantities used in the analysis with graphical interpretation of the contact depth

Therefore, by first making some large indents in a reference material with a known reduced modulus such as fused silica ($E^* = 69.7$ Ga), one can calculate directly the contact area A for large depths from the area function for a perfect indenter without any roundness on the top: $A(h_c) = 24.5h_c^2$ for the perfect Berkovich indenter and $A(h_c) = 2.6h_c^2$ for the perfect cubic indenter. By plotting C versus A^{-2}, one can directly measure the value of C_f from the intercept of the plot, while one can obtain the value of E_r from the inverse of the slope. Next, to determine the area function $f(h_c)$ of imperfect indenters with some roundness on the top, some small indents have to be made in the fused silica. The contact area A for a small depth can be calculated from the reduced modulus E_r and the specimen compliance C_s can be obtained by subtracting C_f from C, by rearranging (36.13) to

$$A = \frac{\pi}{4} \frac{1}{E_r^2} \frac{1}{(C - C_f)^2}, \quad (36.14)$$

where E_r and C_f are obtained from the large indents. Thus, several pairs of (A, h_c) can be given over a range of depths from (36.14), and the initial area function can be found by fitting the A versus h_c data to the form

$$A(h_c) = C_0 h_c^2 + C_1 h_c^1 + C_2 h_c^{1/2} + C_3 h_c^{1/4} + \ldots + C_8 h_c^{1/128}, \quad (36.15)$$

where C_0 is a constant that varies depending on the indenter geometry; $C_0 = 24.5$ for the Berkovich indenter and $C_0 = 2.6$ for the cubic indenter. C_1 through C_8 are fitting constants. Because the initial area function is obtained from the perfect area function in a large depth range and the exact form of the area function influences the values of C_f and E_r, the procedure should be iterated a few times using the new area function.

Consequently, according to the approach of Pharr et al., the reduced modulus of specimens can be found from the measured stiffness S of the unloading curve and the contact depth h_c of the indentation load–displacement curve by rewriting Sneddon's stiffness equation (36.9) as

$$E_r = \frac{\sqrt{\pi}}{2} \frac{S}{\sqrt{A(h_c)}} \quad (36.16)$$

and the hardness can be obtained by dividing the peak load P by $A(h_c)$ as shown in (36.11).

36.4.2
Practical Application Aspects

The nanoindentation technique based on the approach of Pharr et al. is widely applied for evaluating changes in the hardness and reduced modulus of small-volume materials due to miniaturization processes. Although rapid progress has been made in the past 10 years in upgrading the level of this technique and in expanding its application to various fields, there is still a lack of knowledge and experience concerning this method that needs to be overcome. A better understanding is needed regarding the size effect at the nanometer scale, the measurement scale effects due to the indentation process, reasonable error level, effect of the substrate and the relationship with the conventional Vickers hardness measurement, among other aspects.

36.4.2.1
Quantum Size Effect and Measurement Scale Effects

Recent nanoindentation techniques have made it possible to indent specimens within a depth of less than 10 nm. It is necessary to confirm if there is any quantum size effect and if there are any measurement scale effects on the mechanical properties owing to the smaller measurement area. Ye et al. [38] made indents in fused silica by nanoindentation, microindentation and macroindentation techniques using diamond Berkovich indenters with different radii of 100–500 nm and with different loads ranging from 10 μN to 50 N, and determined the hardness and reduced modulus based on the approach of Pharr et al. As shown in Fig. 36.3, it was found that the hardness and modulus in a contact depth range from 10 nm to 10 μm have almost constant values, with a small measurement error of less than 5–10%. The results revealed that there was no quantum size effect in the nanoindentation measurement for fused silica, which possesses a single domain exhibiting a homogeneous and stable glass phase from the nanometer scale to the macroscale. If there is no size effect from the microstructure of the specimen in a nanoindentation measurement, in terms of estimating the mechanical properties on a nanometer scale, the specimen can be simulated as a continuum using the finite-element method (FEM), similar to bulk materials. Moreover, the results shown in Fig. 36.3 also indicate that the approach of Pharr et al., which was developed for evaluating hardness and modulus on a nanometer scale, can also be generally applied and extended to the bulk range by standardizing this method from the nanometer scale to the macroscale.

As for measurement scale effects, Xu and Li [39] have found that a significant effect occurs when the sample size is comparable to the indent size and suggested that nanoindentation must be done within a certain critical ratio of indent size to sample size. They recommended that the ratio of the indent radius to the specimen radius should be less than 0.3 for accurate determination of hardness within an error of 5%, and that the ratio of the indentation displacement to the thickness of the specimen

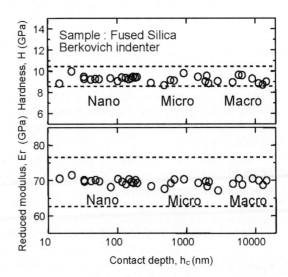

Fig. 36.3. Hardness and reduced modulus versus contact depth from the nanometer scale to the micrometer scale with indentation load ranging from several micronewtons to several tens of newtons. (From Ye [38], with permission)

should be less than 0.03–0.06. On the other hand, some materials have been found to exhibit changes in mechanical properties depending on the indenter geometry and indentation depth [40–42]. In this case, therefore, some factors such as plastic deformation and work hardening can be considered in the FEM simulation [40–42]. The relative mechanical properties such as yield stress and stress–strain curve are described in the following sections.

36.4.2.2
Tip Radius and Surface Roughness Effect

On the basis of the Hertzian solution in (36.6), the indentation hardness depends on the tip radius, reduced modulus, and indentation displacement in elastic contact. The hardness can even be zero at the surface. It increases from zero with increasing displacement and reaches a constant value in the elastoplastic contact range. Therefore, an indenter with a small tip radius is preferred for making shallow indents to obtain hardness independent of the depth. On the other hand, with a small tip radius, hardness and reduced modulus measurements can be easily affected by the surface roughness. Usually, the tip radius has to be at least 10 times larger than the surface roughness.

36.4.2.3
Pile-Up and Sink-In Effects

Equation (36.12) indicates that the contact depth h_c is lower than the total indentation displacement h under load. In some materials with low yield strain or that are difficult to work-harden, however, pile-up may occur, causing the contact depth h_c to become greater than the total indentation displacement h [13,23]. For materials such as work-hardened Cu and Al, the calculated contact area is much smaller than the real value, resulting in an overestimation of hardness by as much as 20–30% and of modulus by as much as 10–15%. Loubet et al. [43] and Hochstetter et al. [44] have taken into account pile-up and sink-in effects and proposed a different method for estimating the contact depth:

$$h_c = \alpha \left(h - \frac{P}{S} \right), \qquad (36.17)$$

where α is a constant that differs depending on the indenter geometry and $\alpha = 1.2$ for the Berkovich indenter [43,44]. Another way of measuring the contact area is from the residual imprint using atomic force microscopy or some other microscopy method.

36.4.2.4
Error Level and Substrate Effect

Just a few years ago, a reasonable level of error in nanoindentation measurements was considered to be around 10%. However, more advanced nanoindentation systems have been developed recently, including commercial ones, that feature greater rigidity, higher loading resolution, improved sensitivity for detecting displacement and

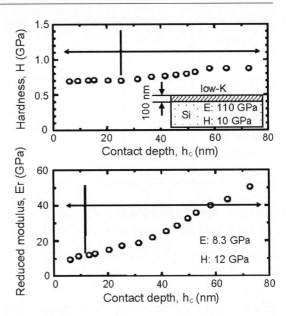

Fig. 36.4. Hardness and reduced modulus versus contact depth for an ultrathin low-k film having a thickness of 100 nm. (From Ye [38], with permission)

less thermal drift. Some measurements can now be made in a controlled environment or in a room with constant temperature and constant humidity. The indenting load, displacement noise flows and thermal noise flows are designed and controlled within a range of less than 50 nN, 0.1 nm and 0.05 nm/s, respectively. This higher performance has made it possible to reduce nanoindentation measurement error to an ultralow level, with standard deviation of less than 3% and accuracy to within 5% [22].

Generally, the indenting displacement has to be controlled to less than 10% of the film thickness to prevent an overestimate of the modulus because of stiffening from the substrate. The ultralow level of measurement deviation and higher accuracy allow specimens to be indented with a displacement of several nanometers [11]. This makes it possible to estimate the hardness and modulus of a thin film that is less than 100 nm thick [11, 38]. Figure 36.4 shows the hardness and reduced modulus versus contact depth for a softer film of low-dielectric materials with a thickness of 100 nm on a harder Si substrate. The measured reduced modulus value increased owing to the influence of the substrate when the contact depth exceeded 10% of the film thickness, while the hardness value increased when the contact depth exceeded 20% of the film thickness.

36.4.2.5
Relationship with Conventional Vickers Hardness

Conventional Vickers hardness (HV) is defined in terms of mass per unit contact surface area and is typically expressed in kilograms per square millimeter [45]. If the real contact area remains constant during indentation, indentation hardness H (GPa) can be converted geometrically to HV (kg/mm^2) as [23, 45]

$$HV = 94.59H \ . \tag{36.18}$$

Sample	H (GPa)	HV0.1
Gold	0.36	32
Copper	0.55	42
Brass	1.3	104
Cu-Be Alloy	2.5	193
Fused silica	9.5	700
Tool Steel	12.0	906
Silicon Nitride	22.9	1760

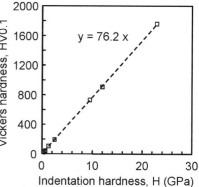

HV0.1 = 76.2 × H

H = 1.3E-02 × HV0.1

Fig. 36.5. Convertible relationship between indentation hardness H (GPa) and Vickers hardness HV (kg/mm^2) based on indentation test results. (From Ye [46], with permission)

However, as mentioned already, the indentation hardness is defined by dividing the maximum applied load P_{max} by the contact area under load A. A is obtained from the stiffness of the unloading curve and it is not equal to the residual area in the measurement of Vickers hardness. Thus, it is meaningless to obtain HV from H by just considering the geometric difference between the projected area and the surface area of the Vickers indenter. Because there is no residual area remaining after indenting in the range of elastic deformation, Vickers hardness is not especially meaningful in this range except as an indication of indentation hardness with respect to elastic indentation. Ye et al. made some large indents with a Berkovich indenter under the standards used for the HV measurement [46]. As shown in Fig. 36.5, they found that the relationship between HV and H is linear for deep indentations, in which the microstructure did not show any influence on the value of the hardness. A coefficient of 76.2 instead of 94.59 is a reasonable value for converting HV to H as follows

$$\mathrm{HV} = 76.2 H , \qquad (36.19)$$

when a Berkovich indenter is used at high load.

36.4.3
Recent Applications

36.4.3.1
High-Temperature Nanoindentation

Generally, nanoindentation measurements for evaluating the mechanical properties of thin films are restricted to room temperature. The rigorous accuracy required for loading in the micronewton range and displacing the indenter on a nanometer scale makes it difficult to perform nanoindentation measurements at high temperature. In recent years, however, several attempts have been made to evaluate the temperature dependence of the hardness and modulus of thin films [47–49]. Schuh et al. [49] have presented a technique for high-temperature nanoindentation studies at elevated temperatures up to 400 °C in the millinewton loading range. They measured the hardness and Young's modulus of fused silica as a function of temperature from 23 to 405 °C and demonstrated quantitative agreement with literature data for these properties. In the ultralow micronewton loading range, Ye et al. [48] undertook the challenge of indenting thin films on a nanometer scale at elevated temperatures up to 200 °C. In their indentation system, a foam insulator plate and a water-cooling system were used against heat convection from the heating stage; and a Macor insulating holder was used to prevent heat conduction from the diamond stylus to the transducer used for loading and detecting the displacement of the indenter. Measurement reliability, examined by using thermomechanically stable fused silica and a SiO_2 film, revealed no thermal load drift or noise that affected measurement accuracy at high temperature and also no thermal stress that affected the hardness and modulus values. These reliability estimation results indicate that the high-temperature nanoindentation measurement method can become a powerful tool for estimating the thermomechanical strength of thin films.

High mechanical strength as well as good thermal and chemical stability at high temperature is a critical property for film materials. In the semiconductor field, low-k dielectrics have attracted widespread interest for use as intermetal dielectric materials to reduce interconnect resistance in ultra-large-scale integrated devices. However, moisture absorption, thermal decomposition and other factors such as thermostructural changes or thermal stress may lower the mechanical properties of low-k films at high temperature, resulting in thermal deterioration or fracture [48].

Polyarylene ether (PAE) is one well-known organic low-k material. As shown in Fig. 36.6a and b, thermogravimetry–differential thermal analysis revealed that a low-k PAE film (500-nm thickness) grown on a Si (100) wafer had good heat resistance, maintaining a stable temperature up to 400 °C. However, thermal desorption spectroscopy showed that the PAE film had some unstable components in a temperature range below 200 °C; traces of H_2, O_2 and H_2O gases began to evolve from the specimen with increasing temperature from 50 °C and traces of hydrocarbon gases such as CH_3, C_2H_4, C_3H_7 and C_4H_9 evolved from 110 °C. In contrast, Raman and IR spectra revealed no significant difference in the composition and configuration between the specimens before and after heating; the PAE low-k film was thermally stable and suffered no thermal deterioration at temperatures below 200 °C.

These analytical results led researchers to investigate if these physical absorbents in the low-k film affect thermomechanical properties. Figure 36.6c and d shows the temperature dependence of the hardness and modulus of the PAE low-k film and a stable SiO_2 film determined simultaneously under heating and cooling conditions in a temperature range from room temperature to 200 °C. Measurement reliability was confirmed by obtaining constant hardness and modulus values for the SiO_2 film. During heating, the hardness and modulus of the low-k film increased slowly from room temperature and then sharply from 80 °C, reaching their maximum values at 115 °C, after which they abruptly decreased and finally reached their minimum values at 200 °C. These results indicate that the variations in hardness and modulus during heating were attributable to moisture absorption in the lower temperature range (from

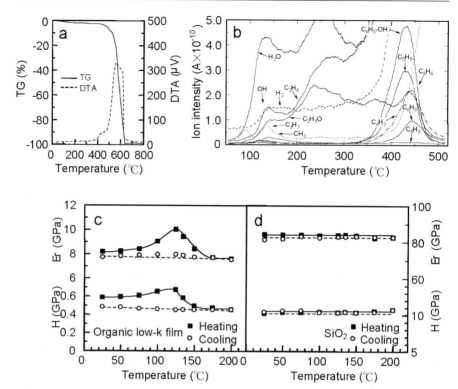

Fig. 36.6. High-temperature nanoindentation results for polyarylene ether low-k film. **a** Thermogravimetry (*TG*)–differential thermal analysis (*DTA*) profiles, **b** thermal desorption spectroscopy profiles, **c** temperature dependence of the hardness and reduced modulus under heating and cooling conditions and **d** simultaneously measured hardness and reduced modulus values of a stable SiO_2 film. (From Ye et al. [48], with permission)

room temperature to 115 °C) and physical absorption of some organic residuals in the higher temperature range (115–200 °C). They suggest that absorbed water generally acts as a plasticizer to lower the mechanical properties of low-k polymers and that the heating process likely desorbed moisture, resulting in the sharp increase in hardness and modulus, while the heating process also evolved hydrocarbon gases, causing the abrupt decrease in hardness and modulus. In contrast, the hardness and modulus remained almost constant during cooling. This suggests that no moisture absorption occurred in the low-k film.

From another point of view, Ye et al. [48] focused on the fact that the hardness and modulus of the low-k film during cooling and the SiO_2 film during heating and cooling remained virtually constant. They suggested that mechanical strength is less influenced by thermal stress. Generally, an organic low-k film possesses a larger thermal expansion coefficient than a Si substrate, so restraint from the substrate induces a tensile stress distribution in the as-grown film at room temperature. Hardness and modulus ordinarily increase under heating owing to a decrease in tensile stress or, inversely, they decrease under cooling. In this experiment, however, a decrease in hardness and modulus during cooling was not observed, although the evolution of the hydrocarbon gases may have facilitated tensile stress relief at 200 °C.

36.4.3.2
Continuous Stiffness Measurement

According to the approach of Pharr et al., the reduced modulus has to be obtained from the measured contact stiffness. In a single indentation experiment, the contact stiffness is obtained only from the unloading curve at the maximum depth of the indent; thus, many indents are needed to evaluate the depth distribution of hardness and modulus. In continuous stiffness measurement (CSM), a small oscillated load is superimposed upon ramp loading and the stiffness S and damping D_s of the contact at all displacements are given by

$$S = \left(\frac{1}{\frac{P_0}{h_0} \cos\phi - (K_s - m\omega^2)} - K_f^{-1} \right)^{-1}, \quad \text{and} \quad D_s\omega = \frac{P_0}{h_0} \sin\phi - D_i\omega,$$

(36.20)

where P_0 is the amplitude of the load oscillation, h_0 is the resulting displacement amplitude, ϕ is the phase angle between the load and displacement, K_s is the stiffness of the indenter support springs, K_f is the stiffness of the load frame, m is the mass of the indenter column and D_i is the damping of the indenter [10, 23, 50, 51]. The system's mechanical parameters K_f, K_s, D_i and m can be determined with the area function or a dynamic calibration procedure. By measuring P_0, h_0 and ϕ, one can simultaneously measure the contact stiffness during loading in an indentation test, making it possible to measure the depth distribution of hardness and modulus with one indent. By changing the radial frequency of the load oscillation, one can also apply CSM to evaluate creep and viscoelastic properties and other mechanical characteristics such as fatigue, which are described in Sects. 36.7 and 36.8.

The thin-film rigid disks that are widely used as magnetic media have a multilayered structure, consisting of an ultrasmooth and flat disk substrate on which 25–50-nm-thick metallic magnetic films are deposited along with a 3–5-nm-thick diamond-like carbon overcoat and a 1–2-nm-thick bonded perfluoropolyether lubricant [29]. The diamond-like carbon coating and absorbed organic lubricant layer are used not only to protect the magnetic layer but also to improve tribological performance. The mechanical properties in the depth distribution of this nanometer-scale multilayered structure can affect its magnetic and tribological performance. Li and Bhushan [29] used the CSM technique to evaluate the hardness and modulus of a magnetic rigid disk. Figure 36.7 shows the contact stiffness, elastic modulus and hardness as a function of the contact depth for the magnetic rigid disk with a multilayered structure. The elastic modulus values of the different layers are comparable, resulting in low interfacial stresses. The underlayer exhibited a lower elastic modulus than the magnetic layer and the Ni–P layer on either side. The hardness values decrease with increasing contact depth. They reported that the hardness and modulus of ultrathin magnetic multilayered structures can be easily measured by the CSM technique and that the values are in good agreement with the results obtained by conventional nanoindentation methods.

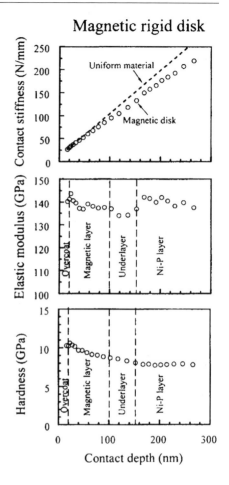

Fig. 36.7. Contact stiffness, elastic modulus and hardness versus contact depth of a magnetic rigid disk having a multilayered structure as determined by continuous stiffness measurement. (From Li and Bhushan [29], with permission)

36.5
Yield Stress and Modulus Measurements

36.5.1
Analysis Method

The tip shape of pyramidal indenters is generally useful for making elastoplastic indents to determine hardness independent of the indentation depth in the surface range. In contrast, spherical indenters easily maintain elastic contact with specimens before the indentation displacement enters the elastoplastic deformation range. It is considered that spherical indentation may make it possible to obtain a critical yield point from the loading curve where the elastic deformation of the contact area changes to elastoplastic deformation. From this viewpoint, Ye et al. [11,12] presented a conventional spherical indentation method for determining the yield stress of ultrathin films based on the Hertzian contact solution in the elastic deformation range and the Tresca yield criterion.

It is well known that ductile materials such as metals yield when the maximum shear stress reaches its critical limit according to the Tresca yield criterion [2,27]. In cases where contact between two bodies of revolution occurs along their symmetry axis, the largest principal shear stress τ_1 is given by

$$\tau_1 = \frac{1}{2}|\sigma_z - \sigma_r|, \tag{36.21}$$

where $\sigma_z, \sigma_r = \sigma_\theta$ are the principal stresses along the axis. By replacing these principal stresses in (36.21) with the stresses along the z-axis given by Johnson for Hertzian contact and letting the derivative of τ_1 with respect to z equal zero, the maximum shear stress τ_{max} and its position z at Poisson's ratio $\nu = 0.3$ are

$$\tau_{max} = 0.31 p_m, \tag{36.22}$$
$$z = 0.47 R, \tag{36.23}$$

where p_m is the maximum contact pressure of indentation and R is the tip radius of a special indenter. If indentation displacement h is much smaller than the tip radius R, p_m can be expressed by

$$p_m = \frac{P_m}{A} = \frac{P_m}{\pi R h}, \tag{36.24}$$

where A is the contact area and P_m is the maximum contact load.

Therefore, by calibrating the tip radius R and measuring the maximum contact load P_y and displacement h_y at the critical yield point from the loading curve, one can determine the shearing yield stress τ_y by substituting (36.24) into (36.22), and the reduced modulus E_r in the elastic deformation range can obtained by rearranging (36.6):

$$\tau_y = 0.31 p_y = 0.31 \frac{P_y}{\pi R h_y}, \tag{36.25}$$

$$E_r = \frac{3 P_y}{4 h^{3/2} R^{1/2}}. \tag{36.26}$$

As for the calibration of the tip radius R, by making some indents in a reference material with a known reduced modulus such as fused silica ($E_r = 69.7\,\text{Ga}$) in the elastic deformation range, one can determine R from the Hertzian solution using (36.6). Otherwise, the tip radius R can also be directly measured by atomic force microscopy. With regard to detecting the critical yield point from the loading curve, the values of P_y and h_y can be obtained by extrapolating the indentation pressure from the load–displacement hysteretic loop energy, which is defined as the energy U_r enclosed within the indentation loading and unloading curves. As shown in Fig. 36.8a, in the elastic deformation range, the unloading curve overlaps the loading curve and the hysteretic loop energy $U_r = 0$. When the load is increased to the elastoplastic deformation range, the unloading curve deviates from the loading curve and the hysteretic loop energy $U_r > 0$. U_r is the irreversible energy consumption associated with plastic deformation. Figure 36.8b shows a plot of U_r versus p_m for

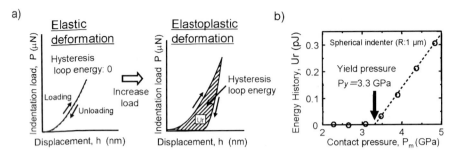

Fig. 36.8. a Change in nanoindentation load–displacement curve from elastic deformation to elastoplastic deformation showing the hysteretic loop energy as irreversible energy consumption associated with plastic deformation. **b** Hysteretic loop energy versus maximum contact load of fused silica for determining critical yield contact load by extrapolation of hysteretic loop energy

fused silica. The critical yield point is found by extrapolating U_r in the range $U_r > 0$ and the yield contact load p_y is the intercept of the plot. In this case, the yield contact load of fused silica is $p_y = 3.3$ GPa, resulting in shearing yield stress of $\tau_y = 1.1$ GPa.

36.5.2
Recent Applications

The spherical nanoindentation method based on Ye's approach is used for evaluating the reduced modulus in the elastic deformation range and the yield stress from the initial plastic deformation. Since this approach induces elastic deformation by making some small indents within a depth of less than several nanometers, it is possible to apply this method to ultrathin films with a thickness of about 20 nm [11, 12]. According to (36.23), the depth of the maximum shear stress can be changed by varying the tip radius of a spherical indenter. It is feasible to evaluate the yield stress relative to the specimen depth by using various indenters with different tip radii. This advantage has gradually attracted attention not only from the standpoint of industrial utility, but also with regard to scientific understanding.

In the automobile field, for example, improvement of engine oil performance is particularly important because of its critical role in reducing friction, which translates directly into better fuel economy. One effective method of modifying engine oil containing the widely used zinc dialkylsithiophosphate (ZDDP) additive is to add the molybdenum dithiocarbamate (MoDTC) friction modifier [11]. These engine oil additives form tribofilms with a thickness of less than several tens of nanometers on steel surfaces. A MoDTC/ZDDP tribofilm originating from both the MoDTC and the ZDDP additives possesses a much lower friction coefficient than a ZDDP tribofilm formed only from the ZDDP additive. This difference in friction behavior was not well known before, and two main explanations for it were suggested in previous studies. One explanation attributed it to a difference in mechanical properties near the tribofilm surface, and the other to a difference in surface roughness.

Ye et al. [11] applied a spherical diamond indenter with a tip radius of 150 nm to these ultrathin films. They examined the difference in yield stress between these tribofilms and found that the low friction behavior does not originate from a difference in surface roughness but from a difference in shear stresses near the surface. Figure 36.9 shows the indentation load–displacement curves of the MoDTC/ZDDP and ZDDP tribofilms, where the dark and gray circles denote the loading and unloading data. It was found that the initial plastic deformation began to appear at $P_y = 6\,\mu N$ for the MoDTC/ZDDP tribofilm and at $P_y = 11\,\mu N$ for the ZDDP tribofilm. The loading and unloading data were traced as $P \sim h^{3/2}$ and $P \sim (h - h_p)^{3/2}$ by the least-squares method, where P, h and h_p are the indentation load, displacement and plastic depth, respectively. Reliability factors for all the approximated curves were in the range 0.95–0.99. As a result, the MoDTC/ZDDP tribofilm was calculated to possess shearing yield stress of $\tau_y = 2.3\,\text{GPa}$ at a depth of $z_y = 9.3\,\text{nm}$; while the ZDDP tribofilm had $\tau_y = 3.3\,\text{GPa}$ at $z_y = 10.4\,\text{nm}$. Thus, the MoDTC/ZDDP tribofilm was demonstrated to possess lower yield stress than the ZDDP tribofilm. By combining this finding with other chemical analysis results allowed the conclusion to be drawn the friction reduction due to the MoDTC/ZDDP additives originates from an inner skin layer formed by MoS_2 nanostrips just below the surface at a depth of about 10 nm.

The yield stress and reduced modulus on a nanometer scale are also well known to be sensitive to changes in microstructures in the miniaturization processes for ob-

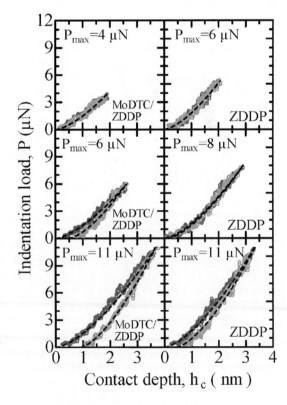

Fig. 36.9. Indentation load–displacement curves for molybdenum dithiocarbamate (*MoDTC*)/zinc dialkylsithiophosphate (*ZDDP*) and ZDDP tribofilms at different loads under various conditions from elastic deformation to elastoplastic deformation, where the *circles* denote the loading and unloading data and the *dotted lines* are plotted for $P \sim h^{3/2}$ and $P \sim (h - h_p)^{3/2}$. (From Ye et al. [11], with permission)

taining high-performance materials. However, macroscale mechanical data obtained from bulk materials have still been applied in developing various new nanoscale materials in industry. In the semiconductor field, for example, although copper is one of the most important materials for reducing interconnect resistance in ultra-large-scale integrated devices, mechanical and thermomechanical reference data of bulk copper materials are still used in computer simulations for device design and life prediction. Shimizu et al. [12] made spherical indents in single-crystal Cu(100) and Cu(111) by using a spherical diamond indenter with a tip radius of 1 μm. They found that the Cu single crystals possessed an anisotropic reduced modulus and yield stress owing to the different crystallographic orientations.

Figure 36.10 shows $P^{2/3}$ versus h plots and U_r versus p_m plots for these two different kinds of single crystals. The $P^{2/3}$ versus h plots were confirmed to have a linear relationship in accord with the Hertzian contact theory. The U_r versus p_m plots show that the Cu(111) plane exhibited initial plastic deformation at $p_m = 0.94$ GPa and that the Cu(100) plane exhibited initial plastic deformation at $p_m = 1.78$ GPa. As a result, the Cu(111) plane was estimated to have a reduced modulus of $E_r = 99$ GPa and shearing yield stress of $\tau_y = 554$ MPa, both of which were smaller than the values of $E_r = 68$ GPa and $\tau_y = 291$ MPa estimated for the Cu(100) plane. These anisotropic mechanical properties agreed with the metallurgical considerations of an active face-centered-cubic slip system. As shown in Fig. 36.10d, when the conventional compression test of bulk materials is assumed at a very low penetration depth

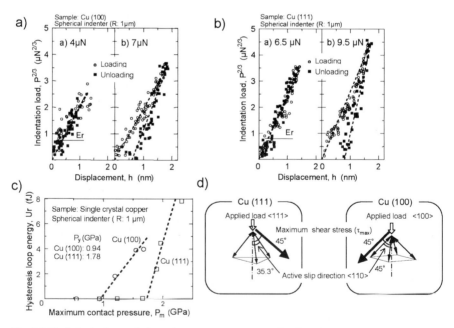

Fig. 36.10. Spherical nanoindentation measurement results for single-crystal Cu(100) and Cu(111). **a, b** $P^{2/3}$ versus h, **c** U_r versus p_m and **d** anisotropic nature of mechanical properties showing the relationships among applied load, maximum shear stress and active slip system

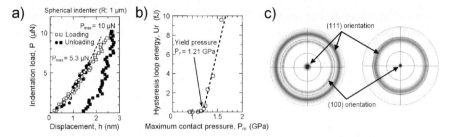

Fig. 36.11. Spherical nanoindentation measurement results for Cu film with a thickness of 200 nm. **a** $P^{2/3}$ versus h, **b** U_r vs. p_m and **c** pole figures of X-ray diffractions along the [111] and [100] directions. (From Shimizu et al. [12], with permission)

compared with the radius of a spherical indenter, the direction of maximum shear stress is oriented at 45° with respect to the direction of the applied load. The active slip direction for the Cu(100) and Cu(111) single crystals was oriented at an angle of 45° and 35°, respectively. This anisotropic nature of yield stress is generated from the gap between the active slip direction and the maximum shear stress direction. Such anisotropic mechanical properties have never been seen on a macrometer scale.

On the basis of these results, Shimizu et al. [12] also applied the spherical nanoindentation method to a copper thin film with a thickness of 200 nm. Figure 36.11 shows the $P^{2/3}$ versus h curves and the plots of U_r and P_m. Elastic behavior was exhibited under a maximum indentation load of 5.3 µN. The copper thin film exhibited initial plastic deformation in the hysteretic loop energy versus contact pressure curve at 1.21 GPa. This copper film displayed a reduced modulus of $E_r = 66.5$ GPa and shearing yield stress of $\tau_y = 375$ MPa. The reduced modulus value was nearly the same as that obtained for the Cu(100) plane, and shearing yield stress was found to be between that of the Cu(100) plane and that of the Cu(111) plane. The X-ray diffraction pattern was used to investigate these mechanical properties relative to the crystallographic orientation in the copper film. They found that this copper film was grown with [100] and [111] orientations as shown in Fig. 36.11c. The elastic deformation of the copper film was dominated by the grains with the [100] orientation such that it occurred in the most compliant direction. In contrast, the plastic deformation of the film was restrained by the grains with the [111] orientation, thus resulting in the highest yield stress. These results indicate that it is necessary to replace the mechanical properties of bulk materials with the reduced modulus and yield stress of thin films or small volumes of materials for developing new materials on a nanometer scale.

36.6
Work-Hardening Rate and Exponent Measurements

36.6.1
Analysis Method

Tabor [2] demonstrated that hardness can be related to plastic stress for a given representative strain. As shown in Fig. 36.11, however, hardness represents the mean value of the contact pressure between the indenter and the material. It is not an intrinsic property of materials. For determining the plastic properties of small-volume materials, it is necessary to find a way to determine the true stress–strain relationship from the indentation load-displacement curve. Recently, Bucaille et al. [13], Chollacoop et al. [14] and Ogasawara et al. [15] developed a dual-indenter method for determining the stress–strain curve based on a finite-element analysis of the influence of the included angle of conical indenters. Cheng and Cheng [16] presented a basic idea for using dimensional analysis and finite-element calculations to extract closed-form universal functions. Dao et al. [17] proposed a representative plastic strain that can be applied as a strain level to make a dimensionless form for indentation loading response independent of the strain hardness exponent. Bucaille et al. [13] and Chollacoop et al. [14] found that this representative strain changes linearly with the apex angle of the indenter. Thus, the stress–strain curve can be obtained by determining the work-hardening rate and exponent using dual sharp indenters with different apex angles. Lately, Ogasawara et al. [15] further advanced the approach suggested by Bucaille et al. [13] and Chollacoop et al. [14], and proposed a new formulation of representative strain that exhibits a stronger physical basis than the others. Because of its wide application range to engineering materials, the approach proposed by Ogasawara et al. is described here.

Figure 36.12 schematically shows the typical stress–strain curve of power-law materials [53]. If the elasticity observes Hooke's law and the plasticity follows the von Mises yield criterion and power-law hardening, the equivalent stress and strain under equibiaxial loading of axisymmetric conical indentation are given by

$$\sigma_a = E(2\varepsilon_a) \qquad \text{for } 2\varepsilon_a \leq \frac{\sigma_y}{E}, \qquad (36.27)$$

$$\sigma_a = R(2\varepsilon_a)^n \qquad \text{for } 2\varepsilon_a \geq \frac{\sigma_y}{E}, \qquad (36.28)$$

where σ_a is the equibiaxial stress and ε_a is the equibiaxial strain, when E is Young's modulus, σ_y is the initial yield stress, n is the work-hardening exponent and R is the work-hardening rate. n is zero for an elastic–perfectly plastic material. For most

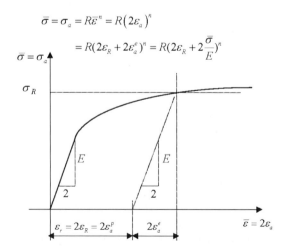

Fig. 36.12. The equibiaxial elasto-plastic stress–strain curve of power-law materials and the representative strain defined as plastic strain. (From Ogasawara et al. [53], with permission)

metals and alloys n is between 0.1 and 0.5. For continuity at yielding, the following condition must hold true:

$$\sigma_y = E\varepsilon_y = R\varepsilon_y^n . \tag{36.29}$$

The equibiaxial strain can be expressed as the summation of axisymmetric elastic stress ε_a^e and plastic stress ε_a^p:

$$\varepsilon_a = \varepsilon_a^e + \varepsilon_a^p . \tag{36.30}$$

Most importantly, (36.27) and (36.28) reveal that the plastic relationship for a given work-hardening material is identical to the uniaxial stress–strain curve when the equibiaxial σ_a is plotted as a function of $2\varepsilon_a$. Ogasawara et al. defined the equibiaxial plastic strain as a representative strain ε_R expressed as

$$\varepsilon_R \equiv \varepsilon_a^p = \varepsilon_a - \varepsilon_a^e , \tag{36.31}$$

and thus the corresponding representative stress σ_R at ε_R can be expressed as

$$\sigma_R \langle \varepsilon_R \rangle = R(2\varepsilon_a^e + 2\varepsilon_R)^n , \tag{36.32}$$

where elastic stress ε_a^e, as shown in Fig. 36.12, can be obtained from [15]

$$\varepsilon_a^e = \frac{\sigma_R \langle \varepsilon_R \rangle}{2E} . \tag{36.33}$$

On the other hand, indentation by a sharp indenter with an included angle α at a load P into a power-law elastoplastic specimen (E_r, ν, σ_y and n) can be expressed as

$$P = P(h, E_r, \sigma_y, n, \alpha) , \tag{36.34}$$

where h is the indentation displacement and E_r is the reduced Young's modulus. By using the stress $\sigma_R \langle \varepsilon_R \rangle$ at the representative strain ε_R, one can rewrite the general indentation loading response on constitutive properties as

$$P = P(h, E_r, \sigma_R \langle \varepsilon_R \rangle, n, \alpha) . \tag{36.35}$$

Based on dimensional analysis, the relationship in (36.35) can be expressed as

$$P = \sigma_R \langle \varepsilon_R \rangle h^2 \Pi_\alpha \left(\frac{E_r}{\sigma_R \langle \varepsilon_R \rangle}, n, \alpha \right) , \tag{36.36}$$

where Π_α is a dimensionless function. According to Mayer's law in (36.5), the indentation displacement responds to loading in the case of conical indentation as

$$P = Ch^2 , \tag{36.37}$$

where C is the loading curvature. Thus, the relationship between the loading curvature C and the dimensionless function can be written as

$$C = \frac{P}{h^2} = \sigma_R \langle \varepsilon_R \rangle \Pi_\alpha \left(\frac{E_r}{\sigma_R \langle \varepsilon_R \rangle}, n, \alpha \right) . \tag{36.38}$$

For determining a representative strain ε_R that can construct a function Π_α independent of the strain-hardening exponent n, the finite element method (FEM) is generally applied to several kinds of conical shape indents with a different included angles α. Ogasawara et al. used material parameters over a large range with $E_r/\sigma_R\langle\varepsilon_R\rangle = 3-3300$ and $n = 0-0.5$, which can cover essentially all engineering materials. A rigid contact surface is assumed to simulate a rigid indenter. Coulomb friction between contact surfaces as a minor factor of indentation is assumed to be 0.15 [54]. Poisson's ratio is fixed at 0.33 [55]. As a result, for a conical Berkovich indenter with an included angle α of 70.3°, the representative stain ε_R is 0.0115, making the relationship between $C_{70.3°}/\sigma_R\langle 0.0115\rangle$ and $E_r/\sigma_R\langle 0.0115\rangle$ independent of the work-hardening exponent n as shown in Fig. 36.13. Thus, (36.38) for $\alpha = 70.3°$ becomes

$$\frac{C_{70.3°}}{\sigma_R\langle 0.0115\rangle} = \Pi_{70.3°}\left(\frac{E_r}{\sigma_R\langle 0.0115\rangle}\right), \tag{36.39}$$

$$\Pi_{70.3°}(\Omega_{70.3°}) = -0.6596(\ln \Omega_{70.3°})^3 + 8.4058(\ln \Omega_{70.3°})^2 \\ - 12.3088(\ln \Omega_{70.3°}) + 9.2102, \tag{36.40}$$

where $\Omega_{70.3°} = E_r/\sigma_R\langle 0.0115\rangle$. In the same way, the representative stain ε_R for $\alpha = 63.14°$ is 0.0162 and the relationship between $C_{63.14°}/\sigma_R\langle 0.0162\rangle$ and $E_r/\sigma_R\langle 0.0162\rangle$ is

$$\frac{C_{63.14°}}{\sigma_R\langle 0.0162\rangle} = \Pi_{63.14°}\left(\frac{E_r}{\sigma_R\langle 0.0162\rangle}\right), \tag{36.41}$$

$$\Pi_{63.14°}(\Omega_{63.14°}) = -0.3093(\ln \Omega_{63.14°})^3 + 3.6164(\ln \Omega_{63.14°})^2 \\ - 2.5183(\ln \Omega_{63.14°}) + 2.3622, \tag{36.42}$$

where $\Omega_{63.14°} = E_r/\sigma_R\langle 0.0162\rangle$. For $\alpha = 75.79°$, the representative stain ε_R is 0.0079 and the relationship is

$$\frac{C_{75.79°}}{\sigma_R\langle 0.0079\rangle} = \Pi_{63.14°}\left(\frac{E_r}{\sigma_R\langle 0.0079\rangle}\right), \tag{36.43}$$

$$\Pi_{75.79°}(\Omega_{75.79°}) = -1.3157(\ln \Omega_{75.79°})^3 + 17.945(\ln \Omega_{75.79°})^2 \\ - 34.8958(\ln \Omega_{75.79°}) + 25.458, \tag{36.44}$$

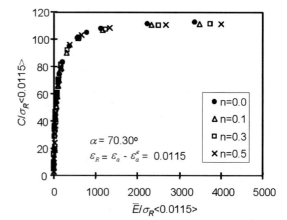

Fig. 36.13. Relationship between $C_{70.3°}/\sigma_R\langle 0.0115\rangle$ and $E_r/\sigma_R\langle 0.0115\rangle$ independent of the work-hardening exponent. (From Ogasawara et al. [53], with permission)

where $\Omega_{75.79} = E_r/\sigma_R \langle 0.0079 \rangle$. These FEM analytic results show that the representative stain ε_R can vary linearly with $\cot \alpha$ as follows:

$$\varepsilon_R = 0.0319 \cot \alpha \ . \tag{36.45}$$

In the plastic region, as shown in (36.27) and (36.28), the work-hardening exponent n and the work-hardening rate R are independent parameters. Two relative equations between $C_\alpha/\sigma_R \langle \varepsilon_R \rangle$ and $E_r/\sigma_R \langle \varepsilon_R \rangle$ made by dual conical indentations with different included angles are enough to obtain the stress–strain curve from the indentation load–displacement curves for power-law materials.

On the basis of the dual-indentation approach explained here, in order to determine the true stress–strain curve, several indents are first made to obtain the loading curvatures C_α from the load–displacement curves using two conical indenters with a different included angle α. The reduced Young's modulus E_r can be determined by Ye's approach using spherical indentation as explained in Sect. 36.5 or by the approach of Pharr et al. using Berkovich indentation as described in Sect. 36.4. By substituting two sets of C_α–E_r pairs into their relative equations between $C_\alpha/\sigma_R \langle \varepsilon_R \rangle$ and $E_r/\sigma_R \langle \varepsilon_R \rangle$ such as (36.40), (36.42) and/or (36.44), one can calculate the representative strains $\sigma_R \langle \varepsilon_R \rangle$ at the corresponding representative strain ε_R for each conical indentation by reverse analysis. Finally, the two sets of $\sigma_R \langle \varepsilon_R \rangle$–$\varepsilon_R$ pairs thus obtained can be transformed into stress–strain (σ_R–ε_R) curves by determining the work-hardening exponent n, work-hardening rate R, and yield stress σ_y from (36.29), (36.32) and (36.33).

36.6.2
Practical Application Aspects

The dual-indenter method has been developed on the basis of materials whose properties follow an exact power-law relationship as expressed in (36.27) and (36.28). Ogasawara et al. [15] applied this method to four materials: gold, aluminum, work-hardened copper and annealed copper. Figure 36.14 shows their σ–ε curves obtained by reverse analysis in comparison with the original input data. They found that the errors of the σ–ε curves between the results of the indentation measurements and the original data were less than 3% for the power-law materials gold and aluminum. In contrast, for real engineering materials such as work-hardened copper and annealed copper, their σ–ε curves measured in uniaxial tensile tests deviated slightly from the ideal power-law hardness. Thus, some inconsistency was observed between the reverse analysis data and the original data. Recently, some studies have been made of non-power-law materials with the aim of further improving the consistency between the original input data and reverse algorithm [56, 57]

As for the influence of the friction coefficient on the indentation load–displacement curve, friction between the indenter and the surface of the material is usually discussed in terms of the Coulomb friction coefficient μ. The classical value of the friction coefficient between metal and diamond is 0.15. The effect of friction on the normal load is

$$P(\mu) = P(0)(1 + \mu \cot \alpha) \ , \tag{36.46}$$

where $P(0)$ is the normal load without the effect of friction and α is the included angle of the conical indenter [2]. Bucalille et al. [13] simulated the indentation of an

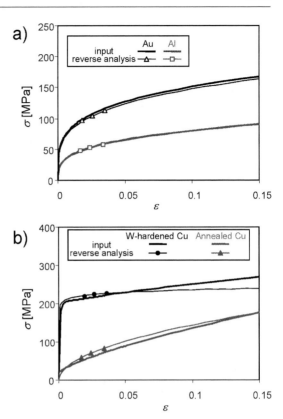

Fig. 36.14. σ–ε curves for gold, aluminum, work-hardened copper and annealed copper obtained by the dual-indenter method in comparison with original input data. (From Ogasawara et al. [15], with permission)

aluminum alloy made with four conical indenters having different included angles from 42.3° to 70.3° for varying values of the Coulomb friction coefficient from 0 to 0.3. As shown in Fig. 36.15, they found that a higher friction coefficient decreased the load slightly at $\alpha = 70.3°$ and increased it at $\alpha = 60°$. The normal load varied less than 3% as the friction coefficient was increased from 0 to 0.3 for included angles of more than 60°. They also suggested that the accuracy of determining the work-hardening exponent n is dependent on the included angle of the indenter and the magnitude of the work-hardening exponent. Better accuracy is obtained for a smaller included angle and for higher values of the work-hardening exponent. Therefore, the influence of friction is a minor factor when the included angle of the indenter is larger than 60°, but it has to be considered when an indenter with a small angle is used for obtaining high accuracy in evaluating the work-hardening exponent.

Another advantage of the dual-indenter method is that the stress–strain curves derived from the indentation data can be evaluated without having to measure the projected contact area. For work-hardened metals, piling-up or sink-in may form around the indenter, resulting in changes in the contact depth and area. Thus, the influence of piling-up and sink-in can affect hardness and modulus estimations made by using the approach of Pharr et al. With the dual-indenter method, however, only the loading curvature C of the indentation load–displacement is taken into account.

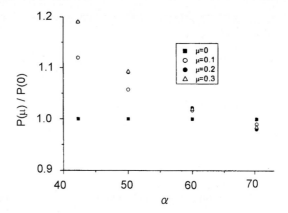

Fig. 36.15. Influence of the Coulomb friction coefficient μ on applied indentation load P as a function of the included angle α for an aluminum alloy. (From Bucaille et al. [13], with permission)

The dual-indenter method is still a little difficult to use, although this approach can determine plastic properties with high accuracy. Certain disadvantages of this method are expected to be overcome through further improvement. One is that this method requires two kinds of indents with different included angles on the same specimen. It is inconvenient for practical use in industry, and sometimes this method cannot be applied when only one small local area or point is of interest. Another disadvantage is that ideally shaped conical indenters must be used in the massive FEM calculations for obtaining the relationship between the loading curvature C and the dimensionless function. Indenters invariably have some roundness on the top. The dimensionless function in the case of a spherical indenter has to be calculated when small indents are made for evaluating the plastic properties of ultrathin films. Recently, some new methods have been proposed that involve the use of single indentation measurements instead of dual indentations, though further improvements are still needed [53].

36.6.3
Recent Applications

The dual-indenter method makes it possible to measure elastoplastic properties and tensile strength in ultrasmall regions and has the potential for broad application in various fields. However, only a few applications have been reported to date. Yonezu et al. [58] used two Berkovich indenters with different tip geometries of 100° and 115° to test various kinds of materials, including pure aluminum, aluminum alloys, titanium alloys, brass, stainless steel and high-strength steel, having a wide range of mechanical properties. They estimated not only the work-hardening rate R and the work-hardening exponent n, but also the yield stress σ_y and the tensile strength σ_B, and compared the values with the results of tensile loading tests. As shown in Fig. 36.16, the values of R, n, σ_y, and σ_B were almost equal to the actual values measured in the tensile loading tests for most of the materials. These results demonstrated that the dual-indenter method is a reliable tool for determining the stress–stain curve and ultimate tensile strength on a microscale. In their experiments, they made large indents to reduce or remove size effects from the specimen microstructure. For practical use, however, making small indents is important in obtaining elastoplastic

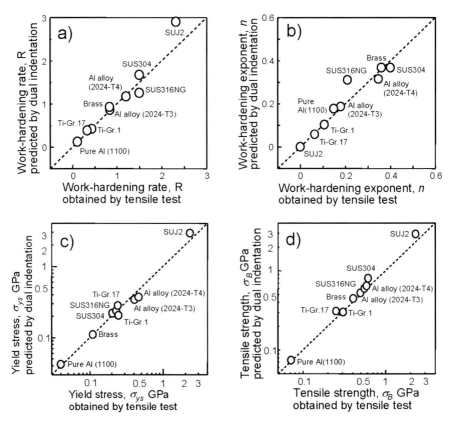

Fig. 36.16. Work-hardening rate, work-hardening exponent, yield stress and tensile strength of pure aluminum, aluminum alloy, titanium alloy, brass, stainless steel and high-strength steel obtained by the dual-indenter method in comparison with tensile test data. (From Yonezu et al. [58], with permission)

properties on a nanometer scale. Instead of macroscale mechanical data obtained from bulk materials, the nanoscale values of R, n, σ_y, and σ_B can be used to predict the strength and life of various new nanoscale materials in industry.

36.7
Viscoelastic Compliance and Modulus

36.7.1
Analysis Method

The methods and approaches mentioned earlier have been thoroughly validated for time-independent materials. However, experimental problems will arise when applying these methods and approaches directly to viscoelastic materials. Tang and Ngan [59] observed that the load–displacement curve shows a nose-shaped pattern during unloading owing to creep effects at the onset of unloading and that such a nose-shaped pattern can affect the measurement accuracy of the contact stiffness and contact area. Hardness, modulus and other mechanical properties can change with the loading rate and loading frequency, being independent of experimental time. In previous studies, efforts were made to evaluate viscoelastic properties using nanoindentation measurement. Oliver and Pharr [10] used load-sensing and displacement-sensing indentation techniques to determine Young's relaxation modulus. Cheng et al. [60] used a flat-punch indentation method to measure viscoelastic properties in the range of linear viscoelastic deformation. Shimizu et al. [61] and Lu et al. [18] proposed methods for measuring creep compliance as a function of elapsed time using spherical and Berkovich indenters. For soft and rheologic materials, it is preferable to express viscoelasticity in terms of complex properties as a function of frequency. Loubet et al. [62] applied an excited dynamic load or displacement to a specimen with a continuous stiffness modulus (CSM) and tried to obtain the complex modulus of viscoelastic materials. They proposed the uniaxial storage modulus in the same formula as in Sneddon's solution as shown in (36.8), which is related to the contact stiffness and contact area. The uniaxial loss modulus is represented in proportion to the damping coefficient and frequency, but in inverse proportion to the root of the contact area. However, this approach produced a result far from the value measured by conventional dynamic mechanical analysis (DMA) under the same specimen conditions. Recently, Huang et al. [19] developed a method for measuring the complex compliance in a frequency-dependent function using the Hertzian solution in combination with a hereditary integral operator proposed by Lee and Radok [20]. Because this approach exhibits better agreement with the conventional DMA method, the method of Huang et al. is described in this section.

The formulas for the computation of complex compliance and modulus are proposed for linear viscoelastic materials using spherical indentation measurement. When a rigid spherical indenter indents into a half-space composed of a homogenous, isotropic and linearly elastic material, the relationship between the applied indentation load P and the displacement h for the spherical indenter in elastic contact follows the Hertzian solution in (36.6) and (36.7), resulting in

$$P = \frac{8\sqrt{R}}{3(1-\nu)} G h^{3/2} , \qquad (36.47)$$

where G is the shear modulus, R is the tip radius and ν is Poisson's ratio. Since Poisson's ratio does not change significantly for most polymers in the glassy state, a constant Poisson's ratio is assumed. If the contact area between the indenter and the specimen is nondecreasing, the hereditary integral operator proposed by Lee and Radok can be applied to (36.47) and the relationship between the indentation load P and the displacement h can be expressed as

$$h^{3/2}(t) = \frac{3(1-\nu)}{8\sqrt{R}} \int_{-\infty}^{t} J(t-\theta) \frac{\mathrm{d}P(\theta)}{\mathrm{d}(\theta)} \mathrm{d}\theta , \qquad (36.48)$$

where $J(t)$ is the creep compliance function of elapsed time in shear [20]. If a sinusoidal nanoindentation load is superimposed in a step loading process as

$$P(t) = [P_\mathrm{m} + \Delta P_0 \sin(\omega t)] H(t) , \qquad (36.49)$$

where $H(t)$ is the Heaviside unit step function [$H(t < 0) = 0$, $H(t = 0) = 1/2$, $H(t > 0) = 1$], P_m is the carrier load and ΔP_0 is the amplitude of the harmonic load, the total displacement as output from a nanoindenter can be expressed as

$$h(t) = h_m(t) + \Delta h_0 \sin(\omega t - \delta), \qquad (36.50)$$

where $h_m(t)$ is the carrier displacement, Δh_0 is the amplitude of harmonic displacement and δ is the out-of-phase angle between the applied harmonic force and displacement. Usually, because $\Delta h_0 \ll h_m(t)$, (36.50) can be rewritten as

$$h^{3/2}(t) = h_m^{3/2}(t) + \frac{3}{2} h_m^{1/2}(t) \Delta h_0 \cos\delta \sin(\omega t) - \frac{3}{2} h_m^{1/2}(t) \Delta h_0 \sin\delta \cos(\omega t), \qquad (36.51)$$

where the high-order terms of Δh_0 are negligible. By substituting (36.49) into (36.48), one can express $h^{3/2}(t)$ as

$$h^{3/2}(t) = \frac{3(1-\nu)}{8\sqrt{R}} \left(P_m J(t) + \omega \Delta P_0 \int_0^t J(t-\theta) \cos(\omega\theta) \, d\theta \right), \qquad (36.52)$$

Considering that the complex compliance is defined after the harmonic response has reached a steady state when t tends to infinity, (36.48) can be transformed into

$$h^{3/2}(t) = \frac{3(1-\nu)}{8\sqrt{R}} \{ P_m J(t) + \Delta P_0 [J'(\omega) \sin(\omega t) - J''(\omega) \cos(\omega t)] \}, \qquad (36.53)$$

$$J'(\omega) = \omega \int_0^\infty J(t) \sin(\omega t) \, dt, \text{ and } J''(\omega) = \omega \int_0^\infty J(t) \cos(\omega t) \, dt, \qquad (36.54)$$

$$J^*(\omega) = J'(\omega) - i J''(\omega), \qquad (36.55)$$

where $J'(\omega)$ and $J''(\omega)$ are the storage compliance and loss compliance in shear and $J^*(\omega)$ is the complex compliance in shear.

By comparing (36.51) with (36.53), h_m, $J'(\omega)$ and $J''(\omega)$ are

$$h_m^{3/2}(t) = \frac{3(1-\nu)}{8\sqrt{R}} P_m J(t), \qquad (36.56)$$

$$J'(\omega) = \frac{4\sqrt{R} h_m^{1/2}(t) \Delta h_0}{(1-\nu)\Delta P_0} \cos\delta, \text{ and } J''(\omega) = \frac{4\sqrt{R} h_m^{1/2}(t) \Delta h_0}{(1-\nu)\Delta P_0} \sin\delta. \qquad (36.57)$$

The uniaxial complex compliance $D^*(\omega)$ can be expressed as

$$D^*(\omega) = D'(\omega) - i D''(\omega) = \frac{J'(\omega) - i J''(\omega)}{2(1+\nu)}, \qquad (36.58)$$

where $D'(\omega)$ and $D''(\omega)$ are the uniaxial storage compliance and loss compliance, which can be expressed as follows from (36.57):

$$D'(\omega) = \frac{2\sqrt{R} h_m^{1/2}(t) \Delta h_0}{(1-\nu^2)\Delta P_0} \cos\delta, \text{ and } D''(\omega) = \frac{2\sqrt{R} h_m^{1/2}(t) \Delta h_0}{(1-\nu^2)\Delta P_0} \sin\delta. \qquad (36.59)$$

The uniaxial complex modulus $E^*(\omega)$, uniaxial storage modulus $E'(\omega)$ and loss modulus $E''(\omega)$ can be represented as

$$E^*(\omega) = E'(\omega) - \mathrm{i}E''(\omega) , \tag{36.60}$$

$$E'(\omega) = \frac{(1-\nu^2)\Delta P}{2\sqrt{R}h_\mathrm{m}^{1/2}(t)\Delta h_0} \cos\delta, \quad \text{and} \quad E''(\omega) = \frac{(1-\nu^2)\Delta P}{2\sqrt{R}h_\mathrm{m}^{1/2}(t)\Delta h_0} \sin\delta . \tag{36.61}$$

On the other hand, if a small sinusoidal load is superimposed upon ramp loading,

$$P(t) = v_0 t + \Delta P_0 \sin(\omega t) , \tag{36.62}$$

where v_0 is the loading rate. Huang et al. also demonstrated that the formulas for determining complex compliance can also be derived under the loading condition of $h_\mathrm{m}(t) \gg \Delta h_0$.

Therefore, for linear viscoelastic materials with less difference in Poisson's ratio, spherical indentation measurement under oscillatory loading in a nondecreasing contact area can be used to determine the complex compliance and the complex modulus in shear or the uniaxial complex compliance and uniaxial complex modulus by detecting the carrier displacement, the amplitude of harmonic displacement and the out-of-phase angle.

36.7.2
Practical Application Aspects

The approach of Huang et al. [19] was based on the assumption of a nondecreasing indentation contact area. In actual measurements, however, the contact area may decrease under an oscillatory loading condition, causing the Lee–Radok integral operator to lose its boundary condition, with the result that residual surface traction occurs outside the the current contact region. It is necessary to examine the approach of Huang et al. in the range of an increasing or a decreasing contact area. Huang et al. [19] suggested that for a small harmonic load superimposed with ramp loading, when the loading rate is $v \geq \Delta P_0 \omega$ in (36.62), a nondecreasing load leads to a nondecreasing contact area during the nanoindentation process. For harmonic loading superimposed with step loading, if the loading frequency $\omega \leq \mathrm{d}(h_\mathrm{m}/\Delta h_0)/\mathrm{d}t$ in (36.49), the contact area is nondecreasing during the whole loading process. When the loading frequency $\omega > \mathrm{d}(h_\mathrm{m}/\Delta h_0)/\mathrm{d}t$ in (36.49), the contact area increases and decreases with loading time. In previous studies, Ting [63, 64] proposed a solution of axisymmetric viscoelastic indentation by a rigid indenter when the contact area varies during the loading condition. However, Ting's approach leads to similar results to those obtained with the hereditary integral operator proposed by Lee and Radok. Huang et al. [19] demonstrated that the solutions derived with their method are very close to Ting's solution, even though their approach is not justified in the range of $\omega > \mathrm{d}(h_\mathrm{m}/\Delta h_0)/\mathrm{d}t$. Because the approach of Huang et al. is a closed-form solution compared with Ting's approach, the formulas of the former approach are convenient for estimating the complex viscoelastic compliance and modulus in the regime of linear viscoelasticity even when $\omega > \mathrm{d}(h_\mathrm{m}/\Delta h_0)/\mathrm{d}t$.

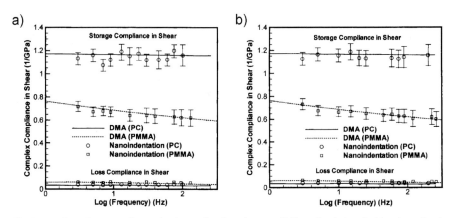

Fig. 36.17. Complex compliance in shear of polycarbonate (*PC*) and poly(methyl methacrylate) (*PMMA*) polymer materials under a harmonic load superimposed on **a** ramp loading and **b** step loading in comparison with dynamic mechanical analysis (*DMA*) results. (From Huang et al. [19], with permission)

Huang et al. [19] also examined their approach by testing the same materials with nanoindentation and conventional DMA apparatuses. In the nanoindentation measurement, they used CSM loading to apply dynamic excitation in the frequency range 3–260 Hz on a flat specimen surface with a spherical indenter having a tip radius of 3.4 μm. The indentation depth was a few hundred nanometers, but the amplitude of harmonic loading was controlled to obtain harmonic displacement with an amplitude between a fraction of a nanometer and a few nanometers. The thermal drift level was typically below 0.05 nm/s. In the DMA measurement, temperature–frequency trade-off was applied to extend the frequency range to 0–260 Hz for comparing the results with the nanoindentation measurement. Figure 36.17 shows the complex compliance in shear of polycarbonate and poly(methyl methacrylate) materials obtained by nanoindentation under a harmonic load superimposed on ramp loading and on step loading, respectively, in comparison with the DMA results. Huang et al. [19] found that the nanoindentation measurements in both types of loading process were in good agreement with the DMA results. The average error for the storage compliance of polycarbonate and poly(methyl methacrylate) at these discrete experimental data was less than 6.2%. The maximum error for the storage compliance of polycarbonate and poly(methyl methacrylate) was 9.1 and 5.1%, respectively.

For other time-independent materials such as metals and ceramic materials, measurement of creep compliance as a function of elapsed time by changing the indentation conditions with respect to temperature, load and loading rate instead of dynamic loading may be an easy way to evaluate viscoelastic properties of these kinds of materials. Shimizu et al. [61] proposed the following creep formula for obtaining uniaxial creep compliance $D(t)$ from the indentation contact depth $h_c(t)$ as a function of time using a constant load P_0:

$$D(t) = \frac{C_0 \tan\beta}{2(1-\nu^2)P_0} h_c^2(t) , \qquad (36.63)$$

where C_0 is a constant that varies depending on indenter geometries; $C_0 = 24.5$ for a Berkovich indenter. β is the inclined face angle. They made some indents in an amorphous selenium specimen with the same constant load but different temperatures ranging from 10 to 42 °C, including the glass transition, and then at the same temperature but with different constant loads, respectively, using a Vickers indenter. The creep curves in Figure 36.18 show that the indentation depth was dependent on time in both cases. Usually, creep under a constant applied stress exhibits a steady-state linear characteristic with respect to time in a viscous regime. In the constant-load indentation measurement, however, as shown in Fig. 36.18a, stress decreased with increasing indentation depth, resulting in a progressive decrease in the creep rate dh_c/dt with time even when the temperature was set high enough for steady-state viscous flow. On the basis of their approach shown in (36.63) and the well-known time–temperature superposition rule for the creep compliance function [65], they calculated the uniaxial creep compliance of α-selenium from the time dependence of the indentation depth and composed a master curve $\log D(t)$ versus $\log t/a_T$ at a standard temperature of 36 °C, as shown in Fig. 36.18b, where a_T is the shift factor for the time–temperature superposition. These results confirmed the self-consistency between the creep formula in (36.63) and the experimental framework, and suggested that other indentation methods should be applied to determine creep compliance for characterizing the viscoelasticity of ceramic, metal and polymer films.

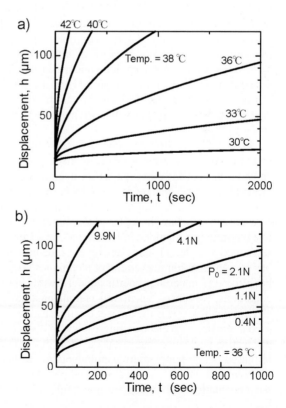

Fig. 36.18. Constant-load indentation creep curves of amorphous selenium at **a** the same constant load but different temperatures and at **b** the same temperature but different constant loads. (From Shimizu et al. [61], with permission)

36.8 Other Mechanical Characteristics

This chapter has described the analysis methods, practical application aspects and the latest application examples of the nanoindentation techniques used most frequently for determining mechanical properties, including hardness and modulus, yield stress, stress–strain curve and viscoelasticity. Besides these properties, nanoindentation techniques can also be used to evaluate many other mechanical properties or characteristics, such as fracture toughness, fatigue, surface residual stress, adhesion/cohesion and friction, among others [29, 66–75]. Morris and Cook [66] proposed an indentation wedging model to estimate radial fracture during indentation by an acute indenter instead of the often-used Vickers indenter, making it possible to extend the fracture toughness estimation method to a very small length scale. Li and Bhushan [29] used the CSM technique to measure contact stiffness as a function of the number of cycles under an oscillated indentation load and found that the number of critical cycles at an abrupt decrease in contact stiffness can be used for estimating the fatigue properties of ultrathin films. Suresh and Glannakopoulos [69] proposed a step-by-step method to determine preexisting residual stresses and residual plastic strains in elastoplastic solids on the basis of continuous, quantitative shape indentation. They demonstrated that this method can be applied to thin films, structural coatings and engineered surfaces containing an equibiaxial residual stress field. Ye et al. [72, 73] applied a nanoscratch technique to measure nanofriction coefficients as a function of ramp loading and found that the critical load at an abrupt decrease in the nanofriction coefficient can be used for estimating the adhesion or cohesion strength of ultrathin multilayered structures. They also applied the nanoscratch technique under constant loads to measure nanofriction coefficients within an area less than 100 nm and found that tribological performance can be characterized on the basis of nanofrictional properties, which are independent of the surface roughness but originate from the friction nature of the material's surface [74, 75].

In recent years, simultaneous nanoindentation measurement techniques using an acoustic emission sensor or magnetic/electric sensors have been developed. By embedding the acoustic emission sensor in an indenter, one can detect acoustic signals originating from yield point phenomena or fractures during the indentation process [76–78]. This technique makes it possible to understand fracture mechanisms by evaluating fracture strength and toughness. On the other hand, by using a conductive boron-doped diamond indenter, one can observe the change in electrical contact resistance at different loads during the indentation process [79]. Thus, simultaneous nanoindentation measurement techniques not only yield mechanical properties but also other material properties under elastic, elastoplastic or viscoelastic contact conditions. Moreover, an in situ nanoindentation technique using a transmission electron microscope is the subject of all researchers' attention. Minor et al. [80–82] used a transmission electron microscope to observe dislocation plasticity, including dislocation nucleation and metal-like flow in single-crystal silicon. Schuh et al. [83] also used a high-temperature nanoindentation technique to examine dislocation nucleation quantitatively in single-crystal platinum. These studies imply that ultralow nanoindentation techniques are powerful tools not only for determining mechanical properties for material engineering purposes, but also for understanding the atomic-level origins of mechanical deformations from a physical viewpoint.

36.9 Outlook

This chapter has demonstrated that elastic, elastoplastic and viscoelastic contact solutions permit nanoindentation load–displacement curves to be used to evaluate many kinds of mechanical properties on a nanometer scale. Although some methods still need to be improved further for practical use, we believe that in the near future all kinds of bulk-scale mechanical properties or characteristics will be easily determined on a nanometer scale by using suitable nanoindentation methods. We also believe that novel nanoindentation techniques not only have broad material applications for estimation of nanomechanical properties, but that they can also be used to characterize nanoscale physical phenomena for material analysis of many descriptions, including phase transformation, thermal stability and time dependence in nanometer volumes, nanoscale electrical, magnetic and optical phenomena, nanoscale complex structures and surface phenomena. We can prognosticate that dramatic progress in nanoindentation techniques will be achieved soon for both nanomechanical estimation and physical phenomenal analysis.

Acknowledgements. The author would like to thank A. Yonezu and S. Shimizu for many helpful discussions and N. Ogasawara, B. Bhushan, A. Yonezu and G. Huang for providing figures from their work. J. Ye would also like to thank K. Ueoka for technical assistance.

References

1. Meyer E (1908) Z Ver Dtsch Ing 52:645
2. Tabor D (1951) The hardness of metals. Oxford University Press, London
3. Sneddon IN (1965) Int J Eng Sci 3:47
4. Harding JW, Sneddon IN (1945) Proc Camb Philos Soc 41:12
5. Hertz H (1881) J Reine Angew Math 92:156
6. Hertz H (1896) In: Schott J (ed) Miscellaneous papers. Macmillan, London
7. Timoshenko S (1934) Theory of elasticity. McGraw-Hill, New York
8. Davies RM (1949) Proc R Soc Lond Ser A 197:416
9. Pharr GM, Oliver WC, Brotzen FR (1992) J Mater Res 7:613
10. Oliver WC, Pharr GM (1992) J Mater Res 7:1564
11. Ye J, Kano M, Yasuda Y (2002) Tribol Lett 13:41
12. Shimizu S, Kojima N, Ye J (2005) Mater Res Soc Symp Proc 863:B8.15
13. Bucaille JL, Stauss S, Felder E, Michler J (2003) Acta Mater 51:1663
14. Chollacoop N, Dao M, Surech S (2003) Acta Mater 51:3713
15. Ogasawara N, Chiba N, Chen X (2005) J Mater Res 20:2225
16. Cheng YT, Cheng CM (2000) Surf Coat Technol 133:417
17. Dao M, Chcolacoop N, Van Vliet KJ, Venkatesh TA, Surech S (2001) Acta Mater 49:3899
18. Lu H, Wang B, Ma J, Huang G, Viswanathan H (2003) Mech Time-Depend Mater 7:189
19. Huang G, Wang B, Lu H (2004) Mech Time-Depend Mater 8:345

20. Lee EH, Radok JRM (1960) J Appl Mech 27:438
21. Bulychev SI, Alekhin VP, Shorshorov MKH, Ternovskii AP, Shnyrev GD (1975) Zavod Lab 41:1137
22. Ye J, Shimizu S, Sato S, Kojima N, Noro J (2006) Appl Phys Lett 89:1913
23. Hay J (1997) Mechanical testing by indentation, course notes. Nano Instruments, Oak Ridge
24. Pharr GM (1998) Mater Sci Eng A 253:151
25. Harding DS, Oliver WC, Pharr GM (1995) Mater Res Soc Symp Proc 356:663
26. Morton WB, Close LJ (1922) Philos Mag 143:320
27. Johnson KL (1985) Contact mechanics. Cambridge University Press, Cambridge
28. Bhushan B (ed) (1999) Handbook of micro/nanotribology, 2nd edn. CRC, Boca Raton
29. Li X, Bhushan B (2002) Mater Character 48:11
30. Li X, Diao D, Bhushan B (1997) Acta Mater 45:4453
31. Li X, Bhushan B (1998) Thin Solid Films 315:214
32. Li X, Bhushan B (1999) Thin Solid Films 315:330
33. Hay J, Bolshakov A, Pharr GM (1999) J Mater Res 14:2296
34. Hay J, Bolshakov A, Pharr GM (1998) Mater Res Soc Symp Proc 522:263
35. Hay J, Pharr GM (1998) Mater Res Soc Symp Proc 522:39
36. Ye J, Kano M, Yasuda Y (2004) Tribol Lett 16:107
37. Ye J, Kojima N, Ueoka K, Shimanuki J, Nasuno T, Ogawa S (2004) J Appl Phys 95:3704
38. Ye J (2005) Tribology 219:24
39. Xu ZH, Li XD (2006) Acta Mater 54:1699
40. Chen SH, Lui L, Wang TC (2004) Acta Mater 52:1089
41. Tho KK, Swaddiwudhipong S, Hua J, Liu ZS (2006) Mater Sci Eng A 421:168
42. Choi Y, Van Vliet KJ, Li J, Suresh S (2003) J Appl Phys 94:6050
43. Loubet JL, Bauer M, Tonck A, Bec S (1993) Mechanical properties and deformation behavior of materials having ultrafine microstructures. Kluwer, Norwell
44. Hochstetter G, Jimenez A, Loubet JL (1999) J Macromol Sci Phys B 38:681
45. Sava T, Tanaka K (2000) J Mater Res 16:3084
46. Ye J (2006) Paper presented at novel techniques of nanoindentation and their applications, 67th Nissan ARC materials and analysis seminar. Nissan ARC, Yokosuka
47. Volinsky AA, Moody NR, Gerberich WW (2004) J Mater Res 19:2650
48. Ye J, Kojima N, Shimizu S, Burkstrand JM (2005) Mater Res Soc Symp Proc 863:B1.5
49. Schuh CA, Packard CE, Lund AC (2006) J Mater Res 21:725
50. Pethica JB, Oliver WC (1989) Mater Res Soc Symp Proc 130:13
51. Syed Asif SA, Pethica JB (1997) Mater Res Soc Symp Proc 436:201
52. Shimizu S, Kojima N, Ye J (2006) Paper presented at the 2006 international conference on solid state devices and materials, Yokohama, Japan
53. Ogasawara N, Chiba N, Chen X (2006) Scr Mater 54:65
54. Bowden FP, Tabor D (1950) The friction and lubrications of solids. Oxford University Press, London
55. Mesarovic SD, Fleck NA (1992) Proc R Soc Lond Ser A 455:2707
56. Wang L, Rokhlin SI (2005) Int J Solids Struct 42:3807
57. Wang L, Ganor M, Rokhlin SI (2005) J Mater Rec 20:987
58. Yonezu A, Ogawa T, Takemoto M (2006) In: Proceedings of the Asian Pacific conference for fracture and strength, Hainan Island, China, p 319
59. Tang B, Ngan AHW (2003) J Mater Rec 18:1141
60. Cheng L, Xia X, Yu W, Scriven LE, Gerberich WW (2000) J Polym Sci B Polym Phys 38:10
61. Shimizu S, Yanagimoto T, Sakai M (1999) J Mater Res 14:4075
62. Loubet JL, Lucas BN, Oliver WC (1995) International workshop on instrumental indentation, San Diego

63. Ting TCT (1966) J Appl Mech 33:845
64. Ting TCT (1968) J Appl Mech 35:248
65. Ferry JD (1980) Viscoelastic properties of polymers. Wiley, New York
66. Morris DJ, Cook RF (2005) Int J Fract 136:237
67. Li XD, Diao DF, Bhushan B (1997) Acta Mater 45:4453
68. Volinsky AA, Vella JB, Gerberich WW (2003) Thin Solid Films 429:201
69. Suresh S, Glannakopoulos AE (1998) Acta Mater 46:5755
70. Tayer CA, Wayne MF, Chiu WKS (2003) Thin Solid Films 429:190
71. Xu ZH, Li XD (2005) Acta Mater 53:1913
72. Ye J, Kojima N, Ueoka K, Shimanuki J, Nasuno T, Ogawa S (2004) J Appl Phys 95:3704
73. Ye J, Ueoka K, Kojima N, Shimanuki J, Shimada M, Ogawa S (2004) Mater Res Soc Symp Proc 812:F5.6
74. Ye J, Kano M, Yasuda Y (2004) Tribol Lett 16:107
75. Ye J, Ueoka K, Kano M, Yasuda Y, Okamoto Y, Martin JM (2005) World tribology congress III, Washington
76. Daugela A and Wyrebek JT (2000) IEEE Trans Magn 581
77. Tymiak NI, Daugela A, Wyrobek TJ, Warren OL (2003) J Mater Rec 18:784
78. Tymiak NI, Daugela A, Wyrobek TJ, Warren OL (2004) Acta Mater 52:553
79. Ruffell S, Bradby JB, Williams J (2006) NanoECR. Hysitron application note. Hysitron, Minneapolis
80. Minor AM, Lilleodden ET, Jin M, Stach EA, Chrzan DC, Morris JW (2005) Philos Mag 85:323
81. Minor AM, Morris JW, Stach EA (2001) Appl Phys Lett 79:1625
82. Stach EA, Freeman T, Minor AM, Owen DK, Cummings J, Wall MA, Chraska T, Hull R. Morris JW, Zettl A, Dahman U (2001) Microsc Microanal 7:501
83. Schuh CA, Mason JK, Lund AC (2005) Nat Mater 4:617

37 Applications to Nano-Dispersion Macromolecule Material Evaluation in an Electrophotographic Printer

Yasushi Kadota

Abstract. In this chapter, applications of analytical methods using SPM to improve the electrophotographic processes used in laser printers, copy machines and so on are described from an industrial viewpoint. Many components for the processes work in a well-controlled manner with macromolecules in high electric field under pressure at high temperature. The achievable print quality and reliability crucially depend on their properties on a microscale or nanoscale. It is noted that the cross-sectional phase imaging by AFM is very powerful to evaluate the compositions and interfacial structures of toner particles on the components as well as charging states and adhesive force.

Key words: Electrophotographic print, Macromolecule, Toner particle, Charge, Cross-sectional observation, Nano-dispersion

37.1 Introduction

At the present time, as personal computers and the Internet have spread all over the world, the technology of printing information on paper has become more indispensable. Information on paper is practically convenient to carry and read, though electronic paper media and other forms are being developed with the aim of taking over from paper. For example, plain paper copiers, electronic printers and normal paper facsimiles are typical machines widely used in an office and at home to print documents as well as images on paper. These machines consist of electronically controlled hardware and software that possess advanced processing functions based on digital high technologies. On the other hand, there are still classical and analog elements such as paper and ink in the machines. In printing machines used widely, an electrophotographic system and an ink-jet system work well with many analog controls in printing processes. The electrophotographic system in the printing machines uses a photoconductive drum capable of holding an electrically charged image depicted on its surface, decorating the image with oppositely charged particles (the so-called toner) and transferring the image onto a sheet of paper.

A laser printer is one of the most popular printing machines which includes the electrophotographic system; the laser beam makes an image to be printed on the drum by scanning its surface. The representative features of the electrophotographic process are an achievable high quality of printing and a low running cost. Monochromatic laser beam printers started to be used in offices from the 1980s. Traditional models of laser printers were very expensive, and they were large, occupying a sizeable area in an office. Inexpensive models of full-color laser printers for personal use came on the market in the early 2000s, and the machines became very compact. A new application of office printers for display labels (e.g., bar-code printing) has been developed on the basis of the high quality and high productivity of printing devices. Major manufacturers are making a push to develop new products such as ink-jet printers. Since they are continuing to develop new printers, in the near future it is expected that novel functionalized compact printers will appear at much lower prices.

Most components of the printing processes in the electrophotographic system use macromolecule materials (e. g., elastomers, resins). Furthermore, the functions in the printing processes were realized by successfully utilizing the characteristics of nanocomposite domains and surfaces. This indicates that the achievable print quality and its reliability crucially depend on microscale or nanoscale characteristics of the macromolecule materials. In this chapter, applications of analytical methods mainly using scanning probe microscopy (SPM) to industrial development of printing processes in electrophotographic systems are described.

37.2
Electrophotographic Processes

37.2.1
Principle and Characteristics of an Electrophotographic System

To characterize the printing processes in electrophotographic systems, the following items are crucial from an engineering point of view:

1. High electric field and electrostatic force. An electric potential as high as several volts to 1000 V is applied between the components used for printing. In addition, electrostatic charging and discharging phenomena are utilized.
2. Pigmented particles used as ink, called toner. The toner particles are triboelectrically charged with mechanical processes. Note that the electrostatic force due to the charge is the dominant driving force exerted on toner particles.
3. Organic photoconductors (OPC). Printing images are drawn on the surface of an OPC by a scanning laser beam.
4. Surface energy of the toner particles. The particles are fixed on paper with heat under pressure. Thus, it is important for the toner particles to have lower surface energies on the paper.

The printing process consists of the following five steps:

1. Charging. The surface of a photoconductor is charged negatively (or positively) by a corotoron using corona discharge with a fine wire or by contact charge with an electrically biased roller.
2. Exposure. The surface of the photoconductor is exposed to a scanning laser beam, which produces a contrast image of the document. This exposure causes photodecay, which is charging state release owing to the increase in photoconduction on areas irradiated with a laser beam, resulting in a charging potential change as shown in Fig. 37.1.

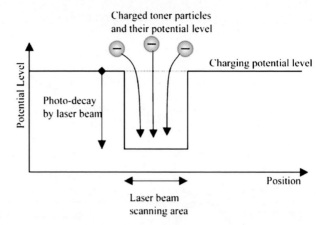

Fig. 37.1. Charging, exposure and development in eletrophotographic processes

3. Development. Negatively (or positively) charged toner particles are supplied over the exposed surface of the photoconductor. They electrostatically adhere to the positive (or negative) potential areas, leading to a visible toner image.
4. Transfer. A sheet of plain paper is placed over the surface of the photoconductor and charged positively (or negatively). The negatively (or positively) charged toner particle image on the surface is electrostatically transferred onto the positively (or negatively) charged paper.
5. Fusing. The toner particle image is fused to the paper by heating under pressure. A fusing roller surface is coated with materials with very small surface energies such as tetrafluoroethylene perfluoroalkoxy vinyl ether copolymer (PFA) or polytetrafluoroethylene (PTFE) which prevent toner particles from fusing with the roller surface.

After the photoconductor surface has been cleaned with a urethane blade, the above-mentioned printing process can be repeated. Figure 37.2 shows a typical cross section of an electrophotographic system.

37.2.2
Microcharacteristics and Analysis Technology for Functional Components

In the printing processes many components, such as rollers and belts, are used. Here, the required characteristics and their evaluation techniques are discussed for functional components and materials in the processes, including many kinds of macromolecule elastomers and resins. For example, degradation of component surfaces is one of the most important issues to be solved from the viewpoint of the reliability of the printing process. The degradation is mostly attributed to the change in the properties at the surfaces of the components.

37.2.2.1
Charging Roller

A charging roller is composed of an electroconductive elastomer (the roller is called an elastomer type) or nanocomposite macromolecule resin (a hard type). Silicone

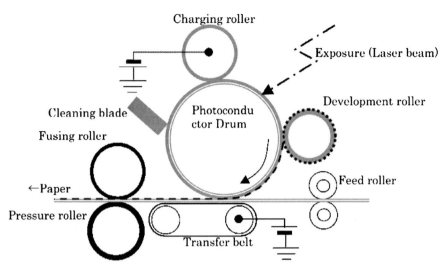

Fig. 37.2. Cross section of an electrophotographic system

or polyurethane is used for the elastomer type, in which conductive particles such as carbon black are dispersed. In this type, the elastomer is required to have good electrical contact to the OPC. Thus, the particle dispersion has to be uniform in a microcosm for both lateral and horizontal directions. A critical problem is "bloom": antioxidants in the elastomer migrate by thermal stress, so electrical conductivity deteriorates therein. On one hand, the hard-type roller is composed of microcomposite macromolecule resins having good electrical conductivity. In respect to mechanical strength and precision acrylonitrile–butadiene–styrene resin or polycarbonate (PC) resin is generally selected as the main component of the resin. In addition, it is noted that the toner or paper particles firmly adhere to the roller surface owing to the electrostatic field and thermal stress.

37.2.2.2
Photoconductor Drum

A photoconductor drum is mainly composed of macromolecules having semiconductive properties (e. g., an azo compound such as phthalocyanine) and high mechanical strength resins (e. g., PC resin). An active layer is needed with a suitable charge carrier concentration with a uniform carrier profile in lateral and horizontal directions on the drum. The mechanical strength of the resin is a central issue as well as the dispersion of semiconducting macromolecules. Surface abrasion with toners and foreign additives in cleaning processes is also critical, and surface adhesion may occur with a charging roller.

37.2.2.3
Fusing Roller

The surface layer of a fusing roller is mainly composed of macromolecules having low surface potential energy such as fluorine-containing macromolecules such as

PFA and PTFE. Furthermore, in order to improve the electrical conductivity and mechanical strength some chemicals are usually added in the surface layer. Since the layer should release toner melted at high temperature, the surface potential energy of the melted toner is designed to be different from that of the surface layer. Adhesion of toner to the surface by discharge, thermal and mechanical stress may be critical. If this is not the case, the surface energy would be increased close to that of the toner. It is also noted that the surface abrasion with foreign additives and paper should be avoided.

37.2.2.4
Toner

Toners are mainly made of three kinds of materials: for example, polyester or styrene–acrylate with internal additive wax (e.g., made from calnauba palm), pigment and dyestuff as a charge control agent (e.g., organic metal) and external additives (e.g., nanoparticles of TiO_2 and SiO_2). In addition, internal additives are important; internal additives should have a suitable concentration in the main resin, and disperse slightly inside the surface for tolerance in the fusing processes.

37.2.2.5
Analysis Methods

Dispersion

Visualization of macromolecules and their distribution is a key to characterize their dispersion. In respect to analysis of functional groups in macromolecules, we applied microscopic Fourier transform infrared (FTIR) spectroscopy and imaging FTIR spectroscopy with a few micrometer resolution. This resolution is enough to apply the methods to charging rollers and photoconductors. When higher resolution is needed, or when the dispersion inside toner particles is analyzed, focused beam analytical methods such as transmission electron microscopy, scanning electron microscopy (SEM) and scanning transmission X-ray microscopy are employed. In the focused beam methods, however, it is difficult to obtain information on the characteristics of organic molecules. Thus, the characterization is carried out by combining the results obtained by other methods for organic analysis (e.g., FTIR spectroscopy).

In recent years, atomic force microscopy (AFM) has been applied to analysis of the distributed states of macromolecules. For example, the following procedure is frequently used. Firstly, AFM imaging is performed with attention to the difference between the dynamics characteristics of macromolecules. For samples with micropolished surfaces, phase images corresponding to viscosity and elasticity are taken by the atomic force microscope as well as topographic images. The images obtained are of help in characterizing and controlling the quality of products. However, it is difficult to apply this method to samples containing not less than three kinds of macromolecules. In addition, the most suitable polishing condition for individual samples should be determined in advance. Secondly, visualization of the difference between electrical characteristics of macromolecules is conducted; it is critical to prepare a thin-film sample.

Surface Adhesion

To analyze materials that adhere to the surfaces, the microscopies and surface analysis methods (e. g., X-ray photoelectron spectroscopy, FTIR spectroscopy) are employed. However, organic macromolecules are frequently damaged by electron beam irradiation in the focused beam methods. From that aspect, SPM including phase imaging is very powerful for achieving nanoscale resolution of organic materials. Phase images can show the difference in local mechanical properties even on rough surfaces. They are very effective for observing cross sections of samples in addition to surface planes of samples, leading us to fruitful analysis of their interfaces with adhesive materials, as mentioned later.

Surface Charge Density

The quantity of triboelectrical charge on a toner particle is one of the most important parameters for designing the material of a toner, though nanoscale visualization of charged sites on a toner is a challenging subject. Although voltage contrast imaging by SEM was applied, we have not succeeded in the quantitative analysis because of the surface roughness of toner. This technique, however, is applicable to measurement of the electrical potential distribution over the OPC. The potential distribution obtained can be compared with those measured by SPM based on AFM (e. g., electric force microscopy, surface potential microscopy, SpoM). SPM is promising for samples with complicated three-dimensional structures, since the potential information can be separated from the topographic information if we take great care.

37.3
SPM Applications to Electrophotographic Systems

37.3.1
Measurement of Electrostatic Charge of Toner

The toner particles with triboelectric charge are supplied onto a developing roller; the quantity of toner charge is crucial for high-quality print using electrophotographic

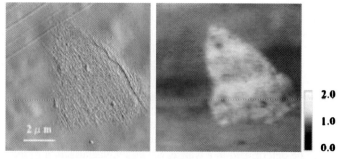

Fig. 37.3. Topography (amplitude image taken by atomic force microscopy, *left*) and surface potential image (*right*) of a cross section of a pulverized toner particle

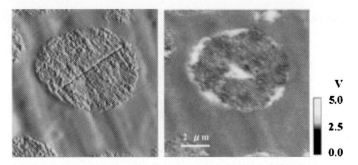

Fig. 37.4. Topography (amplitude image taken by atomic force microscopy, *left*) and surface potential image (*right*) of a cross section of a pulverized toner particle after being positively charged

systems. Although it is hard to measure where charges are accumulated in a drop of toner particles, SPM has the possibility to provide a clue to characterize them [1]. In general, since the toner particles have complicated structures, some mechanical preprocessing is necessary. An example of the procedure is as follows:

1. Toner particles are embedded in resin (epoxy, polyester, etc.). A piece is cut with abrasives and the cut plane is polished. This sample is subjected to cross-sectional observations.
2. The surface is electrically charged and kept in air.
3. Surface potential images are observed by AFM implemented with SpoM.

Charging Toner Image

Toner particles have three-dimentional structures. Thus, a sample is cut to have a flat plane suitable for measurement. Recently, three-dimentional manipulators developed for protein analysis that are easy to use are available, and began to be applied to the sample preparation of toner particles. It is required to examine whether all domains have similar charging properties or not by further refined measurements just after cut sample has been prepared.

37.3.2
Measurement of the Adhesive Force Between a Particle and a Substrate

In an electrophotographic process, toner particles that are electrically charged on a developing roller are transferred to a substrate of the OPC, and afterwards onto a sheet of paper. The toner movement is controlled by electrostatic force induced by applied bias voltage. In general, adhesive force F acting between a particle and a substrate is represented as follows [2–6]:

$$F \propto [F_{\text{van}} + F_{\text{Po}} + F_{\text{ES}}] \times S . \tag{37.1}$$

Here, F_{van} is the van der Waals force, F_{Po} the force from contact potential difference, F_{ES} the electrostatic force and S the contact area. In general, it is supposed that F_{Po}

is very weak between a toner particle and a macromolecule substrate of the OPC. Thus, F_{van} and F_{ES} are the main factors to be taken into account. According to a report from a group at Clarkson University, F_{ES} between a spherical particle and a substrate can be represented as follows:

$$F_{ES} = \pi R^2 \sigma^2 / \varepsilon_0 . \tag{37.2}$$

In addition, F_{van} can be represented as follows:

$$F_{van} = AR/6z_0^2 . \tag{37.3}$$

Here A is the Hamaker constant, R the radius of a spherical toner particle, σ the surface charge density (in coulombs per square meter), ε_0 the dielectric constant of a vacuum and z_0 the separation distance. For a particle attached to a substrate, z_0 is typically about 0.4 nm, ranging up to 1 nm.

When defining R_{crit} as R under the condition of $F_{van} = F_{ES}$, we obtain the following relation:

$$R_{crit} = \frac{A\varepsilon_0}{6\pi z_0^2 \sigma^2} . \tag{37.4}$$

This formula tells us that when $R > R_{crit}$, electrostatic force is dominant, and when $R < R_{crit}$, the van der Waals force is dominant. For instance, if $A = 10^{-19}$ J, $\sigma = 3 \times 10^{-5}$ C/m² and $z_0 = 0.4$ nm, then $R_{crit} = 0.5$ mm. Currently the particle size of toner is not more than 0.01 mm. This indicates that it is almost impossible to drive a toner particle by electrostatic force. Thus, to decrease the van der Waals force, nanometer-sized particles are commonly added to toner as additives, resulting in the control of toner movement by electrostatic force. The decrease in adhesive force with the additives is shown in Fig. 37.5 [7].

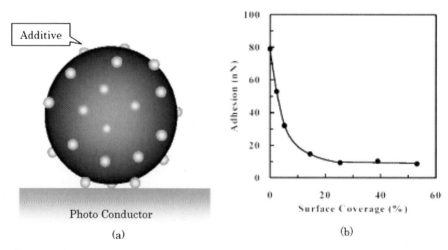

Fig. 37.5. a, b Contact model and average adhesion change [7]. **a** Contact model of additives covering toner on a photoconductor, and **b** dependence of average adhesion on the coverage over a surface of a nontriboelectrically charged toner particle

Fig. 37.6. Cantilever for measuring adhesive force

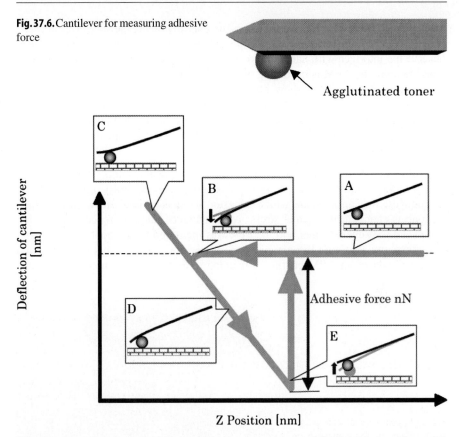

Fig. 37.7. Example of a typical force–distance curve with the deflection of a cantilever due to the tip–substrate interaction. The adhesive force is evaluated from the cantilever deflection multiplied by a cantilever spring constant

It is necessary to measure the adhesive force between a toner particle and a substrate of the OPC. AFM was used to measure the force [8]. With a toner particle agglutinated to the tip of a cantilever (Fig. 37.6), the adhesive force is evaluated from a force–distance curve as shown in Fig. 37.7.

This measurement has contributed to the development of our electrophotographic processes; we have found the influence of not only the surface roughness and materials but also of temperature and humidity.

37.3.3
Observation of a Nanodispersion Macromolecule Interface
—Toner Adhesion to a Fusing Roller

The reliability of an electrophotographic system often depends on adhesive states of toner or paper to the surfaces of the main components such as the OPC and the fusing roller. To improve the reliability, it is important to analyze those adhesion mechanisms, although it is not straightforward even to observe the adhesion interface. Both materials of the toner and the component surfaces are organic; the technique to observe the interface with nanoscale resolution had been not affordable. Nowadays phase imaging by AFM is becoming one of the most effective methods for this.

Figures 37.8 and 37.9 show phase images of cross sections of adhered and fresh toner particles. In general, toner has a different solubility from that of PFA as a substrate. Thus, it is not likely that the toner adheres to the PFA surface physically and chemically. On the basis of this presumption and through the analysis, the following conclusions are drawn [9]:

1. Adhered toner and fresh toner have different compositions; this statement comes from the contrast difference in the phase images. By carefully observing phase images, we can estimate which adhesion toner is of inherent value or which one is changed by thermal stress. As a reference, the difference in phase between the adhered toner and the PFA was measured to be about 20° using our conventional atomic force microscope under the condition that the oscillation amplitude of an AFM cantilever damped to 60%. This means that a 10° dif-

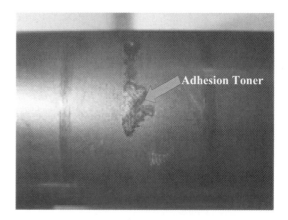

Fig. 37.8. Example of cross-sectional atomic force microscopy observation of a toner particle adhered to a fusing roller on an inherent failure in the electrophotographic process

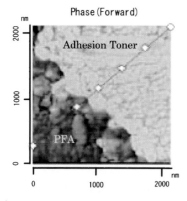
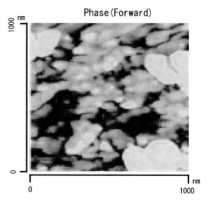

Fig. 37.9. Phase image of the cross section of a toner particle adhered to a fusing roller (*right*), and a phase image of a fresh toner particle (*left*)

ference in phase is of great value for the analysis; the improvement of phases sensitivity would give us more valuable information on nanomaterial analysis. Recently we have found that AFM in a frequency modulation mode or with Q control improves the reproducibility and quantification of the phase measurements.
2. The adhesion area at the interface is less than 50 nm. This implies that mixing of macromolecules at the interface does not take place.
3. Cutting and polishing of a sample containing the toner particles is a highly valuable sample preparation method for AFM observation aiming at the interface between macromolecule materials.

37.4
Current Technology Subjects

Since evaluation technologies using SPM were put to practical use, physical phenomena in an electrophotographic process have been becoming clear with nanoscale resolution. Consequently, the quality of print images and the reliability of the process have been improved; the contribution of SPM to the field of the electrophotographic systems has been of great significance from the viewpoint of industrial applications. However, the following approaches are required for further improvements in electrophotographic systems on the basis of nanotechnology, for example, to measure physical phenomena concerning the electrostatic behavior in the systems quantitatively with better precision on the nanoscale:

1. Sensitivity improvement of measuring force to a piconewton level
2. Precise measurement of viscosity and elasticity without any influence of surface roughness
3. Dynamic range improvement of measuring force, e.g., on electric discharge between nanoscale materials.

References

1. Otomura S (2004) SPM analysis for material properties application of surface potential microscopy for microstructure imaging of toner particles. Ricoh technical report. http://www.ricoh.co.jp/about/business_overview/report/30/pdf/A3003.pdf
2. Zhou H, Götzinger M, Peukert W (2004) Tailoring particle-substrate adhesion. International Congress on Particle Technology PARTEC 2004, Nuremberg, 16–18 March. http://www.lfg.uni-erlangen.de/index_english.html
3. Li Q, Rudolph V, Peukert W (2006) Powder Technol 161(3):248–255
4. Ounis H, Ahmadi G, McLaughlin JB (1992) In: Shen HH et al. (eds) Advances in micromechanics of granular materials. Elsevier, Amsterdam, pp 433–442
5. Soltani M, Ahmadi G, Bayer RG, Gaynes MA (1995) J Adhes Sci Technol 9:453–473
6. Ahmadi G (2005) Particle adhesion. http://www.clarkson.edu/projects/crcd/me637/notes/particle_adhesion/index_particle_adhesion.html
7. Iimura H (2000) Study of toner adhesion—effect of an external additive. Ricoh technical report. http://www.ricoh.co.jp/about/business_overview/report/26/pdf/A04.pdf

8. Mizuguchi Y, Miyamoto T (2004) Measuring non-electrostatic adhesive force between solid surfaces and particles by means of atomic force microscopy. Konica Minolta technology report vol 1.
http://konicaminolta.jp/about/research/technology_report/2004/pdf/treatise_003.pdf
9. Nanbu M, Kadota Y (2005) The availability structure on multi functional product and laser printer with maintenance. International conference of quality 05

38 Automated AFM as an Industrial Process Metrology Tool for Nanoelectronic Manufacturing

Tianming Bao · David Fong · Sean Hand

Abstract. Scanning probe microscope (SPM) techniques, invented 20 years ago, act as eyes for nanotechnology and nanoscience research and development, for imaging and characterizing surface topography and properties at atomic resolution. Particularly for the past decade, atomic force microscopy (AFM, one member of the SPM family) has evolved from laboratory research instrumentation to an industry metrology tool for geometric dimension control in nanoelectronic device manufacturing on production floors. This chapter gives an overview in great technical detail of state-of-the-art AFM applications in process characterization and inline monitoring for semiconductor manufacturing. Use of AFM equally applies for topography, dimension, and sidewall shape metrology in photomask and hard disk recording head processing.

Key words: Dimension metrology, Process control, Sidewall profile, Semiconductor manufacturing, Data storage, Photomask

38.1
Introduction

Integrated microelectronic devices are the backbones for today's technology revolution, to name a few, computers, the Internet, telecommunication, and consumer electronics, in our everyday life. Cell phones and digital cameras use semiconductor microchips, while data storage relies on hard disks and memory devices. Each of these devices contains millions of integrated circuits made of basic functioning elements such as transistors, capacitors, and many other units on the microscale or nanoscale. The basic elements are made by a series of complex fabrication process steps, sequentially layer by layer on a substrate. The fabrication normally starts by transferring circuit design patterns from a mask template to a device layer (microlithography), followed by addition of desired materials (deposition, plating, implant, diffusion), removal of unwanted materials (etch, milling, clean, polish), and heating (anneal, reflow). The final devices are shipped after final packaging and testing. To achieve the goal of high quality, low scrap, and low cost, the manufacturing process engineers insert various process metrology and inspection steps within the manufacturing line to monitor defect density, feature pattern geometry, dimension and topography, film thickness, and material composition to ensure proper process control and prevent catastrophic product yield loss. Nanoelectronics deals with devices with sub-100-nm feature dimension. The feature geometry is often on the micron and the nanometer scale (the transistor gate width is approximately 35 nm at the 45-nm technology node). The current semiconductor devices at 90-, 65-, 45-, and 32-nm technology nodes have marched into the nanotechnology regime. The manufacturing process for such small features requires special metrology instrumentation capable of characterizing the nanoscale geometry dimension to detect any deviation from the nominal process specification and ensure conformity to the design specification [1, 2].

Since such tiny patterned features and small topography are involved in the manufacturing process of integrated circuits for nanoelectronic devices, the metrology tool must be precise and accurate to nanometers, or even angstroms [3, 4]. Deviations from the target geometry cause failures to meet the final product performance specification.

The constant push for higher performance and lower power consumption introduces an even smaller feature geometry with a higher pattern density, new materials, and novel device structures. The tolerance for process variation is shrinking with the advanced transistor dimension. The threshold roughness of 10 Å will not pose an integration issue for a 130-nm node device, but may kill a 45-nm node device. Again, an integrated circuit geometry that small needs special instrumentation [2].

The scanning probe microscope is used in nanotechnology and nanoscience for structural, mechanical, magnetic, topographical, electrical, chemical, biological, engineering basic research, and industrial applications [5]. Atomic force microscopy (AFM) is one branch of the scanning probe microscopy families. AFM is used for cutting-edge research in the emerging field of nanotechnology, which is poised to dramatically affect virtually every aspect of our economy [6]. The analytical research atomic force microscope is the instrumentation for general purposes in biology, nanoscience, nanotechnology, medical science, and material science, while the industrial atomic force microscope is the automated recipe-driven equipment for inline production metrology capable measurements. Measurements are programmed in a recipe for automated wafer handling, alignment, probe handling, site registration, image capture, and image data analysis to output the final measurement data, without operator intervention. In particular, AFM has been widely used in semiconductor fabrication as a dimension metrology tool for the advanced geometry control at the 130-nm technology node and below for etching and chemical mechanical polishing (CMP) characterization [7]. With similar process technologies to those used in the semiconductor industry, the photomask and thin film head industries have also adopted AFM for process metrology control.

There are several equipment suppliers worldwide who make a mix of industrial AFM platforms, each tailored for specific applications. AFM measures the surface topography, 3D dimension and geometrical shape, horizontal surface profiling, and perpendicular sidewall shape profiling [8]. The measurement area can be a small (less than 50 μm) or a long (less than 10 cm) range. On the small scale, the measured variables are height or depth, linewidth, linewidth variation, line edge roughness (LER), pitch, sidewall angle, sidewall roughness (SWR), cross-sectional profile, and surface roughness [9]. On the long range, AFM is used for surface topographical profiling for CMP processes.

This chapter is intended to give an overview of industrial AFM metrology applications across several nanoelectronic industries. Less published literature is found on AFM than on traditional electron-beam and optical metrology techniques [10, 11]. The chapter cites more relevant literature on applications of AFM [12]. The objective is to explain the role of AFM metrology with actual process data in process control and yield enhancement efforts, and to stimulate the use of AFM for advanced process control across fabrication plants. All data discussed are intentionally scaled and are generic in nature across the industry and are only used for the purpose of illustrating the AFM metrology technology.

38.2
Dimensional Metrology with AFM

38.2.1
Dimensional Metrology

Besides AFM, critical dimension (CD) scanning electron microscopy (SEM), cross-sectional SEM (X-SEM), transmission electron microscopy (TEM), dual beam, optical scatterometry, the optical profiler, and the stylus profiler are all examples of other dimensional metrology technologies available for integrated microelectronic process characterization and monitoring. Metrologists need to understand their benefits and limitations to make educated decisions for metrology selection.

There are apparent advantages of AFM over other metrology technologies owing to its unique characteristics. The most trustworthy 3D analysis seems to be using TEM or X-SEM. But the drawback of X-SEM or TEM is sample preparation, instrument operation, time, and cost. X-SEM and TEM require destroying the wafer, but only provides a single cleavage into the feature. TEM cannot be used for resist. CD SEM charges, shrinks, or even damages resist [13, 14]. Like optical microscopy, CD SEM gives little 3D shape information. CD SEM also suffers from the pattern density and proximity effect. Scatterometry is fast and precise, but only works on designated grating structures. Scatterometry cannot work on any arbitrary feature and provides no LER and linewidth roughness (LWR) data. It is often difficult and time-consuming to develop a reliable scatterometry library for the specific film stacks. X-ray, optical thickness, or profiler tools are often limited by spatial resolution and spot size. They only work in the scribe region, not anywhere in the die.

The atomic force microscope simply works in an ambient environment. No sample preparation or vacuum is needed. AFM is a surface force sensitive microscopy, and provides a nondestructive, direct, absolute, and 3D measurement rather than simulation, modeling, or inferring. AFM allows a quick survey of the cross-sectional profile or surface topography to examine if the dimension is in specification, without destroying a product by TEM. The atomic force microscope performs many lines of a scan along a feature at many sites across many wafers, collecting enough statistics to evaluate feature-to-feature, die-to-die, wafer-to-wafer, and lot-to-lot variations. The atomic force microscope has no spot size limitation, and its probe has much higher resolution than optical or stylus profilers for CMP planarity applications.

AFM measures inline samples of any materials used in today's nanoelectronics industry, regardless of film stack, optical properties, or composition. AFM is insensitive to new materials emerging from the newest advanced processing and material integration (strained SiGe, high-k, metal gate, or low-k). With device miniaturization, the circuit geometry and density are scaled down, and the pattern fidelity and dimension depend on the immediate surroundings. However, AFM is free from the bias free of feature proximity or pattern density effects. These are all important requirements outlined in the metrology section of ITRS 2005 [15].

A versatile nanoscale metrology tool, the atomic force microscope scans anywhere within the die, regardless of the specific device structure or function design. The only limitation for use of AFM is the size of the feature space. AFM only works when there is a space wide enough for the tip to step in to perform scanning. The atomic force microscope would not scan a trench space narrower than the tip diameter. Obviously, AFM would not profile a via hole already filled with copper, whereas X-SEM or TEM can. As AFM probe technology advances, smaller tips are already available for very tight space and very high aspect ratio structures [16].

Therefore, AFM is the most accurate, nondestructive, and 3D inline dimension metrology tool among all, independent of material, density, or proximity, and with superior precision and linearity. AFM is fast and inexpensive compared with X-SEM or TEM. Although it is slower than scatterometry or CD SEM, AFM still offers many advantages over them. Consequently, AFM has gained prevalence in the worldwide semiconductor industry and its presence is increasing for the 90-nm node and beyond. Table 38.1 lists typical applications using automated AFM metrology in terms of applicable product types and process steps. In terms of application objectives, AFM can be used for inline monitoring for depth, CD, and profile [9]; replacement of TEM for engineering analysis for cross-sectional profile [8, 17], CD bias determination, and focus exposure matrix (FEM) studies; complementary (not replacement) reference metrology for inline scatterometry and CD SEM [10, 18, 19]; CD targeting, traceability, and correlations among all dimensions tools across fabrication plants or technology generations [20, 21].

38.2.2
AFM Scanning Technology

The atomic force microscope raster-scans a rectangular region anywhere from a few nanometers up to approximately $50\,\mu m$. The scan is highly localized and can be anywhere for in-die metrology, as long as the space permits the tip size. The measurement is free of bias arising from the target shape, pattern proximity, density, and material.

Within a feedback control loop, the atomic force microscope scanner controls a tiny probe to perform scanning motion in x (or y) and z directions to maintain a close proximity between the probe and sample surface, acquiring high-resolution positional data in all x, y, and z axes. The 3D topographic raw image is constructed from the $x/y/z$ spatial data, shown in Fig. 38.1. Then, offline software analysis deconvolutes the tip shape from the AFM images and extracts important geometric parameters about the measured target: depth, linewidth at top/middle/bottom locations, sidewall angle and profile shape, or surface topography.

In addition to the traditional contact and tapping scan modes, recent atomic force microscopes incorporate two improved scanning algorithms: deep trench (DT) and critical dimension (CD) modes [22]. In tapping mode, the scan speed is constant regardless of the feature topography. The probe moves at a constant speed over a feature and skips over sidewalls. The resultant profile wrongfully measures the

Table 38.1. Common atomic force microscope (AFM) applications for surface topography, dimension, and geometry shape control in nanoelectronic manufacturing process control

Process metrology	Vertical Depth/height	Lateral Linewidth/pitch sidewall shape	Surface Long-range profiling small-scale topography
Semiconductor—transistors for logic or NAND/NOR flash memory			
STI resist pattern	Resist height	CD (at any height), profile angle	
STI etch	Trench depth	CD, profile angle	
STI CMP	Recess step height, divot		Post-CMP topography, divot
Gate resist	Resist height	CD, sidewall profile, etch bias	
Gate etch	Line height	CD, sidewall profile, LER, LWR, SWR	
Multigate etch	Gate height	3D shape, CD, sidewall profile	
Gate spacer etch	Line height	CD, sidewall thickness, profile, LWR	
Contact/via resist	Resist hole depth	Resist pattern CD, profile, LER, LWR	
Contact/via etch	Hole depth hole CD, profile		
Contact/via CMP	W plug recess/protrusion		Dishing and erosion, total run-out
Metal resist pattern	Resist depth	CD, profile	
Metal trench etch	Low-*k* etch depth	Trench CD, profile	
Metal CMP	Local topography		Dishing and erosion, total run-out
Dual damascene etch	Multiple trench depths	Multiple trench CD, profile	Low-*k* topography
PMD/IMD CMP	Local topography		Surface topography, global planarity
Films	Microroughness, grain, surface texture or damage for any material, post CMP, implant, or anneal		

Table 38.1. (continued)

Process metrology	Vertical Depth/height	Lateral Linewidth/pitch sidewall shape	Surface Long-range profiling small-scale topography
Semiconductor—memory			
DRAM poly recess	Plug recess etch depth		
DRAM contact etch	Plug depth		
FeRAM capacitor etch	PZT etch depth	Film stack CD, profile	
MRAM magnetic junction	Photopattern, etch, and CMP for magnetic tunnel junction film stacks		
Semiconductor—photomask			
Resist pattern	Resist depth	Resist pattern CD, profile	
Cr/MoSi/quartz etch	Etch depth	CD, profile, LER, undercut, footing	
Phase-shifter	Phase shift depth	Sidewall profile	
OPC development		CD verification, profile	
Defect review/repair	Defect full 3D volume	Defect cross section, bottom shape	Defect topology

Table 38.1. (continued)

Process metrology	Vertical Depth/height	Lateral Linewidth/pitch sidewall shape	Surface Long-range profiling small-scale topography
Data storage—thin-film recording head for hard disk (LMR, PMR, TMR)			
Reader resist pattern	Resist depth	Resist 3D profile, CD	
Reader pole etch	Etch/milling depth	Pole 3D profile, CD	
Writer resist pattern	Resist depth	Resist 3D profile, CD	
Writer pole etch	Etch/milling depth	Pole 3D profile, CD, LWR, LER, SWR	
Slider—after dicing, slicing, and lapping	ABS), PTR, PWP		Slider surface roughness
MEMS/NEMS			
Contact image sensor—CIS or CCD			
Microlenses	Depth	Radius, pitch, width, angle	3D geometry shape and curvature
DLP DMD	Metal thickness	Mirror spacing	Mirror planarity

STI shallow trench isolation, *CMP* chemical mechanical polishing, *CD* critical dimension, *LER* line edge roughness, *LWR* linewidth roughness, *SWR* sidewall roughness, *PMD* premetal dielectric, *IMD* intermetal dielectric, *FeRAM* ferroelectric RAM *PZT* lead zirconate titanate, *OPC* optical proximity correction, *LMR* longitudinal magnetic recording, *PMR* perpendicular magnetic recording, *TMR* transverse magnetic recording, *ABS* air-bearing surface, *PTR* pole tip recess, *PWP* perpendicular writer protrusion, *MEMS* microelectromechanical system, *NEMS* nanoelectromechanical system, *CIS* complementary metal oxide semiconductor image sensor, *DLP* digital light processor, *DMD* digital micromirror device

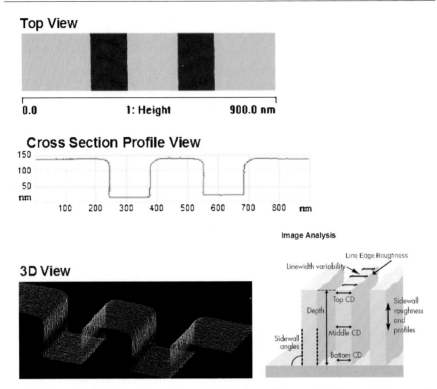

Fig. 38.1. Typical atomic force microscopy (AFM) images and data analysis for a typical nanoscale semiconductor line/space feature

sidewall angle. Unlike the tapping mode, CD and DT modes use adaptive scanning. The scanner slows down in the horizontal direction when sensing a sidewall, allowing the probe to move down along the slope. In DT mode, the scanner only moves in the vertical axis until it reaches the feature bottom and then resumes the horizontal scan. Only a small amount of data is collected along the sidewall, but plenty of data points are collected along the line top and bottom flat levels, making the DT mode ideal for depth measurement. In CD mode, the scanner moves 45° toward the sidewall surface, collecting 2D data about the sidewall profile. The CD mode traces and reproduces the full shape of the flat and sidewall surfaces. Therefore, the CD mode allows a full 3D shape and linewidth measurement, while the DT mode is tailored for depth measurement for high aspect ratio DT features. In addition, there is an AFM profiler mode that performs a single line trace scan up to 10 cm on a sample for surface topography profiling. Advanced atomic force microscopes combine all these scanning modes into the tool platform, giving users versatility and flexibility for different measurement requirements.

The probe scanning algorithm and probe design are core components of AFM. An optimal AFM measurement is achieved by proper selections of hardware platform, scanning mode, probe, scan setting, and image analysis based on feature dimension, topography, and desired measurement outputs.

38.2.3
AFM Probe Technology

One of the core components of AFM is the probe design and geometry. Typically, an AFM probe comprises a small tip affixed to a long cantilever arm attached to a lager substrate. The substrate is generally mounted to a tip mount holder on the scanner head by vacuum suction. Figure 38.1 shows a side view of the probe assembly.

Atomic force microscope tips are application-specific. Tip selection depends on the specific feature geometry and the dimension parameters intended to be measured. The common tip shape and specification are summarized in Table 38.2. Conical, cylindrical, or pyramidal tips accommodate vertical depth or height measurements. A conical tip is slender like a needle with a very small tip radius and small diameter, and is best suited for depth measurements in the DT mode for high aspect ratio narrow DTs or deep holes. A cylindrical post tip diameter remains constant as the tip becomes short, providing a better precision for void-type DTs or deep holes [23]. A pyramidal tip is sharp at the apex with a strong base, and is ideal for surface microroughness characterization or a long-range profiler in the tapping mode.

The tip that looks like an inverted mushroom is the most advanced design. The tip with flared overhangs on the bottom edge is used in the CD mode for feature sidewall scanning and linewidth measurements, where the sidewall profile/angle/slope, cross-sectional profile shape, LER, LWR, and SWR can all be characterized from the same CD AFM image (Fig. 38.2). The protruded overhang allows the tip to work on sidewalls even with reentrant profiles. All tips used in AFM applications are some sort of variation on the abovementioned basic tip designs and shapes. Most of the AFM probes are made of silicon or carbon materials compatible with modern nanoelectronic silicon-based materials.

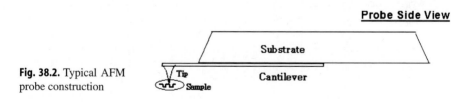

Fig. 38.2. Typical AFM probe construction

38.2.4
AFM Metrology Capability

The following performance metrics are evaluated to assess the fundamental metrology capabilities of AFM. The short-term precision and accuracy are listed in Table 38.3. In Fig. 38.3, the atomic force microscopy results are shown to be well correlated with those from TEM and CD SEM for gate resist and gate poly CD, with $R^2 > 98\%$ [13]. The long-term stability for height, linewidth, sidewall is shown in Fig. 38.4. The measurement linearity is shown in Fig. 38.5. Like other

Table 38.2. Nominal specifications for tips commonly used in an industrial AFM in (nanometers)

Measurement	Tip Shape	Tip Material	Overall Width	Effective Length	Vertical Edge Height	Lateral Overhang	Edge Radius	Apex Radius	Lateral Stiffness (N/m)
Vertical	Conical	Si Needle by Ion Beam	30.0	1000–8000	–	–	–	15.0	0.1
		Sharp Cone by e-Beam	20.0	300–500	–	–	–	10.0	0.5
	Cylindrical	Si Post	55–200	400–800	–	–	–	–	100
		Carbon Coated	50–200	400–1000	–	–	–	15.0	1–5
		Carbon Nanotube	20.0	200–700	–	–	–	20.0	0.5
	Pyramidal	Si Pyramid	–	2000–8000	–	–	–	15.0	1000
		Si Spike for HAR	–	2000–8000	–	–	–	15.0	1000
		Carbon or Diamond Coated	–	2000–8000	–	–	–	15.0	1000
		Co/Cr Coated (MFM)	–	2000–8000	–	–	–	15.0	1000
Lateral	Flared Bottom	Circular Section	30–850	200–1000	10–40	10–35	8–15	–	1–500
		Triangular Section	50, 70, 850	200–1000	10–40	16.4	5–10	–	3.5
		Square Section	75.0	400.0	17.0	15.0	20.0	–	3.0
		SiN Capped	70, 100, 140	400–550	10–20	15–40	8–15	–	20–30
		Carbon Coated	32, 50, 70	400–550	10–20	15–40	8–15	–	5–10
		Si Trident	100–130	300–500	5.0	40.0	5.0	–	5.0

Table 38.3. AFM precision and accuracy data (nanometers) for height and linewidth measurements on NIST traceable standards

Standards	Height				Linewidth	Sidewall angle
Target value	8.70	15.40	188.20	973.90	70.30	90.00
Measurement repeat	10	10	10	10	10	10
Average 8.79	15.50	187.69	969.88	70.28	90.03	
Bias (%)	1.05	0.62	−0.27	−0.41	−0.02	0.04
Precision (3σ)	0.18	0.37	0.37	1.16	0.77	0.05

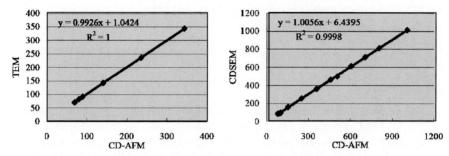

Fig. 38.3. Correlation of critical dimension (CD) AFM with transmission electron microscopy (TEM) results for gate resist and gate poly CD (nanometers). *SEM* scanning electron microscopy (Courtesy of [13])

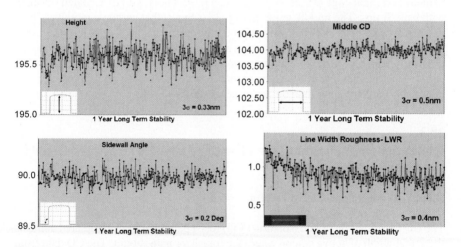

Fig. 38.4. Long-term measurement stability for CD AFM

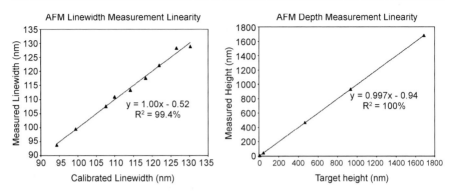

Fig. 38.5. Measurement linearity of AFM for linewidth and height

microscope-based metrology tools, the atomic force microscope first takes a scanning image of the feature of interest (shown in Fig. 38.1). Then the offline image analysis splits out measurement data that process engineers use for process control.

The performance of the atomic force microscope tip is one of the most important factors in AFM metrology [24]. Robust tip design and tip management are the cornerstones for precise and accurate AFM metrology. More systematic studies on the tip shape characterization, tip qualification, tip wear, and tip lifetime have been reported elsewhere [25, 26]. Here we only show the tip wear trend for some typical atomic force microscope tips. The top graph in Fig. 38.6 shows a comparison between SiN-coated and bare Si CD atomic force microscope tips. One of the key tip shape variables (vertical edge height) behaves much more stably with SiN-coated tips than with bare Si tips. The middle graph in Fig. 38.6 shows that carbon nanotube (CNT) tips barely wear even after being used 2000 times on etch silicon or oxide structures. The bottom graph in Fig. 38.6 shows that a SiN-coated CD atomic force microscope tip slowly wears down over 2000 measurement sites on gate poly samples. Owing to the weak interactions between the tip and soft polymers, the tip lifetime on resist materials is much longer than on nonresist materials.

38.3
Applications in Semiconductors—Logic and Memory Integrated Circuits

38.3.1
Shallow Trench Isolation Resist Pattern

The CD for a shallow trench isolation (STI) resist pattern after development controls the actual width of active silicon where the transistor gate is built. The CD data from the grating lines obtained by optical scatterometry are usually used for monitoring the CD. However, CD AFM is the only inline reference metrology tool that can calibrate and validate scatterometry models. An AFM scanning STI resist with a flared tip in the CD mode quickly maps out the whole wafer for resist CD, resist height, and sidewall angle. Figure 38.7 shows results for a flared tip scanning over

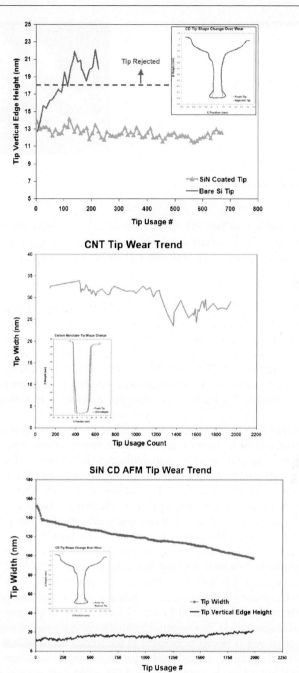

Fig. 38.6. Tip wear trends for some typical atomic force microscope tips used in depth and CD measurements. *CNT* carbon nanotube

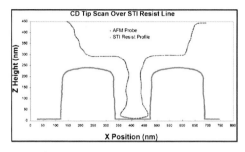 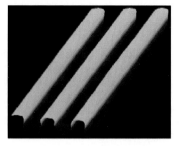

Fig. 38.7. *Left*: A flared CD tip scans over resist lines. *Right*: 3D rendering of a shallow trench isolation (*STI*) resist pattern

the resist lines and a resultant 3D AFM view of the STI resist lines. Figure 38.8 shows a study on a FEM wafer for lithographic process optimization using AFM, where resist height, bottom CD, and sidewall angle are measured in less than 1 h across the whole wafer.

38.3.2
STI Etch

AFM finds its unique place in STI etch depth, linewidth CD, and sidewall profile measurements. Figure 38.9 shows the AFM profile compared with the TEM cross section of the same wafer. The comparison indicates that AFM replaces lengthy and costly TEM to characterize the full 3D geometrical shape of narrow and deep STI trenches.

The STI trench film consists of a nitride hard mask layer on top of active silicon. It is difficult for CD SEM to measure exactly at the nitride/silicon transition for silicon CD. A high-resolution AFM scan delineates the transition (Fig. 38.10, left). The image analysis can be programmed at the right transition location to calculate the nitride bottom CD or silicon top CD. The silicon top CD across a FEM wafer after STI etch is shown on the right in Fig. 38.10. With a move–acquire–measure time of 1.5 min per site, AFM allows a fast and nondestructive mapping across the whole wafer for 45 fields. The task would be otherwise prohibitive with X-SEM or TEM. AFM allows a simple and fast inline examination of the whole wafer during process development before a working scatterometry library is built. AFM is the chosen reference metrology tool for scatterometry or CD SEM. In developing the models for the CD library, AFM is used to validate and maintain calibration/accuracy for optical scatterometry. Figure 38.11 shows correlation plots between AFM and scatterometry for depth and CD, respectively. Studies also showed that the STI trench depth at the scribe box is poorly correlated to the trench depth within circuits such as SRAM cells. Traditional scribe-based measurements are no longer sufficient to monitor the real circuits within a die where AFM is the only nondestructive inline metrology solution.

Etch engineers often desire to monitor a minimum geometry space feature within a die (not in the scribe line) near actual transistors, because its width is a good indicator of potential resist scumming and STI seam voids. The space could be

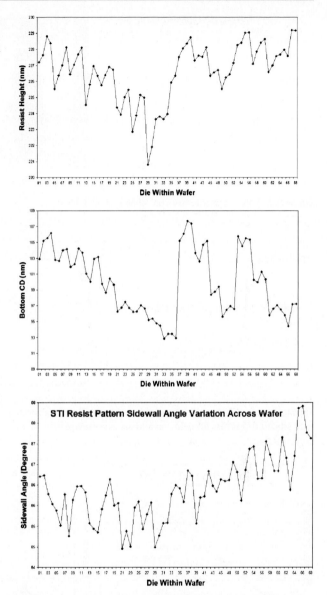

Fig. 38.8. Resist height, bottom CD, and sidewall angle data generated from fast and nondestructive CD AFM scans

as small as 40 nm at the bottom, which is too small for a regular atomic force microscope tip to scan. Recently, atomic force microscope tips made of multiwalled CNTs have adopted for narrow and deep trenches. A comparative study with CNT and conventional focused ion beam (FIB) tips is shown in Fig. 38.12. The 40-nm STI trench is 450 nm deep (10:1 aspect ratio). Figure 38.12 shows that the CNT tip measures the depth correctly, while the FIB tip cannot reach the trench bottom. CNT tips are gaining more presence in industry in semiconductor metrology since its first adoption in AFM [3, 27, 28].

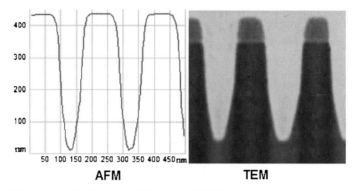

Fig. 38.9. STI trench profile from an AFM scan and TEM

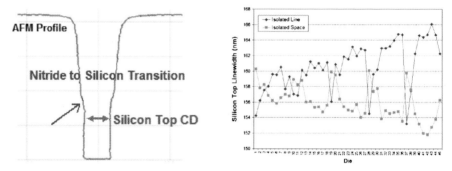

Fig. 38.10. *Left*: AFM profile shows the nitride-to-silicon transition in the STI trench. *Right*: STI etch silicon CD across all fields on a focus exposure matrix wafer

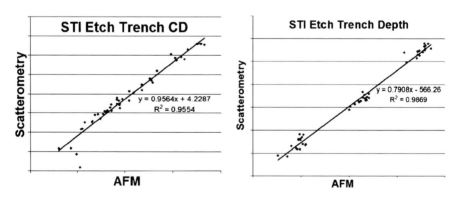

Fig. 38.11. Correlation plots of AFM and scatterometry for linewidth and depth at STI etch

Fig. 38.12. Comparison of CNT and focused ion beam (*FIB*) tips on narrow and deep STI trench depth measurements

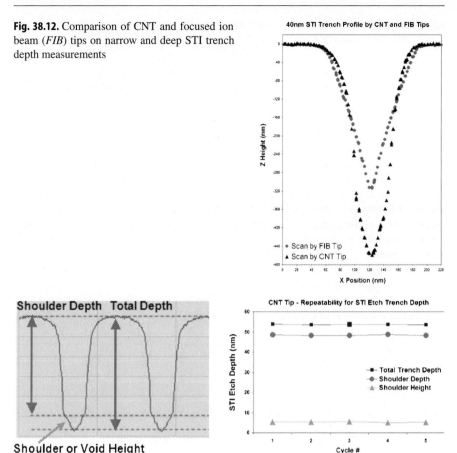

Fig. 38.13. Void-shaped STI trench bottom and the repeatability of AFM depth measurement

In certain DRAM STI etch processes, the trench bottom has a tapered profile with a void shape (Fig. 38.13). The transition from the straight to the sloped sidewall is called a "shoulder". TEM or X-SEM is often used to measure the transition and the shoulder height. AFM is an ideal choice to replace TEM or X-SEM to profile the trench to measure the total trench depth and shoulder height, all with a superior precision. In addition, the nondestructive AFM scans can be performed over many sites across the whole wafer in less than 1 h.

38.3.3
STI CMP

After CMP and nitride strip processes on the STI module, a varying surface topography and height difference between silicon in the active area and the adjacent field oxide is generated (Fig. 38.14). The local topography variation within the actual circuit regions across a wafer is a critical parameter. Transistor electrical failures

Fig. 38.14. STI step height between active silicon and field oxide

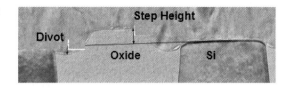

are linked to large or inverted step height difference between the active silicon and the field oxide within dense SRAM or logic circuits. For devices at the 90-nm node and below, the circuit patterns become more complex with the aggressive geometry and density scaling. The topography for STI CMP strongly depends on the feature dimension and pattern density (Fig. 38.15). However, the step height correlation among different features within the die is poor (Fig. 38.16). This implies the traditional ellipsometry and profilometry measurement on a large metrology structure (more than 100 μm) located in the scribe line is no longer sufficient to reflect the real topography of the circuits within the die.

AFM is the only inline metrology technology that enables fast and nondestructive in-die topography monitoring over dense SRAM cells or logic regions (region less than 1 μm) across field and across wafer STI step height monitoring. AFM can detect and measure the reverse silicon/oxide step heights due to the nonuniform polishing

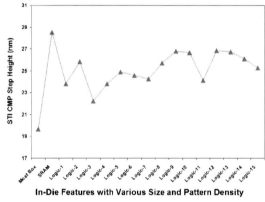

Fig. 38.15. STI chemical mechanical polishing (*CMP*) step heights vary significantly with features of different pattern density within the same die

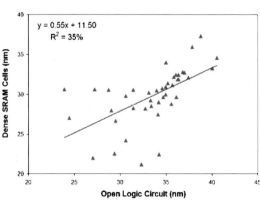

Fig. 38.16. Correlation plot for step height between two distinct features across a wafer

rate in the wafer edges (Figs. 38.17, 38.18). Phase imaging is one way to work around the inverted patterns to center the AFM scans [30]. The oxide divot at the interface between the active area and the isolation area after nitride removal impacts the transistor threshold voltage. AFM is sensitive for profiling the divot (Fig. 38.17) and the divot depth can be monitored.

Some of the primary transistor metrics (such as drive current) are directly impacted by the geometry of the active-to-field interface region. AFM data for the STI recess step height are used to detect process excursions in the production line and to improve STI CMP process uniformity across varied pattern densities, die to die, and wafer to wafer [31]. Figure 38.19 shows a comparison of STI topography for three different CMP splits. Engineers can make a quick decision that process C with a new type of slurry is the worst case and must be reevaluated. AFM offers a fast way to

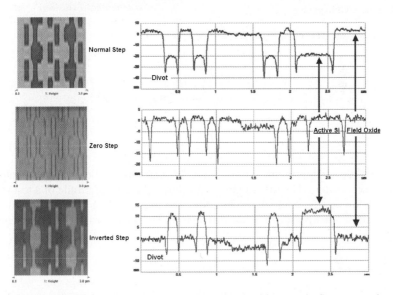

Fig. 38.17. STI CMP and nitride strip processes generate positive, nearly flat, or negative step heights on dense SRAM cells [32]

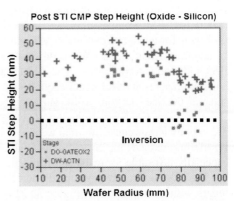

Fig. 38.18. STI CMP step height variation across a 200-mm wafer [32]

Fig. 38.19. STI step heights resulting from three different CMP processes

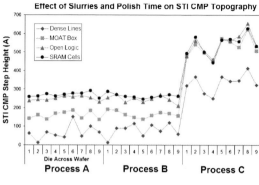

Fig. 38.20. Erosion effect of STI CMP at varied line pitches

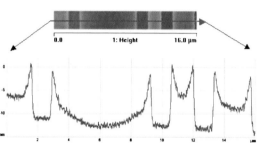

map out the step height uniformly across the whole wafer for dense SRAM cells or any part of the active circuit regions. An AFM scan can also be used to evaluate the erosion effect after STI CMP. Figure 38.20 shows the surface profile of a test structure with varied line pitch after STI CMP. The microloading effect from different line density is clearly shown by the different recess depth across the test structure.

38.3.4
Gate Resist Pattern

The gate photoresist patterning process defines the poly gate dimension and is the most critical step in modern microlithography. Using AFM for dimension characterization introduces no electron-beam damage, curing, or electrostatic charge on resist materials. AFM provides unbiased and absolute CD/sidewall profile data on resist processes for advanced extreme UV, immersion, and double patterning lithography. In addition, resist LER, LWR, and SWR are readily quantified together with the linewidth and sidewall angle. The AFM resist LER data help select types of resist and optimize resist development conditions. CD SEM, the AFM counterpart, cannot separate the line CD from profile variations. An electron beam causes resist charging and shrinking. CD SEM lacks the 3D capability to measure SWR. The soft resist polymer materials cause minimal wear of the atomic force microscope tips. The long tip lifetime reduces the cost of consumables and makes AFM more beneficial and especially favorable for resist applications.

The atomic force microscope tip performs high-resolution scans along the resist length with multiple passes. Figure 38.21 shows the resist bottom linewidth plotted

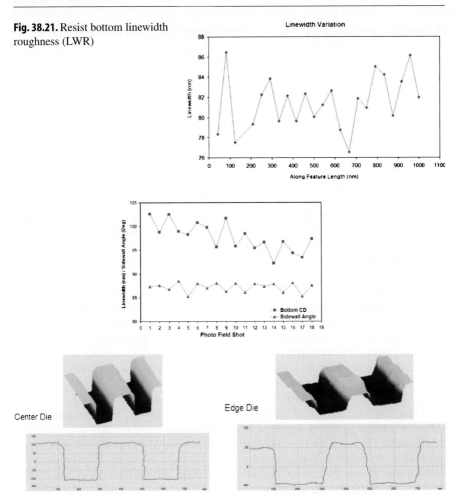

Fig. 38.21. Resist bottom linewidth roughness (LWR)

Fig. 38.22. Resist CD, shape, and profile data from a FEM wafer for lithography process optimization. (Courtesy of [34])

along the length of a resist line to show the LWR data. Figure 38.22 shows the data from a FEM resist wafer. AFM provides a fast and nondestructive resist height, CD, and profile metrology across the whole wafer. The 3D resist shape and profile at edge and center fields is readily attainable for engineers to review. A recent AFM application is to calibrate and verify the overlay metrology tool with CD AFM [33].

38.3.5
Gate Etch

Metal gate or gate poly CD and profile control most critical for defect-free and high-performance transistors. X-SEM and TEM are time-consuming, require wafers to be scrapped, and only provide limited statistics. Thanks to high precision and fast throughput, optical scatterometry is gaining popularity as the preferred CD metrology tool for gate etch. However, scatterometry works on simulation and models from CD data libraries. The accuracy and precision are affected by many process variables, such as poly/moat roughness, oxide thickness, resist, clean, and implant. For complex gate structures, it takes weeks or even months to perfect the scatterometry library. It needs verification and calibration to develop models for specific film stacks. Scatterometry only works on the designated gratings, and cannot be used to characterize any arbitrary features such as memory cells or logic circuits within the die. In addition, scatterometry fails on nonreflecting materials such as antireflection or hard mask coatings. Scatterometry measurement is averaged over a large 50 μm × 50 μm area with no LER, LWR, or SWR data.

AFM offers a bias-free and direct measurement anywhere within the die on any materials, and can serve either as an inline monitoring tool or a reference metrology tool for scatterometry calibration and library optimization. AFM can scan all the fields in a gate photo or etch FEM wafer in hours to obtain linewidth and profile data with sufficient statistics. The LWR and LER data on gate poly help optimize patterning and etch conditions. As a dimension metrology tool with an absolute accuracy, AFM is used to ensure the gate CDs meet design specifications and set design size adjust tables for every technology node and critical devices across multiple fabrication plants. CD AFM is used to measure gate CD offsets between p-type MOS (PMOS) and n-type MOS (NMOS), between isolated and dense grating lines, and between core input and output features. CD AFM is used for the final CD verification for the most critical circuits and parametric test structures, including reticle errors, snapping to grid, optical proximity correction (OPC) models, and OPC tools. Typical gate-etched poly AFM images are displayed in Fig. 38.23.

Shown in Fig. 38.24, the gate poly line sidewall profile variation across two different features (isolated and dense lines) can be easily and quickly determined by AFM scans. Characterized by AFM, Fig. 38.25 shows the bias for CD, LWR, and sidewall angle between n-doped and p-doped ploy scatterometry lines on the same FEM wafer. Figure 38.26 provides a detailed shape profile for NMOS and PMOS features. The difference is apparent from AFM scans without resorting to TEM.

Figure 38.27 shows that the microloading effect on poly etch depth causes a deeper etch over the isolated lines compared with dense lines. The etch offset can be determined in a few minutes by AFM without cleaving the sample. The benefits with AFM metrology are that multiple key geometric measurements of CD and profile are extracted from a single AFM scan image. Table 38.4 lists the linewidth, height, sidewall angle, SWR, LER, and LWR data for two wafers with six sites

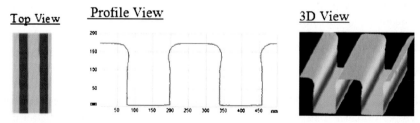

Fig. 38.23. AFM images for poly etch lines

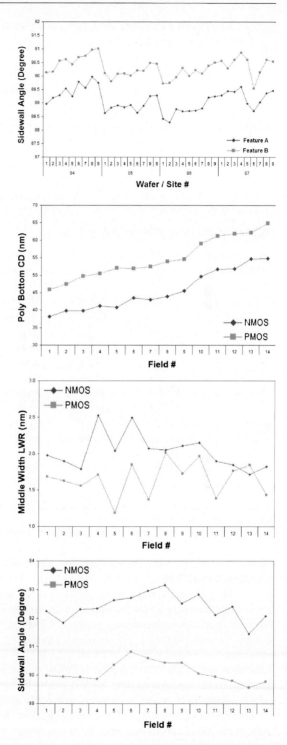

Fig. 38.24. Compaison of the CD and the profile between two features

Fig. 38.25. Bias between n-type MOS (*NMOS*) and p-type MOS (*PMOS*) poly lines

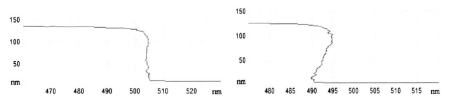

Fig. 38.26. Sidewall profile shape comparison

Fig. 38.27. Microloading effect for poly etch

Table 38.4. AFM metrology data for two wafers across multiple sites on each wafer (nanometers)

Wafer	Site	Poly height	Middle linewidth	SWA	LWR	LER	SWR
A	1	128.73	32.10	89.44	1.16	1.07	1.12
	2	125.65	31.67	90.43	1.32	1.52	1.75
	3	123.61	30.83	90.69	1.68	0.96	1.50
	4	124.48	31.14	90.34	1.41	1.42	1.52
	5	126.85	49.77	90.54	1.41	1.79	1.13
	6	120.73	49.35	90.24	1.38	1.30	1.71
B	1	131.96	51.44	90.72	1.04	1.26	1.15
	2	127.41	48.74	90.90	1.44	1.12	1.49
	3	127.36	52.50	91.58	1.44	1.36	2.13
	4	127.92	49.06	90.68	1.56	1.28	1.41
	5	123.15	48.81	90.14	1.55	1.36	1.83
	6	126.37	48.46	90.85	1.85	1.00	1.22

SWA sidewall angle

measured on each wafer. The task requires less than 1 h with AFM, but is prohibitive with X-SEM or TEM.

At the gate technology level, AFM is often used as an inline monitoring or a reference metrology tool for optical scatterometry. The measurement targets are lines and spaces on scatterometry gratings. AFM can also directly scan the real circuit feature for nondestructive 3D geometry topography failure analysis over memory cells with bit failures. Figure 38.28 shows a high-resolution AFM image for SRAM cells after poly etch. The detailed information on poly line profile, active silicon and field oxide topography, and the STI oxide divots is directly visualized, allowing circuit designers or process engineers to quickly examine for the different features.

During the early stage of gate etch process development, engineers need to understand the effects of etch and photo conditions on the final gate sidewall profile.

Fig. 38.28. 3D AFM scan over SRAM cells

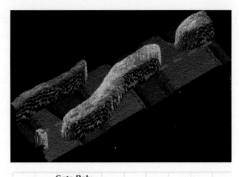

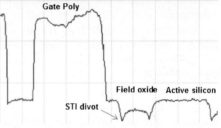

Fig. 38.29. Poly etch profile with reentrant sidewall. *SWA* sidewall angle

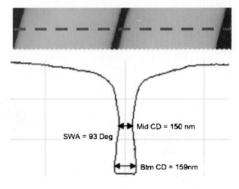

Relying on the X-SEM or TEM feedback adds extra time to the development cycle time. Engineers often desire to continue the experiment on the same set of wafers for subsequent process stages without scrapping the wafers. AFM generates cross-sectional profile scans in the CD scan mode nondestructively, allowing the engineer to quickly assess the poly profile and optimize etch or photo process conditions. Figure 38.29 shows an etch process generates an unwanted undercutting sidewall profile.

38.3.6
FinFET Gate Formation

In the modern complementary MOS (CMOS) semiconductor technology, one of the most superior novel transistor architectures is FinFET, also known as multigate field-effect transistor (FET) or trigate FET, where the transistor gate material (poly or metal) warps around the thin fins of the silicon source/drain. In this 3D FET gate structure, the source/drain vertical fin regions are elevated on the silicon-on-insulator (SOI) substrate. The gate electrodes surround the fin channel region on three sides (left, right, and top). The electric field from the gate is nearly uniform throughout the channel region. The transistor behaves like a fully depleted device, without the extreme short-channel effects of a planar bulk transistor. The switching time of a FinEFT device is significantly faster and the current density is higher than that of the traditional planar CMOS technology. The top chipmakers are already adopting FinFET in both CMOS logic transistors and floating-body flash memory cells at 45 and 32-nm nodes. Most process developers have viewed FinFET as a potential alternative for the 32-nm node and beyond, after all the planar alternatives have been exhausted.

The dimension metrology for the FinFET device is the key enabler for process development and monitoring. The gate length is typically below 50 nm. Figure 38.30 shows the key geometries that are to be characterized. Figure 38.31 shows that high-resolution AFM provides a nondestructive 3D geometrical quantification and characterization for FinFET gate structure. The automated AFM recipe can be set up to take 3D FinFET images. The offline image analysis can output height, linewidth, sidewall profile, and SWR data for gate electrodes or fins as shown in Fig. 38.31.

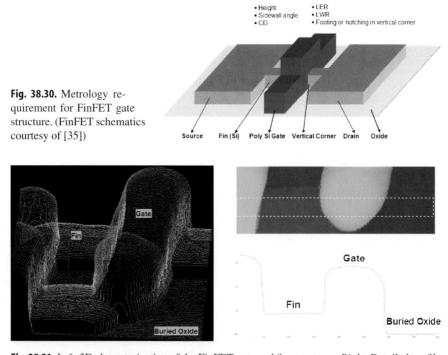

Fig. 38.30. Metrology requirement for FinFET gate structure. (FinFET schematics courtesy of [35])

Fig. 38.31. *Left*: 3D characterization of the FinFET gate and fin structures. *Right*: Detailed profile and CD analysis for FinFET

38.3.7
Gate Sidewall Spacer

The gate spacer is nitride or oxide film deposited on the sidewall of etched gates to provide an offset for the source/drain implantation. The CD, sidewall profile, and sidewall spacer thickness are all important parameters to monitor. The spacer thickness and profile measurement is technologically challenging and continues to be a problematic issue for most available thin-film metrology tools such as X-ray, optical, acoustic sonar, or electron-beam based techniques [36].

Owing to the unique capability of pattern recognition, the atomic force microscopy tip can be precisely placed at the same spot on the same wafer through sequential process steps. Metrologists use AFM to scan the same gate line after gate etch and later after gate spacer etch, to obtain CD and profile data for each process (Fig. 38.32). The difference conveniently gives the dielectric spacer sidewall thickness and profile. The nature of direct measurement with absolute accuracy simply takes the guesswork out of optical metrology. This approach can be extended to the back-end copper seed or atomic layer deposition barrier thickness measurement for trench or via sidewalls.

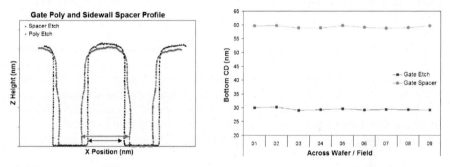

Fig. 38.32. Use of AFM for gate spacer profile and sidewall spacer thickness measurements

38.3.8
Strained SiGe Source/Drain Recess

Over the past 2 years, one of the most successful advancements in pushing the envelope of planar CMOS architectures is strained engineering. The charge carrier mobility is enhanced by introducing strain to silicon crystals within source/drain channels. For PMOS transistors, compressive strain is induced typically in one of the two ways: SiGe embedded in the source/drain region, or compressively strained nitride layer over the gate (dual stress layer). The drive current (CMOS speed) can be boosted by 40% [37]. The strain silicon technology has been implemented in 45-nm-node products by some of the top chipmakers [34].

An embedded SiGe source/drain is typically formed on the SOI substrate as follows. A lateral and vertical etch is performed on source/drain regions after gate

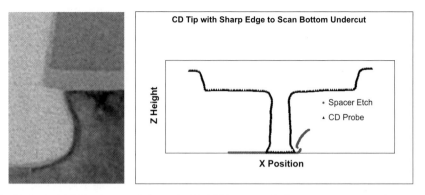

Fig. 38.33. *Left*: TEM micrograph for embedded SiGe source/drain recess structure under a gate spacer. *Right*: A sharp atomic force microscope tip scans the SiGe recess undercut profile. (Courtesy of [34])

spacer formation. The isotropic lateral etch creates a recess in the source/drain channel on either side of the gate. Then in situ doped SiGe is epitaxially grown in the recess. The complete SiGe structure is shown in Fig. 38.33. The recessed SiGe acts as a compressive stressor to enhance the transistor drive current.

The dimension metrology for the recessed SiGe source/drain is critical. Especially, etch engineers desire to monitor and control the recess undercut height and length underneath the spacer. As shown in Fig. 38.33, a sharp flared CD atomic force microscope tip with less than 7-nm edge height can profile the undercut in less than 1 min. The undercut is the difference between the maximum and minimum linewidth along the bottom profile, which is readily available from the offline AFM image analysis. Figure 38.33 shows AFM replaces TEM for the profile and undercut characterization.

38.3.9
Pre-metal Dielectric CMP

Pre-metal dielectric (PMD) or intermetal dielectric CMP is an important process to flatten the dielectric films over the poly gate to ensure defect-free tungsten plug formation. Process engineers adopt AFM to evaluate the local topography or global planarity for both pre- and post-PMD CMP. Figure 38.34 shows the data for 45-nm-node PMD CMP process development. As-deposited PMD film topography over ploy lines is evaluated by AFM scans. The same location is scanned again after the PMD CMP step. The PMD planarity change over ploy lines is clearly displaycd. The PMD step height before and after CMP determines the polishing margin. The surface topography after PMD CMP is important to tune the subsequent contact photolithography process. High-resolution AFM also reveals the PMD divots caused by the higher polishing rate at the poly edge. The divots, if not reduced, trap tungsten metal underneath the PMD layer near the poly gate and cause catastrophic transistor shorting.

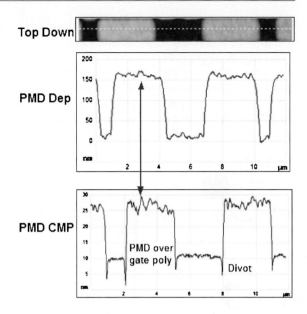

Fig. 38.34. Pre-mela dielectric (*PMD*) topography comparison between as-deposited PMD and after CMP

38.3.10
Contact and Via Photo Pattern

The circular contact or via resist pattern can be characterized by AFM for the hole diameter and sidewall profile. The unusual CD or profiles are quickly detected without destructive cross sectioning. An example is given in Fig. 38.35.

38.3.11
Contact Etch

The contact etch depth and CD measurement is critical to evaluate the etching process. CNT tips provide a solution for depth metrology in the DT scan mode for difficult high aspect ratio contact holes where conventional atomic force microscope tips fail to reach the bottom. Figure 38.36 is for a contact hole with approximately

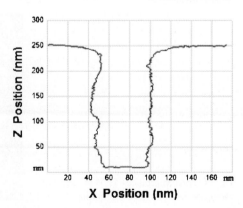

Fig. 38.35. Contact resist profile from an AFM scan

60 nm bottom CD and approximately 520 nm depth, commonly used at 65- and 45-nm-node processing. CNT atomic force microscope tips provide precise depth metrology as shown in Fig. 38.36.

The complex film stack after contact etch often consists of bottom antireflection coating (BARC), tetraethylorthosilicate (TEOS), phosphorus-doped silicate glass (PSG), and silicon nitride over poly silicon. It is often difficult for a CD atomic force microscope tip to reach all the way to the bottom for some high aspect ratio contact

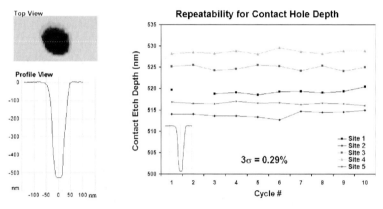

Fig. 38.36. Depth measurement for contact etch and the depth repeatability data

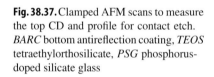

Fig. 38.37. Clamped AFM scans to measure the top CD and profile for contact etch. *BARC* bottom antireflection coating, *TEOS* tetraethylorthosilicate, *PSG* phosphorus-doped silicate glass

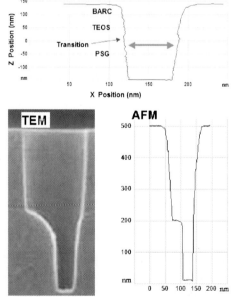

Fig. 38.38. AFM replaces TEM to examine the borderless contact etch profile

38 Automated AFM

holes with 60-nm bottom diameter. Similarly to the STI etch trench case, a clamped AFM scan is performed to scan only the top 300-nm section of the hole, as shown in Fig. 38.37. The transition from the top TEOS layer to the underneath PSG layer is clearly visible and the CD for TEOS or PSG can be properly measured. The contact etch diameter for PSG across all the fields on a wafer for isolated and dense contacts can be measured. Even compared with the clamped scans with AFM, the task would be prohibitive for scatterometry, CD SEM, destructive X-SEM, or TEM.

In the "borderless contact" etch process in some of the DRAM processes, the oxide etch is sensitive to nitride films. As shown in Fig. 38.38, AFM is an effective tool to verify the etch profile.

38.3.12
Contact CMP

While microphotolithography and etch processes create the circuit patterns within a semiconductor device, CMP enables layer stacking. Both local topography and long-range planarity are important for the back-end interconnect integration. AFM

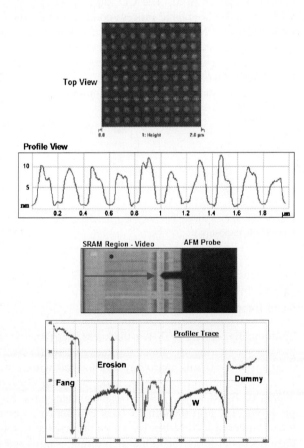

Fig. 38.39. AFM examines the local topography and global planarity after tungsten CMP

has both small-scale (less than 50 μm) and long-range (less than 10 cm) topographic profiling capability anywhere within the die [3, 38, 39]. AFM is the most accurate and direct measurement on any material without modeling. The resolution is less than 1 nm. These are obvious advantages over stylus or optical profilometry.

Figure 38.39 shows the local tungsten plug topography in a 2 × 2 μm area over SRAM cells for a 65-nm-node product. CMP engineers can easily identify if the plugs are protruded or recessed relative to the oxide after polishing. The global planarity over the SRAM region is profiled with a 900-μm AFM profiler mode scan where the SRAM erosion and the fang depth between the dummy area and the SRAM region are apparent. In addition, the data from multiple sites across the field or wafer can be taken in less than 1 h to speed up the process development cycle.

38.3.13
Metal Trench Photo Pattern

Figure 38.40 shows the top 300-nm section of a metal-1 (M1) resist line profile measured by CD AFM. The CD, profile, and SWR can be used for resist type selection and lithographic process condition optimization. Figure 38.41 shows an AFM study on a M1 FEM wafer. It is a nondestructive and accurate metrology tool to allow process engineers to quickly evaluate effects of photo conditions on the resist pattern. Similarly, AFM can be equally applied for resist dimensional analysis for all metal layers (M1 through M9), regardless of materials (resist, dielectric, or metal), feature shape, or feature location in the field.

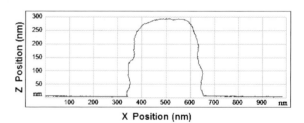

Fig. 38.40. An isolated resist line profile after a metal-1 (M1) photo exposure

38.3.14
Metal Trench Etch

Traditionally, optical or X-ray thickness metrology is used for trench etch depth process monitoring by direct thickness or subtractive thickness methods, using a large (50 × 50 μm) metrology box located within the scribe line. As the technology goes beyond the 65-nm node, a typical dielectric trench film stack may consist of BARC, SiC, TEOS, organosilicate glass, SiC, TEOS, and PSG multiple layers. The film stack in the back-end trench becomes too complex for optical tools to interpret. AFM becomes a natural choice to replace optical or destructive cross-sectional tools for inline monitoring. Additional advantages include in-die

Fig. 38.41. Profile view and 3D view for a M1 resist trench pattern. Also shown are the sidewall angle and CD data across a FEM wafer

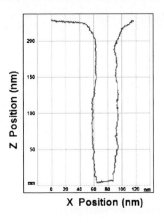

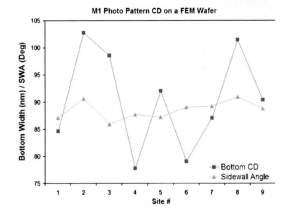

measurements anywhere in the circuits by AFM, with no spot size limitation, with no modeling or calibration needed. Figure 38.42 shows AFM inline monitoring data from M2 and M6 trench depth data from the metrology film box for a 45-nm-node wafer.

Like for any trench-type features, AFM can characterize CD and profiles for the back-end dielectric trench anywhere within the die. Figure 38.43 shows a 420-nm-tall trench with a 60-nm bottom CD for the first M1 trench (M1 trench with the minimum pitch). For the back-end trenches, AFM is typically used as a calibration and validation tool for scatterometry simulation or its use is limited for inline monitoring. Porous low-k dielectrics used in the transistor interconnect are susceptible for electron-beam damage from CD SEM. The material properties are too complex to develop a reliable scatterometry model in a timely fashion. AFM can be used to directly measure the low-k trench profile. Figure 38.44 shows a clear transition from the top antireflection coating layer to the underneath low-k dielectric layer. The CD and profile data at the transition for the low-k layer are readily available from CD AFM profiles. By the same token, AFM equally applies to all metal layers (M1 through M9) for trench etch depth, CD, and profile analysis with CD, CNT, or FIB tips (Table 38.2).

Fig. 38.42. M2 and M6 trench etch depth data across the same wafer

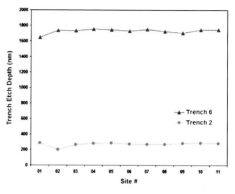

Fig. 38.43. M1 trench cross-sectional profile by AFM

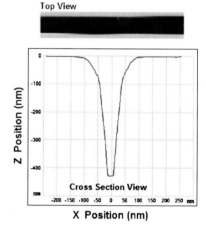

Fig. 38.44. Low-k trench etch profile by AFM. *ARC* antireflection coating

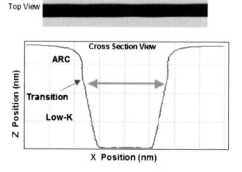

38.3.15
Via Etch

Similarly to the contact photo and contact etch, AFM is an applicable metrology tool for via photo pattern or etch characterization. A via hole normally has the highest aspect ratio within a device, which creates a great difficulty for CD SEM or optical tools [41]. Destructive TEM, X-SEM, or dual beam is often the only way to analyze the via etch profile. Figure 38.45 displays an AFM scan across an isolated via hole with a 10:1 aspect ratio, with a superior metrology repeatability. The via hole depth and profile measured by AFM is a fast and effective process diagnostic tool [42]. Figure 38.46 shows a case of three different etching conditions. A flat bottom indicates an open via (desired). A triangular or bullet-shaped via bottom often indicates an underetched or closed via (defect).

For some of the high-aspect-ratio holes or trenches with unusually small bottom dimension, even AFM shows a limitation to reach the hole bottom with conventional probes. Figure 38.47 shows the advantage of CNT tips over conventional conical FIB tips. The CNT tip diameter is constant at all depths, whereas the FIB tip diameter rises rapidly as it goes deeper into the hole, preventing the FIB tip from reaching the feature bottom.

Dual inlaid damascene etch CD and profile is also a unique application for AFM. Figure 38.48 shows a dual via/trench etch profile. Automated AFM image analysis software within the recipe readily outputs the multiple step heights and CD within a speed of approximately 1 min per site.

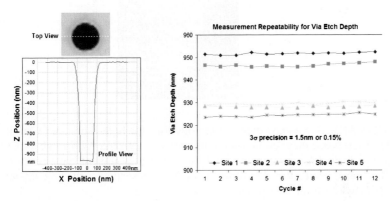

Fig. 38.45. High-aspect-ratio (10:1) via etch profile by AFM and the etch depth precision data (Courtesy of [34])

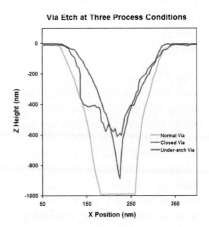

Fig. 38.46. Use of via etch profile and depth by AFM to diagnose an open or a closed via

Fig. 38.47. Comparison of the CNT and conical FIB tip shapes

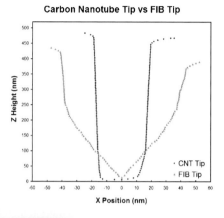

Fig. 38.48. Dual inlaid damascene etch profile by AFM

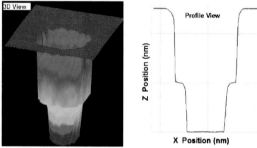

While AFM is a robust depth metrology tool for high-aspect-ratio via holes, the AFM CD mode may run into some limitations regarding the tip size to perform repeatable linewidth measurements on the very narrow bottom of via or contact etch. A way to work around this is to scan the top 250-nm section of the hole to measure the top linewidth at the transition.

By the same concept, process engineers use AFM for fast and absolute depth or CD measurements for all via layers (V1 through V9), regardless of materials (resist, dielectric, or metal), feature shape, or feature location in the field.

38.3.16
Via Etch

The CMP process on a wafer generates surface topography variation and height difference between dissimilar materials. Illustrated in Fig. 38.49, erosion refers to the height difference between a material in a large contiguous area and the same material in an area of dense features of alternating materials such as dielectrics and copper. Dishing refers to the height difference between neighboring areas of dissimilar materials (copper and dielectrics). The overshoot near the edge of the intersection of two dissimilar materials is called total indicated run-out, or fang.

In back-end copper CMP, the local topography variation is not as important as the global planarity. Film thickness tools (optical, X-ray, or acoustic sonar) and

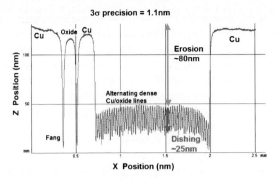

Fig. 38.49. Typical dishing, erosion, and total indicated run-out (fang) measurements by AFM

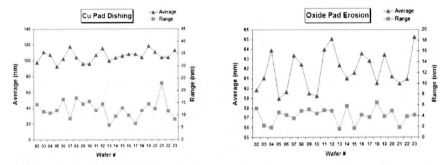

Fig. 38.50. Typical dishing and erosion measurement by AFM

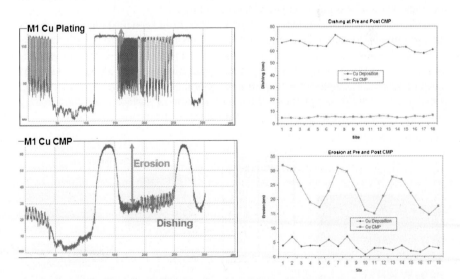

Fig. 38.51. Copper CMP marginality study before and after CMP using the AFM profiler mode

surface profilers (optical or stylus) are available for measurements on a large metal test pad for copper process control. However, AFM also holds its place in copper CMP metrology owing to its many advantages. AFM is an absolute and accurate topography measurement method without spectra fitting or interpretation. AFM, with a lateral resolution of less than 1 nm, resolves densely spaced alternating copper and oxide lines for dishing measurement. AFM can be used for in-die measurements to examine the local topography in the actual circuits after CMP to avoid the correlation error between the test pad and in-die topography.

Figure 38.50 shows dishing and erosion data for M1 copper CMP across 11 fields on a wafer. The average and range of dishing and erosion within a wafer are for inline process monitoring. Figure 38.51 shows the polish marginality is evaluated by the topography change from copper plating (befor CMP) to copper polish (after CMP). The dishing value reduces, while the erosion increases after the copper polish.

38.3.17
Roughness

The traditional use of an atomic force microscope as an imaging and roughness analysis tool still provides many insights for modern nanoelectronic thin-film materials. It is a popular and powerful characterization tool for today's 45-nm node and below for technology development for emerging materials and new processes [43]. Microroughness for incoming SOI substrates, strained Si or SiGe films used in strain engineering, poly roughness and surface texture, contact tungsten deposition, and post-CMP microscopic surface inspection are all powerful process diagnostic tools. There is a great uncertainty with optical-based (laser or X-ray) roughness characterization owing to simulation and modeling.

All atomic force microscope tips have a finite size and generate biased roughness measurements (lower). It is recommended to compare roughness data only with the same tip. The tip sharpness characterization and maintaining the sharpness are critical for production-worthy microroughness applications. Figure 38.52 shows AFM surface scans for a tungsten film with a repeatability of $3\sigma = 0.7$ Å. Figure 38.52 (left) shows that AFM provides a direct comparison of surface grain morphology and microroughness for SiGe films on SOI substrates from two vendors.

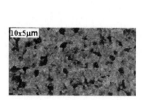
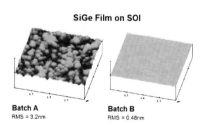

W Roughness (nm)	
Cycle	RMS Rq
1	0.704
2	0.712
3	0.687
4	0.672
5	0.648
6	0.639
7	0.676
8	0.681
9	0.657
10	0.652
3 sigma	0.07

Fig. 38.52. *Left*: Microroughness measurement and the repeatability on tungsten films by AFM. *Right*: Microroughness comparison between two batches of SiGe substrates. *SOI* silicon on insulator

38.3.18
LWR, LER, and SWR

As the device geometry shrinks, the feature LWR, LER or asperity, and SWR start to have significant impacts on device performances, on the wafer level and mask plate level. In the past 2 years, many research papers have discussed these issues. LER impacts the interconnect resistance and LWR affects the transistor leakage current [4]. Good discussions on LER and LWR measurements and their impact on device functions are given in [44–48]. A good review on LWR measurement is given in [49]. For SWR, one can refer to [3, 16, 50]. The capability of high spatial resolution AFM can provide LWR, LER, and SWR data for resist or etch features. The details were discussed in previous respective sections. The limitation of AFM is the tip resolution owing to the tip sharpness and edge radius. For 45 nm and beyond, CNT tips may provide better resolution for LER and SWR. In addition to vertically aligned CNT, Liu et al. [16] evaluated tilted CNT tips for SWR and LER characterization. The small radius of curvature (approximately 20 nm) offers significantly higher imaging resolution than the traditional flared CD probes.

38.3.19
DRAM DT Capacitor

DRAM is the dominant form of computer memory that runs today's PC desktop applications. Nonvolatile flash memories are ubiquitous in modern consumer electronics: portable data drive, digital camera, and music players, to name a few. The AFM applications in DRAM and flash NOR/NAND memory manufacturing are similar to those in the logic device processing, because all device features are of a similar dimension, and they share common manufacturing process technology. Some unique applications with DRAM are the capacitor polysilicon recess and DT etch depth measurements. Figure 38.53 shows a high-aspect-ratio (20:1) deep contact hole for the DRAM capacitor.

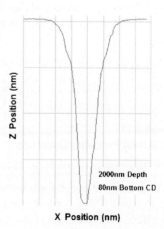

Fig. 38.53. AFM deep trench mode scan for a 20:1 contact hole

38.3.20
Ferroelectric RAM Capacitor

Ferroelectric RAM (FeRAM) is a nonvolatile computer memory. FeRAM consumes less power but reads/writes faster than DRAM. FeRAM consists of one capacitor stacked on a transistor (1C-1T). The FeRAM capacitor cell includes a hysteresis ferroelectric dielectric film material, typically, lead zirconate titanate (PZT). The PZT film is deposited over the CMOS contact plug layer and is sandwiched between iridium electrodes. AFM plays an important role in dimensional monitoring for pattering and etching processes during fabrication of ferroelectric capacitors. Figure 38.54 displays a 3D rendering of FeRAM capacitor cells after stack etch. AFM reveals the true profile and linewidth data for hard mask etch before final patterning for the electrode stack. AFM can replace nondestructive dual-beam cross-sectional analysis. In addition, Fig. 38.55 shows that AFM offers multiple inline measurements for partial depth, bottom, or middle linewidth at the layer transition across multiple wafers. All can be automated by the versatile AFM image analysis software.

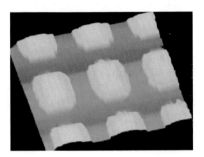

Fig. 38.54. 3D view of ferroelectric RAM (FeRAM) capacitors

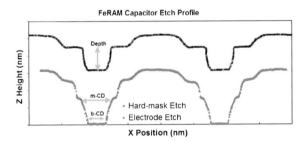

Fig. 38.55. CD AFM profile for hard mask etch and stack etch

38.3.21
Optical Proximity Correction

OPC on a photomask is used to minimize the optical proximity effect originating from the exposure equipment. Without using OPC, the actual pattern on the wafers would appear different from the design pattern. In order to minimize the optical proximity effect, pattern design is purposely altered during the mask-making process in order to achieve the desired pattern. There are different kinds of OPC patterns for different target design patterns, such as scatter bars, assistant slots, hammerheads, jogs, serifs, lines, spaces, line ends, and space ends. The effect of the OPC feature on the final printed CD on a wafer must be verified. Unlike for CD SEM, neither geometry density nor the pattern proximity affects AFM scans. Scatterometry cannot be used on OPC features since OPC patterns are often not in the form of a grating. AFM is a better choice for OPC design validation on the printed wafer [52].

38.4
Applications in Photomask

38.4.1
Photomask Pattern and Etch

A photomask is an essential semiconductor component that contains the detailed blueprint of circuit designs. With use of the photomask in the photolithography process, specific images of detailed device design are transferred onto silicon wafers. The photomask requires precise and accurate depth, profile, and CD targeting to achieve the desired phase shift and imaging conditions. There are three critical processes that make a mask template [53]. The first is feature patterning with lithography, where the

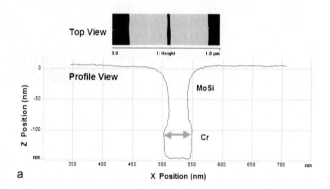

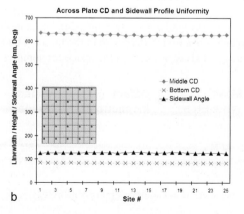

Fig. 38.56. a AFM profile of an etched MoSi layer. b CD uniformity data by AFM across a whole plate

feature CD control and layer-to-layer overlay is vital [54]. The second is quartz etch after patterning, where the etch depth targeting and sidewall angle control is critical. The final process is defect inspection and repair. AFM finds its use in all these critical steps for all today's masks: binary, chromless, embedded attenuated phase shift mask (PSM), alternating-aperture PSM, high-transmission PSM, or nanoimprint template. The Cr or MoSi etch depth or step height obtained by AFM improves the phase shift control [55, 56]. The CD and sidewall profile data after electron-beam writing or dry etch help maintain the CD targets. Figure 38.56a shows that CD AFM can profile the sidewall for etched mask features. The across-plate depth, step height, CD, and profile uniformity are critical. Figure 38.56b shows that AFM can quickly scan many sites and examine the CD uniformity across the whole plate. CD and profile data obtained by AFM can also be correlated to CD SEM for calibration and validation to ensure the mean to target unaffected by electrostatic charges or feature proximity effects. More discussions on AFM photomask applications can be found in [26, 57–59].

38.4.2
Photomask Defect Review and Repair

Photomask defect repair is to eliminate defects on the mask generated during previous patterning and etch processes. Otherwise, the reticle defect reproduces itself on the final wafer products as a repeater defect in each field. As illustrated in Fig. 38.57, besides depth and CD control, AFM provides high-resolution scan images of the defects found by an inspection tool [60]. The 3D AFM image contains $x/y/z$ volume information about the defect, based on which repair tools adjust the beam (electron beam, FIB, or laser) dose and repair time to properly repair the defects. Depending on the nature of the defect, the repair either adds or removes materials from the defect site. In addition, the lithographic simulation software can use the volume information from AFM for defect printability study.

Figure 38.58 shows a 3D view of extra Cr deposit on a Cr line. AFM provides $x/y/z$ 3D information for the defect, while CD SEM only gives a top-down view with no sidewall, depth, or topographical information. Figure 38.59 shows more examples of defect review images by AFM for typical mask defects.

AFM can be used to optimize the repair process conditions by reimaging the postrepair site. Figure 38.60 shows two defect sites after electron-beam and FIB removal repairs, respectively. The electron beam generates overetch for the bridging defect site and causes a trench deeper than the target. The repair dose or time needs to be reduced. The FIB repair causes the space broadening and deepening.

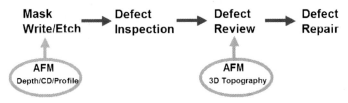

Fig. 38.57. The role of AFM in photomask process metrology

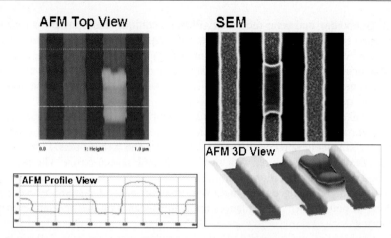

Fig. 38.58. 3D view of a defect by AFM. *SEM* scanning electron microscopy

Fig. 38.59. More review images by AFM for various defects

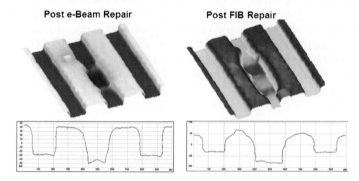

Fig. 38.60. Using AFM to examine the defect site after two different repair processes

38.5 Applications in Hard Disk Manufacturing

38.5.1 Magnetic Thin-Film Recording Head

A hard disk (or a hard drive) is the primary mass data storage solution for today's information technology. The core component of a hard disk is the magnetic thin-film transducer head that reads and writes digital bits in the magnetic recording media platter (disk). The demand for a higher areal recording density results in the perpendicular magnetic recording (PMR) technique and constantly shrinking head dimensions (less than 200-nm writer pole tip). This, in turn, creates daunting challenges for optical-based dimension metrology used in process control for head manufacturing. A thin-film head is normally made of a reader pole and a write pole in sequential process steps on a round AlTiC metal substrate, similarly to silicon-based logic or memory semiconductor wafer processing.

A recent advance in AFM applications is to apply AFM as a dimensional metrology tool for wafer processing in magnetic thin-film head manufacturing, especially for the resist and etch processes for making reader and writer poles. Dual beam or TEM is destructive in nature, with slow turnaround, and is costly. The benefit of AFM is a unique combination of nondestructive and high-resolution measurement capabilities for 3D pole geometry and LWR. Pinching and other geometry excursions resulting in poor device performances are readily identified during routine process monitoring.

Figure 38.61 shows the AFM profile for a reader resist line with 230-nm height and 70-nm bottom CD. AFM can be used a routine monitor of the reader resist or postetch CD and profile.

The PMR write pole is normally fabricated by either metal plating or metal etching. In plating, a resist trench is created by the photolithography process. Nickel iron metal is then electrochemically plated into the trench to form a pole. In the etching alternative, a thick metal layer is deposited first, then the PMR resist pattern is created, and this is followed by plasma etch to create a PMR pole. During the resist pattern process for the PMR write pole, engineers are often interested in measuring the trench profile and understanding the effect of antireflective coating on the resist profile. Figure 38.62 shows an AFM trench profile scanned by a flared CD tip. The repeatability for the resist trench middle CD is $3\sigma = 2$ nm (Fig. 38.62).

After plating, the pole is trimmed by an ion milling process to the final dimension. The overlay of the as-plated and trimmed PMR profiles is shown in Fig. 38.63.

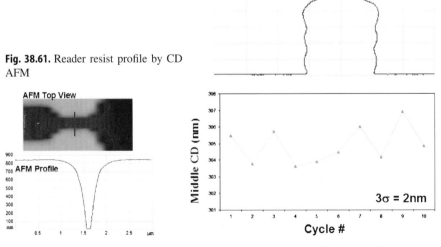

Fig. 38.61. Reader resist profile by CD AFM

Fig. 38.62. AFM used to profile writer pole resist trench (*left*) and the CD data (*right*)

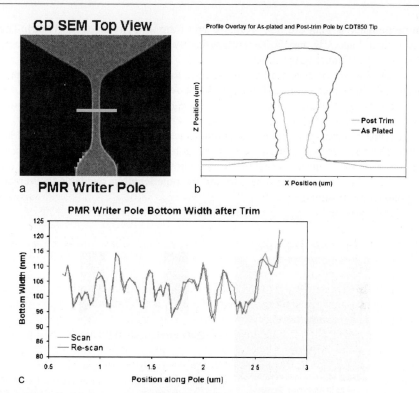

Fig. 38.63. a AFM scan location. **b** Profile overlay for as-plated writer pole with trimmed pole. **c** Pole LWR data. *PMR* perpendicular magnetic recording

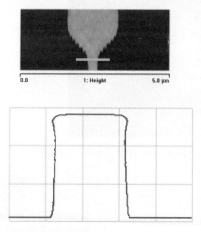

Fig. 38.64. Writer pole resist profile by AFM

The bottom LWR data are also available from the AFM scans (Fig. 38.63). The geometrical shape, CD, and profile data are readily available for engineers to evaluate the plating and trimming processes.

In the etch-based pole formation process, the resist pattern profile is important for the final pole etching. Figure 38.64 shows a writer pole resist profile obtained by AFM. The CD and profile data are readily attainable for inline monitoring.

The PMR writer pole profile after final etching can be examined by AFM. Figure 38.65 shows the 3D scan view for the pole yoke portion of the writer. AFM provides a capable metrology tool for the pole dimension analysis. In Fig. 38.66, the linewidth variation along the yoke part and the pole length is available to evaluate the yoke corner profile.

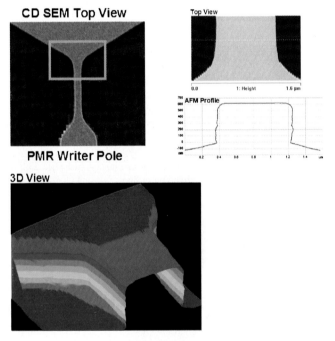

Fig. 38.65. Writer pole yoke profiles by AFM after final etching

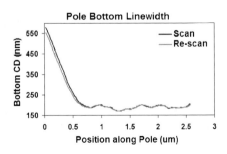

Fig. 38.66. Writer pole LWR data along the pole edge

38.5.2
Slider for Hard Drive

Similar to the integrated circuit manufacturing process, data storage recording heads are currently manufactured in mass on wafers typically 5–6 in. in diameter. Although the magnetic materials used are quite different, they also use a variety of etch, lithography, and deposition processes to create recording heads whose CDs are currently below 100 nm. After wafer processing, the manufacturing steps deviate from those of typical semiconductors. Sliders are mechanically assembled with the thin-film heads sliced from a completed AlTiC wafer. The heads are diced into individual sliders which are ultimately attached to gimbal assemblies. The head–gimbal assembly hovers at a fly height less than 5 nm above the storage media disk, which typically spins at upwards of more than 7000 rpm. In order for the read/write head to magnetically interact successfully with the media, the fly height must be held to a narrow tolerance (Fig. 38.67). The key to this height is the pole-to-tip recess (PTR), defined as the distance the writer pole is recessed above the slider air-bearing surface. AFM is a proven metrology tool in thin-film head slider processing. The PTR, perpendicular writer protrusion, cavity sidewall measurement, and ABS microroughness are all critical geometrical parameters that must be monitored and controlled.

With diameters typically in the sub-20-nm range, AFM probes can quantify PTR with a precision on the order of 1 Å(1σ). In addition, a magnetic imaging mode, magnetic force microscopy (MFM), can be utilized to discertain pole and shield regions from the ABS. This is critical as there are often no fiducial references on the sliders from which to locate the various magnetic regions of interest. Other metrology techniques, such as SEM or interferometry, either require an atomic force microscope to validate models, or cannot quantify this small height from a top-view image.

An example of a typical slider PTR AFM scan is shown in Fig. 38.68. On A complementary MFM image is shown on the right. Using the MFM image to locate the pole tip (arrow), one can use customized algorithms to precisely locate the pole and shield regions, and one can then measure these regions using the topography image in Fig. 38.68 (left).

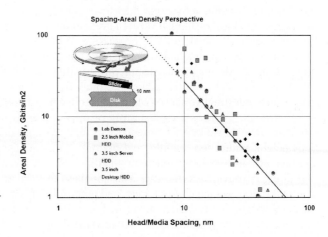

Fig. 38.67. The hard drive technology roadmap. (Courtesy of Hitachi Global Storage Technologies Web site; San Jose Development Center)

Fig. 38.68. *Left*: AFM image for a pole-to-tip recess (PTR). *Right*: Magnetic force microscopy image fo PTR for pattern recognition

38.6
Applications in Microelectromechanical System Devices

38.6.1
Contact Image Sensor

CMOS image sensors (CIS) are a relatively recent technological innovation in the video and camera technology that are rapidly replacing charge coupled device (CCD) sensors in low-power and portable imaging applications. A CIS typically consists of a small linear array of light detectors, covered by a layer of red, blue, and green color filters and microlenses. To generate an image, light must be accurately focused by the lenses, through the color filters and down onto the light detectors.

Microlenses are fabricated by creating blocks of lens material by familiar photolithography methods. These rectangular blocks are then heated and melted. As they melt, they form the convex surface required to focus light. As devices such as cell phones, Webcams, and digital cameras try to pack electronics into tighter and tighter boxes, the optimal distance between the camera lens and the image sensor may be compromised. The image sensor is then called upon to maximize its light-collecting ability. In order to do this, the lens shape is precisely tuned to permit on-axis and of-axis focusing and limit crosstalk as light is stretched to the peripheral collectors [66].

To further complicate the issue, typically the color filters that lie between the microlenses and the light detectors are at slightly different heights with respect to each other. To properly focus incoming light onto the light detector, the pixel-to-pixel height variation at the color filter level must be accurately known. The surface roughness and local dishing are also of great importance. Figure 38.69 describes the general metrology needs for CIS processing. High-resolution AFM scans offer an automated metrology method to characterize the geometrical shape and curvature for the microlenses during the fabrication process. Illustrated in Fig. 38.70, the measurements include depth, width, radius, and pitch for the lenses arrays. AFM provides a fast and nondestructive across-wafer measurement to monitor the process variation.

Figure 38.71 shows a typical array of microlenses as imaged by a dimension AFM in TappingMode®. Nanoscope CIS analysis may then be used to measure lens height, width, top, middle, and bottom angles for the left and right sides as well as for a fit to either a sphere or an ellipse. A root mean square deviation from the fitting is provided to the user as a goodness-of-fit metric. Figure 38.72 shows a TappingMode® image of red, blue, and green color filters. The surface roughness of the filters is clearly visible along with the peak-to-valley dishing for each filter and the relative height differences between the filters.

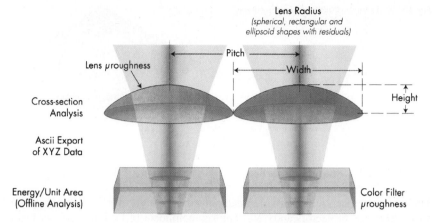

Fig. 38.69. Complementary MOS image sensor (CIS) metrology needs

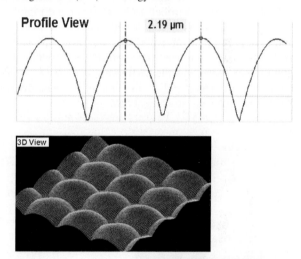

Fig. 38.70. AFM used to characterize the shape and curvature for CIS microlenses

Fig. 38.71. Another 3D view of a CIS array

Fig. 38.72. 3D view of color filter roughness

38.6.2
Digital Light Processor Mirror Device

Digital light processors (DLP) are inside much of today's display technology, such as video/movie projectors and TVs. The key compoent of DLP technology resides in digital micromirror devices (DMD) integrated on a CMOS chip. Millions of micromirrors are controlled to be repositioned rapidly to reflect light through the lens. The light intensity can be varied to create shades of gray in addition to white and black. A prism or color wheel is used to create colors. 3D nondestructive AFM has long been used for geometry and topography characterization of DMD pixel superstructures in the DMD wafer manufacturing process [67]. AFM measures hinge or yoke planarity (Fig. 38.73a), mirror planarity (Fig. 38.73b), mirror spacing, and metal thickness, to name a few.

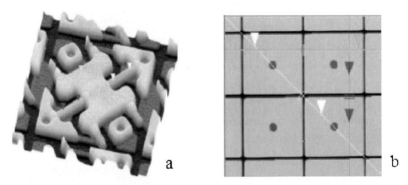

Fig. 38.73. a Digital micromirror device hinge and yoke planarity. **b** Micromirror planarity or roughness by AFM. (Courtesy of [67])

38.7
Challenge and Potential Improvement

The probe size limits the minimal space feature that AFM can scan (Table 38.2). CNT tips have shown promise for deep hole/DT depth metrology. CNT tips with various dimension designs and aspect ratios have to become available in mass production with reasonable costs. Novel types of small and rigid AFM probes for CD and profile metrology must be developed to meet the need for technology at the 32-nm node and below [15]. The CD atomic force microscope tip lifetime needs to be further improved to increase the metrology precision and accuracy, and to reduce the cost of consumables. CD atomic force microscope tips coated with silicon nitride are a recent type of a long-lasting probe (Fig. 38.6), but another type of design or coating needs to be developed for further improvements. Novel techniques to sharpen and recondition tips on the fly are also necessary to reduce tip wear and to preserve the lifetime [61, 62]. Basic research to understand the tip and surface interaction kinetics would be helpful to reduce tip wear. The sidewall profile accuracy by CD AFM is limited by the radius of curvature of the tip (approximately 5 nm). The last 5–10 nm of the feature bottom has to be verified with TEM. Therefore, CD atomic force micropscope tips with sharp edge radius and low edge height are necessary to improve the sidewall profile and bottom corner profile.

An improved scanning algorithm for AFM is being developed to reduce the tip sticking, sidewall image noise, and profile fidelity. A recent implementation of a high-bandwidth actuator is able to pull away CD atomic force microscope tips much faster from sticking and significantly improves the scan profile with an additional transitional rescan. New and innovative AFM scanning modes continue to be explored for improving tip wear, scan speed, and measurement capability.

AFM is a much faster and more economical replacement for X-SEM or TEM. Despite the many advantages discussed previously, the current AFM throughput is much slower than that of CD SEM or scatterometry, especially in the CD scanning mode. The speed of AFM needs to be improved to a level to compete with CD SEM and optical tools. Progress has been reported on fast AFM scanning 2–3 orders of magnitude faster than conventional AFM [63, 64]. This yet needs to be transferred to industrial AFM. In addition to fast scanning, increasing the speed for automated tip exchange and the tip engaging on the surface is part of improvement programs. A recent improvement had reduced the tip engagement time from 25 to 5 s.

The AFM image is always a convolution of the real feature geometry with the tip shape. The image is actually an artifact and the real feature shape is unknown. Accurate and precise tip shape characterization and image reconstruction algorithms are necessary to recover the real feature shape. The current tip characterization scheme can ensure a precise and accurate measurement of depth, linewidth, sidewall angle, and all other critical geometrical parameters [25]. For the purpose of image and true shape visualization, it is highly desirable to remove the tip artifacts in the full three dimensions and extract the high-fidelity image close to TEM resolution.

38.8
Conclusion

There are significant challenges for dimension metrology for the 45-nm node and beyond for semiconductor process control [3, 11]. Metrologists have to resort to a combination of scanning probe, electron beam, and optical based technologies.

Owing to its unique characteristics, AFM offers beneficial solutions for either inline or complementary metrology for linewidth, sidewall profile, geometry shape, and surface topography measurements for process monitoring in semiconductor (logic and memory), photomask, data storage, and microelectromechanical system (MEMS) industries. This chapter has discussed advantages and limitations, and state-of-the-art AFM applications in nanoelectronics manufacturing processes. AFM is being used for manufacturing inline monitoring, product and process characterization during technology development, or complementary reference metrology. In some cases, AFM is the only available metrology tool for nondestructive 3D dimension and geometry measurements.

AFM is used for inline monitoring in the routine production for the depth, height, step, linewidth, profile, or topographical variations within any location of the circuits and on any type of material. No modeling or guesswork is involved in interpreting the data. AFM demonstrates the unique capability to profile very high aspect ratio trench or hole features. In some cases, AFM replaces X-SEM, TEM, or dual beam for cross-sectional profile and shape analysis [17]. It serves a fast and nondestructive shape metrology tool for engineering analysis to speed up technology development and troubleshooting.

AFM is immune to the new materials emerging for advanced semiconductor devices such as SOI, SiGe incorporation, novel resist materials, metal gate, fully silicided (FUSI) polysilicon (nickel silicide), high-k gate dielectrics, and back-end low-k dielectric. Unlike optical scatterometry, AFM needs no film stack modeling, no guesswork, no modification for recipe settings, and simply provides a direct dimension measurement. AFM is calibrated to national standards with regard to height, linewidth, and pitch. Independent of material properties and pattern proximity, AFM equally works on STI etch, gate poly, metal gate, or gate spacer without any modifications or rebuilding models. Owing to the absolute accuracy, AFM can be used as the golden reference standard to calibrate other dimension metrology tools to establish the traceability chain and develop uncertainty budgets [65]. AFM can be used to establish CD targets for CD SEM, scatterometry and electric CD, and to establish correlation between X-SEM, TEM, CD SEM, scatterometry, and electric CD. AFM helps perform the CD metrology tool matching and the design rule matching across different factories around the world, to ensure the 30-nm CD in one fabrication plant is really the same as the 30-nm CD in another fabrication plant. CD AFM is often used as a complementary CD metrology tool to maintain calibration for inline optical scatterometry and CD SEM, and to help speed up the CD library development for scatterometry.

In summary (refer to Table 38.1), automated AFM is a versatile industrial metrology tool for process control and characterization in nanoelectronic manufactuing industries ranging from semiconductor, photomask, and data storage to MEMS technologies.

Acknowledgements. The authors are extremely grateful for inspiring inputs from colleagues, including Marc Osborn, Ingo Schmitz, Max Ho, Qi Chen, Damian Conseur, Moon Keun Lee, Joe Hsu, Kevin Chen, Melvin Lim, Misao Suzuki, Shu-rui Wang, Brian Chang, Tony Hare, Fuad Hassan, and many others.

References

1. Serry FM (2005) Metrology at the micrometer and nanometer scale. OnBoard Technol Sept 38–40
2. Schattenburg M, Smith H (2002) Proc SPIE 4608:116–124
3. Braun B (2006) Semicond Int, 15 June 2006, p 45
4. Diebold A (2005) AIP Conf Proc 788:21–32
5. Serry FM (2006) Photonics Spectra Dec, p 123
6. Magonov S (2005) Proc SPIE 6002:145–153
7. Braun A (2002) Semicond Int, July, p 78
8. Foucher J, Sundaram G, Gorelikov D (2005) Proc SPIE 5752:489–498
9. Miller K, Geiszler V, Dawson D (2004) Proc SPIE 5375:1325–1330
10. Ukraintsev V et al (2005) Proc SPIE 5752:127–139
11. Rice B et al (2005) AIP Conf Proc 788:379–385
12. Bao T, Fong D (2007) In: Proceedings of the 6th international semiconductor technology conference (ISTC 2007), Shanghai, China, 18–20 March 2007, p 517
13. Perng B et al (2006) Proc SPIE 6152:61520Q
14. Wu C et al (2001) Proc SPIE 4345:190–199
15. International Technology Roadmap for Semiconductors (2005) International Technology Roadmap for Semiconductors 2005 edition. Metrology. http://www.itrs.net/Links/2005ITRS/Metrology2005.pdf
16. Liu H et al (2006) Proc SPIE 6152:61522Y
17. Foucher J, Miller K (2004) Proc SPIE 5375:444–455
18. Dixson R, Guerry A (2004) Proc SPIE 5375:633–646
19. Ukraintsev V (2006) Proc SPIE 6152:61521G
20. Dixson R, Orji N et al (2006) Proc SPIE 6152:61520P
21. Dixson R et al (1999) Proc SPIE 3677:20–34
22. Martin Y, Wickramasinghe H (1994) Appl Pys Lett 64(19):2498
23. Liu H et al (2005) J Vac Sci Technol B 23(6)3090–3093
24. Cottle R (2002) Proc SPIE 4562:247–255
25. Dahlen G et al (2005) J Vac Sci Technol B 23(6):2297–2303
26. Bao T, Zerrade A (2006) Proc SPIE 6349:63493Z
27. Dai H et al (1996) Nature 384:147–150
28. Yenilmez E et al (2002) Appl Phys Lett 80(12):2225
29. Larsen T, Moloni K, Flack F, Black C (2002) Appl Phys Lett 80(11):1996
30. Lagus M, Hand S (2001) In: Proceedings of the 18th international VLSI multilevel interconnection conference, 28–29 November 2001, p 239
31. Bao T, Romani R, Ercole M (2006) In: Proceedings of the international conference on planarization/CMP technology (2006 ICPT), Foster City, CA, USA, 12–13 October 2006, p 165
32. Caldwell M, Bao T (2007) CMP-MIC (in press)
33. Lipscomb W, Allgair J, Bunday B et al (2006) Calibrating optical overlay Proc SPIE 6152:615211
34. Caldwell M, Bao T et al (2007) Proc SPIE (in press)
35. Ozturk A, Liu J (2005) AIP Conf Proc 788:222–231
36. Vachellerie V, Kremer S et al (2005) AIP Conf Proc 788:411–420
37. Peters L (2006) Semicond Int, Jan, p 96
38. Heaton M, Ge L, Serry S (2001) Atomic force profilometry for chemical mechanical polishing metrology. In: Semiconductor FABTECH, 11th edn, p 97
39. Cunningham T, Todd B, Cramer J et al (2000) Solid State Technol, Jul, p 105
40. Strausser Y, Hetherington D (1996) Semicond Int, Dec, p 75

41. Prakash V et al (1999) Proc SPIE 3677:10–17
42. Lagus M, Marsh J (1999) Proc SPIE 3677:2–9
43. Borionetti G, Bazzali A, Orizio R (2004) Eur Phys J Appl Phys 27:101–106
44. Walch K et al (2001) Proc SPIE 4344:726–732
45. Patterson K et al (2001) Proc SPIE 4344:809–814
46. Yamaguchi A et al (2004) Proc SPIE 5375:468–476
47. Villarrubia J (2005) AIP Conf Proc 788:386–393
48. Orji N, Raja J, Vorburger T (2002) Proc Am Soc Precis Eng, Oct, p 821
49. Villarrubia J, Bunday B (2005) Proc SPIE 5752:480–488
50. Jang J et al (2003) Appl Phys Lett 83(20):4116–4118
51. Foucher J (2005) Proc SPIE 5752:966–976
52. Miller K et al (2004) Proc SPIE 5446:720–727
53. PKL (1997) http://www.pkl.co.kr/english/product/product14.html
54. Yoshida Y et al (2004) Proc SPIE 5446:759–769
55. Todd B et al (2001) Proc SPIE 4344:208–221
56. Miller K et al (2002) Proc SPIE 4689:466–472
57. Muckenhirn S, Meyyappan A (1998) Proc SPIE 3332:642–653
58. Tanaka Y et al (2004) Proc SPIE 5446:751–758
59. Miller K, Todd B (2001) Proc SPIE 4186:681–687
60. Muckenhirn S, Meyyappan A et al (2001) Proc SPIE 4344:188–199
61. Martinez J, Yuzvinsky T et al (2005) Nanotechnology 16:2493–2496
62. Wendel M, Lorenz H, Kotthaus J (1995) Appl Phys Lett 67(25):3732–3734
63. Schitter G et al (2005) In: Proceedings of the 2005 IEEE/ASME international conference on advanced intelligent mechatronics, Monterey, CA, USA, July 2005, pp 24- 28
64. Schitter G et al (2001) Rev Sci Instrum 72(8):3320–3327
65. Bunday B et al (2005) Proc SPIE 5878:58780M
66. Boettiger U, Li J (2006) US Patent 7,068,432, 27 June 2006.
67. Serry F, Nagy P, Horwitz J, Oden P, Heaton M (2002)3D MEMS metrology with the atomic force microscope. Veeco AFM applications notes. http://www.veeco.com

Subject Index

accumulation *VIII* 410
acoustic emission *X* 342
acrylonitrile–butadiene–styrene resin *X* 350
actin *X* 303
β-actin *IX* 164
actuation
  bimetallic *VIII* 211
  electrostatic *VIII* 210
  piezoelectric *VIII* 211
  thermal bimetallic *VIII* 210
adhesion *VIII* 160, *IX* 228
adhesion force *VIII* 151, *VIII* 162, *VIII* 166, *VIII* 168, *IX* 208, *IX* 225, *IX* 226, *IX* 229, *IX* 230, *IX* 238, *IX* 263, *IX* 278
admittance *VIII* 380
AFM *IX* 90, *IX* 287, *X* 272, *X* 274, *X* 276–278, *X* 352, *X* 355–357
  micro/nanotribology *IX* 284
  noncontact *X* 237
AFM probes *VIII* 429, *X* 367
  conductive *VIII* 429
  conductive coating *VIII* 431
AFM–SEM systems *VIII* 283
aging (piezoelectric) *VIII* 300
Al unmodified *IX* 254
alkylphosphonic acid *IX* 244
alkylphosphonic acid SAMs on Al
    nonperfluorinated *IX* 248
Amonton *IX* 285
amplifier *VIII* 432, *X* 78
  bandwidth *VIII* 432
  current-to-voltage *VIII* 432
  input capacitance *VIII* 434
  integrator–differentiator scheme *VIII* 433
  noise *VIII* 432
  transimpedance *VIII* 432
amplitude oscillations *IX* 97
amyloid *IX* 177–179, *IX* 181, *IX* 182, *IX* 184, *IX* 186, *IX* 189–193, *IX* 199

diseases *IX* 177, *IX* 178, *IX* 180, *IX* 182, *IX* 183, *IX* 189, *IX* 193
fibrils *IX* 177–181, *IX* 183–193, *IX* 196–199, *IX* 201
peptides *IX* 178, *IX* 192–195, *IX* 199
proteins *IX* 178–180, *IX* 184, *IX* 185, *IX* 189, *IX* 190, *IX* 193–195, *IX* 197, *IX* 198
protofibrils *IX* 189–191
protofilaments *IX* 179–181, *IX* 184–186, *IX* 188, *IX* 189, *IX* 199
amyloidosis *IX* 178, *IX* 191, *IX* 192
angle beveling *X* 67
anodization *X* 219–221
  anodic oxide *X* 220
antenna *VIII* 117, *VIII* 118, *VIII* 120, *VIII* 122, *VIII* 123
anti-stiction *VIII* 64
antibody *VIII* 238
antigen *VIII* 239
aperture probe *VIII* 82
  chemical etching *VIII* 98
  fiber probes *VIII* 84, *VIII* 93
  heating *VIII* 97
  metallic tips *VIII* 84
  pulling *VIII* 97
  tube etching *VIII* 100
aperture SNOM *VIII* 83
apertureless probes *VIII* 118
  dielectric tips *VIII* 119
  electrochemical etching *VIII* 119
  metallic tip *VIII* 83
apex radii *IX* 326
AR-XPS *IX* 240
area function *X* 315, *X* 316, *X* 323
artifacts *VIII* 14, *VIII* 15
atomic force microscopy (AFM) *VIII* 81, *VIII* 185, *VIII* 190, *VIII* 191, *VIII* 195, *VIII* 211, *VIII* 213, *IX* 23, *IX* 76, *IX* 77, *IX* 79–81, *IX* 89, *IX* 111, *IX* 334, *X* 185,

X 189, X 192–194, X 199, X 200, X 204, X 286, X 351
attractive regime   IX 76
Au   X 91
Auger   IX 257
Auto Tip Qual   VIII 43

Babinet's reciprocity principle   VIII 2
backbone chain   IX 278
background removal   VIII 17
ballistic electron emission microscopy (BEEM)   X 88
Bell–Kaiser formalism   X 89
bending experiment   VIII 272
bending rigidity   IX 199, IX 200
Berkovich indenters   X 313–318, X 320, X 332, X 335, X 337, X 341
bias   VIII 60
  DC   X 65
  modulating AC   X 65
biexciton   IX 363
binding energy   IX 248
biological samples   X 285
biomaterials   IX 329, IX 347
BioMEMS   IX 283
biomolecules   VIII 437, X 285, X 287
bipolar devices
  net base   X 68
blind region   VIII 65, VIII 66
blind tip reconstruction   VIII 43
boiling water treatment   IX 276
boundary element method (BEM)   VIII 356
boundary lubrication   IX 329, IX 333, IX 347
bow-tie   VIII 115, VIII 123–125
bradykinin   IX 139
breakdown   X 93–95, X 97, X 100
  dielectric   X 100
  extrinsic   X 100
  field   X 97
  hard   X 93
  intrinsic   X 100
  kinetics   X 97, X 99
  soft   X 93
  spots   X 97, X 98
  weak   X 98
brush model   IX 263
buffer
  bootstrapped   VIII 383

c-fos   IX 166
CAFM   X 90, X 98

calibration
  dC/dV X 70
  N X 70
  radiation pressure   VIII 292
calibration grating   IX 326
calibration samples   X 81
  GaAs   X 81
  InP   X 81
CAM analysis   IX 254
  PFMS/Cu   IX 261
cantilevers   VIII 290, IX 89, IX 334, IX 335
  effective mass   VIII 296
  force sensors   VIII 257
  mass sensors   VIII 255, VIII 258, VIII 276
  silicon nitride   IX 338, IX 344
  spring constants   VIII 290, IX 334, IX 335
  tips   IX 91
  V-shaped   VIII 292
  vibration   IX 316
capacitance   VIII 379, X 64
  apex   VIII 441
  attofarads   X 64
  dC/dV X 65
  gain   X 65
  local   VIII 440
  measurement   VIII 439
  resolution   VIII 440
  sensor   X 65
  stray   VIII 440
capacitance–distance curves   VIII 440
capacitor structure   VIII 53
capillary forces   VIII 72
carbon   X 225
  amorphous carbon   X 225
  diamond   X 225, X 240
  graphite   X 219, X 225
carbon nanotubes   VIII 437
carrier
  density   X 67
  intrinsic   X 67
  spilling   X 69
CD AFM   VIII 31, VIII 32
CD mode   VIII 34, VIII 52, X 362
CD probes overhang   VIII 72
cell
  endothelial   IX 127
  epithelial   IX 127
cell adhesion   IX 89, IX 98, IX 102
cell mechanics   IX 50, IX 89
cell membrane   IX 217, IX 227, IX 228, IX 230

cell surfaces   *IX* 111
cell viscoelasticity   *IX* 93
cell–cell junctions   *IX* 132
characterizer   *VIII* 48
charge density   *VIII* 358
charge transfer   *IX* 15
charge trapping   *X* 94
charge writing   *VIII* 202
charges   *X* 95
  fixed   *X* 66
  interface   *VIII* 403
  space   *VIII* 404
  trapped   *X* 66, *X* 96
charging   *X* 349
chemical bond   *IX* 12
chemical contrast   *X* 248
chemical manipulation   *X* 248
chemical mechanical planarization (CMP)   *X* 132, *X* 134
chemical stability   *IX* 268
chemical state   *IX* 273
chemical vapor deposition (CVD)   *VIII* 140, *VIII* 142
chemisorbs   *IX* 236
cholesterol   *IX* 207, *IX* 220–222, *IX* 224
chromatin fibers   *X* 290
chromium oxide   *IX* 344
chromosomes   *X* 290, *X* 295, *X* 296, *X* 299, *X* 305
cilium   *IX* 132
circuit
  lumped element   *VIII* 385
  resonant   *VIII* 384
clamped beams   *IX* 323
CMOS gate   *VIII* 68
CMP   *X* 132, *X* 389
CMP pre-metal dielectric   *X* 386
CNT   *VIII* 46, *VIII* 48, *VIII* 73
CNT probes   *VIII* 69
CNT-across-trench   *VIII* 47, *VIII* 48
cobalt–chromium (CoCrMo)   *IX* 343
CODY mode   *IX* 26, *IX* 48, *IX* 49
collagen   *IX* 191, *X* 289, *X* 290
collection efficiency   *IX* 355, *IX* 357
collector   *X* 88
colloidal probe   *IX* 151
complex compliance   *X* 337–340
composite modulus   *IX* 337
compositional imaging   *IX* 24
compression experiments   *VIII* 274

compression-free force spectroscopy   *IX* 156, *IX* 164
condensation dropwise   *IX* 238, *IX* 279
condensed phase   *IX* 5
conductive AFM (CAFM)   *VIII* 422, *VIII* 434, *X* 63, *X* 189, *X* 190, *X* 205–208, *X* 211
  applications   *VIII* 435
  experimental setup   *VIII* 434
conical indenters   *X* 311, *X* 313–315, *X* 330, *X* 333–335
contact
  ohmic   *X* 78
  permanent   *VIII* 154, *VIII* 156
contact angle   *IX* 338, *IX* 340, *IX* 342
  hydrophilic   *IX* 340–342
  hydrophobic   *IX* 329, *IX* 333, *IX* 340–343, *IX* 347
contact CMP   *X* 389
contact etch   *X* 387
contact image sensor   *X* 406
contact mechanics   *IX* 336
contact mode   *VIII* 291, *VIII* 432, *IX* 335, *X* 78
contact models   *IX* 95
  Derjaguin–Muller–Toporov (DMT)   *IX* 29, *IX* 38
  Dugdale   *IX* 37
  Dupré's work of adhesion   *IX* 38
  Hertz model   *IX* 33, *IX* 36
  Johnson–Kendall–Roberts (JKR)   *IX* 29, *IX* 38
  Maugis–Dugdale (MD)   *IX* 37, *IX* 38
  Sneddon's extensions   *IX* 36
contamination   *IX* 245
continuous stiffness measurement   *X* 309
contrast   *VIII* 392
converter
  current-to-voltage   *VIII* 382
convolutions   *VIII* 354
copper oxide   *IX* 240
correction factor (CF)   *X* 84
correlation coefficient   *VIII* 58
correlation spectrum analyzer   *VIII* 427
  resolution   *VIII* 427
correlator   *VIII* 410
corrosion inhibition   *IX* 238
cost of ownership   *VIII* 39
Coulomb   *IX* 285
covalent bond   *IX* 171
  Si–C bond   *IX* 172

CPD *VIII* 353
creep compliance *X* 337, *X* 340, *X* 341
critical dimension (CD) modes *X* 362
critical surface tension *IX* 241, *IX* 255, *IX* 261, *IX* 277, *IX* 278
cross-linker *IX* 170
crystallinity *IX* 329, *IX* 338, *IX* 340, *IX* 343, *IX* 347
  cross-linking *IX* 332
  recrystallization *IX* 335
  semicrystalline *IX* 332
CS-AFM *VIII* 422
cube-corner indenter *X* 314
current *X* 91
  nanometric localization *X* 91
current transport *X* 93
  direct tunnel *X* 93
  Fowler–Nordheim *X* 93
  percolation *X* 93
current-sensing atomic force microscopy (CS-AFM) *VIII* 422
  conductive probes *VIII* 429
  current detection instrumentation *VIII* 432
  experimental setup *VIII* 429
  operating modes *VIII* 431
cutoff *VIII* 86, *VIII* 93, *VIII* 95, *VIII* 109, *VIII* 115
CVD
  hot filament *VIII* 144

1D nanostructures *IX* 311, *IX* 312
2D SPM Zoom *VIII* 39
3D imaging *IX* 25
deep level transient spectroscopy (DLTS) *VIII* 408
deep trench (DT) *X* 362
deformations
  plastic *X* 79
delocalization *IX* 368
depletion *VIII* 401, *X* 66
Derjaguin–Muller–Toporov (DMT) *IX* 336
Dexel *VIII* 49, *VIII* 51
diamond-like carbon *VIII* 67
dielectric *X* 94
dielectric constant nonlinear *X* 108
diffusion *VIII* 204–207, *X* 75
digital light processor mirror device *X* 408
digital pulsed force mode (DPFM) *IX* 23, *IX* 32
dilated scan *VIII* 47

dimensional metrology *X* 361
  AFM *X* 361
  dual beam *X* 361
  optical profiler *X* 361
  scatterometry *X* 361
  SEM *X* 361
  stylus profiler *X* 361
  TEM *X* 361
  X-SEM *X* 361
dimyristoylphosphatidylcholine (DMPC) *IX* 72
  phase separation *IX* 73
dip-pen nanolithography (DPN) *VIII* 191, *VIII* 193–195, *VIII* 200, *VIII* 204, *VIII* 206, *VIII* 207, *VIII* 209
dipalmitoylphosphatidylcholine (DPPC) *IX* 76
dipole *VIII* 356
dishing, erosion, and total indicated run-out *X* 395
disordered systems *IX* 360, *IX* 367
dissipation *VIII* 259
DNA *VIII* 184, *VIII* 186, *VIII* 188, *VIII* 191, *VIII* 198, *VIII* 200–202, *VIII* 213, *VIII* 437, *IX* 64, *IX* 71, *IX* 72, *IX* 76, *IX* 78, *X* 287
  elastic behaviour *IX* 72
dopant *VIII* 403
DOPC *IX* 209, *IX* 210, *IX* 212, *IX* 216–218
doping *VIII* 430, *IX* 8
Doppler velocimetry *VIII* 303
DP SAM *IX* 271
DPFM *IX* 23, *IX* 39, *IX* 50
DPN
  electrochemical (e-DPN) *VIII* 203
DPPC *IX* 76, *IX* 209–220, *IX* 224
DRAM DT capacitor *X* 397
drift *VIII* 299
DT mode *VIII* 34, *VIII* 52
dual-indenter method *X* 330, *X* 333–336
dynamic force microscopy (DFM) *X* 237
dynamic force spectroscopy (DFS) *IX* 89, *IX* 100
dynamic mechanical analysis (DMA) *X* 309, *X* 337
dynamic methods (of calibration) *VIII* 292, *VIII* 295
dynamic mode *VIII* 291
dynamic range *X* 64
dynamic technique *IX* 315

electrostatic resonant-contact AFM *IX* 315
mechanical resonant-contact AFM *IX* 315

effective modulus *IX* 36
effective radius *IX* 34
eigenmodes
  higher-order *IX* 24
elastic modulus *IX* 338
elasticity *VIII* 151, *VIII* 158, *VIII* 160, *VIII* 168, *IX* 89, *IX* 94
elastoplastic *X* 79
electric dipole *VIII* 89
electric field built-in *X* 68
electrical activation *X* 74
electrical noise
  cross-correlation spectrum *VIII* 445
electrical noise microscopy (ENM) *VIII* 422, *VIII* 443
  experimental setup *VIII* 444
  thermal noise *VIII* 444
electrical transport *VIII* 424
electrochemical reaction *IX* 80
electroless plating *X* 242
electron band *IX* 6
electron correlation effect *IX* 7
electron injection *IX* 80
electron mean free path *X* 88
electron microscopy *IX* 177, *IX* 181, *IX* 183, *IX* 184
electrophotographic *X* 345, *X* 347
electrophotographic process *X* 347, *X* 353, *X* 355
  charging *X* 348, *X* 351
  development *X* 349
  exposure *X* 348
  fusing *X* 349, *X* 350, *X* 355, *X* 356
  organic photoconductors (OPC) *X* 348, *X* 352–355
  phase image *X* 356
  photoconductors *X* 348–351
  transfer *X* 349
electrophotographic systems *X* 347–349
electrospinning *IX* 318
electrostatic *IX* 208, *IX* 212, *IX* 216–219, *IX* 222–225
electrostatic discharge (ESD) *VIII* 62
electrostatic force microscopy *VIII* 351
eletrophotographic processes *X* 349
ellipsometry *IX* 65

elongation *IX* 323
energy gap *IX* 6
environmental control *X* 226
  CO_2 *X* 228
  ethyl alcohol *X* 228
  EtOH *X* 228
  hydrocarbon liquid *X* 229
enzyme-linked immunosorbent assay (ELISA) *VIII* 188–190, *VIII* 213
epithelial cells *IX* 224, *IX* 225, *IX* 229, *X* 293, *X* 301
equipotential *VIII* 356
erosion *VIII* 42
erosion algorithm *VIII* 49
ESD *VIII* 62
etching *X* 242, *X* 243, *X* 247, *X* 248
evanescent field *IX* 353
exciton *IX* 361, *IX* 362
exciton excited state *IX* 364

failure probability *X* 100
far-field *VIII* 92
fast dither tube actuation *VIII* 71
fast scan control *VIII* 71
F/D curve *IX* 313
F/D curve nonlinear *IX* 325
FDTD *VIII* 116
ferroelectric data storage *X* 114
FFM analsyis *IX* 261
FIB *VIII* 104, *VIII* 113
FIB-SEM *VIII* 54
fiber-based optical probes *IX* 354
fibroblast *X* 303
fibroblasts living *X* 300
fibronectin (FN) *IX* 156
fiducial mark *VIII* 57
field enhancement *VIII* 84, *VIII* 93, *VIII* 120
filter membranes *IX* 139
FinFET *VIII* 32
FinFET gate formation *X* 383
fingerprinting *VIII* 54
finite element analysis (FEA) *VIII* 63, *VIII* 293, *X* 309
finite-difference time domain *VIII* 88
finite-element modeling *VIII* 223
flash memory *X* 109
flat-band *X* 65, *X* 94
fluid *IX* 137
fluorescence *VIII* 78, *VIII* 88

fluorocarbon   *IX* 237, *IX* 252, *IX* 257,
  IX 265, *IX* 278
focused electron/ion beam induced processing
  deposit density   *VIII* 282
  deposition   *VIII* 251, *VIII* 253, *VIII* 280
  etching   *VIII* 251, *VIII* 253
  milling   *VIII* 253, *VIII* 281
  principle   *VIII* 251
  process control   *VIII* 276
force   *IX* 89, *IX* 209, *IX* 210, *X* 78
  electrostatic   *VIII* 380
  lateral   *VIII* 309
force curve   *IX* 92, *IX* 153
force mapping mode   *X* 296
force measurements   *IX* 92
force mode analog pulsed   *IX* 32
force mode pulsed   *IX* 26, *IX* 47
force modulation mode   *X* 298
force spectroscopy   *IX* 113, *IX* 116, *IX* 187,
  IX 199, *IX* 200
force standards   *VIII* 292
force volume imaging   *IX* 24
force–distance curve   *IX* 27
force–extension curve   *IX* 153, *IX* 172
Fowler–Nordheim
  injection   *X* 96
  tunnel   *X* 97
fracture   *VIII* 269
  strain   *VIII* 269, *VIII* 272
  stress   *VIII* 269
frequency domain   *IX* 97
frequency modulation (FM)   *VIII* 154,
  VIII 291
  damping   *VIII* 154, *VIII* 156
  frequency shift   *VIII* 154, *VIII* 155
frequency modulation atomic force
    microscopy   *VIII* 315, *VIII* 318,
    VIII 327
  applications   *VIII* 326
  biological systems   *VIII* 327
  instrumentation   *VIII* 318
  liquids   *VIII* 315
  magnetic activation   *VIII* 316
  nonbiological systems   *VIII* 326
  phase detuning   *VIII* 343
  theoretical foundations   *VIII* 317
fretting   *IX* 329, *IX* 332, *IX* 347
friction   *IX* 80, *IX* 329, *IX* 332–337, *IX* 340,
  IX 342, *IX* 343
  adhesive   *IX* 332, *IX* 336, *IX* 342, *IX* 344
friction force   *IX* 262, *IX* 263

friction mapping   *IX* 301
friction measurement   *IX* 47
functionalization   *IX* 99

GaNAs   *IX* 368
gate etch   *X* 379
gate resist pattern   *X* 378
gate sidewall spacer   *X* 385
geometrical filter   *VIII* 41
graphite-like layer   *IX* 303
gray-scale   *VIII* 47
green fluorescent protein (GFP)   *IX* 161,
  X 303

Hall
  mobility   *X* 86
  scattering factors   *X* 86
halothane   *IX* 211–215, *IX* 218, *IX* 219
hardness   *IX* 315, *IX* 338
height fluctuations   *IX* 136
Hertz   *IX* 94
Hertzian contact model   *IX* 155, *X* 311,
  X 312, *X* 324, *X* 325, *X* 328
  spherical indenter   *IX* 155
heterodyne   *VIII* 1, *VIII* 3, *VIII* 8, *VIII* 19,
  VIII 20, *VIII* 23
  pseudo-heterodyne   *VIII* 8, *VIII* 13
heterostructure
  Si/SiGe/Si   *X* 76
high-resolution   *IX* 252, *IX* 255, *IX* 257,
  IX 259
higher bias   *IX* 4
higher harmonics   *IX* 24
higher-order SNDM (HO-SNDM)   *X* 105,
  X 110
highest occupied molecular orbital (HOMO)
  VIII 370
hollow cantilevers   *VIII* 85, *VIII* 106,
  VIII 127
homodyne   *VIII* 1, *VIII* 3, *VIII* 5, *VIII* 7,
  VIII 8, *VIII* 12, *VIII* 13, *VIII* 20
humidity   *VIII* 193, *VIII* 204, *VIII* 206,
  VIII 209, *VIII* 210, *X* 226, *X* 242
  adsorbed water   *X* 226
  humidity effect   *X* 226, *X* 228
hydrodynamic drag   *IX* 98
hydrophilicity   *X* 226, *X* 228, *X* 242
hydrophobicity   *VIII* 67, *IX* 255, *IX* 261,
  X 226, *X* 228
hydroxylated   *IX* 251
hysteresis (piezoelectric)   *VIII* 300

Subject Index

ICTS *VIII* 416
image *VIII* 174
imaging *IX* 112, *IX* 113, *IX* 119, *IX* 123
immobilization *VIII* 184, *VIII* 189
impedance *VIII* 386
 electrical *VIII* 425
 measurement *VIII* 425
 resolution *VIII* 426
 spectroscopy *VIII* 425
impingement *IX* 324
in-line metrology *VIII* 39
indentation *IX* 89, *IX* 95
indentation measurements *IX* 25, *IX* 29
integrins *IX* 157, *IX* 159
interaction mechanism *IX* 10
interface states *X* 66
interfacial shear strength *IX* 336, *IX* 342
interfacial surface energy *IX* 241
interferometric techniques *VIII* 3, *VIII* 13
intermittent contact *VIII* 154, *VIII* 156, *VIII* 170, *IX* 24
intermittent-contact AFM (IC-AFM) *X* 237
intrinsic noise *VIII* 263
IQTS *VIII* 413
ITRS *VIII* 37
IVPS *VIII* 46, *VIII* 47

joint simulators *IX* 334
JT distortion *IX* 10
jumping mode *VIII* 432, *IX* 26, *IX* 51
junction *X* 68
 EJ *X* 68
 MJ *X* 68

Kelvin *VIII* 388
Kelvin probe force microscopy (KFM) *VIII* 351, *X* 231, *X* 236
Kelvin probe microscopy (KPM) *X* 147

Langmuir–Blodgett *IX* 56, *IX* 60, *IX* 61, *IX* 72, *IX* 73, *IX* 76, *IX* 84
 amphiphiles *IX* 60
 horizontal deposition *IX* 61
 Langmuir–Schaefer *IX* 61
 vertical deposition *IX* 61
lateral force microscopy (LFM) *X* 234
lateral resolution *X* 67
lateral stiffness *VIII* 62
layers
 epithelial *IX* 129
LER *X* 397

leukocytes *IX* 144
lift-off *X* 242
light–matter interaction *IX* 361
line edge roughness (LER) *VIII* 55
line width variation (LWV) *VIII* 55
living cells *IX* 113, *IX* 123, *X* 292, *X* 302
loading rates *IX* 101
local chemical reactions *IX* 79
local material properties *IX* 26
local mechanical properties *IX* 23, *IX* 50
localization *IX* 368
lock-in amplifier *X* 65
lock-in detection *VIII* 426
loop energy
 hysteretic *X* 311, *X* 325, *X* 326, *X* 329
loss compliance *X* 338
loss modulus *X* 337, *X* 339
lubrication *IX* 278
LWR *X* 397

macromolecule material *X* 345, *X* 350
 PFA *X* 351, *X* 356
 photoconductor *X* 350
 polycarbonate (PC) *X* 350
 polyester *X* 351
 polytetrafluoroethylene (PTFE) *X* 349, *X* 351
 styrene–acrylate *X* 351
 tetrafluoroethylene perfluoroalkoxy vinyl ether copolymer (PFA) *X* 349
macromolecule materials *X* 348
macromolecules *X* 350–352, *X* 355
magnetic thin-film recording head *X* 401
manipulation *X* 304
mass method (of calibration) *VIII* 296
mathematical morphology *VIII* 40
 dilation *VIII* 40
 erosion *VIII* 40
Mayer's law *X* 312, *X* 331
mechanical properties *IX* 311, *IX* 312
 elastic modulus *IX* 312
 elongation *IX* 311
 tensile strength *IX* 311
 toughness *IX* 311
 Young's modulus *IX* 311, *IX* 312
median filter *VIII* 51
membrane proteins *IX* 112, *IX* 124
membranes *IX* 139, *IX* 193, *IX* 195–197, *IX* 225
 cell *IX* 191–193
 model *IX* 177, *IX* 194

MEMS   *VIII* 64
MEMS/NEMS   *IX* 283
meniscus   *VIII* 204, *VIII* 206, *VIII* 207
meniscus bridge   *IX* 302
meniscus force   *VIII* 61
16-mercaptohexadecanoic acid (MHA)   *VIII* 192
messenger RNAs (mRNAs)   *IX* 150
metal   *X* 223
  Al   *X* 240
  Mo   *X* 243
  Nb   *X* 240
  Ni   *X* 240
  NiFe   *X* 240
  Ti   *X* 223, *X* 240, *X* 243
metal–insulator–semiconductor
  nanoMIS   *X* 64
metal-coating   *VIII* 103
metal nanostructures   *VIII* 18
metal trench etch   *X* 390
metal trench photo pattern   *X* 390
metallic nanostructures   *VIII* 4, *VIII* 19
metrology   *VIII* 31
MHA   *VIII* 194, *VIII* 199, *VIII* 200, *VIII* 206, *VIII* 207
micro-contact-printed   *IX* 50
microarrays   *VIII* 184, *VIII* 185, *VIII* 188–190
microbead   *IX* 151
  carboxylated polystyrene beads   *IX* 153
microcontact printing   *VIII* 185
microelectromechanical system (MEMS)   *VIII* 61
microelectronic device   *IX* 238
microfabricated devices   *VIII* 302
microfluidics   *VIII* 197, *VIII* 198
microscopy   *VIII* 379, *IX* 64
  scanning near-field optical microscopy (SNOM)   *IX* 65
microspotters   *VIII* 196
microsteps   *VIII* 34
microtribology/nanotribology   *IX* 284
middle line width (MCD)   *VIII* 58
miniaturization   *IX* 237
Minkowski addition/subtraction   *VIII* 50
mobility   *X* 85
  drift   *X* 85
  Hall effect   *X* 85
modal analysis   *VIII* 258
modelling   *VIII* 403
modification   *IX* 80

molecular dynamics (MD) simulations   *X* 149
molecular orientation   *IX* 12, *IX* 323
molecular packing   *IX* 250
molecular recognition   *IX* 112, *IX* 116, *IX* 122
molecular-beam epitaxy   *X* 242
Moore's law   *VIII* 37
morphology   *X* 184, *X* 187, *X* 188, *X* 190, *X* 205
MOS   *X* 94
MOSFET   *X* 239
  effective channel length   *X* 68
mRNAs   *IX* 164
multi wall nanotube (carbon) (MWCNT)   *VIII* 141
multiple-bond attachments   *IX* 103

N clusters   *IX* 368, *IX* 370
nanoelectrochemical techniques
  oxidation   *IX* 79
nanoelectronics   *IX* 62, *IX* 64
nanofibers
  polymeric   *IX* 318
nanofibers oriented   *IX* 319
nanofountain probes (NFP)   *VIII* 191, *VIII* 199, *VIII* 200, *VIII* 202, *VIII* 209, *VIII* 211, *VIII* 212
nanografting   *VIII* 202
nanoindentation   *IX* 313, *IX* 338, *X* 309
nanolithography   *IX* 56
  addition   *IX* 56
  constructive   *IX* 56
  elimination   *IX* 56
  nanopatterning   *IX* 56
  soft lithography   *IX* 67
  substitution   *IX* 56
nanomanipulation   *VIII* 256
nanomanipulator   *IX* 144, *IX* 164
nanomechanics   *VIII* 268
nanoMOS   *X* 98
nanoparticle   *VIII* 83
nanopatterning   *IX* 76
nanoscale capacitance microscopy   *VIII* 423
nanoscale impedance microscopy (NIM)   *VIII* 422, *VIII* 437
  bandwidth   *VIII* 440
  calibration   *VIII* 443
  experimental setup   *VIII* 438
  parasitic contribution   *VIII* 437
  stray contribution   *VIII* 437

Subject Index 421

nanoscale spectroscopy   *VIII* 24
nanoscope   *VIII* 45
nanoscratch technique   *X* 342
nanosource   *VIII* 83, *VIII* 84
nanosystem   *IX* 55
nanotechnology   *IX* 55, *IX* 62, *IX* 283
nanotemplates   *IX* 84
nanotube   *VIII* 135, *VIII* 138, *VIII* 149
 mechanical properties   *VIII* 149, *VIII* 150, *VIII* 158, *VIII* 160, *VIII* 166
nanowear mapping   *IX* 308
near-field   *VIII* 78, *VIII* 89
 resolution   *VIII* 92
 reverse tube etching   *VIII* 102
 spatial frequency   *VIII* 91
 spectrum   *VIII* 91
near-field optical microscopy (SNOM)   *X* 189, *X* 201–205
near-field scanning optical microscopy (NSOM)   *IX* 351, *X* 144
Nioprobe   *VIII* 44
NIST-traceable standards   *VIII* 57
nitric acid treatment   *IX* 257, *IX* 273
nitride   *X* 223, *X* 244
 TiN   *X* 223
 ZrN   *X* 223
noise   *VIII* 359, *VIII* 382, *VIII* 383, *VIII* 443
 correlated instrument   *VIII* 428
 electrical   *VIII* 426
 measurement   *VIII* 427
 nanoscale   *VIII* 444
 Nyquist formula   *VIII* 444
 resolution   *VIII* 427
 spectroscopy   *VIII* 427
noise spectrum   *IX* 316
nondestructive   *VIII* 56
notching   *VIII* 68
NSOM   *IX* 353, *X* 144
NSOM aperture   *IX* 353
nuclear magnetic resonance (NMR)   *IX* 177, *IX* 181, *IX* 185

OCD   *VIII* 35
ODP SAM   *IX* 270
OP SAM   *IX* 272
optical efficiency   *VIII* 84, *VIII* 85
 throughput   *VIII* 78, *VIII* 86, *VIII* 88, *VIII* 94, *VIII* 99, *VIII* 106
optical lever   *VIII* 290
 calibration   *VIII* 290
optical microscope   *IX* 182

optical proximity correction   *X* 398
optical resolution   *VIII* 78, *VIII* 102
optical selection rules   *IX* 361
optical waveguides   *VIII* 19
optimal estimator   *VIII* 360
organic thin-film transistors (OTFTs)
   VIII 368
oscillation amplitude   *VIII* 168
oscillator   *VIII* 380
oscillatory measurements   *IX* 98
oxidation   *X* 218, *X* 219
 anodic   *X* 98
 cathodic oxidation   *X* 219, *X* 222
 electrochemical oxidation   *X* 219, *X* 220
 field-enhanced oxidation   *X* 222
 low-temperature   *X* 72
 UV/ozone   *X* 72
 wet   *X* 72
oxide thickness   *X* 70

parallel beam model   *VIII* 293
PBS   *IX* 344
PCR   *IX* 164
PDMS   *VIII* 185, *VIII* 194
percolation   *X* 96, *X* 100
perfluoroalkyl chain   *IX* 256
perfluoroalkylphosphonate   *IX* 237
perfluoroalkylsiloxy   *IX* 237
perfluorodecyldimethylsiloxy–copper
   IX 255
perfluorosiloxy   *IX* 252
permeability   *IX* 127
permittivity   *VIII* 378
persistence length   *IX* 197, *IX* 198
PFDP SAM   *IX* 244, *IX* 270
PFM   *IX* 50, *IX* 51
PFMS SAM   *IX* 252, *IX* 271
PFMS-based SAM on Cu   *IX* 255
phase measurements   *X* 357
phase transformation   *IX* 303, *IX* 308
phase transition   *IX* 209–212, *IX* 218, *IX* 219
phase-locked loop (PLL)   *VIII* 232
phosphate-buffered saline (PBS)   *IX* 340
phospholipids   *IX* 209, *IX* 210, *IX* 217–220
phosphorus   *IX* 250
photolithogaphy   *X* 248
photolithography   *VIII* 186–188
photomask defect review and repair   *X* 400
photomask pattern and etch   *X* 399
physisorbtion   *IX* 240, *IX* 257

piezo actuator *VIII* 71
piezo stage *IX* 287, *IX* 288, *IX* 292
  large amplitude *IX* 288
  ultrasonic linear piezo drive *IX* 292
piezoresistive cantilever *VIII* 227
piezoresistive detection *VIII* 267
piezoresistive sensor *VIII* 227
pin-on-disk *IX* 334
pin-on-flat *IX* 334
pinocytotic organelles *IX* 140
pipette *VIII* 198, *VIII* 199
planarization efficiency (EFF) *X* 141
point-mass model *VIII* 259
Poisson *X* 85
Poisson equation *VIII* 355, *IX* 338
Poisson's ratio *VIII* 293
polariton resonances *VIII* 3, *VIII* 18, *VIII* 22
polarization *VIII* 108, *VIII* 110, *VIII* 114, *VIII* 116
poly(dimethylsiloxane) *VIII* 185
polymer *IX* 252
polymer nanofibers *IX* 318
polymer solar cells *X* 182–184, *X* 189
polymers
  binary mixture *IX* 42
  mixing entropy *IX* 43
  poly(methyl methacrylate) (PMMA) *IX* 29
  styrene–butadiene rubber (SBR) *IX* 29, *IX* 31, *IX* 41, *IX* 42, *IX* 45, *IX* 46
  surface energy *IX* 43
process control *VIII* 256
procollagen *X* 289
profile
  carrier *X* 85
  dopant *X* 85
  resistivity *X* 85
prostaglandin (PG) *IX* 162
proteins *VIII* 437, *IX* 60, *IX* 75, *IX* 76, *IX* 78, *IX* 177–179, *IX* 182, *IX* 187–189, *IX* 191–193, *IX* 195, *IX* 196, *IX* 329, *IX* 333, *IX* 338, *IX* 340–343, *IX* 347
  bovine serum albumin (BSA) *IX* 340
  height reduction *IX* 71
  plastic deformation *IX* 75
  streptavidin *IX* 71, *IX* 72
proteins denatured *IX* 342
PSF *VIII* 358
pull-off *IX* 335

quality factor *VIII* 236
quantum dots *IX* 351
quantum wells *X* 76
quercetin-3-O-palmitate (QP) *IX* 72
  fibrelike *IX* 73
  self-organized spiral *IX* 73

radius *VIII* 391
RAM capacitor
  ferroelectric *X* 398
Raman *VIII* 78, *VIII* 119, *VIII* 121, *VIII* 128, *VIII* 129
raster scanning *VIII* 38
real area of contact *IX* 336, *IX* 337
receptor *IX* 150
reconstruction *VIII* 359
reduction *X* 230
  cathodic reaction *X* 230
reentrant image reconstruction *VIII* 41
reference metrology *VIII* 39
reference metrology system (RMS) *VIII* 35
reinforcement *IX* 321
reliability *IX* 323
repassivation *IX* 346
repeatability *VIII* 47, *VIII* 299
repulsive regime *IX* 76
residual stress *IX* 343, *IX* 345–347
resist *VIII* 55, *X* 244, *X* 247, *X* 248
resistance *X* 78
  electrical *VIII* 424
  measurement *VIII* 424
  probe R_{tip} *X* 78
  spreading *X* 78
  tip–semiconductor nanocontact *X* 78
resistivity *X* 81, *X* 91
  bulk *X* 91
  polycrystalline Au films *X* 91
resolution *VIII* 386
  effective contact radius *X* 81
  energy *X* 91
  spatial *X* 81, *X* 91
resonance experiment *VIII* 274
  clamping *VIII* 276
  vibration amplitude *VIII* 274
resonance frequency *VIII* 221–223
resonant imaging *IX* 24
resonator *VIII* 384
RMS *VIII* 31, *VIII* 32
rotational degrees of freedom *IX* 5
roughness *IX* 340, *IX* 343, *IX* 344, *X* 396
rupture force *IX* 100

Subject Index

SA *IX* 84
SAM coatings *VIII* 67
SAM formation *IX* 240
sample cross section *VIII* 60
sample preparation *X* 69
  passivation *X* 69
  polishing *X* 69
sample properties
  adhesion force *IX* 28
  deformation ratio *IX* 45
  detachment energy *IX* 31
  elasticity *IX* 26
  electrical double layer *IX* 47
  electrostatic interactions *IX* 30
  energy diffusion *IX* 46
  hysteretic loss *IX* 30
  indentation depth *IX* 44
  local adhesion *IX* 26, *IX* 29
  local stiffness *IX* 29
  plastic deformation *IX* 26
  Poisson's ratio *IX* 33, *IX* 46
  postflow *IX* 31, *IX* 32
  real surface topography *IX* 45
  relaxation time *IX* 31
  sample compliance *IX* 29
  stiffness *IX* 30
  topography *IX* 28, *IX* 30, *IX* 44
  viscoelastic *IX* 26
  Young's modulus *IX* 33, *IX* 46
SAMs comparison between *IX* 273
satellite *IX* 242
scan mode
  CD mode *VIII* 33
  contact mode *VIII* 33
  DT mode *VIII* 33
  noncontact mode *VIII* 33
  tapping mode *VIII* 33
scanning capacitance *VIII* 352
scanning capacitance microscopy (SCM)
  VIII 377, *VIII* 423, *VIII* 443, *X* 63,
  X 236
  scanning capacitance spectroscopy (SCS)
  X 69
scanning electron microscopy (SEM) *X* 286
scanning force microscopy (SFM) *IX* 56,
  IX 68, *IX* 73, *IX* 80
  attractive regime *IX* 56, *IX* 69, *IX* 71–73
  DSFMA *IX* 84
  dynamic *IX* 68, *IX* 69, *IX* 84
  dynamic SFM *IX* 56
  fibres *IX* 74

  force modulation *IX* 76
  hard tapping *IX* 73–75
  phase-lag *IX* 76
  Q-control *IX* 70, *IX* 76
  quality factor *IX* 69, *IX* 71
  repulsive regime *IX* 71
  tapping mode *IX* 68–71, *IX* 73, *IX* 76,
  IX 84
scanning Kelvin probe microscopy (SKPM)
  VIII 388
scanning near-field optical microscope
  (SNOM) *VIII* 78, *X* 294
  aperture *VIII* 78
  apertureless *VIII* 78
  illumination mode *VIII* 82
scanning nonlinear dielectric microscopy
  (SNDM) *X* 105
scanning probe lithography (SPL) *IX* 67,
  IX 76, *IX* 79, *IX* 82, *IX* 84
  addition nanolithography *IX* 76
  constructive nanolithography *IX* 76
  dip-pen nanolithography (DPN) *IX* 76,
  IX 77, *IX* 84
  elimination nanolithography *IX* 76
  nanoelectrochemical *IX* 84
  nanopipet *IX* 76
  nanotemplates *IX* 82
scanning probe microscopy (SPM) *IX* 56,
  IX 76, *IX* 78, *IX* 80, *IX* 84, *IX* 89,
  IX 334, *X* 63, *X* 64, *X* 257, *X* 261,
  X 348, *X* 360
  atomic force microscopy (AFM) *X* 360
  dip-pen nanolithography (DPN) *IX* 78
  elimination nanolithography *IX* 78, *IX* 79
  elimination substitution *IX* 79
  force modulation microscopy *IX* 72
  scanning capacitance microscopy (SCM)
  X 64
  scanning spreading resistance microscopy
  (SSRM) *X* 64
  substitution nanolithography *IX* 78
scanning spreading resistance *VIII* 352
scanning spreading resistance microscopy
  (SSRM) *VIII* 402, *X* 63
scanning tunnelling microscopy (STM)
  VIII 366, *VIII* 377, *IX* 67, *IX* 68, *IX* 78,
  IX 79, *X* 88
scattering
  interface charge *X* 85
  interface roughness *X* 85
  local strain fluctuations *X* 85

scatterometer *VIII* 35
Schottky barrier *X* 88
schottky barrier height (SBH) *X* 88
SCM *VIII* 377, *VIII* 443
secretion *IX* 134
self-assembled *IX* 84
self-assembled monolayers (SAM) *VIII* 67,
    IX 57–59, *IX* 62, *IX* 64, *IX* 65, *IX* 67,
    IX 76, *IX* 78–81, *X* 225, *X* 234, *X* 245,
    X 247, *X* 249
  1-alkenes *IX* 57, *IX* 60
  alkylthiols *IX* 57, *IX* 59, *IX* 60
  alkyltrichlorosilane *IX* 60
  alkyltrichlorosilanes *IX* 80
  1-alkynes *IX* 57, *IX* 60
  amino-terminated SAM *X* 231
  aminosilane SAM *X* 226
  chemical maps *IX* 81
  chemisorption *IX* 57, *IX* 58, *IX* 60
  coatings *IX* 62
  friction *IX* 81
  multifunctional nanoparticle *IX* 63
  1-octadecene *IX* 60, *IX* 80, *IX* 81
  organosilanes *IX* 57, *IX* 60, *IX* 80
  physisorption *IX* 60
  sensor *IX* 62
  SH-terminated SAM *X* 230
  1-terminated alkenes *IX* 80
  thiolated *IX* 62
  thiols *IX* 68, *IX* 78, *IX* 79
  trichlorosilane *IX* 78
self-assembly *IX* 56, *IX* 79, *IX* 82
self-organization *IX* 76, *IX* 82, *IX* 84
self-replacement *IX* 79
SEM *VIII* 35, *VIII* 36
semiconductors *VIII* 401, *X* 218, *X* 223,
    X 236, *X* 239
  a-Si *X* 244, *X* 247
  GaAs *X* 218, *X* 223, *X* 239, *X* 240
  III–V *X* 67
  InP *X* 223
  n-type *X* 66
  p-type *X* 66
  Si *X* 218, *X* 237, *X* 239
  Si–H *X* 218, *X* 222, *X* 241, *X* 242, *X* 248
  SiC *X* 67, *X* 223
  SiGe *X* 67, *X* 223
  silicon *X* 218
sensitivity *VIII* 361
separation work *IX* 155
sessile drop *IX* 241, *IX* 252

SFA
  micro/nanotribology *IX* 284
SFM tip *IX* 73
shake-up *IX* 242
shallow trench isolation (STI) *VIII* 52,
    X 136, *X* 370
shape factor *X* 100
shear force feedback *IX* 354
shear modulus *IX* 89, *IX* 96, *IX* 199, *IX* 200
SI measurement system *VIII* 303
sidewall angle *VIII* 54
sidewall roughness *VIII* 53, *VIII* 55, *VIII* 69
SiGe source/drain recess *X* 385
silanization *IX* 242, *IX* 257
silanols *IX* 238, *IX* 259
silicon nitride *VIII* 65
silicon probe *VIII* 69
silicon-on-insulator (SOI) *X* 239
siloxane *IX* 252, *IX* 257, *IX* 259
siloxy *IX* 278
siloxy–copper *IX* 259, *IX* 261, *IX* 278
SIMS *VIII* 403, *X* 75
single molecules *VIII* 436, *IX* 89
single wall nanotube (carbon) *VIII* 141,
    VIII 144
single-electron transistor (SET) *X* 240
single-molecule measurements *IX* 100,
    IX 116, *IX* 120
size dependence *IX* 312, *IX* 322
skin depth *VIII* 103
slider for hard drive *X* 405
sliding velocity *IX* 278
SNDM (NC-SNDM) noncontact *X* 105,
    X 111
Sneddon's stiffness *X* 310, *X* 311, *X* 313,
    X 316
SNOM apertureless *VIII* 83
SNOM/AFM *X* 295
SOCS *VIII* 45, *VIII* 46
sodium hydroxide treatment *IX* 277
solid-supported monolayer (SSM) *IX* 56,
    IX 57, *IX* 61, *IX* 63–65, *IX* 67–69,
    IX 76, *IX* 84
  phenomena *IX* 61
  properties *IX* 61
  technological needs *IX* 61
space charge region (SCR) *VIII* 355
spatial resolution *IX* 358
spectroscopy *VIII* 83, *VIII* 127, *VIII* 408,
    IX 64
  IR *IX* 65, *IX* 80

IR spectroscopy   *IX* 80
  lasers   *IX* 65
  Raman   *IX* 65
  sampling depth   *IX* 64
  secondary ion mass spectrometry (SIMS)   *IX* 65
  X-ray photoelectron spectroscopy (XPS)   *IX* 65
spherical indenter   *X* 312, *X* 314, *X* 324, *X* 326, *X* 329, *X* 335, *X* 337, *X* 340
spherical indenters   *X* 310
SPM   *X* 347, *X* 352, *X* 353, *X* 357
  force–distance curve   *X* 355
  phase imaging   *X* 347, *X* 356
  surface potential microscopy   *X* 352
spreading resistance   *X* 70, *X* 81
spring constants   *VIII* 222, *VIII* 257, *IX* 92, *IX* 325
  calibration   *VIII* 281, *VIII* 292
SSRM   *X* 78
  resolution   *X* 80
stability   *IX* 257
stage drift   *VIII* 70
step height reduction ratio (SHRR)   *X* 141
STI   *X* 142
STI CMP   *X* 375
STI etch   *X* 372
stick–slip   *IX* 285, *IX* 302
stiffness   *VIII* 150, *VIII* 166
STM   *IX* 2
STM interpretation   *IX* 3
storage compliance   *X* 338, *X* 340
storage modulus   *X* 337, *X* 339
strain distribution   *VIII* 223
strength   *VIII* 270, *VIII* 272
  bending   *VIII* 272
  tensile   *VIII* 269
  Weibull statistic   *VIII* 268
streptavidin   *IX* 71, *IX* 72
stress   *IX* 347
  compressive   *IX* 344, *IX* 346
  constant voltage   *X* 94
  electrical   *X* 93, *X* 94
  ramped voltage   *X* 94
  tensile   *IX* 344–346
  time   *X* 98
stress concentration   *VIII* 63
stress–strain curve   *X* 309, *X* 311, *X* 314, *X* 318, *X* 330, *X* 331, *X* 333, *X* 334, *X* 342
structuring element   *VIII* 50

substrates   *IX* 241
superlensing   *VIII* 25, *VIII* 26
surface (heterogeneous)   *VIII* 174
surface band bending   *VIII* 362, *VIII* 363, *VIII* 365, *VIII* 366
surface charge region   *VIII* 365
surface energy   *IX* 263, *IX* 338, *IX* 343
surface local density of states   *IX* 3
surface modification   *VIII* 66
surface patterning tool (SPT)   *VIII* 195, *VIII* 196, *VIII* 208
surface photovoltage spectroscopy (SPS)   *VIII* 363, *VIII* 366
surface plasmons   *VIII* 121, *VIII* 124, *VIII* 129
surface potential   *VIII* 355, *X* 145
surface potential image   *X* 353
surface properties   *IX* 116, *IX* 117
surface reconstruction   *IX* 14
surface tension   *IX* 323
surfactant   *IX* 207, *IX* 208, *IX* 219, *IX* 220, *IX* 222, *IX* 225, *IX* 229, *IX* 230
SWR   *X* 397
synovial fluid   *IX* 330, *IX* 333, *IX* 340, *IX* 347
  hyaluronic acid   *IX* 333
  lubricin   *IX* 333
  phospholipids   *IX* 333
  serum   *IX* 333
  synovia   *IX* 333
system noise   *VIII* 70

tangent plane   *VIII* 49
tangent slope algorithm   *VIII* 49
taper angle   *VIII* 88, *VIII* 106, *VIII* 110
tapping   *X* 362
technology node   *VIII* 73
TEM   *VIII* 36, *VIII* 60
temperature stability   *VIII* 265
tensile experiment   *VIII* 269
tensile test   *IX* 316
thermal fluctuations   *IX* 93
thermal method   *VIII* 292, *VIII* 297
thermal noise   *VIII* 236
thermoionic emission   *X* 91
thin oxides   *VIII* 435
three-point bending   *IX* 314
threshold voltage   *X* 96
time-lapse AFM   *IX* 189
tip–sample interaction   *VIII* 60
TipCheck   *VIII* 44

tips   *VIII* 135, *VIII* 148, *VIII* 391, *IX* 89, *X* 72
 apex   *VIII* 362
 artifacts   *IX* 324
 aspect ratio   *VIII* 72
 characterizer   *VIII* 42
 Co/Cr   *X* 72
 conductive diamond   *X* 72
 diamond   *X* 80
 geometries   *IX* 94
 lateral stiffness   *VIII* 72
 lifetime   *X* 370
 mechanical properties   *VIII* 149, *VIII* 158, *VIII* 160, *VIII* 166
 on aperture   *VIII* 125, *VIII* 126
 overhang   *VIII* 44
 Pt/Ir   *X* 72
 qualification   *X* 370
 radius   *X* 67, *X* 70
 tip reconstruction *in situ*   *VIII* 41
 shape characterization   *X* 370
 shape profile   *VIII* 65
 wear   *VIII* 61, *VIII* 64, *X* 370
 width   *VIII* 44
tissue section   *X* 292
TJRs   *IX* 334
tool-to-tool matching   *VIII* 37
total joint replacements (TJRs)   *IX* 329, *IX* 330, *IX* 332
 alumina   *IX* 333
 ceramic   *IX* 330
 femoral head   *IX* 330, *IX* 332
 hip joint   *IX* 329
 metallic alloy   *IX* 330
 revision surgeries   *IX* 330
 UHMWPE   *IX* 329, *IX* 338, *IX* 347
 zirconia   *IX* 333
transducer   *VIII* 379, *VIII* 382
transient   *VIII* 378
transition   *IX* 209, *IX* 210, *IX* 214
transmission coefficient   *IX* 357
transmission efficiency   *IX* 355, *IX* 357
transmission-line   *VIII* 386
trap density
 critical   *X* 96
traps   *X* 93
trench structure   *IX* 319
Tresca criterion   *X* 310, *X* 324
tribology   *IX* 332
trident probe   *VIII* 68
tunnel transmission probability   *X* 89

tunneling current   *IX* 3
TUT   *VIII* 36
type-B measurement error   *VIII* 47

UHMWPE   *IX* 332
ultrahigh vacuum (UHV)   *VIII* 361
ultrastructure   *IX* 116
ultrathin film   *IX* 56
uncertainty   *VIII* 56, *VIII* 60, *VIII* 70

vacuoles   *IX* 142
vacuum levels   *X* 94
van der Waals   *IX* 208, *IX* 212–214, *IX* 225, *IX* 250, *IX* 273
variational principle   *VIII* 151
VEH   *VIII* 44
vesicles   *IX* 136
Vickers indenter   *X* 313, *X* 314, *X* 320, *X* 341, *X* 342
VideoDisc pickup   *VIII* 379
villi   *IX* 130
viscoelasticity   *IX* 89, *IX* 96, *X* 296, *X* 309–311, *X* 337, *X* 339, *X* 341, *X* 342
vitronectin (VN)   *IX* 159
voltage   *VIII* 198
volume   *IX* 134
von Mises criterion   *X* 310, *X* 330

W-mapping   *IX* 158, *IX* 161, *IX* 162
water balance   *IX* 127
water molecule adsorption   *VIII* 233
wavefunction   *IX* 360
waveguide   *VIII* 85, *VIII* 89, *VIII* 103, *VIII* 116
wear   *VIII* 65, *IX* 332, *IX* 334, *IX* 336, *IX* 340, *IX* 343–347
 adhesive   *IX* 332, *IX* 342
 depth   *IX* 347
 groove depths   *IX* 336
 mapping   *IX* 303
 trench depth   *IX* 344
 wear depth   *IX* 347
 wear volume   *IX* 344
wear-mechanism mapping   *IX* 304
wear resistance   *IX* 267
wedge method   *VIII* 311
Weibull
 distribution   *X* 99
Wiener filter   *VIII* 360
work function   *VIII* 362, *X* 94
work of adhesion   *IX* 103

work-hardening   *X* 329, *X* 330, *X* 333, *X* 335, *X* 336

X-ray diffraction   *IX* 177, *IX* 180, *IX* 181, *IX* 185

yield stress   *X* 309–311, *X* 314, *X* 318, *X* 324–330, *X* 333, *X* 335, *X* 336, *X* 342

Young's modulus   *VIII* 62, *VIII* 73, *VIII* 269, *VIII* 272–274, *VIII* 293, *IX* 89, *IX* 94, *IX* 134, *IX* 198, *IX* 200, *IX* 227–229, *IX* 231, *X* 296

Zisman method   *IX* 261
Zisman plot   *IX* 241, *IX* 252

Printing: Krips bv, Meppel, The Netherlands
Binding: Stürtz, Würzburg, Germany